FUNCTIONAL POLYMER COATINGS

WILEY SERIES ON POLYMER ENGINEERING AND TECHNOLOGY

Richard F. Grossman and Domasius Nwabunma, Series Editors

Polyolefin Blends
Edited by Domasius Nwabunma and Thein Kyu

Polyolefin Composites
Edited by Domasius Nwabunma and Thein Kyu

Handbook of Vinyl Formulating, Second Edition
Edited by Richard F. Grossman

Total Quality Process Control for Injection Molding, Second Edition
M. Joseph Gordon, Jr.

Microcellular Injection Molding
Jingyi Xu

Poly(lactic acid): Synthesis, Structures, Properties, Processing, and Applications
Edited by Rafael Auras, Loong-Tak Lim, Susan E.M. Selke, and Hideto Tsuji

Hyperbranched Polymers: Synthesis, Properties, and Applications
Edited by Deyue Yan, Chao Gao, and Holger Frey

Advanced Thermoforming: Methods, Machines and Materials, Applications and Automation
Sven Engelmann

Biopolymer Nanocomposites: Processing, Properties and Applications
Alain Dufresne, Sabu Thomas, Laly A. Pothan

Polymers for PEM Fuel Cells
Hongting Pu

Polyurethanes: Science, Technology, Markets, and Trends
Mark F. Sonnenschein

Functional Polymer Coatings: Principles, Methods, and Applications
Edited by Limin Wu and Jamil Baghdachi

FUNCTIONAL POLYMER COATINGS

Principles, Methods, and Applications

Edited by
Limin Wu
Jamil Baghdachi

WILEY

Published by John Wiley & Sons, Inc., Hoboken, New Jersey
Published simultaneously in Canada

Library of Congress Cataloging-in-Publication Data:

Functional polymer coatings : principles, methods and applications / edited by Limin Wu, Jamil Baghdachi.
pages cm
Includes index.
ISBN 978-1-118-51070-4 (cloth)
1. Coating processes. 2. Plastic coating. 3. Polymers–Industrial applications.
I. Wu, Limin. II. Baghdachi, Jamil.
TP156.C57F864 2015
668.9′2–dc23

2015003604

Set in 10/12pt Times by SPi Global, Pondicherry, India

10 9 8 7 6 5 4 3 2 1

Contents

Contributors

Jamil Baghdachi Coatings Research Institute, Eastern Michigan University, Ypsilanti, MI, USA

Hao Chen Department of Materials Science and Advanced Coatings Research Center of Ministry of Education of China, Fudan University, Shanghai, P.R. China

and

Engineering Department of Zhejiang Agriculture and Forestry University, Hangzhou Linan, P.R. China

Carlos A. Escobar Department of Chemical and Biomolecular Engineering, Vanderbilt University, Nashville, TN, USA

A.C.C. Esteves Materials and Interface Chemistry Group, Chemical Engineering and Chemistry Department, Eindhoven University of Technology, Eindhoven, The Netherlands

S.J. García Novel Aerospace Materials Group, Faculty of Aerospace Engineering, Delft University of Technology, Delft, The Netherlands

Sergio Granados-Focil Gustaf Carlson School of Chemistry and Biochemistry, Clark University, Worcester, MA, USA

Melissa A. Grunlan Department of Biomedical Engineering, Texas A&M University, College Station, TX, USA

and

Department of Materials Science and Engineering, Texas A&M University, College Station, TX, USA

Yunjiao Gu Department of Materials Science and Advanced Coatings Research Center of Ministry of Education of China, Fudan University, Shanghai, P.R. China

G. Kane Jennings Department of Chemical and Biomolecular Engineering, Vanderbilt University, Nashville, TN, USA

W. (Marshall) Ming Department of Chemistry, Georgia Southern University, Statesboro, GA, USA

Sandro Olveira Department of Chemistry, University of Zurich, Zurich, Switzerland

H. Perez Coatings Research Institute, Eastern Michigan University, Ypsilanti, MI, USA

Marc A. Rufin Department of Biomedical Engineering, Texas A&M University, College Station, TX, USA

Stefan Seeger Department of Chemistry, University of Zurich, Zurich, Switzerland

Zhehong Shen Department of Materials Science and Advanced Coatings Research Center of Ministry of Education of China, Fudan University, Shanghai, P.R. China

and

Engineering Department of Zhejiang Agriculture and Forestry University, Hangzhou Linan, P.R. China

Ana Stojanovic Department of Chemistry, University of Zurich, Zurich, Switzerland

Punthip Talapatcharoenkit Coatings Research Institute, Eastern Michigan University, Ypsilanti, MI, USA

Limin Wu Department of Materials Science and Advanced Coatings Research Center of Ministry of Education of China, Fudan University, Shanghai, P.R. China

Qinhua Xu Coatings Research Institute, Eastern Michigan University, Ypsilanti, MI, USA

Jie Zhao Department of Chemistry, Georgia Southern University, Statesboro, GA, USA

Shuxue Zhou Department of Materials Science and Advanced Coatings Research Center of Ministry of Education of China, Fudan University, Shanghai, P.R. China

Preface

Coatings are used on surfaces of most products to offer decoration, protection, and special functions. Coating science and technology is an old field; however, it has not reached a perfect maturity. In particular, with increasingly strict environmental protection laws and rules enforced in various countries and the demands of continuously developing hi-tech industries, coatings with better or novel performances are highly expected. Generally, polymers and coatings will be evolved to respond to the following major trends: (i) to provide environmentally friendly coatings, which require synthesis of novel resins for waterborne, solvent-free, thermal-insulating, air-purifying coatings, and so on; (ii) to enhance the performances of current coatings, including better scratch and mar resistance, enhanced corrosion-resistance, aging and heat resistance, anti-fingerprint performances, and so on; (iii) to develop multifunctional even smart coatings, including self-cleaning coatings, temperature-controllable coatings, bionic anti-fouling coatings, self-healing coatings, light/heat/electricity switching coatings, sensory coatings, and so on.

These functions of coatings are not easily achievable by traditional synthesis methods and formulation techniques, but they can be possibly realized by application of modern science and technology, that is, controllable/live free-radical polymerization, graft polymerization, and micro-emulsion polymerization for novel binders. And organic–inorganic hybrid, self-assembly and nanotechnology for special coating functions. In addition, the use of new pigments and modification methods and construction of micro- and nanosurfaces can potentially afford coatings with enhanced and multifunctional properties.

This book mainly focuses on some important and hot functional coatings. The authors of various chapters in this book are recognized experts in their specific areas of expertise of the subject. This book begins with the organic–inorganic nanocomposite coatings (OINCs), which are the simplest and widely investigated since nanotechnology. Chapter 1 discusses in details general fabrication principles and performance features of OINCs as well as partially transparent OINCs. In addition, fabrication methods and properties of transparent OINCs with mechanically reinforced, high refractive index, UV shielding, near-infrared light-shielding, barrier, conductive coatings are discussed. Chapter 2 reviews and discusses the recent progress in design, preparation, and typical properties of super-repellent polymeric surfaces, including the concept of surface wettability, various approaches to obtain super-repellent surfaces, and applications of super-repellent polymeric surfaces. Chapter 3 focuses on the important fundamentals and definitions of superhydrophilic and superamphiphilic surfaces, examples of naturally occurring superhydrophilic and/or superamphiphilic surfaces, the most prominent examples of artificial superwetting coatings, the most common techniques used for manufacturing such coatings, applications of superhydrophilic and superamphiphilic coatings, etc. Chapter 4 discusses the self-healing mechanisms and approaches for functional polymeric coatings, and some examples of healable functionalities, referring to potential applications on polymeric coatings. Chapter 5 describes the stimuli–responsive soft materials with special emphasis on stimuli such as temperature changes, electromagnetic radiation exposure, magnetic fields, electrical fields, and selective binding of biochemically relevant molecules. Chapter 6 mainly focuses the basic concepts of self-stratifying polymers and coatings, design and formulation, characterization, as well as their properties.

This book further focuses on some binders and their applications in functional coatings: Chapter 7 presents methods of polymer surface-grafting, characterization of such modified surfaces and utility of surface-grafted polymer coatings for antifouling applications. Chapter 8 discusses surface-initiated ring-opening metathesis polymerization to fabricate partially fluorinated coatings of a few to several micrometers in thickness with ultralow critical surface tensions and dispersive surface energies. Chapter 9 presents a new concept on structural color coatings that are derived from photonic crystals in physics. The last two chapters, Chapters 10 and 11, discuss basic concepts, formulation, properties, utility, and characterization of specific functional coatings of antibacterial polymers and coatings (Chapter 10) and novel marine antifouling coatings (Chapter 11).

This book targets professionals, industrial practitioners, as well as researchers and graduate students in the fields of polymers chemistry and engineering, coatings materials science, and chemical engineering that need to know the most recent developments in coatings science and technology.

October 2014 *Limin Wu and Jamil Baghdachi*

CHAPTER 1

Transparent Organic–Inorganic Nanocomposite Coatings

Shuxue Zhou and Limin Wu
Department of Materials Science and Advanced Coatings Research Center of Ministry of Education of China, Fudan University, Shanghai, P.R. China

1.1 INTRODUCTION

The combination of organic and inorganic ingredients is the most popular strategy to achieve coatings with optimal properties. The two components with different or even opposing intrinsic properties can be mixed at the microscale, nanoscale, and even molecular level. Composite coatings at the microscale actually are conventional pigmented coatings with an opaque appearance. Molecular hybrids were first reported in the 1980s and are an early form of organically modified ceramics (Ormocers) wherein the organic groups act as an inorganic network modifier or network former [1, 2]. These products were further developed in this century as organic phase-dominated materials with an unmatured inorganic phase especially as crystalline inorganics. Nanoscale hybrid coatings based on an organic matrix are actually organic–inorganic nanocomposite coatings (OINCs). The inorganic domain is a dispersed phase with at least one dimension on the nanometer size regime (1–100 nm). In the past 15 years, OINCs have attracted broad research interest both in academics and in industries. Many papers and patents have been published related to OINCs.

Based on Rayleigh scattering theory, the transmission (T) of light through the heterogeneous coatings like OINCs can be calculated according to the following equation:

$$T = \exp\left[-\frac{3\phi_p L r_p^3}{4\lambda^4}\left(\frac{n_p}{n_m} - 1\right)\right] \tag{1.1}$$

where L is the thickness of the coatings, r_p is the radius of the scattering element (namely, the inorganic phase), ϕ_p is the volume fraction of the inorganic phase, λ is

Functional Polymer Coatings: Principles, Methods, and Applications, First Edition.
Edited by Limin Wu and Jamil Baghdachi.

the wavelength of the incident light, and n_p and n_m are the refractive indices of the inorganic phase and the polymer matrix, respectively. It can be clearly seen from Equation (1.1) that the transparency of OINCs depends on the size of the dispersed phase, coating thickness, and the refractive index (RI) difference between the organic matrix and the inorganic phase. The OINCs have a high transparency because the size of the inorganic phase is significantly smaller than the wavelength of light. Normally, 40 nm is an upper limit for nanoparticle diameters to avoid intensity loss of transmitted light due to Rayleigh scattering and thus achieve highly transparent OINCs.

In addition to excellent transparency, OINCs can efficiently combine the advantages of rigidity, functionality (optic, electric, magnetic, etc), durability (to chemicals, heat, light) of the inorganic phase with the softness and processability of the organic phase. They can find wide applications in abrasion- and scratch-resistant coatings, optical coatings, barrier coatings, corrosion-resistant coatings, antibacterial coatings, electrically conductive coatings, self-cleaning coatings (superhydrophilic and superhydrophobic), heat-resistant coatings, flame-retardant coatings, etc. The OINCs are often the best solution especially for those cases that require high coating transparency.

The nanophase of the OINCs can be either simply introduced by blending with *ex situ* nanostructure materials or *in situ* by a sol–gel process or intercalation. The blending method is similar to the fabrication process of conventional organic coatings wherein the inorganic nanostructure materials rather than microparticles are used as the filler. As for the sol–gel method, the inorganic nanophase can be created in the formulating step or the drying step in bottom–up strategies. In most cases, the nanophases precursors are first prehydrolyzed and then blended with a binder. Normally, amorphous metal oxides and metal nanophases in OINCs can be fabricated with this method. The intercalation method is particularly suitable for layered inorganic fillers, for example, clay. In this method, the process is quite analogous to the blending method. However, the inorganic nanophase is *in situ* generated based on a top–down strategy.

In this chapter, the general fabrication principles and performance features of OINCs as well as partially transparent OINCs are presented. Primarily focus is on transparent OINCs with mechanically reinforced, high RI, ultraviolet (UV)-shielding, near-infrared (NIR) light-shielding, barrier, conductive coatings, etc. Because the pigmented OINCs even with the aforementioned performance are opaque, they are beyond the scope of this chapter and not discussed further.

1.2 FABRICATION STRATEGIES

1.2.1 Blending Method

Blending is frequently adopted for inclusion of *ex situ* nanostructure materials into organic coatings. These nanostructures include nanoparticles, nanofibers, nanorods, nanotubes, nanosheets, etc. Among them, nanoparticles are the most

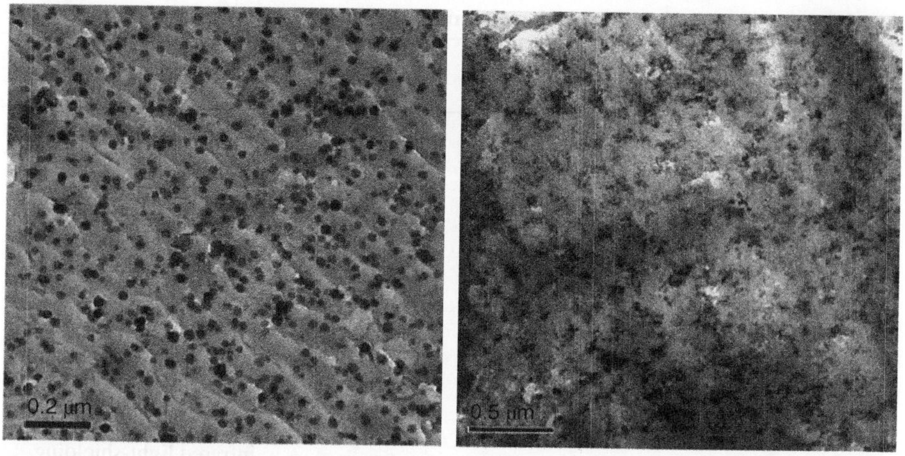

FIG. 1.1 TEM micrographs of nanocoatings filled with 10 wt.% nanoparticles: colloidal nanosilica (left) and pyrogenic nanosilica (right). Reprinted with permission from Ref. 3. © 2011 Elsevier.

common nanofiller for the fabrication of transparent OINCs. The particles can be nanopowders or colloidal. Figure 1.1 shows the typical morphology of colloidal silica and pyrogenic silica in coatings. Colloidal silica particles are spherical and individually dispersed in the organic matrix, whereas pyrogenic silica particles are irregular aggregates. Table 1.1 summarizes some typical nanostructure materials. All nanostructure materials could be possibly used to produce mechanically reinforced OINCs. Nevertheless, the functionality of nanostructure materials determines the functional performance of the resulting OINCs.

The nanoparticles in sols are already nanoscale. Thus, they can be directly mixed with other ingredients [4]. However, these metal oxide nanoparticles in commercial sols are generally amphorous, which is useless for the fabrication of functional OINCs. In recent years, colloidal sols using crystalline oxide nanoparticles from nonaqueous synthesis or controlled hydrolysis have been successfully acquired, opening a new route to obtain transparent functional OINCs.

The nanoparticles can be embedded into coatings during formulation. Sometimes, the incorporation of nanoparticles is moved forward to the stage of resin synthesis, that is, the so-called "*in situ* polymerization" method. This approach enhances the dispersion of nanoparticles and/or the interaction between nanoparticles and the polymer.

1.2.1.1 Deagglomeration of Nanopowder Nanoparticles in the powder state aggregate due to their large surface areas. The aggregates deteriorate the mechanical properties and transparency of OINCs [5]. Therefore, dispersing nanoparticles in resins or coatings is an extremely important task for the field. Various techniques have been developed for dispersing nanopowders into different liquids, including high shear rate mixing, sonication, milling (or grinding), and microfluidic techniques.

TABLE 1.1 The Physical Properties of Some Typical Nanostructure Materials

Type	Density, (g/cm^3)	Mohn's Hardness	Refractive Index	Functionality
SiO_2	2.2	7	1.42–1.46	Mechanical hardness
Al_2O_3	4.0	9	1.7–1.8	Mechanical hardness
ZrO_2	5.6–6.3	6.5	2.13–2.14	Mechanical hardness, high refractive index
TiO_2	3.9	6.0–6.5	2.7	UV absorption, photocatalytic activity, anti-bacterial property
ZnO	5.6	4.5	2.02	UV absorption, photocatalytic activity, anti-bacterial property
ITO	4.3–7.0	—	1.85–1.95	Electric conductivity, near infrared light-shielding
ATO	—	—	—	Electric conductivity, near infrared light-shielding
$CaCO_3$	2.9	3.0	1.6	Mechanical strength
Silver	10.5	2.5–4	0.13	Anti-bacterial property, optical
Boehmite	3.0–3.1	3.0–3.5	1.64–1.67	Anisotropic mechanics
Carbon nanotube	~1.3	25 GPa for single walled[a]	—	Electric conductivity, anisotropic mechanics
Graphene	—	—	—	Electric conductivity, anisotropic mechanics and barrier property

[a] Indentation hardness.

Figure 1.2 summarizes the possible routes for preparation of waterborne or solvent-based nanocomposite coatings from nanopowders. Ultrasonic and microfluidic techniques are usually used in the lab but are infeasible for industrial applications. High shear-rate mixing deagglomerates nanopowders somewhat, but not completely. Bead milling is the most efficient current technique.

The bead milling apparatus is composed of a bead mill, a circulation pump, and a mixing tank equipped with a stirrer. Besides size reduction, loss of crystallinity often occurs during the intensive grinding process. This crystalline change is undesired especially for crystalline nanoparticles application, for example, the use of titania (TiO_2) nanoparticles for photocatalytic self-cleaning applications. Here, the photocatalytic performance is directly related to crystallinity. Smaller bead size and the appropriate induced energy input better destroy nanoparticle aggregates and maintain crystallinity. Beads down to 15–30 µm can reduce TiO_2 nanopowders to a primary particle size of 15 nm [7, 8]. To separate the small beads, a centrifugation bead mill has been developed (Fig. 1.3). The slurry containing agglomerated particles is pumped into the dispersing section of the vessel, where it interacts with the violently agitated beads. Gradually, the slurry reaches the upper part of the dispersing region, where it is

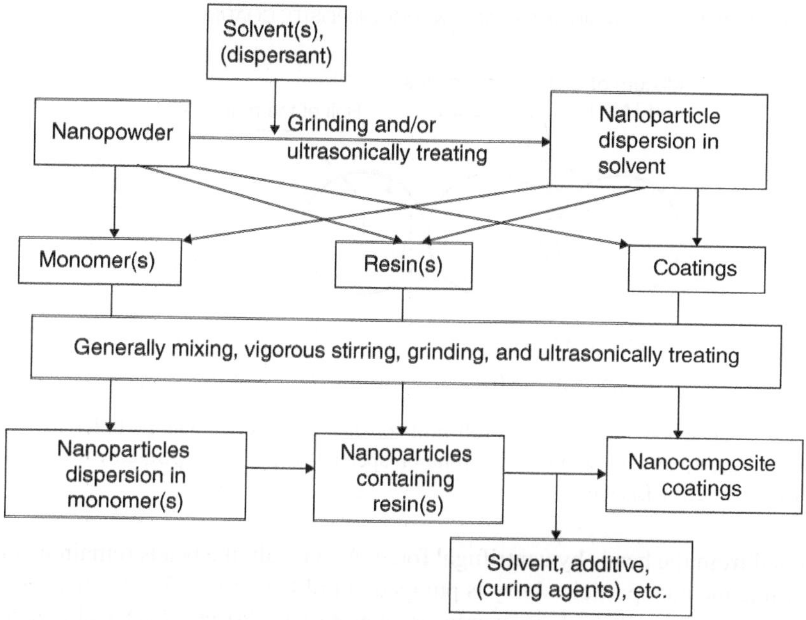

FIG. 1.2 The possible routes for preparation of nanocomposite coatings from nanopowders. Reprinted with permission from Ref. 6. © 2009 American Chemical Society.

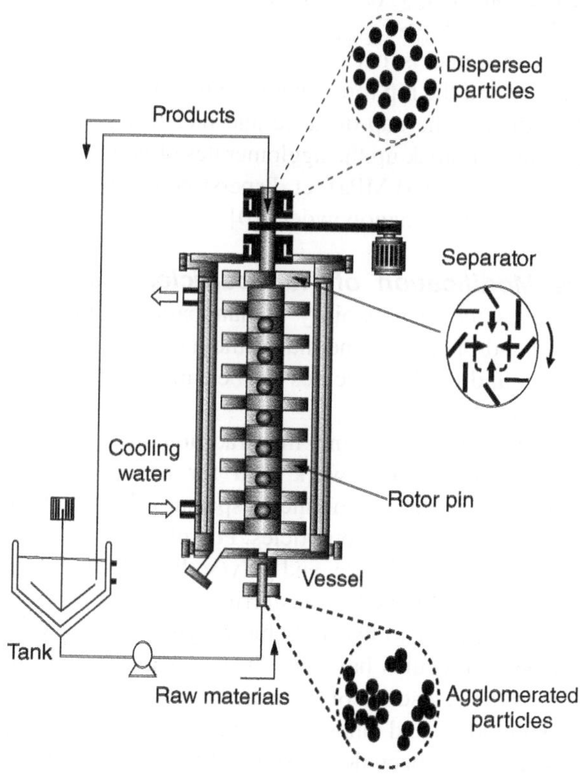

FIG. 1.3 Schematic of the bead mill with centrifugal bead separation. Reprinted with permission from Ref. 7. © 2006 Elsevier.

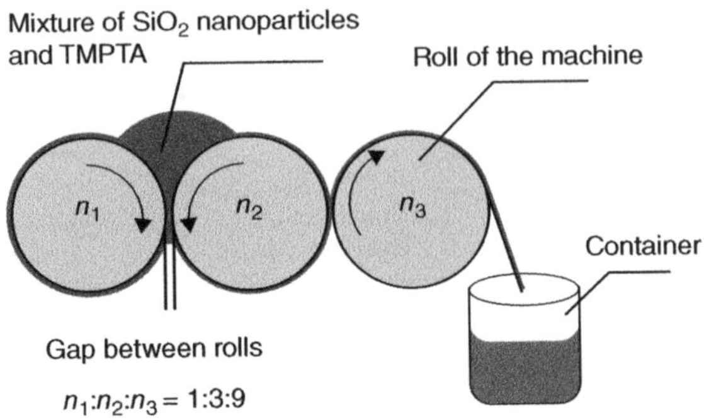

FIG. 1.4 The schematic of a three-roll mill for dispersing silica nanoparticles in TMPTA. The letters (n_1, n_2, and n_3) stand for the rotation speed of the rolls. Reprinted with permission from Ref. 3. © 2011 Elsevier.

separated from the beads by centrifugal force. As a result, the beads remain inside the mill, while the nanoparticle slurry is pumped out of the vessel. We also used a patent describing small beads with an average diameter of 10–70 μm [9]. A stable nanoparticle suspension (D_{50} < 50 nm) with a dry matter content of more than 10 wt.% and a crystallinity loss less than 10% was obtained by controlling the induced energy (E_{kin}) above the deaggregation energy ($E_{de\text{-}aggr}$) but less than the amorphization energy ($E_{amorphous}$), that is, $E_{amorphous} > E_{kin} > E_{de\text{-}aggr}$.

A three-roll mill machine is occasionally used to deagglomerate nanopowder (Fig. 1.4). The distance and the nip forces between the three rolls can be programmatically controlled. Reducing the gap distance and increasing the nip forces generate strong shear force that can break up the agglomerates effectively.

In addition, high pressure (>1 MPa) jet dispersion using at least one nozzle was reported for dispersing of SiO_2 nanopowder [10].

1.2.1.2 *Surface Modification of Nanoparticles* Surface modification of nanoparticles improves the dispersibility of nanoparticles and their compatibility with polymer matrix and/or solvent and makes them reactive with the coating binder. Both macromolecules and small molecules can be employed for surface modification in the physical/chemical bonding.

The commercial polymer dispersants that traditionally are used for the preparation of microparticle slurries also work well for nanoparticle slurries [11–13]. However, much more quantities of polymer dispersants have to be used because of the large specific surface area of nanoparticles. Polyelectrolytes such as polyacrylate sodium, polyallylamine hydrochloride, and poly(sodium 4-styrenesulonate) can also be employed as polymer modifiers for transferring nanoparticles from aqueous phase to nonpolar organic solvent or to hydrophobic polymer matrix without aggregation [14]. Some new macromolecules have also been designed to aid the dispersion of nanoparticles. For instance, a series of hybrid dendritic-linear copolymers (Fig. 1.5) with carboxy-, disulphide-, and phosphonic acid-terminated groups are reported [15]. These copolymers have been demonstrated to be highly efficient for dispersing

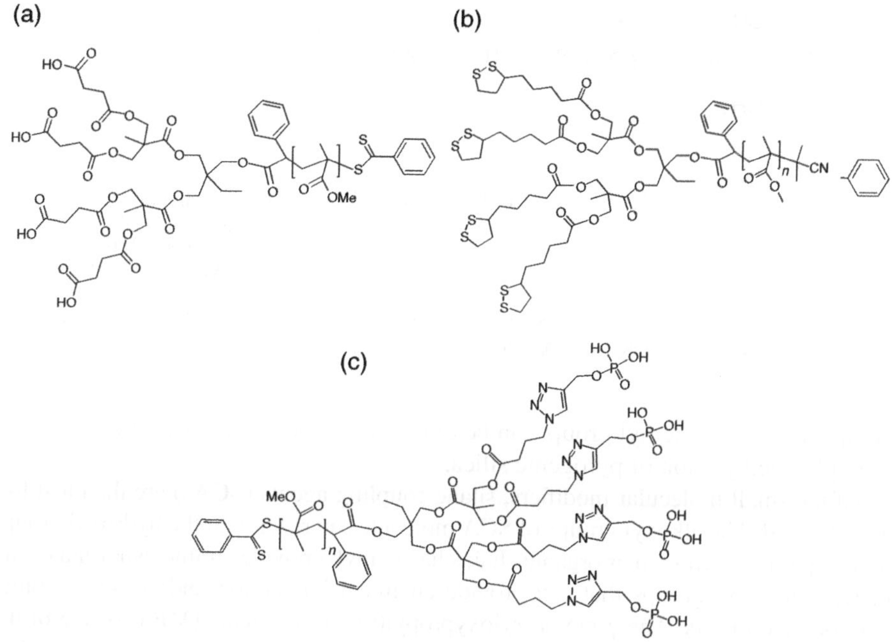

FIG. 1.5 The structures of (a) carboxy-terminated, (b) disulphide-terminated, and (c) phosphonic acid-terminated dendritic-linear block copolymers [15]. Ref. 15. © 2009 Wiley Periodicals, Inc.

TiO_2, Au, and CdSe nanoparticles and are superior to commercial dispersants. Poly(propylene glycol) phosphate ester was synthesized for functionalization of SiO_2 nanoparticles, which are particularly suitable for their application in polyurethane (PU) coatings [16].

The polymer chains chemically attach to nanoparticles through two strategies: "grafting to" and "grafting from." The polymers are directly bonded via the surface hydroxyl groups of nanoparticles in the "grafting to" method. In some cases, chemically reactive organic groups are first attached and then polymers are grafted to nanoparticles chemically. Amici *et al.* even grafted polymer onto magnetite nanoparticles by a "click" reaction between azido functionalized nanoparticles and acetylene end-functionalized poly(ε-caprolactone) or PEG [17]. In contrast, polymer directly propagates from the surface of nanoparticles in the "grafting from" route. In this strategy, an initiator is always attached to nanoparticles in advance. For example, Mesnage's group invented a "Graftfast™" process for functionalization of TiO_2 nanoparticles with poly(hydroxyethyl) methacrylate [18]. In that process, a diazonium salt initiator was first bonded to the surface of nanoparticles.

Besides polymers, organophiliation of nanoparticles with small molecules can be adopted. These short organic segments can attach to the surface of nanoparticles through versatile means. Figure 1.6 gives some possible bonding modes of the grafted organic chains on the nanoparticles. Functionalization of some organic

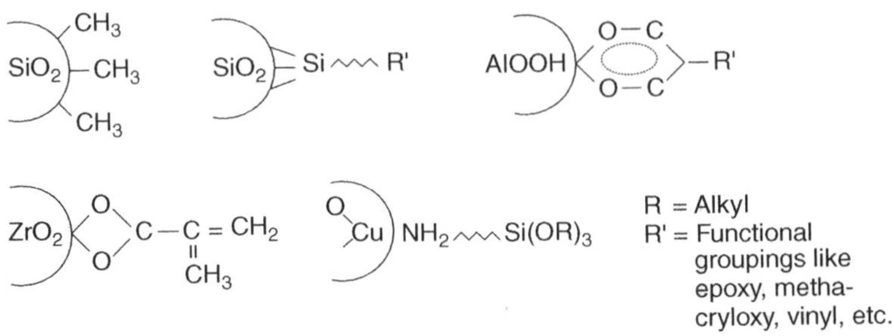

FIG. 1.6 Some principles for surface modification of nanoparticles. Reprinted with permission from Ref. 19. © 1998 Kluwer Academic Publishers.

groups, that is the methyl group, can be done during the nanoparticle synthesis, for example, methylation of pyrogenic silica.

Of the small molecular modifiers, silane coupling agents (SCAs) are the most frequently used. The alkoxyl groups of SCA molecule can react with the hydroxyl groups of nanoparticles while their organic chains have vinyl, epoxide, amine, isocyanate, and mercaptanol end groups that can provide chemical interaction and/or compatibility with organic matrix. The γ-methacryloxypropyltrimethoxysilane (MPS) is one of the most common SCAs for organophilation of nanoparticles because its methacrylate group makes the nanoparticles polymerizable in radical polymerization. The MPS-functionalized nanoparticles have been widely used in the fabrication of UV-curable nanocomposite coatings. Many reports show that MPS molecules bind to nanoparticles via either T^2 or T^3 mode [20, 21]. In most cases, the adsorbed MPS molecules form monolayers with perpendicular and parallel orientations in the absence of catalyst. The parallel orientation might be induced by hydrogen bonding between the MPS-carbonyl and a hydroxyl group of the oxide. With monolayer structure, the amount of MPS bonded could theoretically change in the range of 3.0–6.9 μmol/m^2 [22]. This deviation is due to incomplete coverage or multilayers. If an acidic or basic catalyst is employed during modification, a precondensed MPS structure would be attached to nanoparticles. For an example, a ladder-like arrangement of two linked siloxane chains forming connected eight-membered rings (Fig. 1.7) was demonstrated by Bauer *et al.* [23]. This group used nanosilica (nano-SiO_2) or nanoalumina (nano-Al_2O_3) particles modified with MPS under maleic acid catalyst in acetone. The ladder-like structure was expected to build up a short range of interpenetrating networks with polyacrylate chains during UV or EB curing [24].

To date, many oxide nanoparticles such as SiO_2 [25], TiO_2 [26], ZrO_2 [27], antimony-doped tin oxide (ATO) [28], etc. have been functionalized with MPS. However, MPS-functionalized nanoparticles do not always provide good dispersion in organic solvents, monomers, and oligomers. Modification of highly-dispersible ZrO_2 nanoparticles and deagglomeration of TiO_2 nanopowder with MPS indicate that MPS-functionalized nanoparticles are soluble in THF and butyl acetate [26, 29]. Nevertheless, there is a critical MPS-functionalized nanoparticle load. Above this loading level, phase separation occurs during dispersion in tripropyleneglycol

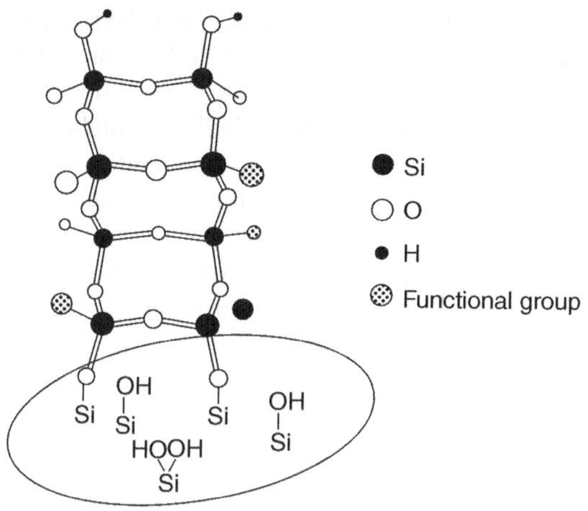

FIG. 1.7 Ladder-like structure of silicon atoms in polysiloxanes grafted on the silica surface [23]. Ref. 23. © 2003 Wiley-VCH Verlag GmbH & Co. KGaA.

diacrylate (TPGDA), 1,6-hexanediol diacrylate (HDDA), trimethylolpropane triacrylate (TMPTA), polyurethane acrylate oligomer, and their mixtures [26, 30–32]. Moreover, as more MPS is attached or higher fraction of PU oligomer in UV-curable coatings is adopted, lower critical MPS-functionalized ZrO_2 load is revealed. This suggests that MPS-functionalized nanoparticles are partially compatible with conventional UV-curable monomers, but poorly compatible with PU oligomer. Therefore, modifying nanoparticles with MPS for UV-curable coatings should be done carefully.

The γ-glycidoxypropylmethoxytriethoxysilane (GPS) and γ-aminopropyltrimethoxysilane (APS) are the other two SCAs for functionalization of nanoparticles [33, 34]. They endow nanoparticles with epoxy and amino groups, respectively, and hence chemical reactivity with the organic binder. The GPS-modified nanoparticles can be readily embedded into epoxy coatings [35], and GPS-based polysiloxane coatings are part of the cross-linking network [36, 37]. More interestingly, the prehydrolyzed GPS is amphiphilic and can modify the aqueous nanoparticle sol [38]. The silylated colloidal particles thus have improved cross-linking ability with themselves or with other polymer binders via the grafted epoxy and/or silanol groups. The APS-modified nanoparticles can be applied to epoxy coatings [39] and PU coatings [40]. It should be noted that nanoparticles grafted with excess APS are unstable in organic solvents due to the high polarity of the amino groups. On the contrary, a high quantity of bonding APS favors the dispersion of nanoparticles in acidic or alkaline water [33]. Therefore, unlike MPS, both GPS and APS can be used to modify nanoparticles for applications not only in solventborne coatings but also in waterborne coatings.

Other SCAs reported for functionalization of nanoparticles include vinyltrimethoxysilane (VTS) [23], *n*-propyltrimethoxysilane (PTS) [23], hexadecyltriethoxysilane (HDTS) [41], *N*-aminoethyl-*N*-aminopropyltriethoxysilane (AEAPS) [42], and even mixtures of SCAs (decyltrimethoxysilane/APS [43]). The VTS has a C=C bond similar to MPS. However, the C=C bond is much less active because of its relatively high rigidity on the nanoparticle surface. The PTS and HDTS are inert SCAs without any terminated groups, and are therefore mainly used for improving the dispersion of nanoparticles in nonpolar solvents [41]. The AEAPS has higher polarity than APS and is an ideal ligand for the fabrication of aqueous nanoparticle dispersions [42]. Mixed SCAs offer more control of the surface wettability to facilitate dispersion of modified nanoparticles in versatile solvents.

Besides SCAs, other small molecule modifiers include acrylic acid [29, 44], 2-acetoacetoxyethyl methacrylate [45], hydroxyethyl methacrylate [46], and cathediol group-containing ligands [29]. Details of their utility with nanoparticles are in the literature.

1.2.2 Sol–Gel Process

The sol–gel process combines inorganic and organic units at the molecular and nano-sized level. In a typical sol–gel process the precursors, that is, metal alkoxides, metal salts, etc., are prehydrolyzed/condensed to form an inorganic sol in the presence of acid or base catalyst. The as-synthesized inorganic sol is then cast on a substrate for further condensation under drying. Baking at a high temperature results in the formation of inorganic coatings, which is very thin (several hundred nanometers) and brittle. To reduce the brittleness, organic-group tethered precursors are always introduced. An inorganic–organic (I/O) hybrid coating (Ormocer) is thus formed at a drying temperature below the decomposition temperature of the organic groups. Besides the inorganic network modifier, some organic groups can react to aid the film formation. If the organic network dominates the film formation, the dried coatings actually transform into organic–inorganic (O/I) hybrid coatings wherein the organic component constitutes the continuous phase [47]. The O/I hybrid coatings can also be prepared by introducing prehydrolyzed inorganic sol into conventional polymer coatings. Nevertheless, it is hard to quantitatively judge the boundary between I/O and O/I hybrid coatings.

The O/I hybrid coatings mixed at the nanosize have distinguished inorganic phases (amphorous or crystalline). Therefore, they are best described as OINCs. The O/I hybrid coatings are generally limited to those sol–gel derived coatings without distinguished inorganic phase. These "hybrid nanocomposites" are often seen in publications [39, 48, 49] and actually represent one important source of OINCs—nanocomposite coatings prepared from a sol–gel process.

Tetraethoxysilane (TEOS) and tetramethoxysilane (TMOS) are the most popular precursors in sol–gel-derived coatings because of their mild hydrolysis and condensation reactions. In most cases, organotrialkoxysilane such as MPS, GPS, APS, and VTS are added to tetraalkoxysilane to endow the inorganic silica sol with reactivity and compatibility with the organic phase. Besides silica precursors, other precursors include titanium *n*-butoxide, titanium tetraisopropoxide (TTIP), aluminum isopropoxide, aluminum *sec*-butoxide, zirconium butoxide, and zirconium tetrapropoxide. These metal alkoxides are highly active in hydrolysis/condensation reactions. To decrease their reactivity, ligands such as ethylene glycol, acetic acid, ethyl acetoacetate, acetylacetone, and their derivates are usually coordinated with them.

Controlling phase separation is very important to achieve sol–gel derived OINCs. Both insufficient and serious phase separations are undesirable. Generally, chemical reactions between an inorganic sol and an organic component are designed to control phase separation. The polymer chains with pendant carboxylic acid groups or triethoxysilyl groups can chemically interact with the inorganic sol to hinder serious phase separation. The growth of the inorganic phase can also be controlled by a limited supply of water. In addition, the hydrolysis/condensation of precursors in dried coatings—aided with moisture from air—is another ideal way to generate OINCs because of the limited space for the growth of inorganic domains.

In comparison with the blending methods, the sol–gel process is an easier route to transparent OINCs. Therefore, transparent sol–gel-derived OINCs are promising as optical (high RI, UV-shielding) coatings and scratch-resistant clearcoats. They are also widely used as corrosion-resistant coatings for metals, in which high transparency is not necessary.

1.2.3 Intercalation Method

The intercalation method is utilized for the fabrication of nanocomposite coatings based on clays or other layered inorganic fillers. These special fillers are incorporated into coatings via high-speed mixing, ball milling, bead milling, and three roll milling similar to nanopowders. However, the nanophases are *in situ* generated through intercalation. Table 1.2 gives the commercial name and suppliers of some layered silicates. Transparent nanocomposite coatings can be readily obtained because the RI of clay (bentonite clay = 1.54) closely matches that of most organic coatings, and at most cases, the clay loading in coatings is very small (usually <5%).

TABLE 1.2 The Type and Supplier of Layered Silicate

Commercial Name	Ingredient	Supplier
Cloisite® Na^+	A natural unmodified montmorillonite clay (cation exchange capacity, 92 mEq/100 g)	Southern clay products (Gonzales, TX)
Cloisite 30B	Organic layered silicate, montmorillonite treated with methyl tallow bis-(2-hydroxyethyl) quaternary ammonium	Southern clay products (Gonzales, TX)
Laponite RD	Synthetic layered silicate, $Na^+_{0.7}$ $[(Si_8 Mg_{5.5}Li_{0.3}) O_{20}(OH)_4]^{0.7}$	Rockwood Additives Ltd. (Cheshire, UK)
Cloisite 20A	Dimethyl dehydrogenated tallow quaternary ammonium modified	Rockwood specialties
Cloisite 20B	—	Rockwood specialties
I.30P	Primary octadecylamine modified	Nanocor Inc.
I.30E	An onium ion surface modified montmorillonite	Nanocor Inc.
Saponite clay	$Ca_{0.25}(Mg,Fe)_3((Si,Al)_4O_{10})(OH)_2 \cdot n(H_2O)$	Kunimine Ind. Co. Ltd., Japan

Quaternary alkylammonium salts or quaternary alkylphosphonium salts are often employed as an intercalating agent. One special ammonium salt, [2-(methacyloyloxy)ethyl] trimethylammonium methyl sulfate, was adopted to modify clay when the clay was used in UV-curable coatings [50]. The clays or related products are first treated with the intercalating agent to cause organophilation of the filler and enlarge the interlayer distance. Sometimes, commercial clays have been already treated with organic groups (see Table 1.2). After that, the organophiliated clay is incorporated into a polymer solution and stirred at a controlled temperature. The polymer chains gradually diffuse into the interlayer of the filler to further increase the interlayer distance and even cause exfoliation. This process is the "solution dispersion" technique [51]. The organically modified clay can also be introduced via "*in situ* polymerization" route in which treated clay is mixed with monomer and then polymerized. The interlayer distance enlarges during the propagation of polymer chains.

The intercalation and exploitation depend on the intercalation degree and are revealed for clay-containing nanocomposites. Actually, the morphology of clay-containing OINCs is also strongly related to the deagglomeration technique due to the initial powder state of clay. Figure 1.8 presents the typical dispersion state of clay in

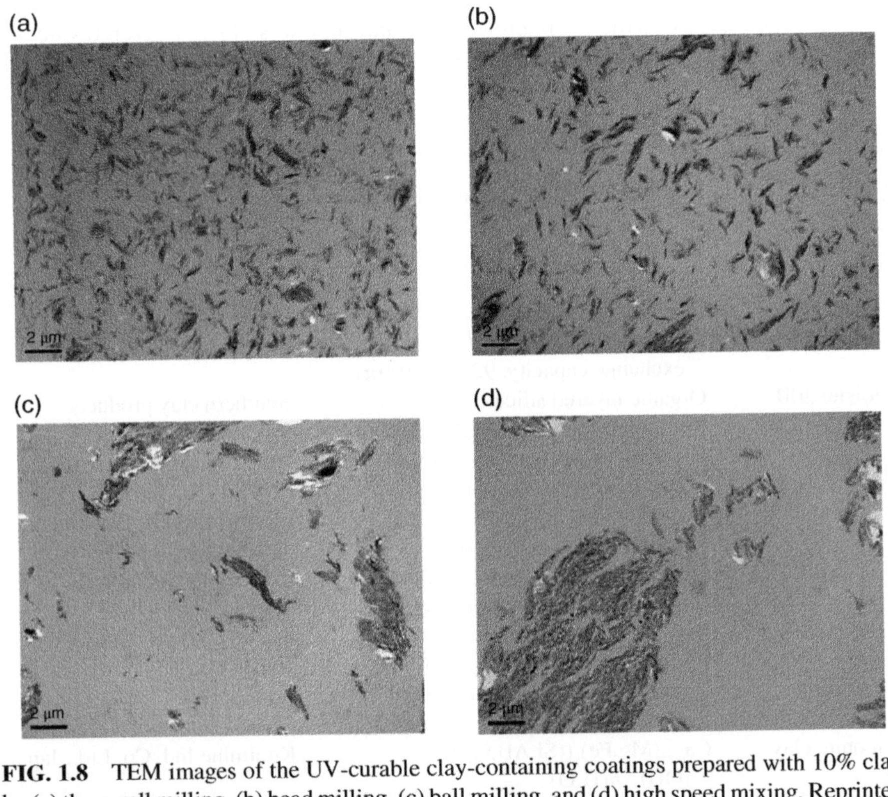

FIG. 1.8 TEM images of the UV-curable clay-containing coatings prepared with 10% clay by (a) three-roll milling, (b) bead milling, (c) ball milling, and (d) high speed mixing. Reprinted with permission from Ref. 52. © 2008 Elsevier.

UV-curable coatings prepared with various mixing techniques. The best dispersion is achieved with a three-roll mill and bead mill. For the ball milling and the high-speed mixing dispersion, large aggregates are observed. This deagglomeration behavior is quite similar to that observed in the deagglomeration of nanopowder.

Clay can improve the hardness, scratch resistance [53], and viscoelastic properties of coatings. Importantly, it can provide corrosion resistance and barrier properties superior to spherical particle fillers [54]. Enhancing corrosion resistance and barrier properties of H_2O and O_2 by clay was observed for different polymer matrices such as polyaniline, poly(*o*-ethoxyaniline), poly(methyl methacrylate) (PMMA), poly(styrene-*co*-acrylonitrile), etc. even at low loading levels (0.5–3%). This excellent barrier performance of polymer-clay nanocomposites is a result of the increased tortuosity of the diffusion pathway for oxygen and water. However, clay does not improve organic coatings with originally excellent mechanical and/or corrosion-resistant properties [53, 54].

Besides mechanical, barrier, and corrosion resistant applications, some clay-containing nanocomposite coatings have been reported in special applications. For example, Majumdar *et al.* developed an aqueous transparent nanocomposite coating consisting of laponite/polyvinylpyrrolidone (PVP)/poly(ethylene oxide) (weight ratio 35:15:50) as a fast drying, high-quality, image-receiving layer for inkjet printing on a variety of substrates such as polyester, polyethylene-coated, or polypropylene-laminated photo paper, GP paper, etc. [55]. Ranade *et al.* prepared a mixed exfoliated and intercalated polyamide–imide nanocomposite with montmorillonite (MMT) as magnetic wire coatings with reduced specific heat and improved Vicker hardness [56].

1.3 MECHANICALLY ENHANCED NANOCOMPOSITE CLEARCOATS

Clearcoats are generally used as topcoats in automobile, wood flooring, wood furniture, and optical plastic applications. High mechanical strength and excellent transparency are two essential properties of these coatings. Traditionally, the mechanical properties of clearcoats are determined by their macromolecular structure and cross-linking density especially for thermoset coatings. Nevertheless, the addition of inorganic nanoparticles provides a new way to improve the mechanical properties of clearcoats [57]. The nanoparticle-embedded coatings can retain the origin transparency of clearcoats because of the small size of the nanoparticle. The rigidity of clearcoats would be naturally enhanced due to the high hardness of inorganic nanoparticles because all inorganic materials have a higher hardness relative to the organic materials. Considering the economic cost and enhanced efficiency, only some inorganic nanofillers—nano-SiO_2, nano-Al_2O_3, nano-ZrO_2 particles, boehmite, and layered silicate—are feasible for the mechanical improvement of clearcoats.

The enhanced mechanical properties can be revealed from the change of the hardness (pencil hardness, pendulum hardness, and micro-indentation hardness) of the coatings upon inclusion of nanofillers. According to the "mixing rule," the hardness (H) of composite coatings could be approximately predicted to be:

$$H = H_p\phi_p + H_m(1-\phi_p) \quad (1.2)$$

where H_p and H_m represent the hardness of organic matrix and inorganic filler, respectively. However, two other aspects are not negligible for nanocomposite coatings. One is the interfacial phase between the organic matrix and the inorganic nanophase. The other is the change in the condensation state of the organic matrix initiated by the embedded nanophase. The volume fraction of the interfacial phase could be high because of the large specific surface area of nanoparticles. Consequently, the properties of the nanocomposite coatings are determined from the organic matrix, nanoparticle filler, and interfacial phase. Equation (1.2) thus becomes

$$H = H_p\phi_p + H_m(1-\phi_p-\phi_i) + H_i\phi_i \quad (1.3)$$

where H_i and ϕ_i are the hardness and volume fraction of interfacial phase.

Besides the specific surface area of nanoparticles, the volume fraction of the interfacial phase is related to the interaction distance of the nanoparticle impacting the matrix, the volume fraction, and the dispersion state of the nanofiller. The interfacial phase may be harder or softer than the organic matrix depending on the interfacial bonding mode between the organic chains and inorganic nanophase. Generally, chemical bonding leads to a hard interfacial phase while weak interactions such as Van der Waal force and hydrogen bonding (low number of anchoring points) produce a soft interfacial phase. Unfortunately, both the volume fraction and the mechanical properties of the interfacial phase are difficult to measure, and thus the mechanical properties of the nanocomposite coatings are difficult to theoretically predict. Nevertheless, Equation (1.3) can theoretically explain the complexity of the mechanical change in nanocomposite coatings or vice versa probe the properties of the interfacial phase.

For cross-linked clearcoats, the addition of nanoparticles will possibly impact the cross-linking density of the matrix. Actually, inorganic nanoparticles can be regarded as cross-linking points if strong interfacial bonding occurs. That is why the solvent resistance of thermoplastic coatings is enhanced via the addition of nanoparticles. On the other side, strong interfacial interactions will hinder the motion of organic chains and deteriorate the cross-linking behavior of the organic matrix. The double-face of the nanoparticle during cross-linking causes a diverse cross-linking structure of matrix. In addition, if the nanophase materials are introduced into a crystalline polymer matrix, the nanoparticles will influence the original crystallinity of the polymer. As a result, the mechanical properties of nanocomposite coatings is possibly dependent on the crystallinity rather than on the nanoparticle material itself. Consequently, variation in the mechanical properties of clearcoats via incorporation of nanoparticles is more complicated than that of microparticles. It must be determined empirically.

To date, many polymer clearcoats have been combined with inorganic nanophases to achieve mechanically-improved clearcoats. These clearcoats include solventborne two-component (2K) PU coatings, waterborne clearcoats, UV-curable coatings, etc. Thereafter, the mechanical improvement of the clearcoats due to the nanophase is introduced based on the type of clearcoats.

1.3.1 Solventborne Polyurethane Nanocomposite Coatings

Nano-SiO_2 particles are most frequently adopted to modify the mechanical properties of 2K PU coatings because they are affordable and available. Nano-SiO_2 particles could more efficiently increase the macro hardness, scratch resistance, elastic modulus of acrylic-based PU coatings than micro-silica particles [58]. They also enhance the microhardness and abrasion resistance of polyester-based PU coatings [59].

Fumed silica (10–40 nm) is one of the nano-SiO_2 particles that was first used in these coatings. These silica nanoparticles are a rheological additive for coatings. Because of the existence of hydroxyl groups in polyol resins, both hydrophilic and hydrophobic fumed silica are quite compatible with polyol resins. Thus, they are easily incorporated into the polyol resins or its solution.

Different groups have reported different results. Zhou *et al.* found that both hydrophilic and hydrophobic fumed silica (Wacker N-20 and Aerosil R972) have the same dispersion in acrylic-based PU coatings with very similar influences on the tribological properties (microindentation hardness, elastic modulus, and the critical force for crack) of acrylic-based PU coatings [60]. Jalili *et al.* also compared the addition effect of hydrophilic Aerosil TT600 and hydrophobic R972 on the 2K PU clearcoat from acrylic polyol (commercial name: Uracrone CY433)/Desmodure N75 [61]. They concluded that the incorporation of 4–8 wt.% of R972 in the 2K PU clearcoat gave optimal rheological, mechanical, and optical properties of the final nanocomposite coatings.

Barna *et al.* produced silica nanoparticles with an average particle size below 100 nm by flame synthesis [62]. These nanoparticles were treated with trimethylchlorosilane (TMCS), dimethyloctylchlorosilane (DMOCS), or APS. The treated or untreated silica nanoparticles were then blended with acrylic polyol (Setalux C-1184 SS-51)/Desmodur N 3300 to form PU nanocomposite coatings. They found that the lacquers containing untreated silica showed the best transmission results. This suggests that the treated silica nanoparticles have poor compatibility with the lacquer. These different results differ in the use of different organic solvents in the acrylic polyol solutions. When fumed silica was added to the polymer solution, the organic solvent imposes considerably on the dispersion of silica nanoparticles besides acrylic polyol. The dispersion of the treated nanoparticles is inferior to the untreated nanoparticles in Barna's case [62] and is because of the extreme nonpolarity (TMCS- or DMOCS-treated SiO_2) or the extreme polarity of silica nanoparticles. Consequently, adequate surface polarity of fumed silica would be desired for their combination with acrylic polyol.

The transparency of PU nanocomposite coatings with fumed silica deteriorates because of nano-SiO_2 aggregates. Completely transparent PU nanocomposite coatings are prepared preferentially using colloidal silica, another type of nanosilica particles. Colloidal silica particles are generated from two ways: ion-exchange of polysilicate and Stöber method with TEOS. The commercial silica sols or alcosols are usually manufactured with the ion-exchange route. Due to the large quantity of hydroxyl groups on the surface of the particles, colloidal silica particles are poorly compatible with the organic solvents that are contained in acrylic polyol solution (butyl acetate, xylene, etc.).

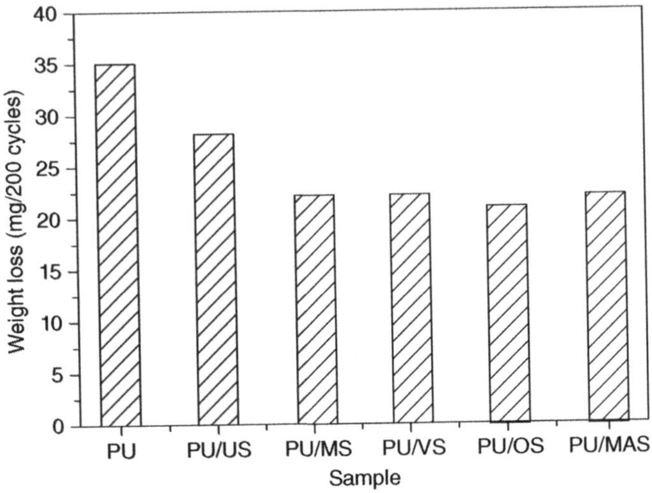

FIG. 1.9 Abrasion resistance of PU/silica composite films. Reprinted with permission from Ref. 63. © 2005 Elsevier. (200 cycles under 1000 g load, US, MS, VS, OS, and MAS represent unmodified, MTES-, VTS-, OTES-, and MPS-modified silica nanoparticles, respectively.)

In addition, the water and/or alcohol in the silica sol are not desired in the curing of polyol with isocyanate. Hence, the silica sol is not allowed to directly mix with acrylic polyol solution. A more complicated process has to be adopted such as the colloidal silica particles should be surface-modified, centrifuged from the sol, and then redispersed in monomers or acrylic polyol solution [63, 64]. Alternatively, the silica nanoparticles are modified in the sol state and then alcohol is substituted with butyl acetate through distillation [65]. Experiments indicate that MPS and octyltriethoxysilane (OTES) are better than methyltriethoxysilane (MTES) and VTS for the redispersion of the functionalized silica nanoparticles in acrylic polyol. The modified silica nanoparticles are superior to unmodified ones in improving the abrasion resistance of acrylic-based PU coatings. However, the type of surface modifier does not obviously influence the abrasion resistance of PU nanocomposite coatings (Fig. 1.9).

Unlike acrylic polyol, silica sol can be directly blended with polyester polyol (blending method) or with dicarboxylic acid and diol monomers and subsequent condensation polymerization [66]. The water and/or alcohol introduced by the silica sol can be removed through evaporation at elevated temperature. The latter *in situ* polymerization caused more polyester segments to chemically bond onto the surfaces of the silica particles than the blending method. This lowers the viscosity of the nanocomposite resins and increases the critical silica load for sharp increases in viscosity [67]. Better abrasion resistance is achieved by polyester-based PU nanocomposite coatings via *in situ* polymerization regardless of silica content and diameter (Fig. 1.10) [68].

Some other nanofillers improve the mechanical properties of solventborne PU coatings. Ahmadi *et al.* studied the effect of nanolayered silicates (Sud Chemie) on the properties of an automotive refinish clear coat based on 2K PU [69]. They found

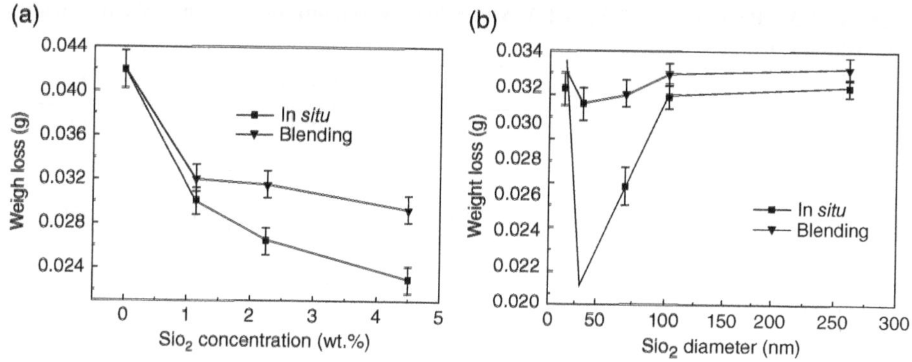

FIG. 1.10 Change of weight loss of PU/nano-SiO_2 composites as a function of (a) silica concentration (silica particle size 66 nm) and (b) silica diameter (silica content 2.25 wt.%) [68]. Ref. 68. © 2005 Wiley Periodicals, Inc.)

a significant improvement in scratch and mar resistance. Meanwhile, there was no gloss reduction due to nanofiller. On the contrary, gloss retention was increased by about 10% in the presence of only 3 wt.% nanofiller. Sabzi *et al.* embedded APS-modified TiO_2 nanoparticles into 2K PU coatings. The nanoparticles increase the mechanical strength and UV protection of PU coatings [70]. However, the transparency of the coatings is unavoidably decreased. Colloidal silica nanoparticles are the best candidate for mechanical improvements in solventborne PU coatings without scarifying its original transparency.

1.3.2 Waterborne Nanocomposite Clearcoats

Waterborne clearcoats are increasingly popular because they are environmentally friendly. However, they are usually weaker in mechanical performance and water resistance versus with their solventborne counterparts. In recent years, the addition of nanofillers was demonstrated to compensate their weakness, which promotes the development of waterborne coatings.

Because of the intrinsic hydrophilicity of inorganic nanofiller, they are very compatible with water and can be directly incorporated into waterborne resins, that is, polymer latex and aqueous polymer dispersion. However, when the aqueous nanocomposite resins are dried, the compatibility between the polymer chains and the inorganic nanofiller dominates the dispersion state of the nanofillers. Poor compatibility generally leads to aggregates of nanofillers during drying. It is fortunate that waterborne resins generally have a hydroxyl or carboxyl groups on their chains. These groups can interact with nanofiller via hydrogen bonding and providing compatibility of nanofillers with polymer matrix. Nevertheless, serious aggregation of nanofillers still exists at high nanofiller load because polymer chains cannot encapsulate the nanofiller during drying.

Prior encapsulation of nanofiller within polymer latex is another way to achieve waterborne nanocomposite coatings [71]. In this strategy, the preparation of a

TABLE 1.3 Properties of P(St-BA-AA)/Silica Nanocomposite Films With Various Nanosilica Contents[a]

	Run 15	Run 6	Run 7	Run 8	Run 9	Run 10
Nanosilica content[b] (wt.%)	0	9.1	16.7	23.1	28.6	33.3
Pencil hardness	2B	H	2H	2H	3H	4H
Pendulum hardness (s)	41	78	95	120	108	66
Flexibility (mm)	0	0	0	0	1	2
Impact strength (kg·cm)	40	50	50	50	40	40
Gel[c] (%)	0	35(28.5)	43(31.6)	56(42.8)	64(49.6)	77(65.6)

[a] Reprinted with permission from Ref. 75. © 2009 Society of Chemical Industry.
[b] Based on the total weight of nanocomposite film.
[c] The data in parentheses represent the weight fraction of insoluble polymer, which is calculated on the assumption that the weight loss after immersion in acetone results entirely from dissolution of the polymer chains.

polymer/nanofiller composite latex is crucial. To date, there are many reports on this topic. The methods include emulsion polymerization using a cationic initiator, 2,2′-azobis(2-methylpropionamide) dihydrochloride [72], seed emulsion polymerization [73], miniemulsion polymerization [74], etc. These will not be described in detail for space limitations. The nanocomposite latex can be used as the new filler for waterborne coatings or can directly form a film. Because the nanofillers in the composite particles generally have a strong interaction with the polymer matrix, aggregation does not happen during drying.

Silica sol is frequently used to improve the performance of polymer latex coatings. Table 1.3 gives the typical influence of colloidal silica nanoparticles on the mechanical properties of poly(butyl acrylate-*co*-styrene-*co*-acrylic acid) [P(BA-St-AA)] latex film. The pencil hardness of latex film increases from 2B to 4H when 33.3% silica nanoparticles are incorporated. Interestingly, the solvent resistance of the polymer latex is also improved [75].

Silica sols can be introduced into waterborne coatings via blending. However, it should be noted that the same surface charge of the silica nanoparticle with polymer latex is essential to avoid destabilization of the nanocomposite latex. The morphology of the polymer/silica nanocomposite latex film is rather complex depending on the size and volume fraction of the colloidal silica nanoparticles, the size ratio of polymer latex/silica, and the T_g of polymer latex [76]. Generally, transparent nanocomposite film can be achieved at low silica load. Because of the enrichment of inorganic nanoparticles at the interfaces, high silica content leads to silica frameworks or silica aggregates and thus may yield an opaque film [75]. Nevertheless, the silica framework can aid the formation of structure color films if monodispersed polymer latex is used.

Although transparent crack-free films can be prepared at silica loads as high as 70 wt.% based on 14 nm silica particles [76], the film thickness is limited to several micrometers due to its brittleness. This does not match the film thickness of the traditional latex paint. Silica nanopowders can also be blended with polymer latex by bead milling [77], but do not usually meet the optical transparency requirements for a clearcoat.

Beside nanosilica, nano-$CaCO_3$ is introduced into the waterborne coatings. Yao *et al.* incorporated nano-$CaCO_3$ into water-soluble PU resin via an *in situ* biomineralization process using CaO solution and CO_2 [78]. A transparent PU nanocomposite film was obtained at 2 wt.%. The storage modulus increased from 441 MPa of neat PU matrix to 1034 MPa of the nanocomposite film. In addition, the same nanocomposite films displayed a significant improvement in its water resistance. However, the transparency of the film beyond 2 wt.% remarkably declined. Gumfekar *et al.* synthesized nanocomposite latex of P(MMA–BA)/nano-$CaCO_3$ (53 nm) by *in situ* emulsion polymerization [79]. The nanocomposite latex was then added to a water-based coating. Superior coating with pencil hardness of 2H and gloss of 18 was obtained at MMA:BA (1:1), 4 wt.% nano-$CaCO_3$, and 50 wt.% alkyd emulsion. The control sample without nano-$CaCO_3$ had a pencil hardness of 3B and a gloss of 9. The impact strength decreases from 20 to 10 cm.

Needle-shaped boehmite, disc-shaped laponite, and MMT platelets were used to improve the one-component (1K) self-crosslinkable acrylic dispersion. Hydrophilic nanoparticles at the interfaces of the polymer droplets and hydrophobic nanoparticles completely or partially located inside the polymer droplets dramatically increased the cured film stiffness [80]. Nevertheless, the dependence of rheological behavior on the dispersion and location of nanoparticles has not been considered. Nitric acid–treated multiwalled carbon nanotubes (A-MWCNT) were added to a water-soluble PU resin (WBPU) [81]. The initial tensile moduli and tensile strengths of the nanocomposite film with 1.5 wt.% loading of A-MWCNT were enhanced by about 19 and 12%, respectively, compared to the corresponding values for the original WBPU film.

1.3.3 UV-Curable Nanocomposite Coatings

UV-curable nanocomposite coatings can be prepared via three routes: (1) Nanoparticles are *ex situ* modified (see Section 1.2.1.2) and then dispersed into the coatings (the *ex situ* method); (2) nanoparticles and surface modifier are added to the coating formulation and the modification of nanoparticles is finished during fabrication (the *in situ* method); and (3) commercial nanoparticle dispersions in monomer (Table 1.4) are used as additives for UV-curable coatings. In comparison with Route 1, Routes 2 and 3 are easier for industrialization of UV-curable nanocomposite coatings.

Much attention has been given to the curing behavior of UV-curable nanocomposite coatings because the curing behavior strongly determines their mechanical properties. Many studies suggest a decline in the photopolymerization rate and final carbon–carbon

TABLE 1.4 Some Commercial Nanoparticle Dispersions in Monomers

Commercial Name	Composition	Supplier
Highlink OG 100	A 50 nm SiO_2 in HEMA at 30 wt.%	Clariant
Highlink NanO G 103-53	A 50 nm SiO_2 in HDDA at 50 wt.%	Clariant
Highlink NanO G 107-53	A 50 nm SiO_2 in DPGDA at 50 wt.%	Clariant
Nanocryl C 145	A 15 nm SiO_2 in TPGDA at 50 wt.%	Hanse Chemie
Nanobyk 3602	A 45 nm Al_2O_3 in HDDA at 30 wt.%	Byk-Chemie
Nanobyk 3601	A 40 nmAl_2O_3 in TPGDA at 32 wt.%	Byk-Chemie
C150	50 wt.% sol–gel-formed silica nanoparticles and 50 wt.% TMPTA	Nanoresins AG, Germany

double bond conversion after addition of nanoparticles. This alleviates the reinforced efficiency of nanofiller in UV-curable coatings. The negative effects of nanofillers on curing behavior may result from three aspects: First, nanoparticles (from nanopowder) cannot deagglomerate very well in UV-curable coatings. These aggregates will scatter the light resulting in a reduced UV-irradiation intensity. Because light scattering is strongly related to the RI of nanoparticles, better curing is seen in nanocomposite formulations with relatively low-refractive-index nanofillers. For example, Landry *et al.* found that a formulation prepared with alumina nanoparticles (from nanopowder, size: 13 nm) via a *in situ* preparation presented higher levels of curing and a higher rate of reaction [82]. On the contrary, zirconia decreased the conversion and the rate of curing. This may be because of the increase in light scattering caused by the high RI of zirconia, around 2.14. This is quite high compared to alumina and acrylate resin. In another case, UV waterborne nanocomposite coatings containing nano-Al_2O_3 or MPS-functionalized nano-SiO_2 (concentration: 1, 3, and 5 wt.%), differential scanning calorimetry (DSC)-exothermic curves showed that the photopolymerization was less efficient when nanoparticles were introduced, but the UV-curing of coatings based on functionalized nano-SiO_2 were better than those containing nano-Al_2O_3 [83]. Second, increased viscosity of nanocomposite coatings restricts the mobility of acryl radicals [52]. Third, compared with the UV-curable coatings without nanofiller, vitrification occurs early for the UV-curable coatings because these polymerization-active nanoparticles act as multifunctional monomers. Usually, a higher functionality of monomer results in earlier vitrification due to the higher cross-linking efficiency. For example, due to a vitrification effect, decreased photopolymerization rate and epoxy-group conversion are observed for photopolymerization of epoxy resin/GPS-modified Fe_2O_3 nanocomposite coatings with increasing amount of iron oxide in the photocurable formulations [84].

Fast curing rates and a high final conversion are generally seen at low nanoparticle content or for those completely transparent nanocomposite coatings. For instance, a remarkable increase in the UV curing rate and final conversion is seen only for coatings prepared at 1 wt.% clay using bead milling (Fig. 1.11) [52]. Unchanged or deteriorated curing behaviors are demonstrated for other samples due to poor dispersion of clay and high viscosity. For the coatings based on methyltrimethoxysilane (MTMS)/MPS comodified colloidal silica nanoparticles and acrylic resin, the

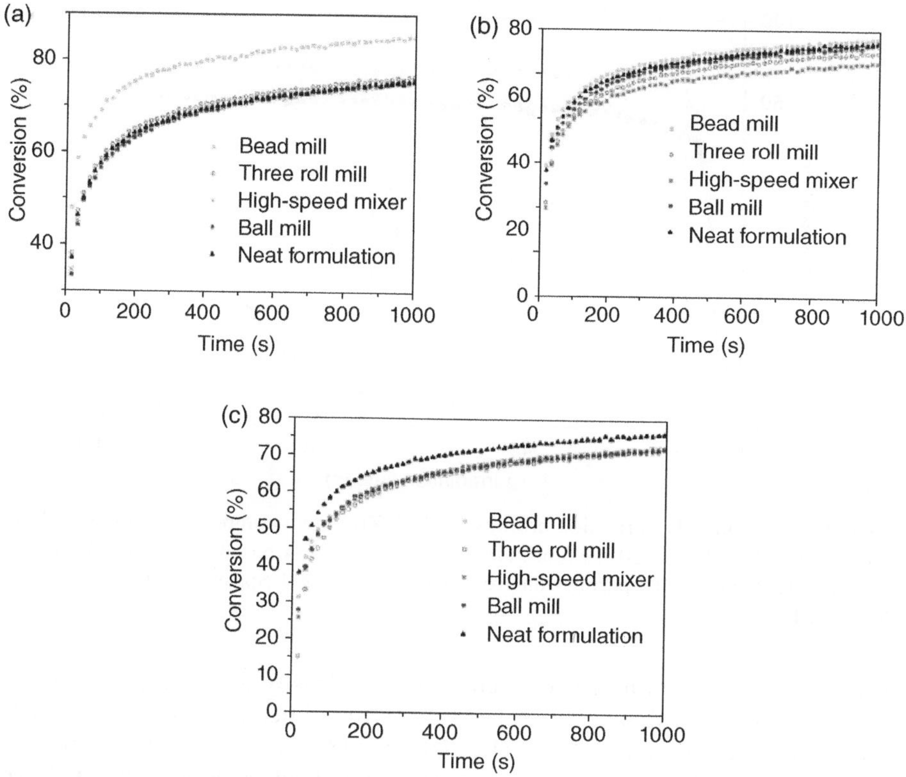

FIG. 1.11 Percent conversion versus time plots of formulations prepared with 1, 3, and 10 wt.% of clay with the roll mill, the ball mill, the bead mill, and the high-speed mixer. Reprinted with permission from Ref. 52. © 2008 Elsevier.

conversion degree of the C=C bond increased up to 85–88% at composition ranges of 0–40 wt.%. However, the conversion degree decreased to 50% at 80 wt.% colloidal silica nanoparticles because of the screen effect of nanoparticles on the polymerization of the acryl or methacryl radicals [85]. In the nanocomposite coatings based on nonaqueous synthesized ZrO_2 nanocrystal (3.8 nm), the final conversion steadily increases from 80.5 to 91.3% as the ZrO_2 content increases from 0 to 20 wt.%. It declines to 68.0% at 25 wt.% of ZrO_2 because of phase separation (Fig. 1.12) [31]. The increased limited conversion of UV-curable nanocomposite coatings may be caused by the reduced oxygen inhibition because inorganic nanoparticles can enhance the barrier property of the coating.

Mechanical improvement of nanoparticles has been widely demonstrated for UV-curable coatings. Bauer *et al.* [86] showed that fumed silica can improve the abrasion and scratch resistance of UV-curable coatings. Soloukhin *et al.* found that nanocolloidal silica particles could enhance the elastic modulus and hardness of acrylic-based UV-curable coatings on a polycarbonate (PC) substrate [87]. Even for those cases with negative effects of nanofiller on UV curing, the addition of

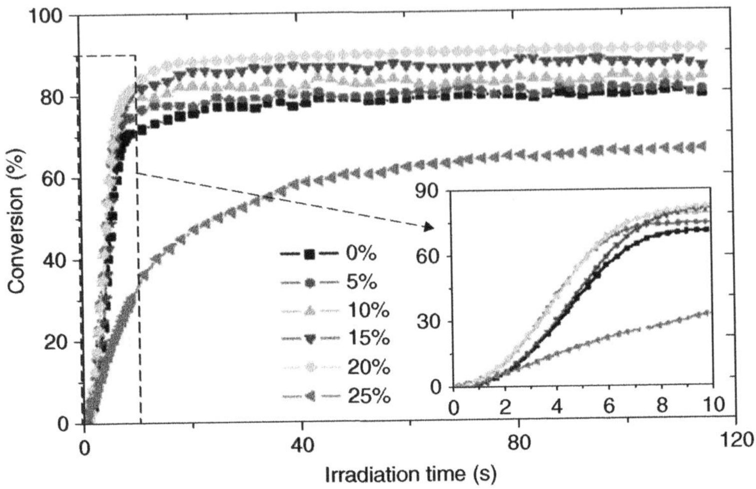

FIG. 1.12 Photopolymerization profiles of PUA/ZrO_2 nanocomposite coatings with different MPS-ZrO_2 loads (10 mW/cm^2, air, 5 wt.% Iragure 184 based on the weight of PUA coating). Reprinted with permission from Ref. 31. © 2009 Springer Science + Business Media, LLC.

nanofiller still results in a rise of cross-linking density and hence an enhanced rigidity of coatings. Kang *et al.* embedded MPS-modified colloidal silica particles (20 nm) into acrylic resin (SK CYTEC, UP 053) [85]. The coating keeps the same transparency as the neat material even with silica loads as high as 80 wt.%. The nanoindentation hardness steadily increases from 0.41 to 0.59 GPa when the silica load increases from 0 to 80 wt.%. Kim *et al.* found that when 10 wt.% MPS-modified ZnO nanoparticles were added, the hardness and elastic modulus of UV-cured PU coating increased from 0.03 to 0.056 GPa and from 2.75 to 3.55 GPa, respectively [88]. However, the reinforced efficiency of nanoparticles in UV-curable coatings strongly depends on the type of nanoparticle, the surface modifier, as well as the polymer matrix [89]. Leder *et al.* incorporated 5–15% treated pyrogenic silica into Laromer PO 43 F oligoether acrylate [90]. An improvement in the mar resistance of 10–35% was seen. This refers to remaining gloss after scrubbing. Meanwhile, the viscosity of the liquid coatings was more stable. Zhang *et al.* mixed C150 sol with a urethane-acrylate oligomer to get a UV-curable nanocomposite coating [91]. They saw a 20% decrease in the coefficient of friction of a coating filled with 40 wt.% nanosilica particles relative to unfilled coating. Under the same fretting test conditions, the wear rate in terms of wear volume of the hybrid coating containing 40 wt.% nanoparticles was about 70 times lower than that of the neat coating, confirming the wear-reduction capability of the nanoparticles. In the UV-waterborne PU-acrylate nanocomposite coatings, the addition of nanoalumina and nanosilica decreases the hardness because of nanoparticle aggregation. Versus the neat coating, the scratch resistance (gloss retention after scratch) of the nanocomposite coatings was significantly improved. Other

TABLE 1.5 Haze and Diamond Microscratch Hardness of Pure SR494/HDDA (1:1) Polyacrylate Film and Nanocomposite Coatings (9 wt.% SiO_2)[a]

	SR494/HDDA	Unmodified Pyrogenic Silica	MPS-Coated Silica	Colloidal Silica
Haze (%)	85	45–65	25–35	50–65
Hardness (N)	1.0	3.0	2.5	1.5

[a] From Ref. 96. © 2002 WILEY-VCH Verlag GmbH & Co. KGaA.

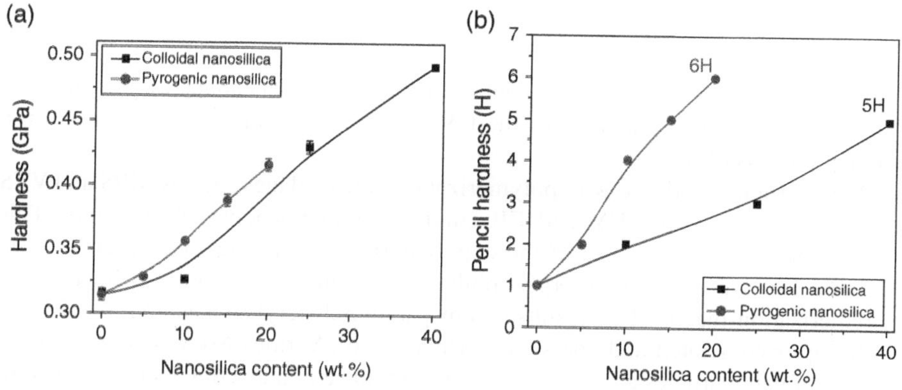

FIG. 1.13 Nanoindentation hardness (a) and pencil hardness (b) of the coatings samples as a function of the silica nanoparticle content. Reprinted with permission from Ref. 3. © 2011 Elsevier.

nanofillers, such as clay [92, 93], boehmite [94], nanotitania [95], have also been employed to enhance the hardness and scratch resistance of UV-curable coatings.

Similar to solventborne or waterborne nanocomposite coatings, pyrogenic silica and colloidal silica are frequently adopted and compared in terms of efficiency and mechanical improvement. In Bauer's report, a distinct improvement in the mechanical properties of a UV-curable SR494/HDDA coating was seen for surface-modified pyrogenic silica compared with colloidal SiO_2, as seen in Table 1.5 [96]. It is assumed that the higher density and hardness of the fumed nanoparticles give better abrasive resistance at the same filler content. Zhang *et al.* also compared the influence of colloidal and pyrogenic silica on the properties of PU coatings [3]. The results are presented in Fig. 1.13. In comparison with colloidal nanosilica-filled coatings, the pyrogenic nanosilica-filled ones had a higher modulus, hardness, wear resistance (in terms of pencil hardness and fretting resistance) as well as slightly lower optical transmittance, gloss, and higher haze. Bautista *et al.* found that nanoalumina did not influence wear resistance [89]. However, nanosilica did modify the elastic and plastic deformation in different ways depending on the polymer matrix and improve the wear resistance of the coating. These different behaviors of nanosilica and nanoalumina in the wear resistance may be due to the chemical nature of the particles or to the surface treatment of the particles to obtain stable commercial suspensions.

TABLE 1.6 Abrasion, Haze, and Diamond Microscratch Hardness of Pure SR494 Polyacrylate and Nanocomposite Coatings (ca. 25 wt.% SiO_2, Modified by Different Silanes)[a]

	Unmodified	Modified With MPS-Coated Silica	Modified With VTS-Coated Silica	Modified With PTS-Coated Silica
Abrasion (mg)	53.1	21.1	16.4	19.8
Haze (%)	85	14.4	3.2	8.5
Hardness (N)	1.0	1.5	2.5	1.7

[a]From Ref. 96. © 2002 WILEY-VCH Verlag GmbH & Co. KGaA.

Nevertheless, enhancement of scratch resistance is independent of the intrinsic hardness of inorganic domains for the UV-curable nanocomposite coatings from a sol–gel process [97].

Nanoparticles modified with polymerization-active silanes such as MPS and VTS form cross-links within UV and EB curable acrylate/nanoparticle systems. This changes the viscoelastic data of the copolymerized composites. However, as seen in Table 1.6, even the simple organophilation of nanosized pyrogenic silica by the polymerization-inactive PTS results in transparent polyacrylate nanocomposite films with improved scratch and abrasion resistance [96]. Various MPS coverage on the surface of nanoparticles also impacts the performance of nanocomposite coatings. In a MPS-modified SiO_2/HDDA/acrylate resin, it was shown that as the MPS/SiO_2 weight ratio increased from 0.2 to 0.6, the dispersion, compatibility, and cross-linking density between the MPS-modified SiO_2 particles and acrylate resin were improved, leading to an increase in pencil hardness from 4H to 6H as well as increased abrasion resistance [98, 99]. Interesting mechanical results were also found in the UV-curable nanocomposite systems using sol–gel derived titania nanoparticles modified with a isopropyl tri(dioctyl)pyrophosphato titanate coupling agent. Both the hardness and flexibility of the photocured nanocomposite films were improved simultaneously, in contrast to neat organic UV-curable formulations [95].

Some studies clearly show that the mechanical increase in nanocomposite coatings is related to the type of polymer matrix. Table 1.7 gives the parameters obtained in wear tests and scratch tests for the nanocomposite coatings based on acrylated polyester resin (A series), acrylated epoxy resin (B series), and acrylated urethane resin (C series) [89]. Improvement in the specific rate of wear was seen for all three coatings, but the increased critical load of scratch start (L_{C1}) with increasing nanosilica load was only seen for the softest polyester coating and hardest epoxy coatings. Figure 1.14 shows the pendulum hardness (Koenig hardness) of the nanocomposites coatings as a function of ZrO_2 content [32]. The efficiency (namely, the slope) for enhancing the hardness of the coating is independent of the inherent hardness of the organic matrix but is strongly related to the amount of PU oligomer in the UV-curable formulation. Lower slopes are seen at higher amounts of PU oligomer because of the low compatibility of PU oligomer with MPS-modified ZrO_2 nanoparticles.

TABLE 1.7 Wear and Scratch Parameters for the Nanocomposite Coatings[a]

		Wear test		Scratch test			
Sample	Nanosilica g/100 g polymer	Specific rate (kg/ mN × 10^9)	Total penetration depth[b] (μm)	Elastic recovery[b] (%)	Residual depth[b] (μm)	L_{C1}[c] (mN) ± 100	L_{C3}[c] (mN) ± 100
AH1	0	7.6	22.2	98.6	0.31	590	—
AH2	5	6.4	17.2	98.6	0.24	640	—
AH3	20	4.7	14.3	98.9	0.16	960	1690
BH1	0	11.0	3.8	77.6	0.85	530	—
BH2	5	9.1	3.3	74.7	0.84	550	—
BH3	20	4.3	5.1	912	0.45	530	—
CHI	0	9.1	7.2	99.0	0.07	400	1620
CH2	5	7.3	10.3	96.9	0.32	400	1540
CH3	20	8.5	9.7	97.9	0.20	210	1670

A, polyester matrix; B, epoxy matrix; C, urethane matrix.
[a]Reprinted with permission from Ref. 89. © 2011 Elsevier.
[b]Determined at 1570 mN.
[c]L_{C1}, critical load of scratch start and L_{C3}, critical load of layer breaking.

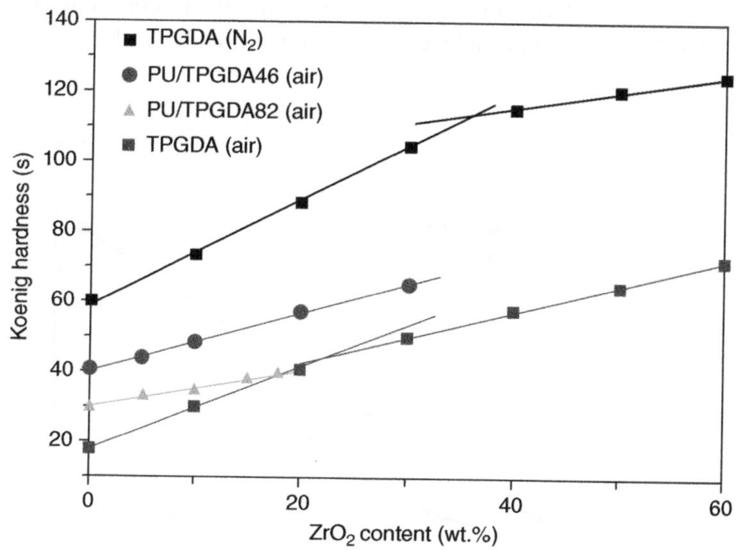

FIG. 1.14 The pendulum hardness of the samples with different ZrO_2 contents. Reprinted with permission from Ref. 32. © 2010 Elsevier. Symbol, the determined value; solid line, linear fitting curve.

1.3.4 Other Mechanically Strong Nanocomposite Coatings

Protection coatings against scratch are critical for transparent plastics such as PC and PMMA. One of the routes for fabrication of these coatings is to mix nanofillers with organoalkoxylsilane-containing hybrid binder. With this route, nanofillers work as dense hard phase, cross-links, and may even serve as the catalyst.

GPS-based hard nanocomposite coatings have been frequently reported since the 1990s. Schmidt *et al.* fabricated a ultrahard coating from acetic acid-modified boehmite powder (8–17 nm) and an I/O hybrid binder that was composed of GPS, aluminum alkoxide, and TEOS [19, 100]. The coatings cast on a PC substrate showed prominent scratch resistance in a modified Vickers test (Fig. 1.15). A scratch resistance below 2% haze was also obtained in 1000 cycles of the Taber abrading test—this is comparable with that of floatglass (1.5% haze after 1000 cycles). The boehmite particles served as the cross-linking catalyst for epoxides in this example also.

Daniels and Francis reported GPS-modified colloidal silica coatings based on the addition of GPS to a commercial acidic silica sol [101]. The GPS/silica weight ratio is crucial to both the critical cracking thickness and the abrasion resistance. The critical cracking thickness of the coatings increased dramatically when the weight ratio of GPS/silica surpassed 0.4. Better abrasion resistance was obtained at a GPS/silica weight ratio of 0.5. Excess GPS content deteriorated the abrasion resistance of the coatings. Mosher *et al.* reported the synthesis of three particles (SiO_2, ZrO_2/CeO_2, CeO_2) that reinforced water-based nanocomposite coatings. The films are sol–gel derived using a nonionic surfactant, aluminum perchlorate [$Al(ClO_4)_3$] catalyst, nanoparticle colloids, and GPS [102]. The silica colloid

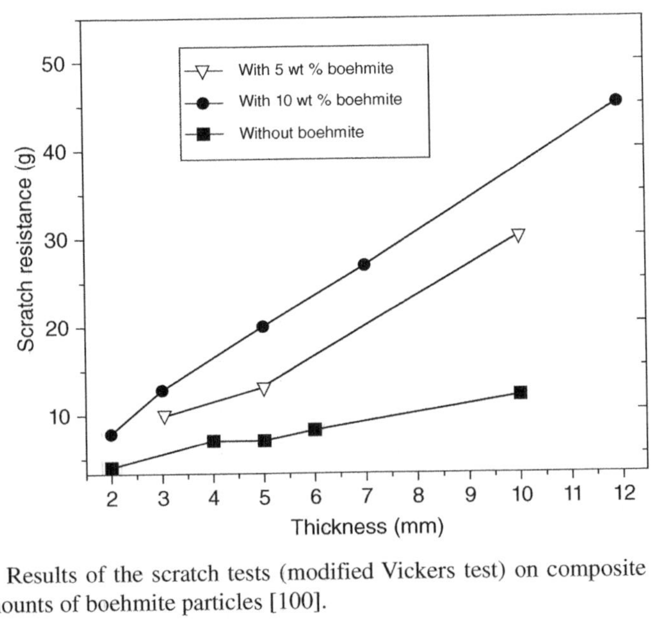

FIG. 1.15 Results of the scratch tests (modified Vickers test) on composite coatings with different amounts of boehmite particles [100].

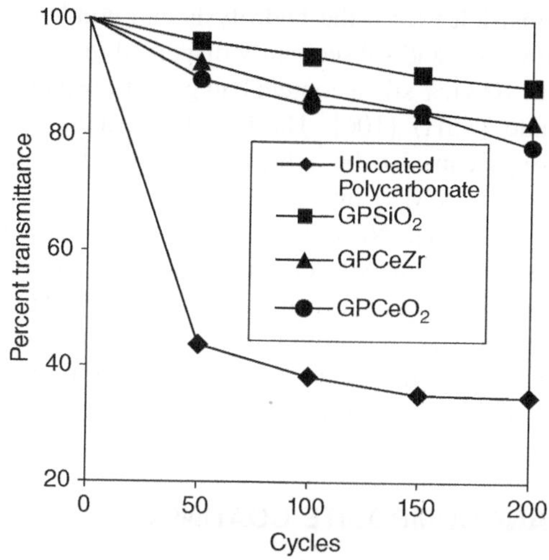

FIG. 1.16 Light transmittance in wear track as a function of wear cycles. Reprinted with permission from Ref. 102. © 2006 Elsevier.

composites ($GPSiO_2$) provide the best mechanical properties without decreasing the unique optical properties of inorganic materials (Fig. 1.16).

Chen *et al.* fabricated transparent GPS composite coatings filled with up to 40 wt.% boehmite nanoparticles and nanorods [103]. The hardness and modulus of the boehmite nanorod filled coatings were slightly lower than those of the same coating filled with boehmite nanoparticles. However, nanorods with an aspect ratio around 20 significantly improve the crack toughness of the GPS composite coating. The authors attributed this observation to the formation of cross-ply structures and the orientation of nanorods in the coatings.

Ultrahard nanocomposite coatings can also be prepared based on MPS and MTMS precursors. For example, nanoscaled AlOOH particles were added to MPS and hydrolyzed in deionized water. The mixture was then diluted with 1-butanol and blended with the UV initiator BYK 306 to produce UV-curable hard coatings. The coatings can be sprayed on plastic substrates (PC, PMMA) to form a transparent film with a thickness of several micrometers. The transparent coatings show excellent adhesion on PMMA and PC (GT/TT=0/0, DIN 53151) even in the absence of primers. Their abrasion resistance after the Taber test was haze values of 10% after 1000 cycles (CS 10F rolls, 5.4 N, DIN 52347) [104]. In another case, scratch-resistant hard coatings were successfully prepared by mixing a basic colloidal silica sol with hydrolyzed MTMS using the sol–gel method. Nanocomposite coatings were applied to PMMA substrates by dip coating and cured in a thermal oven. The presence of SiO_2 nanoparticles in this siloxane-based coating at optimal quantities—40, 50, and 60 wt.% of SiO_2—enabled highly transparent coating films with good hardness and scratch resistance [105].

Besides organoalkoxylsilane-based ultrahard nanocomposite coatings, mechanical protection coatings can also be prepared by mixing nanofillers with polymer. For instance, silica nanoparticles from the hydrolysis and condensation of TEOS were surface modified with MPS and subsequently dispersed into dissolved PMMA solution. The coating with 10 wt.% silica content had good transparency (>90%) and high pencil hardness grade (>5H) [106]. Hard (4H), scratch resistant, and flexible nanocomposite coatings can also be achieved through the introduction of 3 wt.% unmodified nanoclay (Na^+MMT) into a polymer resin based on a hyperbranched polyester Boltorn H30 [107]. Nevertheless, the organoclay prepared by cation exchange between sodium ion MMTs and octadecyl ammonium increases the hardness of alkyd clear coatings much more dramatically. This may be due to the formation of exfoliated structures resulting from the high compatibility of long-chain hydrocarbons of octadecyl ammonium salts with the long-chain hydrocarbons of the fatty acids in the alkyd resin [108].

1.4 OPTICAL NANOCOMPOSITE COATINGS

Because nanocomposite coatings are transparent in visible range, they have high potential as optical coatings including UV-shielding coatings [109], RI-adjustable coatings, NIR-shielding coatings, nonlinear optical coatings [110], etc.

1.4.1 Transparent UV-Shielding Nanocomposite Coatings

Traditionally, choices for protecting coatings from the sun include organic UV absorbers as hindered amine light stabilizers. Organic UV absorbers are colorless or nearly colorless compounds having high absorption coefficients in the UV region. They protect coatings against photoinduced damages by absorbing the harmful solar radiation. The addition of organic UV absorbers protects the substrate, but the effect is not permanent and such stabilizers tend to migrate or decompose during exposure. An alternative is inorganic metal oxide nanoparticles. These do not migrate from the matrix and exhibit excellent photo- and thermostability. Hot topics in inorganic nanoparticles for UV-shielding coatings include nano-TiO_2, nano-ZnO, and nano-CeO_2.

TiO_2 nanoparticles have two main crystalline forms: rutile and anatase. The rutile nano-TiO_2 with high UV opacity and lower photoactivity is an effective UV protector in coatings [111]. Figure 1.17 compares the UV–vis spectra of the cured alkyd films containing organic UV absorber, rutile nano-TiO_2, and anatase nano-TiO_2. The films with organic benzophenone or benzotriazole are excellent at UV-blocking and highly transparent in the visible range regardless of film thickness (0.32 or 0.85 mm). The UV blocking of these films with nano-TiO_2 is incomplete at 300–400 nm because of their inherent wide band gaps. UV-shielding performance can be enhanced by increasing the film thickness. Nevertheless, the visible transmission of thick films via nano-TiO_2 considerably declined. In other words, better UV blocking with TiO_2 nanoparticles usually leads to poorer transparency of the coating [112].

Low transparencies or glosses of UV-block nanocomposite coatings were also demonstrated for APS-modified TiO_2 nanopowder in poly(vinyl alcohol) [113], GPS-modified TiO_2 nanoparticles in polyacrylic clearcoats [114], and heptaisobutyltrisilanol

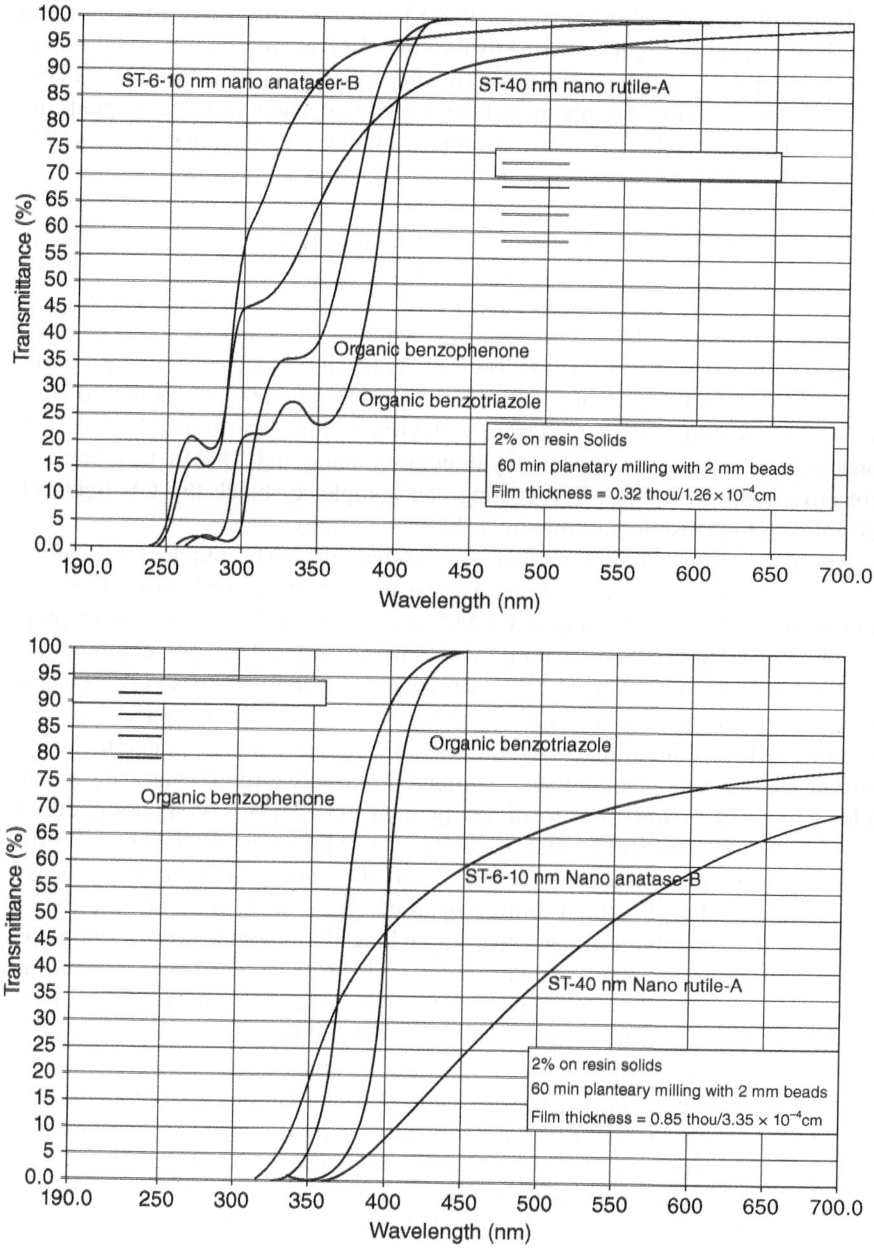

FIG. 1.17 Transmission spectra of Lowilite 24 and Lowilite 26 compared with nanoparticles of anatase-B and rutile-A in a cured alkyd resin film at 2% w/w on resin solids. Reprinted with permission from Ref. 111. © 2002 Elsevier. Film thickness: a, 0.32 mm; b, 0.85 mm.

polyhedral oligomeric silsesquioxanes (POSS)-modified TiO_2 nanoparticles in waterborne coatings [115]. Light scattering by the nano-TiO_2 aggregates is due to the opacity of the coatings. The high RI of the nano-TiO_2 in particular further consolidates the light scattering. In addition, the photocatalytic activity of nano-TiO_2 can degrade most polymer matrices and shorten the durability of the UV-shielding coatings. These two shortcomings seriously restrict the use of TiO_2 nanoparticles as UV absorbers in coatings.

To depress the photocatalytic activity of TiO_2 nanoparticles, they can be encapsulated with SiO_2 or Al_2O_3 to form core/shell composite particles [114, 116] or create rattle-type TiO_2@void@SiO_2 particles [117]. The use of a polymer matrix, that is, polysiloxane that can tolerate photocatalytic degradation, is another solution for fabricating UV-shielding coatings [118]. The opacity of TiO_2-based coatings can be solved using a commercially available colloidal TiO_2 sol. Hwang *et al.* [119] treated HIT-30 M TiO_2 sol (Nissan Chemicals, Japan) with GPS and then performed hydrolysis and condensation reactions with dimethyldimethoxysilane and MTMS to produce UV protective coatings for PC. The coating can completely block the UV light below 350 nm with high visible transmission (>85% at 550 nm).

Transparent TiO_2-based, UV blocking coatings can also be prepared by combining TiO_2 nanophase *in situ* via a sol–gel process using polymer or hybrids. Chen *et al.* employed trialkoxysilane-capped PMMA and titanium(IV) *n*-butoxide to prepare titania-based hybrid optical coatings via an *in situ* sol–gel process combining spin coating and multistep baking [120]. The prepared hybrid films show very high optical transparency in the visible region. The shift in the absorption maximum of the prepared hybrid thin films correlates with the titania content. Sangermano *et al.* introduced TiO_2 nanophase into the cationic UV-curable formulations using nanoparticles *ex situ* or *in situ* from a sol–gel process [118]. The UV-shielding effect of *in situ* TiO_2 is comparable to that of *ex situ* TiO_2. Moreover, the *in situ* TiO_2 offered transparent coatings without interfering with the photopolymerization process or compromising the UV-cured film properties. Mazzocchetti *et al.* reported the synthesis of new polymer-titania hybrids, where the organic phase contains polyesters such as poly(ε-caprolactone), poly(D,L-lactic acid), and poly(L-lactic acid) [121]. Hybrids coated on transparent substrates have their intrinsic optical transparency and the ability to completely block UVB and UV-A2.

When textiles are coated with hybrids, they become radiopaque thus offering new personal protective clothing and equipment. Que and Hu prepared I/O hybrid films using GPS, MTES, and tetrapropylorthotitanate as precursors [122]. The absorption band red shifts as the titanium content in the coatings increases (Fig. 1.18). Complete blocking of UV light below 350 nm was realized when the titanium molar content reached 0.4 M. Xiang *et al.* prepared hybrid materials *via* an aqueous sol–gel technique from tetra-*n*-butyl titanate as the precursor of titania in the presence of MQ silicone resin [123]. The hybrid nanocomposites were colorless and highly transparent in visible light region. It absorbed light below 320 nm.

Although the coatings generated *in situ* with TiO_2 nanophase are highly transparent, they are usually yellow especially those with wide band blocking performance. The colorful appearance will restrict their applications. To avoid phase separation,

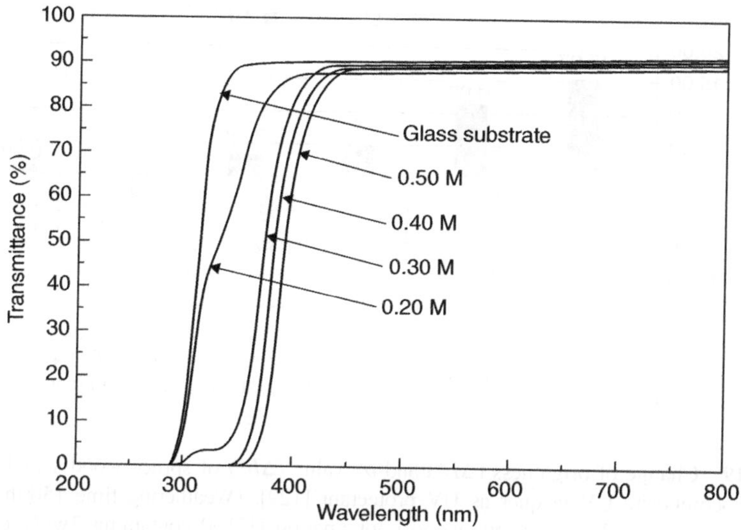

FIG. 1.18 Dependence of the optical transmittance of the films heated at 100°C on titanium content. Reprinted with permission from Ref. 122. © 2013 IOP Publishing Ltd.

chemical bonding between the TiO_2 nanophase and the polymer or hybrid matrix is usually designed in the sol–gel-derived TiO_2-based UV blocking coatings. This chemical bonding design suggests that the coatings have to be 2K coatings—one component for prehydrolyzed TiO_2 sol (or TiO_2 precursor) and the other for the binder (polymer or hybrid). The pot life of coatings thus limits their utility. The sol–gel-derived TiO_2 nanophase in the coatings is usually amphorous. Its photocatalytic activity would be low however, and may not be negligible for long-term service. Unfortunately, the effect of the photocatalytic activity of amphorous TiO_2 on the durability of UV-block coatings is seldom reported and unclear. Nevertheless, similar to coatings based on TiO_2 nanoparticles, the polysiloxane or sol–gel-derived silica hybrids would be a better choice as the binder of UV blocking coatings with TiO_2 nanophase generated *in situ*.

ZnO nanoparticles are a semiconductor with a band gap of 3.37 eV and have been widely used as UV absorbers in sunscreens [124], textile fibers [125], transparent packing materials for UV-light based white light–emitting diodes [126], and especially coatings. For example, Li *et al.* indicated that nanosized ZnO particle could be an UV absorbent for I/O hybrid coatings derived from TEOS and GPS on a PMMA substrate [127]. Khrenov *et al.* incorporated ZnO nanoparticles prepared by inverse emulsion polymerization into a solution of poly(2-ethylhexyl methacrylate), PMMA, and PS [128]. These nanocomposite films demonstrated good UV absorption. Weichelt *et al.* introduced various kinds of commercial ZnO nanoparticles into a UV-curable wood lacquer [129]. Spruce wood coated with the ZnO-based nanocomposites showed comparable brightness change (ΔL^*) but decreased yellowing factor (Δb^*) versus those in the conventional HALS/UVA system (Fig. 1.19). Even better protection can be realized using a combination of lignin protecting impregnation and nano-ZnO in

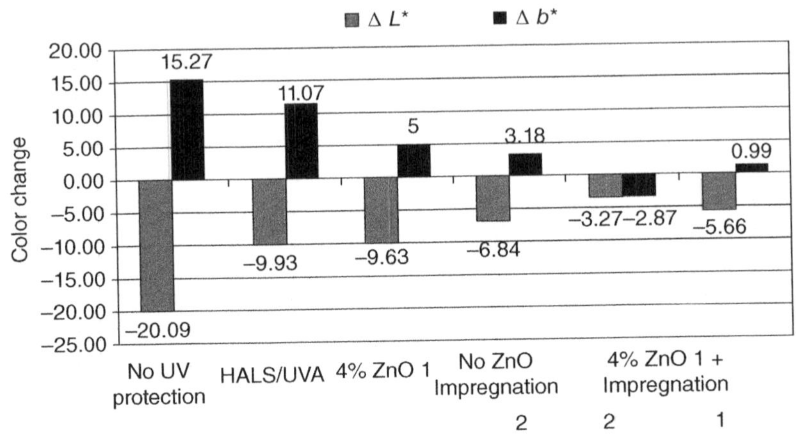

FIG. 1.19 Change of brightness (ΔL^*) and b^* value (Δb^*) of spruce wood samples coated with nanocomposite UV lacquer as UV protectant [129]. (Weathering time 1500h, ZnO 1: 20nm, Impregnation 1: an aqueous wood impregnation H5100 containing 2wt.% of a solid lignin protecting agent 1; Impregnation 2: containing 5% of water-based lignin protecting formulation 2.) Ref. 129. © 2010 WILEY-VCH Verlag GmbH & Co. KGaA.

the top coat. Kim *et al.* found that UV light below 450nm can be efficiently absorbed by incorporating ZnO nanoparticles into a UV-curable PU matrix [88]. However, because of their intense scattering of visible light, the pigmentary grades of ZnO nanoparticles usually produce opaque systems when they are embedded into polymer matrices [88, 129]. This seriously restricts their applications in highly transparent materials.

ZnO quantum dots (ZnO QDs) produce highly transparent ZnO-based UV-shielding coatings. They have very small size (usually <10nm) and retain the excellent UV absorption capability making them good inorganic UV absorbers. Some UV-shielding polymer-based nanocomposites, such as PMMA/ZnO nanocomposites [130], PS/ZnO thin films [131], and poly(butyl methacrylate) (PBMA)/ZnO nanohybrid films [132] have been reported. Figure 1.20 shows the UV–vis absorbance and transmittance spectra of PBMA/ZnO nanocomposite films prepared by bulk polymerization [132]. The nanocomposite films can completely block the UV light below 350nm while maintaining the same transparency as pure PBMA film with greater than 90% visible light transmittance regardless of the nanophase content. Nevertheless, Jeeju *et al.* found that the absorption window shifts toward a shorter wavelength as the size of the ZnO nanocrystals decreases [133]. This implies that the UVA and even part of UVB may not be blocked by nanocomposite coatings containing ZnO QDs. The remarkable photocatalytic activity of ZnO QDs is even higher than that of TiO_2 nanoparticles (P25) [134] and is another disadvantage of ZnO QDs as a UV absorber. Bare ZnO QDs cannot fabricate long-term durable UV-shielding coatings. However, encapsulation of ZnO QDs with inert silica shells can efficiently insulate their photocatalytic activity depressing the ZnO QD–induced decomposition of polymer matrices [135].

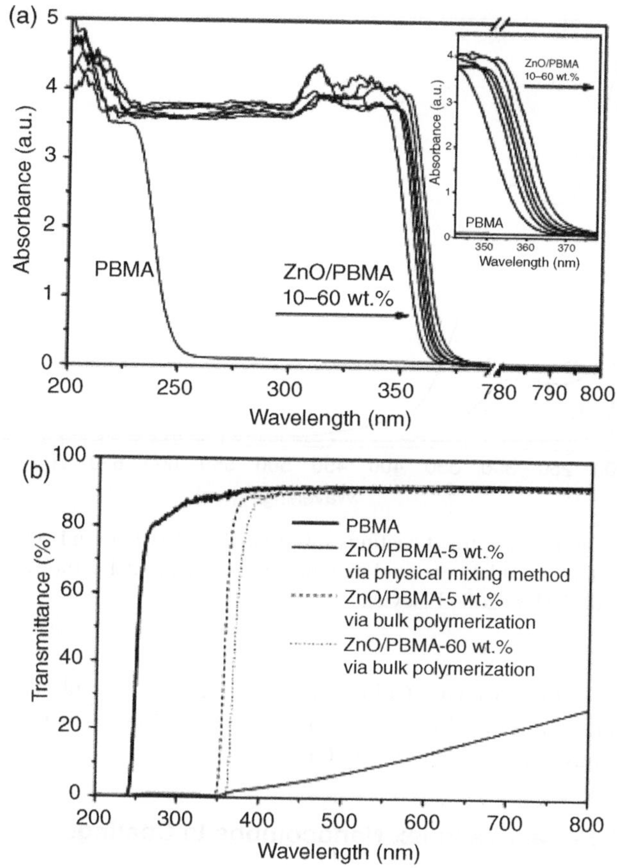

FIG. 1.20 (a) The UV–vis absorbance spectra of pure PBMA film and ZnO/PBMA nanohybrid films synthesized by bulk polymerization containing ZnO NPs from 10 to 60 wt.% in 10 wt.% increments. (b) The UV–vis transmittance spectra of pure PBMA film and ZnO/PBMA nanohybrid films prepared by bulk polymerization and physical mixing method. The thickness of each film was about 100 ± 5 μm. Reprinted with permission from Ref. 132. © 2012 American Chemical Society.

Compared to nano-TiO_2 and nano-ZnO, CeO_2 nanoparticles have minimal photocatalysis. Their disadvantage is oxidation catalytic activity. It can be significantly reduced by doping with metal oxide. CaO-doped cerium dioxide had excellent UV absorption and transparency in the visible ray region even better than undoped cerium dioxide [136]. Figure 1.21 shows the absorption spectra of water-based coatings containing various colloidal particles. The coating with colloidal CeO_2 displays strong UV absorption. In contrast, UV absorption is not present for the coating with colloidal SiO_2 [102]. Protection of organic coatings by UV-induced degradation from CeO_2 nanoparticles was also demonstrated in a water-based PU clearcoat [137].

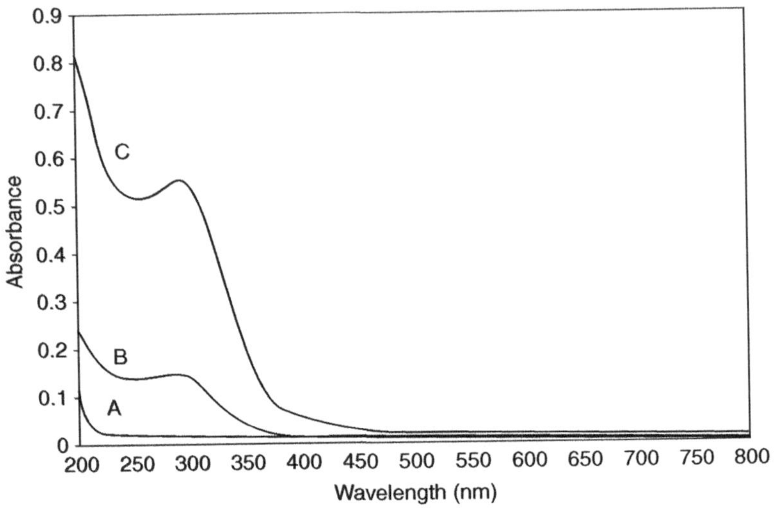

FIG. 1.21 Absorption spectra of coatings sols diluted 10,000 times in H_2O. Coating with (A) colloidal silica, (B) colloidal ceria and zirconia, and (C) colloidal ceria. Reprinted with permission from Ref. 102. © 2006 Elsevier.

Recently, cerium phosphate ($CePO_4$) nanoparticles with low photocatalytic activity and high absorption in the UV region were developed [138]. This provides new route to formulate transparent, UV-shielding OINCs.

1.4.2 High Refractive Index Nanocomposite Coatings

High RI coatings are used as optical coatings of lens, optical fiber coatings, and antireflective coatings. Unfortunately, the refractive indices of conventional organic coatings such as polyacrylate, PU, alkyd, and polysiloxane are lower than 1.50. The introduction of sulfur, bromine, heavy metals, or phenyl and alicylic groups into the polymer chains is one strategy to increase the RI of the polymer. Nevertheless, with development of nanomaterials, embedding the nanofillers with high RI into organic coatings has been increasingly employed to acquire high RI coatings.

The RI (n) of nanocomposite coatings can be approximately calculated by the following equation:

$$n = \phi_{\mathrm{p}} n_{\mathrm{p}} + (1 - \phi_{\mathrm{p}}) n_{\mathrm{m}} \tag{1.4}$$

If the weight fraction of the inclusion nanoparticles is adopted, Equation (1.4) can be changed as follows,

$$n = \frac{\rho_{\mathrm{p}} n_{\mathrm{m}} - w_{\mathrm{p}} (\rho_{\mathrm{p}} n_{\mathrm{m}} - \rho_{\mathrm{m}} n_{\mathrm{p}})}{\rho_{\mathrm{p}} - w_{\mathrm{p}} (\rho_{\mathrm{p}} - \rho_{\mathrm{m}})} \tag{1.5}$$

where ρ_p and ρ_m are the densities of nanoparticle and host matrix, respectively, and w_p is the weight fraction of nanoparticles.

The RI of nanocomposite coatings can also be theoretically calculated with an equation deduced from the Maxwell–Garnett approximation,

$$n = n_m \sqrt{1 + \frac{3\phi_p \beta}{1 - \phi_p \beta}} \tag{1.6}$$

$$\beta = \frac{m^2 - 1}{m^2 + 2} \tag{1.7}$$

where β is a coefficien, and m is the relative RI, namely, n_p/n_m. If the inherent RI of the host matrix is not impacted by the nanoparticles, then the calculated value from Equation (1.6) will equal that from Equation (1.4).

Equation (1.4) clearly shows that high RI and high volume fraction of nanoparticles both favor high RI OINCs and thus a wide modulation range of the RI. To date, the inorganic nanoparticles with high RI, that is, metal sulfides (PbS, ZnS) and metal oxides (TiO_2, ZrO_2, ZnO), have been employed to fabricate high RI nanocomposite coatings.

PbS nanoparticles exhibit an RI on the order of 4 across a wide wavelength range. They were ever combined with gelatin to produce nanocomposite films through aqueous solution mixing and subsequent spinning [139]. Figure 1.22 shows the RI of the obtained films as a function of weight fraction of PbS. The weight fraction of PbS is up to 86.4 wt.%, and RI values as high as 2.5 were achieved. ZnS nanoparticles with RI of 2.36 in the visible range are also used to tune the RI of coatings. Antonello *et al.* synthesized ZnS nanoparticles with a mean diameter of 4.8 nm from Na_2S and

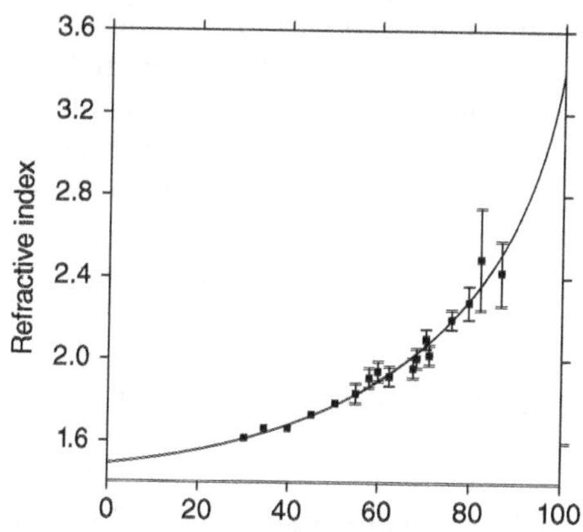

FIG. 1.22 The refractive index of PbS-gelatin nanocomposites as a function of weight fraction of PbS. The solid line represents best-fitting of Equation (1.5). Reprinted with permission from Ref. 139. © 1993 Materials Research Society.

$ZnSO_4$ in the presence of mercapto-group containing capping agents [140]. Then, the nanoparticles were blended in the sol–gel hybrid GZ and DGZ films that were correspondingly synthesized using molar ratios of GPS/zirconium tetraisopropoxide (ZrOPr) = 75:25 and diphenyldymethoxysilane/GPS/ZrOPr = 50:25:25. The RI was adjusted from 1.50 to 1.56 for the GZ film and 1.53 to 1.59 for the DGZ film when the volume fraction of ZnS increased to 0.077%. However, the toxic character of PbS and the photocatalytic activity of ZnS are important considerations.

Among the metal oxides, TiO_2 has the highest RI at 2.7. Thus, high RI nanocomposite coatings commonly use TiO_2 nanophase. The routes to attain high RI TiO_2-based nanocomposite coatings are analogous to those for UV-shielding TiO_2-based nanocomposite coatings. Nevertheless, in order to assure absolute optical transparency, high RI TiO_2-based coatings are mainly fabricated through a sol–gel process. The sol–gel-derived TiO_2-based coatings have two typical morphologies—distinguished TiO_2 nanophase (or nanocrystal) and tianium-oxo network structure. These depend on the sol–gel process. The former morphology is acquired by mixing the TiO_2 nanoparticles, premade from the hydrolysis/condensation of titania precursor, with binder. The TiO_2 content could be high and the TiO_2 nanophase is relatively dense resulting in a high RI of the coatings. The latter is formed by mixing titania precursor with binder and then carrying out a sol–gel process. It is difficult to achieve high Ti loading levels because of the use of a crosslinkable chelating agent (i.e., methylacrylic acid (MAA), 2-(methacryloyloxy) ethyl acetoacetone [141]) and incomplete removal of organic species of the titania precursor. This leads to a narrow range of RI modulation of coatings (generally 1.5–1.6 [122, 141]). Hence, the sol–gel process for high RI TiO_2-based nanocomposite coatings mostly uses advanced synthesis of TiO_2 nanoparticles.

For example, Chau *et al.* synthesized highly dispersed TiO_2 nanoparticles via HCl-catalyzed hydrolysis/condensation reactions in absolute ethanol with TTIP precursor [142]. When these nanoparticles were incorporated into epoxy coatings, an RI of 1.668 was obtained at 30 wt.% TiO_2. Although higher RI values can be obtained by increasing the content of the TiO_2 nanoparticles, cracks appear beyond 50 wt.% TiO_2. Antonello *et al.* prepared TiO_2 nanocrystals with a similar HCl-catalyzed sol–gel process but in a methanol/water mixture [143]. The TiO_2 nanocrystal was mixed with an O/Ihybrid via the cohydrolysis/condensation of GPS/TTIP. The RI values of the coatings are tunable between 1.51 and 1.89. Figure 1.23 shows the RI evolution of the coatings with TiO_2 content (both in volume fraction and in weight fraction).

Chang *et al.* prepared a TiO_2 sol via HCl-catalyzed hydrolysis of tetrabutyl orthotitanate (TBOT) in 1-butanol [144]. The TiO_2 nanoparticles are subsequently coated with MAA or MPS and mixed with dipentaerythritol hexacrylate and photoinitator to yield UV-curable nanocomposite coatings with RI of 1.64–1.77. Liu *et al.* premade TiO_2 nanoparticles by hydrolysis/condensation of TIPP in water, dissociation of gel with nitric acid, and low-temperature (80°C) treatment [145]. After a washing step, the wet TiO_2 nanoparticles were redispersed in a mixture of *N,N*-dimethylacetamide and butanol (1:1, w/w). The TBOT and epoxy resin were then added to the nanoparticle dispersion. As the composition of TBOT/epoxy/TiO_2 film

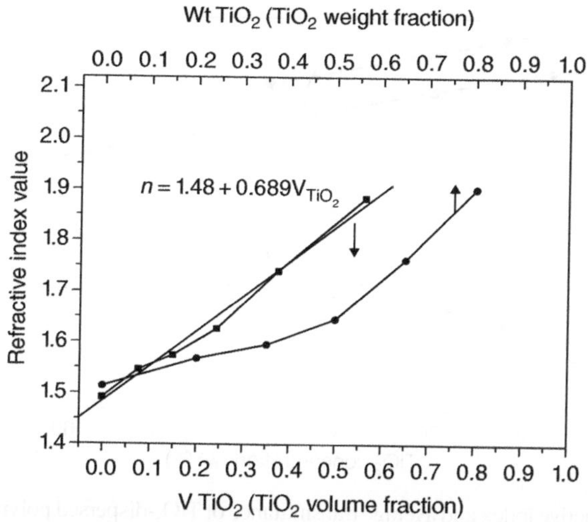

FIG. 1.23 Refractive index values at 630nm plotted versus TiO_2 volume fraction and weight fraction. Reprinted with permission from Ref. 143. © 2010 Springer Science+Business Media B.V.

TABLE 1.8 Compositions and Properties of TBOT/epoxy/TiO_2 Hybrid Films[a]

Codes	TiO_2 Nanoparticles (wt.%)	TBOT(TiO_2[b]) (wt.%)	Epoxy Resin (wt.%)	Residue[c] (wt.%)	Refractive Index[d]
TE75	75	35.2(8.4)	16.6	55.3	1.908
TE79	79	37.2(8.8)	12.2	65.9	1.965
TE84	84	39.5(9.4)	6.6	68.5	1.972
TE87	87	40.9(9.7)	3.3	72.7	1.917
TE88	88	41.4(9.8)	2.2	79.4	1.873

[a]Reprinted with permission from Ref. 145. © 2011 Elsevier.
[b]Calculated based on the conversion of 1 g of TBOT into 0.2347 g of TiO_2.
[c]Determined through TGA at 500°C.
[d]Measured at 633 nm.

changed, the RI became as high as 1.972 (Table 1.8)—close to that of a pure anatase titania thin film fabricated at high temperature.

Besides the sol–gel process, bead milling of TiO_2 nanopowder was adopted to prepare high RI nanocomposite films. In a typical example [146], TiO_2 nanopowder, which has a primary diameter of 15 nm and a needle-like rutile crystal structure, was milled using 50 µm zirconia beads in disperse media, that is, two monomers (either neopentyl glycol dimethacrylate or divinylbenzene) and butylacetate. The TiO_2-dispersed monomers after addition of benzoyl peroxide were polymerized under heating to form the dried film. The TiO_2 content in the film can be adjusted via addition of TiO_2/butylacetate slurry. Figure 1.24 shows the RI value

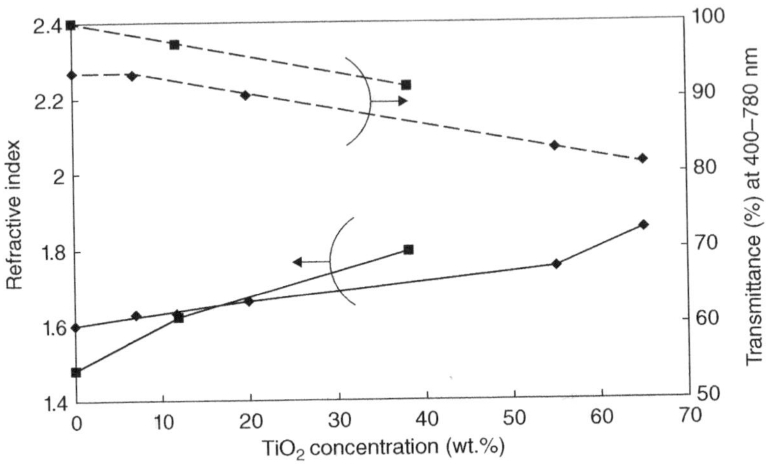

FIG. 1.24 Refractive index and average transmittance of TiO_2-dispersed polymer films fabricated via the bead milling method (◆, neopentyl glycol dimethacrylate; ■, divinylbenzene; straight line, refractive index; dotted line, transmittance (400–780nm) at film thickness of 5μm). Reprinted with permission from Ref. 146. © 2008 The Society of Polymer Science, Japan.

of the nanocomposite film as a function of TiO_2 content. The highest concentration achieved was 65wt.% TiO_2 and gave a nanocomposite film with a RI of 1.85 and transmittance of 81.5% (thickness: 5μm).

The combination of TiO_2 nanophase with a high RI polymer is another way to obtain high RI nanocomposite coatings at relatively low TiO_2 content. Nakayama and Hayashi synthesized TiO_2 nanoparticles from titanium oxychloride and further treated the TiO_2 surface with ZrO_2 to depress its photocatalytic activity [147]. The resulting TiO_2-ZrO_2 nanoparticles were modified with acrylic acid and then added into a mixture of 2-mercapatoethylsulfide diacrylate (MES-DA) and urethane acrylate (UA-306I) along with photoinitiator to formulate UV-curable high RI coatings. The RI value of the cured coatings can be readily tuned from 1.59 to 1.81 with increasing TiO_2-ZrO_2 content from 0 to 80wt.%.

Nakayama and Hayashi further blended the TiO_2-ZrO_2 nanoparticles with mercaptoethylsulfide-thiourethane methacrylate (MES-TUMA) or isophorone diisocyanate-mercaptoethylsulfide-thiourethane methacrylate (IPDI-MES-TUMA) [148]. This UV-curable nanocomposite coating is suitable for hard coating of high RI lens substrates (polythiourethane and PC). Chang *et al.* prepared high RI polyimide/titania nanocomposite thin films from a soluble polyimide, a coupling agent (APS), and TTIP [149]. Transparent hybrid thin films can be obtained at TiO_2 content as high as 40wt.%. The refractive indices at 633nm of the prepared hybrid thin films increase linearly from 1.66 to 1.82 with increasing TiO_2 content. Photosensitive polyimide–nanocrystalline titania optical thin films were synthesized in Chen's research group using TBOT, and the soluble polyimide with residual carboxyl acid groups and pendant methacrylate groups [150]. The carboxylic acid groups provided bound to titania while methacrylate groups offered photosensitivity. The prepared hybrid films

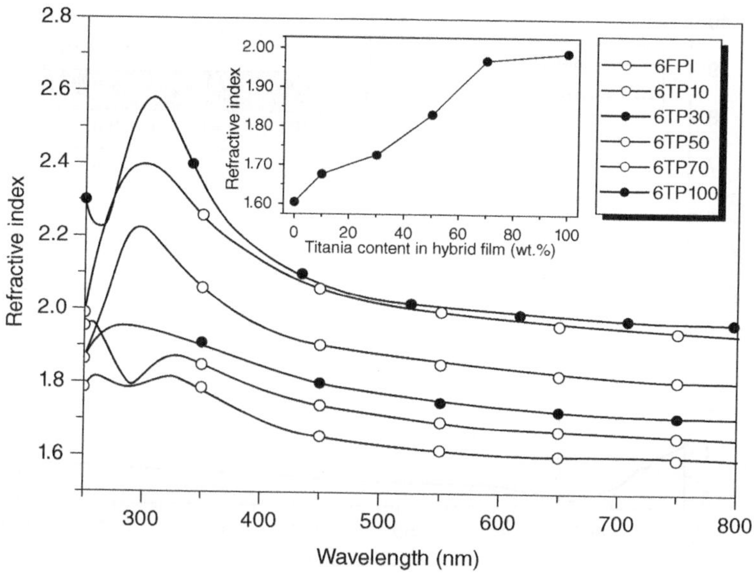

FIG. 1.25 Variation of the refractive index of the 6FPI/TiO_2 nanocomposites with wavelength. The inset shows the variation of refractive index at 633 nm with titania content. Reprinted with permission from Ref. 151. © 2009 The Royal Society of Chemistry.

had tunable and high RI (n = 1.58–2.03), surface planarity, excellent thermal properties, and good transparency in the visible range. Crystalline anatase titania domains in the hybrid films were around 4–7 nm.

Chen's group also employed 6F-poly(*p*-hydroxy-imide) (6FPI) to fabricate polyimide/TiO_2 nanocomposite films [151]. The bulky CF_3 in the polyimide backbones enhanced both the solubility and optical transparency while the attached hydroxyl groups provided the O/I bonding. Figure 1.25 shows the RI of the films at 300–800 nm and the inset is the variation of RI at 633 nm as a function of titania content. These nanocomposite films can be utilized to fabricate a three-layer antireflection coating that revealed a reflectance of less than 0.7% in the visible range (Fig. 1.26).

ZrO_2 nanophase materials are good candidates for the fabrication of high RI coatings because of their chemical inertance, high hardness, and non-photocatalytic activity. ZrO_2 has a RI value (~2.1) lower than TiO_2. Nonaqueous synthesized ZrO_2 nanocrystals with high dispersibility synthesized from a solvothermal reaction of zirconium (IV) isopropoxide isopropyl alcohol complex in benzyl alcohol were even adopted to prepare UV-curable ZrO_2-based nanocomposite coatings [30–32]. Prior to addition to the coatings, the ZrO_2 nanocrystal was treated with MPS. The upper limit of ZrO_2 content for completely transparent nanocomposite coatings changed with the composition of the coatings as well as the amount of MPS attached to ZrO_2 nanoparticles. The highest ZrO_2 content was 60 wt.% when the MPS-modified ZrO_2 nanoparticles were mixed with TPGDA. The cured coatings had a RI value of as high as 1.78 [32].

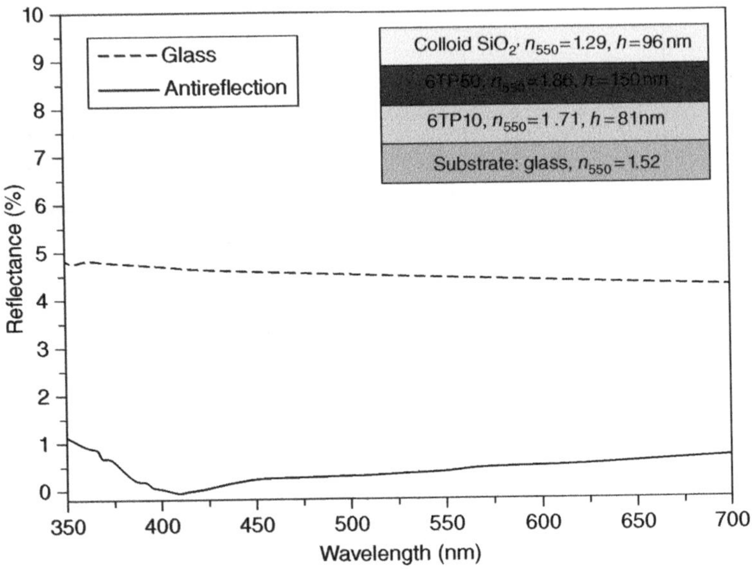

FIG. 1.26 Variation of the reflectance with wavelength: BK7 optical glass and the three-layer anti-reflection coating. The inset shows the structure of the three-layer anti-reflection coating. Reprinted with permission from Ref. 151. © 2009 The Royal Society of Chemistry.

Nonaqueous synthesized ZrO_2 nanocrystals were functionalized with AEAPS and dispersed in water under basic condition (pH 13–14) [42]. The aqueous ZrO_2 nanoparticle dispersion together with the cross-linker 1,4-butanediol digylcidyl ether was formulated as a water-based scratch resistant and high RI coating for PC substrates. Figure 1.27 presents the refractive indices of ZrO_2 nanoparticle films. The absolute refractive indices at 632 nm are 1.70, 1.73, and 1.77 for films prepared at AEAPS-to-ZrO_2 molar ratios of 0.25, 0.18 and 0.12, respectively. Commercial ZrO_2 nanoparticle dispersions in toluene were ever used to prepare high RI nanocomposites with polydimethylsiloxane (PDMS) matrix [152]. The obtained ZrO_2–PDMS nanocomposite had a transparency of 93.3% across the entire visible range via ligand molecule engineering. The RI of the ZrO_2–PDMS nanocomposite could be varied from 1.39 to 1.65 simply by increasing the ZrO_2 content from 0 to 20.8 vol.%.

ZnO nanoparticles are seldom used in the fabrication of high RI coatings due to their low RI value (2.02) and photocatalytic activity. Tsuzuki dispersed ZnO nanoparticles prepared by mechanochemical processing and coated with poly(methylsilsesquioxane) into caprylic capric triglyceride using poly(hydroxystearic acid) in a bead mill at 60 wt.% [153]. The product was 20 μm thick with a RI of 1.44–1.55 and high visible transparency.

Other high RI inorganic nanofillers include Si (n=3.91 at 620 nm, crystalline), Ge (n=5.59 at 620 nm), GaP (n=3.33 at 620 nm), Ta_2O_5 (n=2.1 at 550 nm), indium-doped tin oxide (ITO, n=2.0 at 550 nm), Nb_2O_5 (n=2.3 at 550 nm), $Bi_4Ti_3O_{12}$ (n=2.3 at 520 nm), etc. The details can be found in a prior review [154].

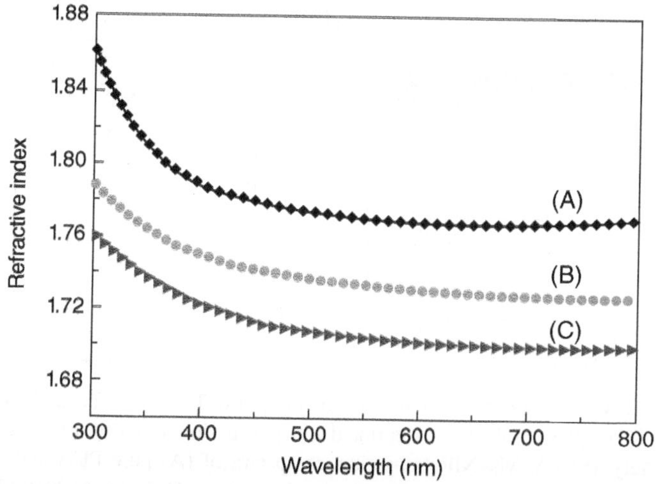

FIG. 1.27 Refractive indices of ZrO_2 nanoparticle films prepared at AEAPS-to-ZrO_2 molar ratio of (A) 0.12, (B) 0.18, and (C) 0.25. Reprinted with permission from Ref. 42. © 2010 Elsevier.

1.4.3 Transparent NIR-Shielding Nanocomposite Coatings

Transparent NIR (700–2600 nm)-shielding coatings are strongly desired for solar control of windows in automotive and architectural applications to provide internal comfort as well as reduce energy consumption. Transparent NIR-shielding coatings can be either in the form of inorganic coatings or as OINCs. Due to low cost and easy *in situ* application to windows, NIR-shielding OINCs are especially developed via inclusion of functional inorganic nanoparticles into the polymer or hybrid matrix. The character (type, size) and introduction method of the nanoparticles are crucial to the performance of the coatings. For practical applications, a stronger and wider-band NIR with high visible light transmittance is required for the inorganic NIR absorbent.

As one of the most common transparent conductive oxides, ITO nanoparticles absorb NIR light by activating surface plasmon polaritons of the free electrons. Thus, they are often used as the functional nanofiller of transparent NIR-shielding nanocomposite coatings. The ITO nanoparticles can be in the form of commercial nanopowders. With this source, adequate deagglomeration of the nanopowder is a key step to achieve nanocomposite coatings with the desired transparency. Chemical synthesis of ITO nanoparticles from ITO precursors under mild conditions is usually reported.

As examples, ITO nanoparticles were prepared by a solvothermal synthesis at 300°C using octadecene as the solvent, indium acetate and tin acetate as precursors, and oleic acid and oleylamine as the capping agents [155]. The ITO nanoparticles were also synthesized at 220°C for 8 h using *N*-methyl-pyrrolidone as both the reaction solvent and surface modifier and indium acetylacetonate and tin bis(acetylacetonate) dichloride as the precursors [156]. The former ITO nanoparticles were grafted with

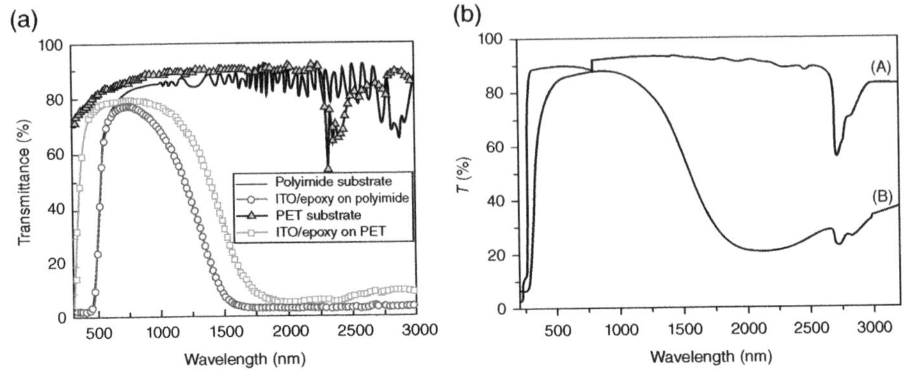

FIG. 1.28 (a) UV–vis–NIR transmittance spectra of epoxy/ITO coating (with thickness of 5 μm) with 35 wt.% ITO on plastic substrates. Reprinted with permission from Ref. 155. © 2010 American Chemical Society. (b) UV–vis–NIR transmittance spectra of (A) pure PUA and (B) PUA/ITO nanocomposite films at 5 wt.%. Reprinted with permission from Ref. 156. © 2010 Elsevier.

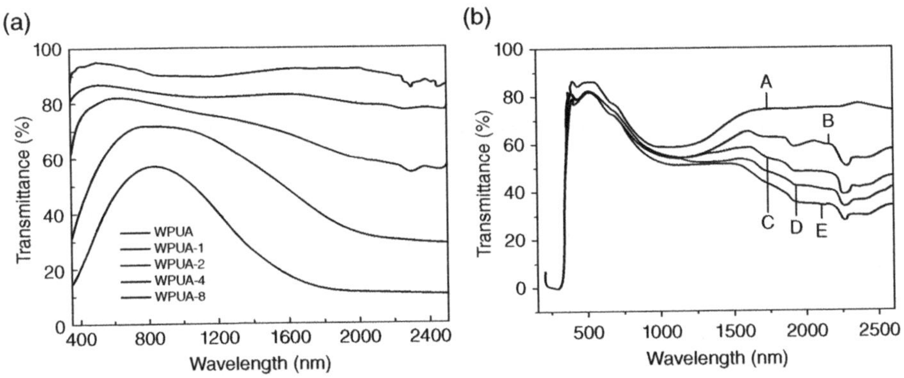

FIG. 1.29 UV–vis–NIR spectra of (a) waterborne UV-cured PUA/ITO coatings (WPUA, WPUA-1, WPUA-2, WPUA-4, and WPUA-8 correspond to the nanocomposite coatings containing 0, 1, 2, 4, 8 wt.% ATO). Reprinted with permission from Ref. 158. © 2010 Elsevier. and (b) poly(MMA-BA)/ATO nanocomposite films with different ATO amounts: (B) 3%, (C) 5%, (D) 7%, and (E) 10%. Reprinted with permission from Ref. 159. © 2011 Springer.

poly(glycidyl methacrylate) and added to epoxy coatings while the latter ones were added to acrylic polyurethane varnish. Figure 1.28 presents the UV–Vis–NIR spectra of the nanocomposites. Both ITO nanoparticles offer similar NIR shielding in organic coatings. The absorption band is wide, yet is limited to above 1200 nm. One limitation is that ITO nanoparticles are relatively expensive despite their commercial availability and routine synthesis.

Alternatives include ATO nanoparticles. Unlike ITO nanoparticles, ATO nanoparticles are mostly from a commercial product [157–159]. Figure 1.29 shows the UV–vis–NIR spectra of waterborne UV-cured PUA/ATO coatings and poly(MMA-BA)/ATO films. The IR shielding performance of ATO nanoparticles is not as good as that

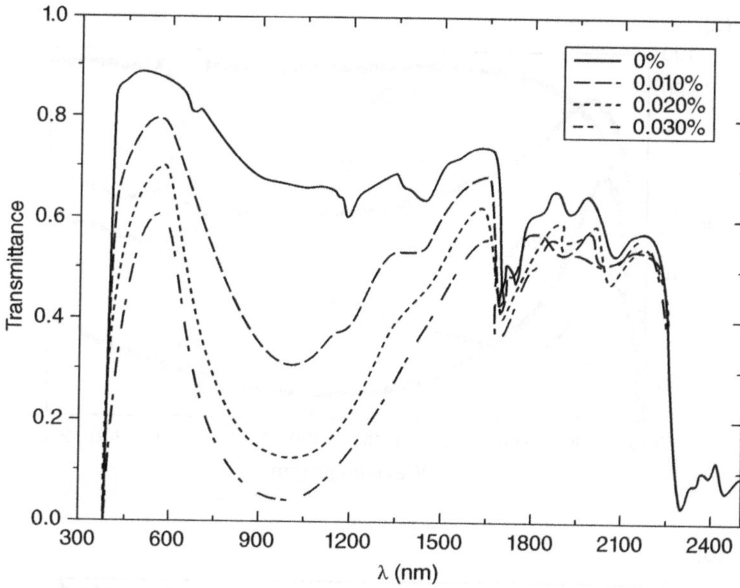

FIG. 1.30 Normal incidence transmission of polyvinyl butyral/LaB_6 nanocomposite laminate sheets for different weight percent of LaB_6 nanoparticles. Reprinted with permission from Ref. 161. © 2003 American Institute of Physics.

of ITO nanoparticles. Therefore, other efficient NIR absorbents have to be added in addition to ATO to enhance the NIR-shielding performance. Furthermore, Al-doped ZnO (AZO)/epoxy composite was fabricated as thermal insulation coating for glass [160]. Compared to ITO- or ATO-based nanocomposite coatings, the AZO-based coating is suboptimal for NIR shielding.

Highly conducting metals and black compounds can also behave as a tinted NIR absorber when dispersed at nanosized dimensions. Metals and black compounds reported include silver, gold, ruthenium dioxides, rhenium trioxides, and lanthanum hexaborides (LaB_6) [161]. The LaB_6 nanoparticles have a characteristic absorption around 1000 nm and a shorter NIR wavelength in contrast to ITO/ATO. Figure 1.30 shows the UV–vis–NIR spectrum of polyvinyl butyral/LaB_6 nanocomposite sheet with a thickness of 0.8 mm. The NIR shielding efficiency of LaB_6 is quite high. Even at 0.030% LaB_6, transmittance at the NIR absorption peak of the nanocomposite approaches zero. Meanwhile, the transmittance in the visible region remains 0.6. However, the tail of the NIR absorption peak extends to the visible range and deteriorates the transparency of the nanocomposite.

Size effect of LaB_6 on the optical properties of LaB_6-based composites was reported by Yuan *et al.* [162]. They found that PMMA/LaB_6 composites prepared via *in situ* polymerization with 70 nm LaB_6 particles had the best performance in the NIR and visible absorption of particle sizes ranging from 50 to 400 nm. To date, preparation of LaB_6-based OINCs is rarely found in publications. Because LaB_6 nanoparticles are prepared via a solid state reaction at high temperature, a stirred bead milling process is necessary to achieve transparent LaB_6-based OINCs.

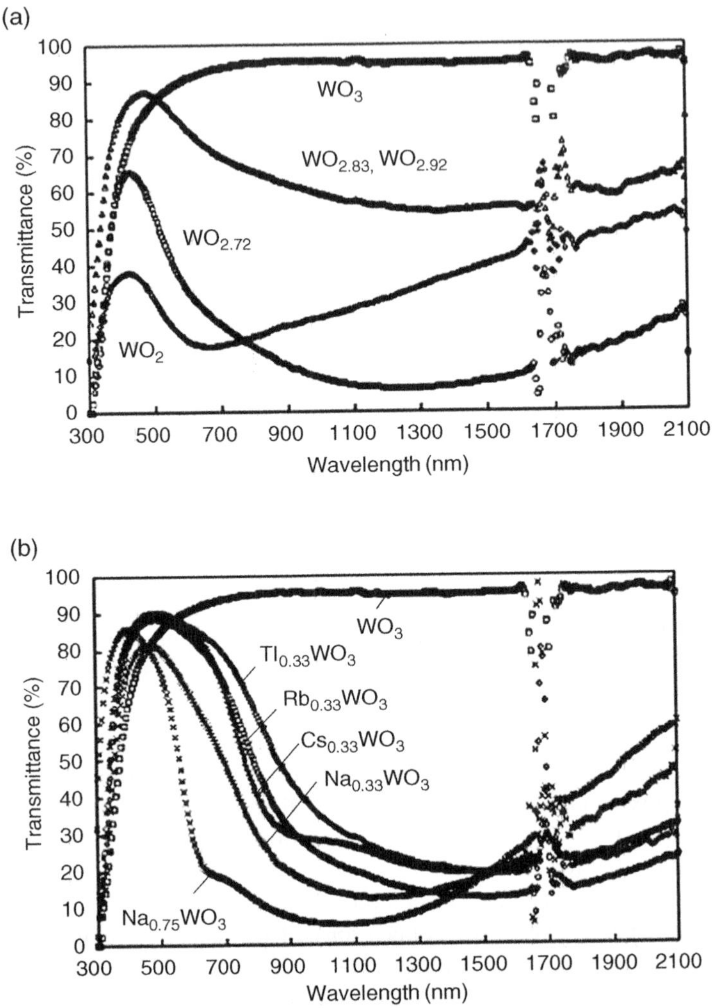

FIG. 1.31 Transmittance spectra of (a) tungsten oxide nanoparticles and (b) tungsten bronze nanoparticles dispersed by 0.01 wt.% in toluene. The disturbance at wavelengths between 1620 and 1790 nm is due to absorption by toluene. Reprinted with permission from Ref. 163. © 2007 The American Ceramic Society.

Tungsten trioxide (WO_3) has a wide band gap of 2.62 eV and is transparent to visible and NIR light. However, a metallic conductivity and a strong NIR absorption can be induced when free electrons are introduced into crystals by either decreasing the oxygen content or by adding ternary elements. Figure 1.31 is the transmittance spectra of tungsten oxide and tungsten bronze nanoparticles dispersed in toluene [163]. The pattern of the spectra strongly depends on the W/O atom ratio as well as the dopant. A remarkable NIR absorption effect was found in $WO_{2.72}$, $Na_{0.75}WO_3$, and $M_{0.33}WO_3$ (M = Na, Cs, Tl, Rb). Particularly, the $M_{0.33}WO_3$ with a hexagonal tungsten

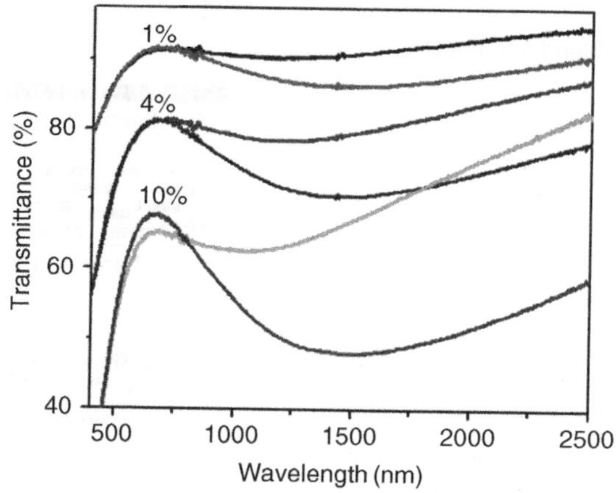

FIG. 1.32 Vis–NIR transmittance spectra of polysiloxane/VO_2(W) nanocomposite coatings with 1, 4, and 10 wt.% VO_2(W) content. Reprinted with permission from Ref. 164. © 2013 Elsevier. (Molar ratio of W:V = 0.15:100; film thickness: ~20 μm; the solid and dashed lines are for the films at 20 and 60°C, respectively.)

bronze structure is highly attractive in solar filter applications because the absorption in the visible range is small enough that their dispersion can yield a lower solar heat gain coefficient than that by ITO nanoparticle dispersion. Therefore, tungsten bronze nanoparticles will be very promising in the preparation of NIR-shielding OINCs.

Vanadium dioxide (VO_2) has a relatively low transition temperature (T_c) at 68°C for the change from a distorted to an undistorted rutile structure. Thus, VO_2 exhibits temperature-dependant reflective properties. At room temperature, VO_2 is almost transparent to NIR light but changes to reflective above the T_c. More interestingly, the transition temperature can be adjusted to near room temperature through doping with W, Mo, etc., demonstrating its positional application as a smart-window coating. Currently, most thermochromic coatings are inorganic VO_2 coatings. Thermochromic OINCs were prepared by blending the deagglomerated (W)VO_2 (T_c = 36°C) nanoparticles with the moisture-curable polysiloxane coatings [164]. Figure 1.32 gives the Vis–NIR transmittance spectra of the corresponding coatings. Luminous transmittance at 550 nm of about 60% and solar modulation at 2500 nm of 23% were achieved at 10% (W)VO_2 load. Nevertheless, VO_2-based OINCs are difficult to envision because W-doped VO_2 nanoparticles are not commercially available and have a reduced crystallinity and thus thermochromic property after deagglomeration process, for example, bead milling. They are also easily oxidized.

1.5 TRANSPARENT BARRIER NANOCOMPOSITE COATINGS

A coating layer is usually cast on a package to enhance the barrier properties or to improve the printability. Traditional barrier layers include vacuum-deposited aluminium or glass-like SiO_x films [165]. Currently, there is a trend toward chilled and possibly

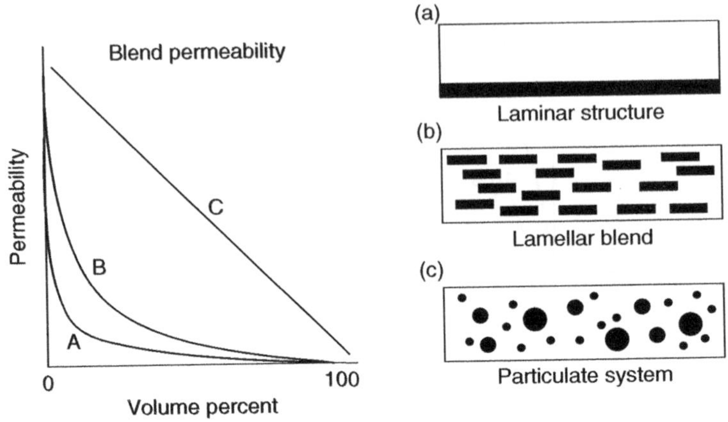

FIG. 1.33 Schematic illustration of the permeability of a blend of a high-barrier material in a low-barrier matrix as a function of the morphology: (a) laminar structure, (b) lamellar blend, and (c) particulate system [165]. Ref. 165. © 2003 John Wiley & Sons, Ltd.

modified atmosphere packaging. Consumers also demand high product visibility, as in food packaging. This requires a coating layer with good and sometimes selective barrier properties in combination with high transparency, good print quality, etc. Unfortunately, an aluminim layer is opaque and unsuitable for microwave heating. It also requires high energy during production. The SiO_x films are transparent, retortable, and microwaveable, but have limited flex and crack resistance. Nevertheless, organic barrier coatings are transparent and flexible. Epoxy-amine coating has been employed to improve the oxygen barrier of PET by a factor of 2 or more. However, the barrier performance of other conventional organic coatings such as PU and polyacrylate coatings is moderate. These coatings have been modified to include barrier inorganic nanofillers in the organic coatings to yield the so-called barrier nanocomposite coating. Besides packaging, barrier nanocomposite coatings are also useful in the anti-corrosion field.

Different modeling approaches have shown that significant barrier improvement factors of the order of 50 or higher can be achieved with high aspect ratio nanofiller particles [166, 167]. Figure 1.33 schematically illustrates the barrier properties of a blend as a function of the morphology. Lamellar, filler-based composites have a barrier performance close to the fully laminar layer, whereas particulate filler is not as efficient at improving the barrier. Using a model with a multiscale hierarchical approach, Xiao *et al.* proposed an empirical equation to predict the barrier properties of polymer nanocomposite coatings [168], as follows:

$$\frac{K_e}{K_m} = \exp\left(-\left(\frac{\alpha\phi_p^n}{1.33}\right)^{0.68} \times \left(\frac{\varepsilon^{pn*}}{1.72}\right)^{0.08}\right) \quad (1.8)$$

where ε^{pn*} and α denote the polymer–nanoparticle interaction strength and nanoparticle aspect ratio, respectively. The K_e and K_m correspond to the permeabilities of the

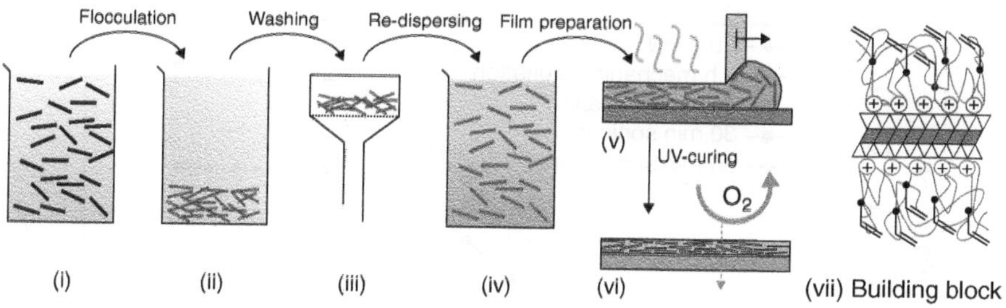

FIG. 1.34 Process steps to obtain UV-curable barrier coatings. Aqueous dispersions of clays (i) are flocculated via the addition of cationic, UV-curable polyurethane (ccPU)-dispersion (ii). Loose aggregates of obtained clay hybrids are washed (iii) and redispersed in THF (iv). Doctor-blading of the clay hybrid dispersions (v) allows a final UV-curing step (vi) to get the homogeneous composite films. An idealized O-HEC hybrid-platelet consisting of a clay lamella with ccPU adsorbed on both sides is shown in (vii) [172]. Ref. 172. © 2012 WILEY-VCH Verlag GmbH & Co. KGaA.

nanocomposite coatings and organic matrix. The equation can guide experimentalists in the rational design of nanocomposite coatings by offering quick screening of candidate formulations, identification of the most promising design spaces, and ideal experimental strategies. It also theoretically demonstrates why higher aspect ratio nanofillers offer better barrier performance.

Clay is currently the most common and most affordable high aspect ratio filler. Different clays have been reported in the fabrication of barrier nanocomposite coatings. The key factor lies in the type of clay as well as the dispersion state (i.e., intercalation and exfoliation) of the matrix filler. In a waterborne polymer latex/clay nanocomposite coating, the samples containing cetyltrimethylammonium bromide (CTAB)-modified saponite clay (Sap-CTAB) outperformed Cloisite 30B due to the much increased surface area of the Sap-CTAB after chemical modification [169]. When clay was incorporated into SILRES BS 1701, a polysiloxane emulsion sealant for concrete structure, reduced water permeability was observed [170]. Moreover, the moisture barrier performance was better for the nanocomposites containing Cloisite 20A clay than those with I.30P clay or the neat silane coating.

In an unsaturated polyester-based UV-curable coating, Choiste 30B is better in barrier performance for water but poorer in optical transparency versus CTAB-modified clay at the same load [171]. Recently, Möller *et al.* demonstrated that high charge, coarse-grained Li-hectorites obtained by melt synthesis delaminate spontaneously to yield platelets with aspect ratios typically larger than 1000 [172]. The synthetic hectorite was embedded into UV-curable coatings via a smart preparation process (Fig. 1.34). With respect to oxygen barriers, the hectorite-based hybrids outperform corresponding materials made from MMT by more than an order of magnitude while showing superb optical properties.

Influence of the preparation process on the barrier properties of nanocomposite coatings was also reported. Heidarian *et al.* prepared 3 wt.% PU/organically modified MMT (OMMT) composites by dispersing Cloisite 30B in castor oil polyol

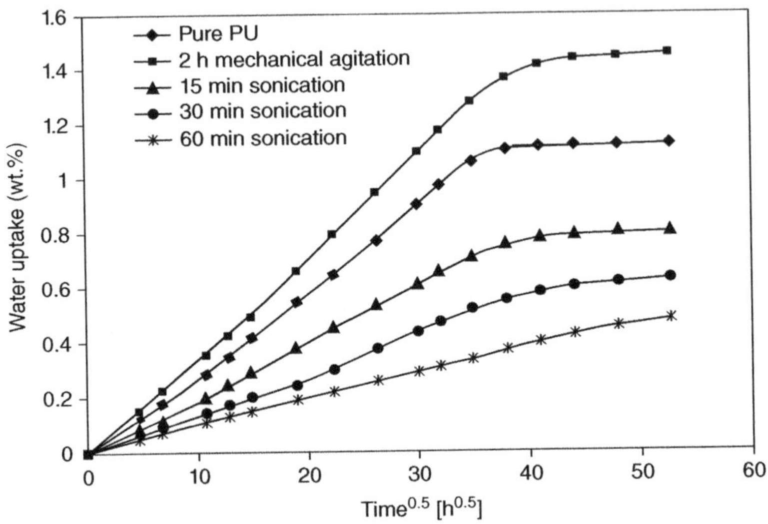

FIG. 1.35 Water absorption curves for pure PU and nanocomposite films after certain time intervals. Reprinted with permission from Ref. 173. © 2010 Springer.

resin through an ultrasonication-assisted technique [173]. Figure 1.35 shows the water uptake profiles of pure PU and 3 wt.% PU/OMMT composites after 2 h of mechanical agitation and 15, 30, and 60 min of sonication. For the composite film prepared by 2 h of mechanical agitation, the maximum water uptake (1.45 wt.%) and diffusivity coefficient (3.09×10^{-7} mm^2/s) were increased compared to that of pure PU (1.12 wt.% and 2.98×10^{-7} mm^2/s). However, the maximum water uptake and diffusivity coefficient is decreased to 0.49 wt.% and 2.23×10^{-7} mm^2/s, respectively, via 60 min sonication.

Pavlacky *et al.* reported UV-curable nanocomposite barrier coatings by dispersion of clay in monomers and then *in situ* polymerization or a sonication technique that synthesized unsaturated polyester and a dispersion of clay in the resin [171]. All the nanocomposite coatings demonstrate reduced water vapor permeability (WVP) and water vapor transmission (WVT), but no concrete changes to oxygen gas permeability. In contrast to those nanocomposite coatings prepared by sonication, the *in situ* nanocomposite coatings have lower WVP and WVT due to the better dispersion in the *in situ* process.

Actually, an ideal clay-based nanocomposite coating should have a fully exfoliated morphology with parallel orientation to the substrate. For conventional techniques (spraying, rolling, dipping, etc.), control of the orientation of the exfoliated clay is a rather difficult task because of the viscous binders. Priolo *et al.* employed a layer-by-layer assembly technique to create a super-gas barrier film where the cationic polyethylenimine (PEI) and anionic MMT clay and poly(acrylic acid) (PAA) are deposited in the sequence of PEI/PAA/PEI/MMT [174]. All the individual clay platelets lie parallel to the substrate to construct an open nano brick wall structure as illustrated in the TEM cross-sectional image (Fig. 1.36).

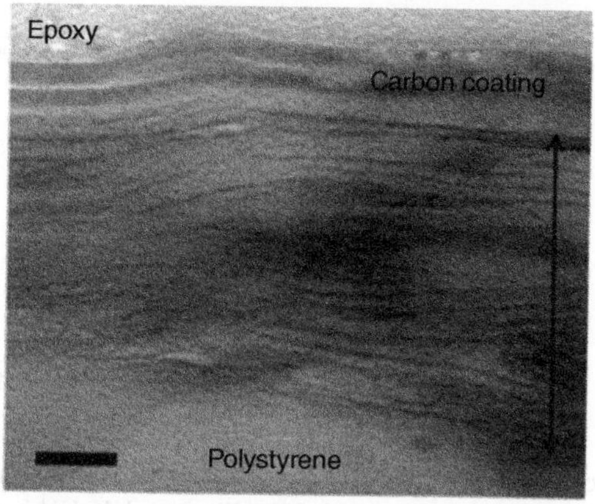

FIG. 1.36 TEM cross-section of a five QL thin film deposited on 250 μm polystyrene. The scale bar is 20 nm, and the double arrow spans the film's 80 nm thickness. Reprinted with permission from Ref. 174. © 2010 American Chemical Society.

At only 51 nm of thickness, these nanocomposite thin films exhibit an oxygen permeability of $\leq 5 \times 10^{-22}$ cm^3(STP)·cm/(cm^2·s·Pa). This is lower than that of SiO_x (8.3×10^{-20} cm^3(STP)·cm/(cm^2·s·Pa)). This product combines high flexibility, transparency, and barrier protection and is thus a good candidate for a variety packaging applications.

1.6 TRANSPARENT CONDUCTING NANOCOMPOSITE COATINGS

Transparent conductive coatings are essential components in numerous applications that require high transmission and high electrical conductivity. Such coatings are therefore used as electrodes in photoelectronic devices, as an IR-reflecting or heatable layer, for electromagnetic shielding, for dissipating static, etc. The most common transparent conducting coatings are inorganic ITO, ATO, and AZO. However, these inorganic coatings are unadapted to applications as flexible plastic substrates or as big objects because of their inherent brittleness and limitations in fabrication techniques (high cost and small scale of physical and chemical vapor deposition, high temperature treatment during the sol–gel process, etc.).

An alternative is to utilize conducting OINCs that are mainly composed of conducting nanofillers and adequate organic (or hybrid) binders. The use of nanofillers leads to low light scattering and high transparency. Because the crystallization step of the conducting material in the OINC was separated from film formation, conducting OINCs can be cured either by a low temperature thermal treatment or by UV light irradiation using polymerizable organic additives. Therefore, the major advantages of OINC are large-scale fabrication, low cost, and adaptation to many different substrates. However, in contrast to inorganic conducting coatings, the conductivity of OINCs is relatively low.

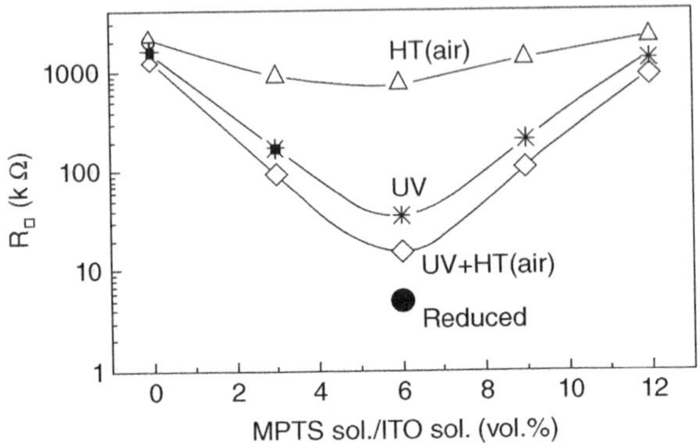

FIG. 1.37 Sheet resistance of 500 nm thick MPS/ITO coatings versus sol composition. Reprinted with permission from Ref. 176. © 2001 Elsevier. UV, UV irradiation 105 mW/cm^2, 110 s; HT (air), heat treatment in air: 130°C, 15 h.

Transparent conducting OINCs were first developed by Aegerter and coworkers [175]. The crystalline ITO or ATO nanoparticles, synthesized via a wet chemical method in the presence of the surface modifying agent β-alanine and subsequent calcination, were redispersed in ethanol or water. Organofunctionalized silanes such as MPS (together with photoinitiator) and GPS were used as the binder. Figure 1.37 shows the effect of annealing on the evolution of the sheet resistance ($R_\square$) with MPS/ITO coatings deposited on PC having sol composition [176]. Similar overall behavior was obtained with GPS/ITO coatings. Regardless of the treatment, the $R_\square$ for pure ITO is high (1 MΩ), and the coating adhesion is poor. Cured MPS/ITO coatings have a minimal sheet resistance for a composition volume ratio of 6%. At or above this ratio, they do not exhibit abrasion damage and only slight scattering. The lowest stable sheet resistances are $R_\square = 5$ kΩ for MPS/ITO coatings and $R_\square = 15$ kΩ for GPS/ITO coatings. This used a 110 s UV irradiation and a 15 h heat treatment at 130°C with further annealing in air. If annealing was conducted in a reducing atmosphere (N_2/H_2), the sheet resistance can be further reduced to as low as 800 Ω, and the transparency in the visible range is higher than 85% for 600 nm thick single layers [177].

Besides organoalkoxysilanes, PVP was also reported to be a binder in ITO nanoparticle coatings. Figure 1.38 presents the specific resistance of PVP/ITO nanocomposite coatings as a function of the volume fraction of PVP [178]. The lowest specific resistance was measured to be 6 Ω·cm at PVP contents of 33–40 vol.%. This further decreased to 0.5 Ω·cm after annealing at 200°C. The decrease in the specific resistance with increasing PVP content can be attributed to a densification of the ITO nanoparticle network due to shrinkage of PVP resulting from solvent evaporation. Unfortunately, the densification of the ITO nanoparticle network caused a transmission of just 50% at 550 nm for a layer thickness of 4 μm at a volume fraction of 40%. For PVP contents above 40 vol.%, the specific resistance increases as a result of the disturbed formation of ITO nanoparticle network.

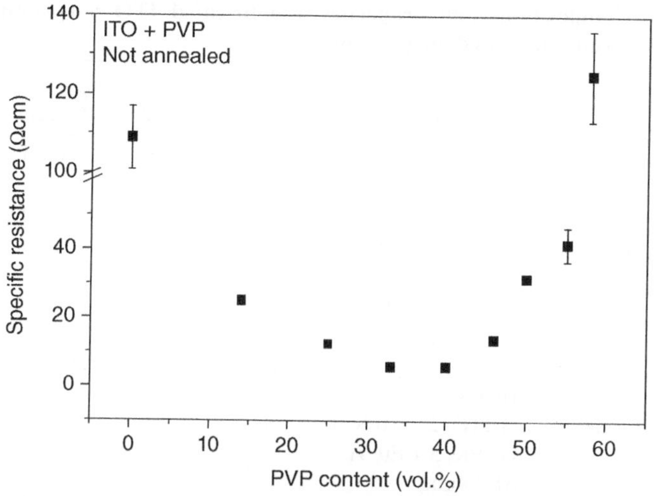

FIG. 1.38 Specific resistance of ITO/PVP nanocomposites as a function of the PVP content. Reprinted with permission from Ref. 178. © 2009 Springer Science+Business Media, LLC.

Maksimenko *et al.* introduced a binder from combination of PVP and MPS [179]. Table 1.9 compares the transparency and specific conductivity of ITO-MPS, ITO–PVP, and ITO–MPS–PVP coatings. In contrast to the ITO–PVP films, the optical properties were not dependent on the solvent even for the ITO–MPTS–PVP coatings. Compared to best pure ITO layers (3.1 Ω^{-1}/cm), the ITO–MPTS–PVP nanocomposite coatings have a conductance of 9.8 Ω^{-1}/cm. Stable sheet resistances of 750 $\Omega_{\square}$ at a coexistent transmittance of 86% at 550 nm for a layer thickness of about 1.3 μm were achieved. The combination effect of MPS and PVP on the conductance enhancement may be due to the filling of smaller cavities that cannot be completely filled with PVP because of the gyration radius (13.8 nm for commercial PVP K-15).

The above conducting OINCs involve the conducting behavior of the coatings with high volume fraction of nanofiller. Their conductivities are dominated by the densification degree of the conducting nanoparticle network. Actually, the conductivity of most composites containing fillers in an insulating matrix can be described by classic percolation theory:

$$\sigma = \sigma_0 \left(V - V_c\right)^t \tag{1.9}$$

where σ is the direct-current conductivity, σ_0 is the proportionality constant, V is the conductive filler volume fraction, V_c is the percolation threshold, and t is the critical conductivity exponent. Another prediction of conductivity is based on the effective medium approximation (EMA) theory and general effective media theory. A modified EMA model, which takes into consideration different types of contacts between the adjacent particles in a particle network, was proposed by Soloukhin *et al.* [180]. However, the modified EMA equation is much more complicated versus the percolation equation.

TABLE 1.9 Compositions and properties of fabricated ITO nanocomposite layers (thickness: 1–1.3 μm) deposited on glass substrates[a]

Coatings		Composition (wt.%)	T^b (550 nm), %	$\sigma_{after\ treatment}$, Ω^{-1}/cm	$\sigma_{after\ reducing}$, Ω^{-1}/cm
Pure ITO	Eth.	→got from Evonik 35.5 ITO, 64.5 EtOH	86	0.6±0.02	3.1±0.04
	Acac.	30.7 ITO, 69.3 Acac	86	0.6±0.02	3.1±0.04
ITO-PVP	Eth.	33.4 ITO, 6 PVP, 60.6 EtOH	43	0.9±0.06	—
	Acac.	33.4 ITO, 5 PVP, 5.2 EtOH, 62.2 Acac	85	1.2±0.02	—
ITO-MPS	Eth.	32.7 ITO, 3.3 MPS, 4.7 HCPK[c], 59.3 EtOH	88	1.5±0.07	4.2±0.15
	Acac.	27.1 ITO, 2.7 MPS, 3.9 HCPK, 5.1 EtOH, 61.2 Acac	88	1.9±0.03	5.3±0.09
OTP-MPS-PVP	Eth.	30.9 ITO, 3.1 MPS, 4.4 HCPK, 5.6 PVP, 56 EtOH	86	4.3±0.17	9.8±0.41
	Acac.	25.9 ITO, 2.6 MPS, 3.7 HCPK, 4.7 PVP, 4.9 EtOH, 58.2 Acac	86	3.7±0.07	8.7±0.34

[a] Reprinted with permission from Ref. 179. © 2010 Elsevier.
[b] T_{max} = 92% due to Fresnel reflection losses at the glass air interfaces.
[c] The photoinitiator, 1-hydroxycyclohexylphenylketone.

The percolation power-law equation was followed in the conductive composite coatings that were fabricated from poly(vinyl acetate-acrylic) (PVAc-*co*-acrylic) copolymer latices (50–600 nm) and nanosized ATO particles (15 nm) with V_c between 0.05 and 0.075 volume fractions of ATO and t ranging from 1.34 to 2.32 [181]. Compared to the PVAc-*co*-acrylic coating, the nanocomposite coatings had lower transparency because of the Rayleigh scattering. However, the transparency of the composite coatings can be improved by a reduction in the coating thickness. The best transparency for the coatings with a direct current conductivity of approximately 10^{-2} S/cm was around 85% at 600 nm. Interestingly, a low percolation threshold was exhibited at 1–2 vol% of ATO volume fraction for the UV-curable ATO-acrylate nanocomposite coatings using an acrylic prepolymer mixture (Ebecryl 745) as the binder and low amount of MPS grafted to the ATO nanoparticles [180]. Figure 1.39 gives the conductivity of the nanocomposite coatings as a function of ATO volume fraction and their modified EMA fitting curves.

Carbon nanotubes (CNTs) with properties ranging from semiconductor to metallic conductivity are the other candidates for fabrication of transparent conducting coatings. Because of the high aspect ratio, the conductivity of the nanocomposite coatings can be remarkably enhanced even at low loading levels. As shown in Figure 1.40, waterborne PU/CNT nanocomposite coatings containing 1.5 wt.% of

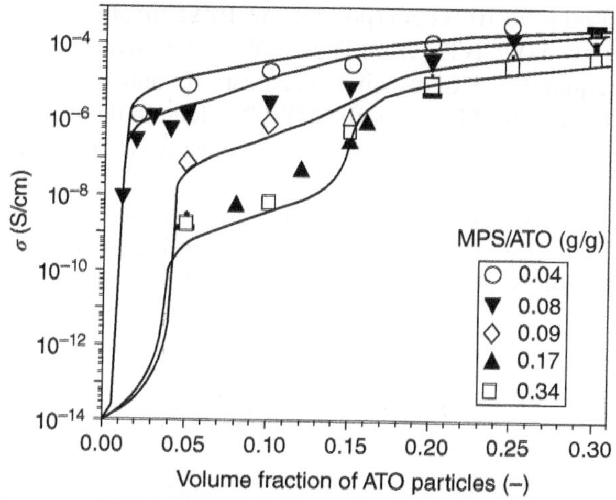

FIG. 1.39 The experimentally determined σ values for each volume faction of ATO for different MPS/ATO ratios (symbols) fitted with the modified EMA model (solid lines) [180]. Ref. 180. © 2007 Wiley Periodicals, Inc.

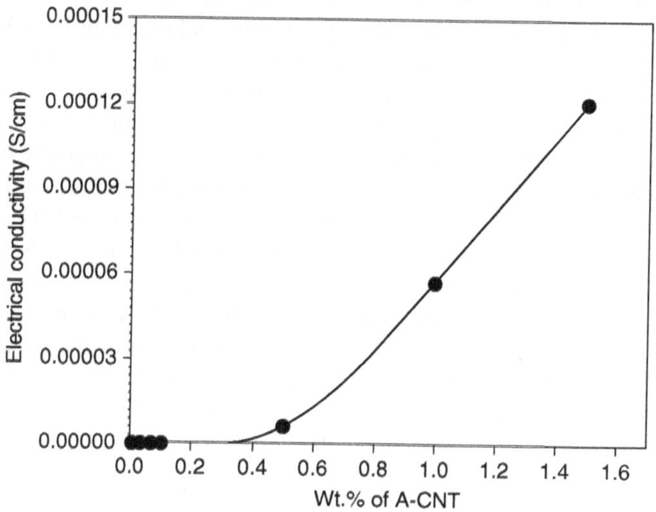

FIG. 1.40 Effect of nitric acid–treated MWCNT (A-MWCNT) content on the electrical conductivity of the WBPU/A-MWCNT nanocomposite films [81]. Ref. 81. © 2005 Wiley Periodicals, Inc.

A-MWCNTs exhibited a conductivity of 1.2×10^{-4} S/cm, which was nearly eight orders of magnitude higher than that of the pure PU film (2.5×10^{-12} S/cm) [81].

In another example, PU/MWCNT nanocomposite coatings were prepared based on a *para*-phenylenediamine-grafted MWCNT and toluene diisocyanate prepolymer with NCO functional groups as well as 4,4′-methylenebis(*o*-chloroaniline). The electrical resistivity decreased from $10^{13}\,\Omega\cdot$cm (pure PU) to $10^{8}\,\Omega\cdot$cm at 0.5 wt.% MWCNTs and $10^{7}\,\Omega\cdot$cm at 3.0 wt.% MWCNTs [182]. Thus, MWCNT-based nanocomposite coatings easily meet the required level of resistivity necessary for antistatic application (volume resistivity: 10^{5}–$10^{10}\,\Omega\cdot$cm). Moreover, the MWCNT load in these antistatic coatings is usually lower than 1.0 wt.%, which gives the coatings reasonable transmission in the visible range.

To meet minimum industry standards, a material whose conductivity is invariant under flexing must have a sheet resistance of $R_s \leq 100\ \Omega_{\square}$ coupled with an optical transparency of $T \geq 90\%$ (550 nm). For thin conducting films, R_s and T are linked through

$$T(\lambda) = \left(1 + \frac{188.5}{R_s}\frac{\sigma_{Op}(\lambda)}{\sigma_{DC}}\right)^{-2} \qquad (1.10)$$

where σ_{Op} and σ_{DC} are the optical (generally quoted at 550 nm) and direct current (DC) conductivities of the materials, respectively.

The single wall CNT (SWCNT) films typically have ~1.5×10^{4} S/m. Thus, to achieve the standard, σ_{DC} requires $\geq 5.3\times10^{5}$ S/m, which is a great challenge [183]. Academic efforts including the use of SWCNTs and a conductive polymer matrix have been made to tackle this challenge. Examples include metallic SWCNTs with a high conductivity that were separated from the arc discharge SWCNTs via 1-docosyloxymethyl pyrene and embedded into poly(3,4-ethylene dioxythiophene): poly(styrene sulfonate) (PEDOT:PSS) matrix [184]. Figure 1.41 demonstrates clearly the enhanced electrical conductivity of the separated metallic SWCNTs in the transparent conductive films. Meanwhile, the optical transparency of PEDOT:PSS is not sacrificed and make the composite films competitive to ITO coatings for transparent electrodes and other applications. Similar (PEDOT:PSS)/SWCNT nanocomposite films were prepared by De *et al.* [183]. DC conductivities of $>10^{5}$ S/m for mass fractions >50 wt.% were obtained. For an 80 nm thick composite filled with 60 wt.% arc discharge nanotubes, the conductivity ratio was maximized at $\sigma_{DC}/\sigma_{Op} = 15$. This translates into a transmittance (550 nm) and sheet resistance of 75% and $80\,\Omega_{\square}$, respectively, which are better than the data shown in Fig. 1.41. These composites were electromechanically very stable with <1% resistance change over 130 bend cycles.

1.7 OTHER FUNCTIONAL NANOCOMPOSITE COATINGS

Other functional OINCs have superhydrophilic, superhydrophobic, antibacterial, corrosion-resistant performance. When these functional coatings are used as clearcoats, high optical clarity is essential. Transparent superhydrophilic and superhydrophobic coatings are highly desired for self-cleaning and anti-frosting applications on windows or when the original surface appearance must not be impacted.

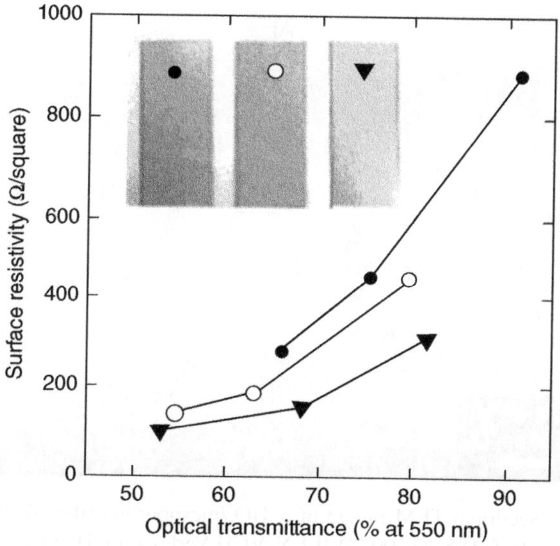

FIG. 1.41 Surface resistivity results of PEDOT:PSS/SWCNT films on glass substrate with the same 10 wt.% nanotube content but different film thickness and optical transmittance at 550 nm (○, preseparation purified sample; ▼, separated metallic SWCNTs, for comparison; ●, blank PEDOT:PSS without nanotubes) Inset shows representative films photographed with tiger paw print as background. Reprinted with permission from Ref. 184. © 2008 Elsevier.

These transparent superhydrophilic nanocomposite coatings are usually fabricated based on photocatalytic TiO_2 nanoparticles. Because of the high photodegradation of TiO_2 nanoparticles, hybrid or inorganic binders, not pure polymer, are generally used for the nanocomposite coatings. These binders are typically synthesized from a sol–gel process of TEOS [185] or TIPP [186]. Copolymer is sometimes added to avoid cracking during film formation [186]. This yields a porous structure via calcination [185] or nano-TiO_2-induced photodegradation [187] or both, for example, the role of *n*-octylamine in TiO_2-SiO_2 self-cleaning coatings for stones [188].

A typical morphology of TiO_2 nanoparticles dispersed in a mesoporous silica matrix is presented in Figure 1.42. The porous structure remarkably enhances the photocatalytic self-cleaning performance of the coatings but also deteriorates the optical transparency due to scattering [187]. Therefore, not only the dispersed size of the TiO_2 nanoparticles but also the pore size and distribution are crucial to successfully achieve transparent superhydrophilic coatings.

Superhydrophilic nanocomposite coatings are also developed using a simple mixture of polymers containing reactive TMOS groups, including quaternized poly(2-(dimethylamino) ethyl methacrylate and poly(MPS), and silica nanoparticles [189]. Superhydrophilic behavior is achieved with a high weight ratio of fumed silica nanoparticles or polymer/fumed silica nanoparticle bilayer coatings.

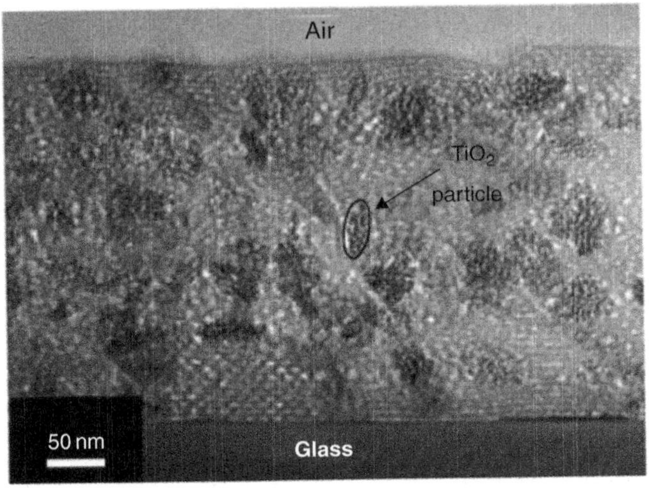

FIG. 1.42 Cross-sectional TEM image of a TiO_2/mesoporous silica film obtained with Ti/(Ti+Si)=0.2 [185]. Ref. 185. © 2007 WILEY-VCH Verlag GmbH & Co. KGaA.

It is well known that a rough surface structure and low surface free energy are essential to acquire a superhydrophobic surface. The preparation of transparent superhydrophobic coatings is difficult because the rough surface can scatter light. The surface roughness and the aggregate size of nanoparticles have to be carefully controlled at the nanoscale level. Thus far, transparent superhydrophobic coatings are seldom reported. Hwang *et al.* did prepare a coating of 50 nm colloidal silica particles and a hybrid binder from a sol–gel process of GPS/TMOS [190]. The coating was deposited on glass or silicon wafer to form a 40 nm thick film. Thereafter, the film was covered in a layer of prehydrolyzed isobutyltrimethoxysilane (IBTMS). The highest water contact angle of 130° was achieved at 7.21 nm thickness of IBTMS layer. However, the coating is very thin and not mechanically strong.

Lin *et al.* prepared a UV-curable transparent superhydrophobic coating using a system containing a fluoroimide acrylate oligomer, neopentyl glycol diacrylate (NPGDA), photoinitiator, and 20–25 nm colloidal SiO_2 nanoparticles [191]. The water contact angle of the coatings increases with silica content and reached 142° at 60 wt.% SiO_2 (NPS60). Figure 1.43 clearly illustrates a rough surface enriched with nanosize aggregates. The NPS60 still exhibits good transparency (96.44% at 400 nm and 100 μm film) because its surface roughness and aggregate size are smaller than 100 nm with minor light scattering. Meanwhile, the nanocomposite coating also shows high thermal stability and mechanical durability.

Transparent antibacterial coatings are prepared by the inclusion of antibacterial nanoadditives. These nanoadditives include nanosilver [192], nanocopper [193], and nano-semiconductive oxide (i.e., ZnO [194]). To assure the transparency of the coatings, the nanometal particles were made *in situ*. For example, an aqueous $AgNO_3$/PVA solution was spin-coated on glass slides and then heated in a hot air oven at 130–190°C to create Ag nanoparticles *in situ* within transparent PVA films. The

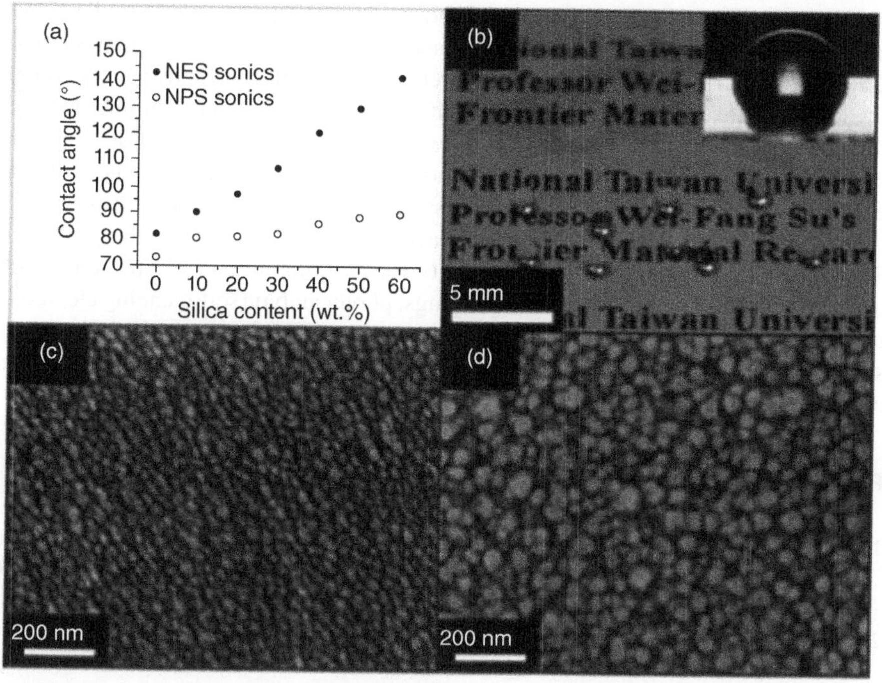

FIG. 1.43 (a) Comparison of contact angles between NES (prepared from ethoxylated bisphenol A diacrylate, NPGDA, and colloidal silica) and NPS (prepared from fluoroimide acrylate oligomer, NPGDA, and colloidal silica) series. (b) Water droplets on the surface of highly transparent NPS60 coated on glass. Inset shows the contact angle of the film is 142°. The SEM images of nanocomposites (c) NES60 and (d) NPS60. Reprinted with permission from Ref. 191. © 2010 The Royal Society of Chemistry.

composite films showed excellent antimicrobial performance toward bacteria such as *Escherichia coli* [192]. Other detailed examples of antibacterial coatings are given in Chapter 10.

1.8 CONCLUSIONS AND OUTLOOK

OINCs have been recognized as the most commonly used solution to exploit new organic coating products as well as to modify the properties of traditional organic coatings. Among the fabrication methods, addition of nanopowder is undoubtedly the simplest way. Nevertheless, the deagglomeration of nanopowder for preparation of transparent OINCs is a great challenge. To fulfill the deagglomeration well, some special deagglomeration machines, for example, grinding machine using bead with size less than 100 μm, have been manufactured. Some knowledge on surface molecular engineering of nanoparticles are also necessary to get fine nanoparticle dispersions. Due to the high-tech level of nanoparticle dispersion, some companies specializing on nanoparticle dispersions are even established, promoting the commercialization of transparent OINCs. However, the dispersions of the nanofillers with high aspect ratio, that are useful in the preparation of transparent barrier OINCs and antistatic (or conductive) OINCs, are still lack. In addition, OINCs with extremely high nanofiller loads are difficult to attain absolute transparency. *In situ* generation of nanophase during drying step of organic coatings would be an alternative way for these cases. But it needs further efforts on exploitation of new drying process (e.g., drying in different atmospheres).

As a whole, transparent OINCs have achieved great progress in the past years. Scratch-resistant nanocomposite clearcoats for automobile and plastic lens, transparent NIR-shielding nanocomposite coatings, photocatalytic self-cleaning clearcoats, transparent antibacterial nanocomposite coatings, etc., are now available in market somehow. Of course, transparent OINCs with better performances and more functions are highly expected in the near future.

REFERENCES

1. Schmidt, H., Seiferling, B. (1986) Chemistry and Applications of Inorganic-Organic Polymers (Organically Modified Silicates). MRS Spring Meeting: Materials Research Society, Palo Alto, CA, USA.
2. Schmidt, H., Wolter, H. (1990) Organically modified ceramics and their applications. *Journal of Non-Crystalline Solids*, 121(1–3): 428–435.
3. Zhang, H., Zhang, H., Tang, L.C., Zhou, L.Y. Eger, C., Zhang, Z. (2011) Comparative study on the optical, surface mechanical and wear resistant properties of transparent coatings filled with pyrogenic and colloidal silica nanoparticles. *Composites Science and Technology*, 71: 471–479.
4. Chen, X.C., You, B., Zhou, S.X., Wu, L.M. (2003) Surface and interface characterization of polyester-based polyurethane/nano-silica composites. *Surface and Interface Analysis*, 35(4): 369–374.
5. Poovarodom, S., Hosseinpour, D., Berg, J.C. (2008) Effect of particle aggregation on the mechanical properties of a reinforced organic-inorganic hybrid sol-gel composite. *Industrial & Engineering Chemistry Research*, 47(8): 2623–2629.
6. Zhou, S.X., Wu, L.M., You, B. Gu, G.X. (2009) Preparation, structure, and properties of organic-inorganic nanocomposite coatings. in *Smart Coatings II*. ed., Theodore Provder, Jamil Baghdachi. Vol. ACS Symposium Series No. 1002 : Washington, DC: American Chemical Society, 193–219.
7. Inkyo, M., Tahara, T., Iwaki, T., Iskandar, F., Hogan, C.J., Jr, Okuyama, K. (2006) Experimental investigation of nanoparticle dispersion by beads milling with centrifugal bead separation. *Journal of Colloid and Interface Science*, 304: 535–540.
8. Joni, I.M., Purwanto, A., Iskandar, F., Okuyama, K. (2009) Dispersion stability enhancement of titania nanoparticles in organic solvent using a bead mill process. *Industrial & Engineering Chemistry Research*, 48: 6916–6922.
9. Iversen, S.B., Rasmussen, H., Christensen, C.A., Jensen, H., Reenberg, T. (2010) Production of Titania Nanoparticle Colloidal Suspensions with Maintained Crystallinity by Using a Bead Mill with Micrometer Sized Beads. Applicants: VÄLINGE INNOVATION AB and PHOTOCAT A/S, International publication number: WO2010/110726.

10. Bock, M., Engbert, T., Groth, S., Klinksiek, B., Yeske, P., Jonschker, G., Dellwo, U. (2000) Transparent coating compositions containing nanoscale particles and having improved scratch resistance. Bayer Aktiengesellschaft, USA and US Patent 6,020,419.
11. Xiong, M.N., Gu, G.X., You, B., Wu, L.M. (2003) Preparation and characterization of poly(styrene butylacrylate) latex/nano-ZnO nanocomposites. *Journal of Applied Polymer Science*, 90(7): 1923–1931.
12. Liu, Y.L., Yu, Z.F., Zhou, S.X., Wu, L.M. (2006) De-aggregation and dispersion of nanofine TiO_2 in an agitator bead mill. *Journal of Dispersion Science and Technology*, 27(7): 983–990.
13. Hazan, Y.D., Heinecke, J., Weber, A., Graule, T. (2009) High solids loading ceramic colloidal dispersions in UV curable media via comb-polyelectrolyte surfactants. *Journal of Colloid and Interface Science*, 337: 66–74.
14. Alkilany, A.M., Thompson, L.B., Murphy, C.J. (2010) Polyelectrolyte coating provides a facile route to suspend gold nanorods in polar organic solvents and hydrophobic polymers. *ACS Applied Materials & Interfaces*, 2(12): 3417–3412.
15. Vestberg, R., Piekarski, A.M., Pressly, E.D., Berkel, K.Y.V., Malkoch, M., Gerbac, J., Ueno, N., Hawker, C.J. (2009) A general strategy for highly efficient nanoparticle dispersing agents based on hybrid dendritic linear block copolymers. *Journal of Polymer Science Part A: Polymer Chemistry*, 47: 1237–1258.
16. Gao, X.Y., Zhu, Y.C., Zhao, X., Wang, Z.C., An, D.M., Ma, Y.J., Guan, S., Du, Y.Y., Zhou, B. (2011) Synthesis and characterization of polyurethane/SiO_2 nanocomposites. *Applied Surface Science*, 257: 4719–4724.
17. Amici, J., Kahveci, M.U., Allia, P., Tiberto, P., Yagci, Y., Sangermano, M. (2012) Polymer grafting onto magnetite nanoparticles by "click" reaction. *Journal of Materials Science*, 47(1): 412–419.
18. Mesnage, A., Magied, M.A., Simon, P., Herlin-Boime, N., Jegou, P., Deniau, G., Palacin, S. (2011) Grafting polymers to titania nanoparticles by radical polymerization initiated by diazonium salt. *Journal of Materials Science*, 46: 6332–6338.
19. Schmidt, H.K., Geiter, E., Mennig, M., Krug, H., Becker, C., Winkler, R.-P. (1998) The sol-gel process for nano-technologies: new nanocomposites with interesting optical and mechanical properties. *Journal of Sol-gel Science and Technology*, 13: 397–404.
20. Bauer, F., Ernst, H., Decker, U., Findeisen, M., Gläsel, H.J., Langguth, H., Hartmann, E., Mehnert, R., Peuker, C. (2000) Preparation of scratch and abrasion resistant polymeric nanocomposites by monomer grafting onto nanoparticles, 1-FTIR and multi-nuclear NMR spectroscopy to the characterization of methacryl grafting. *Macromolecular Chemistry and Physics*, 201(18): 2654–2659.
21. Chen, H., Zhou, S.X., Gu, G.X., Wu, L.M. (2004) Study on modification and dispersion of nano-silica. *Journal of Dispersion Science and Technology*, 25(6): 837–848.
22. Miller, J.D., Ishida, H. (1984) Quantitative monomolecular coverage of inorganic particulates by methacryl-functional silanes. *Surface Science*, 148(2–3): 601–622.
23. Bauer, F., Sauerland, V., Ernst, H., Glasel, H.-J., Naumov, S., Mehnert, R. (2003) Preparation of scratch- and abrasion-resistant polymeric nanocomposites by monomer grafting onto nanoparticles, 4 application of MALDI-TOF mass spectrometry to the characterization of surface modified nanoparticles. *Macromolecular Chemistry and Physics*, 204: 375–383.
24. Bauer, F., Mehnert, R. (2005) UV curable acrylate nanocomposites: properties and applications. *Journal of Polymer Research*, 12: 483–491.

25. Li, Y.L., Chen, Z.X., Li, X.X., Zeng, H.W. (2011) A new surface modification method to improve the dispersity of nano-silica in organic solvents. *Journal of Sol-Gel Science and Technology*, 58: 290–295.
26. Lu, Y.F., Zhou, S.X., Wu, L.M. (2012) De-agglomeration and dispersion behavior of TiO_2 nanoparticles in organic media using 3-methacryloxypropyltrimethoxysilane as a surface modifier. *Journal of Dispersion Science and Technology*, 33(4): 497–505.
27. Scholz, S., Kaskel, S. (2008) Surface functionalization of ZrO_2 nanocrystallites for the integration into acrylate nanocomposite films. *Journal of Colloid and Interface Science*, 323: 84–91.
28. Posthumus, W., Magusin, P.C.M.M., Brokken-Zijp, J.C.M., Tinnemans, A.H.A., van der Linde R.. (2004) Surface modification of oxidic nanoparticles using 3-methacryloxypropyltrimethoxysilane. *Journal of Colloid and Interface Science*, 269: 109–116.
29. Zhou, S.X., Garnweitner, G., Niederberger, M., Antonietti, M. (2007) Dispersion behavior of zirconia nanocrystals and their surface-functionalization with vinyl group-containing ligands. *Langmuir*, 23(18): 9178–9187.
30. Zhou, S.X., Wu, L.M. (2008) Phase separation and properties of UV-curable polyurethane/zirconia nanocomposite coatings. *Macromolecular Chemistry and Physics*, 209(11): 1170–1181.
31. Xu, K., Zhou, S.X., Wu, L.M. (2009) Effect of highly-dispersible zirconia nanoparticles on the properties of UV-curable poly(urethane-acrylate) coatings. *Journal of Materials Science*, 44(6): 1613–1621.
32. Xu, K., Zhou, S.X., Wu, L.M. (2010) Dispersion of γ-methacryloxypropyltrimethoxysilane-functionalized zirconia nanoparticles in UV-curable formulations and properties of their cured coatings. *Progress in Organic Coatings*, 67(3): 302–310.
33. Luo, K.Q., Zhou, S.X., Wu, L.M., Gu, G.X. (2008) Dispersion and functionalization of non-aqueous synthesized zirconia nanocrystals via attachment of silane coupling agents. *Langmuir*, 24(20): 11497–11505.
34. Barna, E., Rentsch, D., Bommer, B., Vital, A., von Trzebiatowski, O., Graule, T. (2007) Surface modification of nanoparticles for scratch resistant clear coatings. *KGK-Kautschuk Gummi Kunststoffe*, 60(1–2): 49–51.
35 Kang, S., Hong, S., II, Choe, C.R., Park, M., Rim, S., Kim, J. (2001) Preparation and characterization of epoxy composites filled with functionalized nanosilica particles obtained via sol-gel process. *Polymer*, 42: 879–887.
36. Luo, K.Q., Zhou, S.X., Wu, L.M. (2009) High refractive index and good mechanical property UV-cured hybrid films containing zirconia nanoparticles. *Thin Solid Films*, 517: 5974–5980.
37. Douce, J., Boilot, J.P., Biteau, J., Scodellaro, L., Jimenez, A. (2004) Effect of filler size and surface condition of nano-sized silica particles in polysiloxane coatings. *Thin Solid Films*, 466(1–2): 114–122.
38. Greenwood, P., Gevert, B. (2011) Aqueous silane modified silica sols: theory and preparation. *Pigment & Resin Technology*, 40(5): 275–284.
39. Huang, T.C., Su, Y.-A., Yeh, T.-C., Huang, H.-Y., Wu, C.-P., Huang, K.-Y., Chou, Y.-C., Yeh, J.-M., Wei, Y. (2011) Advanced anticorrosive coatings prepared from electroactive epoxy–SiO_2 hybrid nanocomposite materials. *Electrochimica Acta*, 56: 6142–6149.
40. Chen, L., Shen, H.X., Lu, Z., Feng, C., Chen, S., Wang, Y.R. (2007) Fabrication and characterization of TiO_2-SiO_2 composite nanoparticles and polyurethane (TiO_2-SiO_2) nanocomposite films. *Colloid and Polymer Science*, 285(13): 1515–1520.

41. Gomathi, A., Rao, C.N.R. (2008) Hexadecyltriethoxysilane-induced dispersions of metal oxide nanoparticles in nonpolar solvents. *Journal of Cluster Science*, 19: 247–257.
42. Luo, K.Q., Zhou, S.X., Wu, L.M., You, B. (2010) Preparation and properties of cross-linked zirconia nanoparticle films on polycarbonate. *Thin Solid Films*, 518(23): 6804–6810.
43. Iijima, M., Kobayakawa, M., Kamiya, H. (2009) Tuning the stability of TiO_2 nanoparticles in various solvents by mixed silane alkoxides. *Journal of Colloid and Interface Science*, 337: 61–65.
44. Deb, B., Kumar, V., Druffel, T.L., Sunkara, M.K. (2009) Functionalizing titania nanoparticle surfaces in a fluidized bed plasma reactor. *Nanotechnology*, 20: 465701.
45. Sayılkan, F., Asiltèurk, M., Burunkaya, E., Ertuğrul, A. (2009) Hydrothermal synthesis and characterization of nanocrystalline ZrO_2 and surface modification with 2-acetoacetoxyethyl methacrylate. *Journal of Sol-Gel Science and Technology*, 51: 182–189.
46. Hu, Y.Q., Gu, G.X., Zhou, S.X., Wu, L.M. (2011) Preparation and properties of transparent PMMA/ZrO_2 nanocomposites using 2-hydroxyethyl methacrylate as a coupling agent. *Polymer*, 52: 122–129.
47. Pandey, S., Mishra, S.B. (2011) Sol-gel derived organic–inorganic hybrid materials: synthesis, characterizations and applications. *Journal of Sol-Gel Science and Technology*, 59: 73–94.
48. Sowntharya, L., Lavanya, S., Chandra, G.R., Hebalkar, N.Y., Subasri, R. (2012) Investigations on the mechanical properties of hybrid nanocomposite hard coatings on polycarbonate. *Ceramics International*, 38: 4221–4228.
49. Sangermano, M., Amerio, E., Epicoco, P., Priola, A., Rizza, G., Malucelli, G. (2007) Preparation and characterization of hybrid nanocomposite coatings by cationic UV-curing and the sol-gel process of a vinyl ether based system. *Macromolecular Materials and Engineering*, 292: 634–640.
50. Huang, H.-C., Huang, S.-P., Hsieh, T.-E., Chen, C.-H. (2012) Characterizations of UV-curable montmorillonite/epoxy nanocomposites prepared by a hybrid of chemical dispersion and planetary mechanical milling process. *Journal of Applied Polymer Science*, 123: 3199–3203.
51. Yeh, J.M., Chen, C.L., Chen, Y.C., Ma, C.Y., Huang, H.Y., Yu, Y.H. (2004) Enhanced corrosion prevention effect of polysulfone-clay nanocomposite materials prepared by solution dispersion. *Journal of Applied Polymer Science*, 92(1): 631–637.
52. Landry, V., Riedl, B., Blanchet, P. (2008) Nanoclay dispersion effects on UV coatings curing. *Progress in Organic Coatings*, 62: 400–408.
53. Shemper, B.S., Morizur, J.-F., Alirol, M., Domenech, A., Hulin, V., Mathias, L.J. (2004) Synthetic clay nanocomposite-based coatings prepared by UV-cure photopolymerization. *Journal of Applied Polymer Science*, 93(2): 1252–1263.
54. Chen, C.G., Khobaib, M., Curliss, D. (2003) Epoxy layered-silicate nanocomposites. *Progress in Organic Coatings*, 47(3–4): 376–383.
55. Majumdar, D., Blanton, T.N., Schwark, D.W. (2003) Clay-polymer nanocomposite coatings for imaging application. *Applied Clay Science*, 23(5–6): 265–273.
56. Ranade, A., D'Souza, N.A., Gnade, B. (2002) Exfoliated and intercalated polyamide-imide nanocomposites with montmorillonite. *Polymer*, 43(13): 3759–3766.
57. Sangermano, M., Messori, M. (2010) Scratch resistance enhancement of polymer coatings. *Macromolecular Materials and Engineering*, 295(7): 603–612.
58. Zhou, S.X., Wu, L.M., Sun, J., Shen, W.D. (2002) The changes of the properties of acrylic based polyurethane via addition of nano-silica. *Progress in Organic Coatings*, 45(1): 33–42.

59. Zhou, S.X., Wu, L.M., Sun, J., Shen, W.D. (2003) Effect of nanosilica on the properties of polyester-based polyurethane. *Journal of Applied Polymer Science*, 88(1): 189–193.
60. Zhou, S.X., Wu, L.M., Shen, W.D., Gu, G.X. (2004) Study on the morphology and tribological properties of acrylic based polyurethane/fumed silica composite coatings. *Journal of Materials Science*, 39(5): 1593–1600.
61. Jalili, M.M., Moradian, S., Dastmalchian, H., Karbasi, A. (2007) Investigating the variations in properties of 2-pack polyurethane clear coat through separate incorporation of hydrophilic and hydrophobic nano-silica. *Progress in Organic Coatings*, 59: 81–87.
62. Barna, E., Bommer, B., Kürsteiner J., Vital, A., Trzebiatowski, O.v., Koch, W., Schmid, B., Graule, T. (2005) Innovative, scratch proof nanocomposites for clear coatings. *Composites: Part A*, 36: 473–480.
63. Chen, G.D., Zhou, S.X., Gu, G.X., Yang, H.H., Wu, L.M. (2005) Effects of the surface property of nanosilica particles on its redispersibility and properties of acrylic based polyurethane/silica composites. *Journal of Colloid and Interface Science*, 281(2): 339–350.
64. Chen, G.D., Zhou, S.X., Gu, G.X., Wu, L.M. (2007) Modification of colloidal silica on the mechanical properties of acrylic based polyurethane/silica composites. *Colloids and Surfaces A—Physicochemical and Engineering Aspects*, 296(1–3): 29–36.
65. Chen, G.D., Zhou, S.X., Liao, H.M., Wu, L.M. (2005) Preparation of solvent-borne acrylic-based polyurethane-modified silica nanocomposites. *Journal of Composite Materials*, 39(3): 215–231.
66. Chen, Y.C., Zhou, S.X., Yang, H.H., Gu, G.X., Wu, L.M. (2004) Preparation and characterization of nanocomposite Polyurethane. *Journal of Colloid and Interface Science*, 279(2): 370–378.
67. Chen, Y.C., Zhou, S.X., Chen, G.D., Wu, L.M. (2005) Preparation and characterization of polyester/colloidal silica nanocomposite resins. *Progress in Organic Coatings*, 54(2): 120–126.
68. Chen, Y.C., Zhou, S.X., Yang, H.H., Wu, L.M. (2005) Structure and properties of polyurethane/nanosilica compoiste. *Journal of Applied Polymer Science*, 95(5): 1032–1039.
69. Ahmadi, B., Kassiriha, M., Khodabakhshi, K., Mafi, E.R. (2007) Effect of nano layered silicates on automotive polyurethane refinish clear coat. *Progress in Organic Coatings*, 60: 99–104.
70. Sabzi, M., Mirabedini, S.M., Zohuriaan-Mehr, J., Atai, M. (2009) Surface modification of TiO_2 nano-particles with silane coupling agent and investigation of its effect on the properties of polyurethane composite coating. *Progress in Organic Coatings*, 65: 222–228.
71. Tiarks, F., Leuninger, J., Wagner, O., Jahns, E., Wiese, H. (2007) Nanocomposite dispersions for water-based coatings. *Surface Coatings International*, 90(5): 221–229.
72. Luna-Xavier, J.-L., Bourgeat-Lami, E., Guyot, A. (2001) The role of initiation in the synthesis of silica/poly(methyl methacrylate) nanocomposite latex particles through emulsion polymerization. *Colloid and Polymer Science*, 279(10): 947–958.
73. Diaconu, G., Asua J.M., Paulis, M., Leiza, J.R. (2007) High-solids content waterborne polymer-clay nanocomposites. *Macromolecular Symposia*, 259(1): 305–317.
74. Qi, D.-M., Bao, Y.-Z., Weng, Z.X., Huang, Z.-M. (2006) Preparation of acrylate polymer/silica nanocomposite particles with high silica encapsulation efficiency via miniemulsion polymerization. *Polymer*, 47(13): 4622–4629.

75. You, B. Wen, N.G., Cao, Y.C., Zhou, S.X., Wu, L.M. (2009) Preparation and properties of poly(styrene-co-butyl acrylate-acrylic acid)/silica nanocomposite latex using an acidic silica sol. *Polymer International*, 58(5): 519–529.
76. Yang, L., Zhou, S.X., Gu, G.X. Wu, L.M. (2013) Film-forming behavior and mechanical properties of colloidal silica/polymer latex blends with high silica load. *Journal of Applied Polymer Science*, 129(3): 1434–1445.
77. Xiong, M.N., Wu, L.M., Zhou, S.X., You, B. (2002) Preapration and characterization of acrylic latex /nano-SiO_2 composites. *Polymer International*, 51: 693–698.
78. Yao, L., Yang, J., Sun, J., Cai, L.F., He, L.H., Huang, H., Song, R., Hao, Y.M. (2011) Hard and transparent hybrid polyurethane coatings using in situ incorporation of calcium carbonate nanoparticles. *Materials Chemistry and Physics*, 129: 523–528.
79. Gumfekar, S.P., Kunte, K.J., Ramjee, L., Kate, K.H., Sonawane, S.H. (2011) Synthesis of $CaCO_3$–P(MMA–BA) nanocomposite and its application in water based alkyd emulsion coating. *Progress in Organic Coatings*, 72(4): 632–637.
80. Nobel, M.L., Picken, S.J., Mendes, E., Waterborne nanocomposite resins for automotive coating applications. *Progress in Organic Coatings*, 58: 96–104.
81. Kwon, J.-Y., Kim, H.-D. (2005) Preparation and properties of acid-treated multiwalled carbon nanotube/waterborne polyurethane nanocomposites. *Journal of Applied Polymer Science*, 96(2): 595–604.
82. Landry, V., Riedl, B., Blanchet, P. (2008) Alumina and zirconia acrylate nanocomposites coatings for wood flooring: photocalorimetric characterization. *Progress in Organic Coatings*, 61(1): 76–82.
83. Sow, C., Riedl, B., Blanchet, P. (2010) Kinetic studies of UV-waterborne nanocomposite formulations with nanoalumina and nanosilica. *Progress in Organic Coatings*, 67(2): 188–194.
84. Sangermano, M., Priola, A., Kortaberriac, G., Jimeno, A., Garcia, I., Mondragon, I., Rizza, G. (2007) Photopolymerization of epoxy coatings containing iron-oxide nanoparticles. *Macromolecular Materials and Engineering*, 292: 956–961.
85. Kang, D.J., Han, D.H., Kang, D.P. (2009) Fabrication and characterization of photocurable inorganic–organic hybrid materials using organically modified colloidal-silica nanoparticles and acryl resin. *Journal of Non-Crystalline Solids*, 355: 397–402.
86. Gläsel, H.-J., Bauer, F., Ernst, H., Findeisen, M., Hartmann, E., Langguth, H.,Mehnert, R., Schubert, R. (2000) Preparation of scratch and abrasion resistant polymeric nanocomposites by monomer grafting onto nanoparticles, 2 characterization of radiation-cured polymeric nanocomposites. *Macromolecular Chemistry and Physics*, 201: 2765–2770.
87. Soloukhin, V.A., Posthumus, W., Brokken-Zijp, J.C.M., Loos, J., With, G. (2002) Mechanical properties of silica-(meth)acrylate hybrid coatings on polycarbonate substrate. *Polymer*, 43: 6169–6181.
88. Kim, D., Jeon, K., Lee, Y., Seo, J., Seo, K., Han, H., Khan, S.B. (2012) Preparation and characterization of UV-cured polyurethane acrylate/ZnO nanocomposite films based on surface modified ZnO. *Progress in organic coatings*, 74(3): 435–442.
89. Bautista, Y., Gonzalez, J., Gilabert, J., Ibanez, M.J., Sanz, V. (2011) Correlation between the wear resistance, and the scratch resistance, for nanocomposite coatings. *Progress in Organic Coatings*, 70(4): 178–185.
90. Leder, G., Ladwig, T., Valter, V., Frahn, S., Meyer, J. (2002) New effects of fumed silica in modern coatings. *Progress in Organic Coatings*, 45: 139–144.

91. Zhang, H., Zhang, H., Tang, L.C., Zhang, Z., Gu, L., Xu, Y.Z., Christian, E. (2010) Wear-resistant and transparent acrylate-based coating with highly filled nanosilica particles. *Tribology International*, 43: 83–91.

92. Gianni, A.D., Amerio, E., Monticelli, O., Bongiovanni, R. (2008) Preparation of polymer/clay mineral nanocomposites via dispersion of silylated montmorillonite in a UV curable epoxy matrix. *Applied Clay Science*, 42: 116–124.

93. Mohamadpour, Sh., Pourabbas, B., Fabbri, P. (2011) Anti-scratch and adhesion properties of photo-curable polymer/clay nanocomposite coatings based on methacrylate monomers. *Scientia Iranica*, 18(3): 765–771.

94. Corcione, C.E., Frigione, M. (2012) UV-cured polymer-boehmite nanocomposite as protective coating for wood elements. *Progress in Organic Coatings*, 74: 781–787.

95. Zhang, X.H., Yang, J.W., Zeng, Z.H., Huang, L., Chen, Y.L., Wang, H.H.(2006) Stabilized dispersions of titania nanoparticles via a sol–gel process and applications in UV-curable hybrid systems. *Polymer International*, 55(4): 466–472.

96. Bauer, F., Sauerland, V., Gläsel, H.-J., Ernst, H., Findeisen, M., Hartmann, E.,Langguth, H., Marquardt, B., Mehnert, R. (2002) Preparation of scratch and abrasion resistant polymeric nanocomposites by monomer grafting onto nanoparticles, 3 effect of filler particles and grafting agents. *Macromolecular Materials and Engineering*, 287: 546–552.

97. Sangermano, M., Gaspari, E., Vescovo, L., Messori, M. (2011) Enhancement of scratch-resistance properties of methacrylated UV-cured coatings. *Progress in Organic Coatings*, 72(3): 287–291.

98. Hsiang H.-I., Chang, Y.-L., Chen, C.-Y., Yen, F.-S. (2010) Silane functional effects on the rheology and abrasion resistance of transparent SiO_2/UV-curable resin nano-composites. *Materials Chemistry and Physics*, 120: 476–479.

99. Hsiang H.-I., Chang, Y.-L., Chen, C.-Y., Yen, F.-S. (2011) Silane effects on the surface morphology and abrasion resistance of transparent SiO_2/UV-curable resin nano-composites. *Applied Surface Science*, 257: 3451–3454.

100. Schmidt, H.K., Krug, H., Sepeur-Zeitz, B., Geiter, E. (1997) Inorganic-organic nanocomposites for optical coatings. Proc. SPIE 3136. In Dunn, B.S., Mackenzie, J.D., Pope, E.J.A., Schmidt, H.K., Yamane, M., eds. *Sol-Gel Optics IV*. SPIE, San Diego, CA.

101. Daniels, M.W., Francis, L.F. (1998) Silane adsorption behavior, microstructure, and properties of glycidoxypropyltrimethoxysilane-modified colloidal silica coatings. *Journal of Colloid and Interface Science*, 205: 191–200.

102. Mosher, B.P., Wu, C.W., Sun, T., Zeng, T.F. (2006) Particle-reinforced water-based organic-inorganic nanocomposite coatings for tailored applications. *Journal of Non-Crystalline Solids*, 352: 3295–3301.

103. Chen, Q. Tan, J.G.H., Shen, S.C., Liu, Y.C., Ng, W.K., Zeng, X.T. (2007) Effect of boehmite nanorods on the properties of glycidoxypropyltrimethoxysilane (GPTS) hybrid coatings. *Journal of Sol-gel Science and Technology*, 44: 125–131.

104. Sepeur, S., Kunze, N., Werner, B., Schmidt, H. (1999) UV curable hard coatings on plastics. *Thin Solid Films*, 351: 216–219.

105. Chantarachindawong, R., Luangtip, W., Chindaudom, P.,Osotchan, T., Srikhirin, T. (2012) Development of the scratch resistance on acrylic sheet with basic colloidal silica (SiO_2)—methyltrimethoxysilane (MTMS) nanocomposite films by sol–gel technique. *The Canadian Journal of Chemical Engineering*, 90: 888–896.

106. Chau, J.L.H., Hsieh, C.-C., Lin, Y.-M., Li, A.-K. (2008) Preparation of transparent silica–PMMA nanocomposite hard coatings. *Progress in Organic Coatings*, 62: 436–439.

107. Fogelstrom, L., Malmstrom, E., Johansson, M., Hult, A. (2010) Hard and flexible nanocomposite coatings using nanoclay-filled hyperbranched polymers. *ACS Applied Materials & Interfaces*, 2(6): 1679–1684.

108. Jiratumnukul, N., Pruthipaitoon, S., Pitsaroup, T. (2006) Nanocomposite alkyd coatings. *Journal of Applied Polymer Science*, 102(3): 2639–2642.

109. Calvo, M.E., Smirnov, J.R.C., Miguez, H. (2012) Novel approaches to flexible visible transparent hybrid films for ultraviolet protection. *Journal of Polymer Science Part B: Polymer Physics*, 50: 945–956.

110. Yuwono, A.H., Xue, J.M., Wang, J., Elim, H.I., Ji, W., Li, Y., White, T.J. (2003) Transparent nanohybrids of nanocrystalline TiO_2 in PMMA with unique nonlinear optical behavior. *Journal of Materials Chemistry*, 13: 1475–1479.

111. Allen, N.S., Edge, M., Ortega, A., Liauw, C.M., Stratton, J., McIntyre, R.B. (2002) Behaviour of nanoparticle (ultrafine) titanium dioxide pigments and stabilisers on the photooxidative stability of water based acrylic and isocyanate based acrylic coatings. *Polymer Degradation and Stability*, 78(3): 467–478.

112. Zhou, S.X., Wu L.M., Xiong, M.N., He, Q.Y., Chen, G.D. (2004) Dispersion and UV-VIS property of nanoparticles in coatings. *Journal of Dispersion Science and Technology*, 25(4): 417–433.

113. Mallakpour, S., Barati, A. (2011) Efficient preparation of hybrid nanocomposite coatings based on poly(vinyl alcohol) and silane coupling agent modified TiO_2 nanoparticles. *Progress in Organic Coatings*, 71: 391–398.

114. Godnjavec, J., Znoj, B., Vince, J., Steinbucher, M., Znidarsie, A., Venturini, P. (2012) Stabilization of rutile TiO_2 nanoparticles with glymo in polyacrylic clear coating. *Materials and technology*, 46(1): 19–24.

115. Godnjavec, J., Znoj, B., Veronovski, N., Venturini, P. (2012) Polyhedral oligomeric silsesquioxanes as titanium dioxide surface modifiers for transparent acrylic UV blocking hybrid coating. *Progress in Organic Coatings*, 74: 654–659.

116. Zhang, Y.H., Yu, L., Ke, S.M., Shen, B., Meng, X.H., Huang, H.T., Lv, F.Z., Xin, J.H., Chan, H.L.W. (2011) TiO_2/SiO_2 hybrid nanomaterials: synthesis and variable UV-blocking properties. *Journal of Sol-Gel Science and Technology*, 58: 326–329.

117. Ren, Y., Chen, M., Zhang, Y., Wu L.M. (2010) Fabrication of rattle-type TiO_2@void@SiO_2 particles with both high photoactivity and UV-shielding property. *Langmuir*, 26(13): 11391–11396.

118. Sangermano, M., Borlatto, E., D'Herin Bytner, F.D., Priola, A., Rizza, G. (2007) Photostabilization of cationic UV-cured coatings in the presence of nanoTiO_2. *Progress in Organic Coatings*, 59: 122–125.

119. Hwang, D.K., Moon, J.H., Shul, Y.G., Jung, K.T., Kim, D.H., Lee, D.W. (2003) Scratch resistant and transparent UV-protective coating on polycarbonate. *Journal of Sol-Gel Science and Technology*, 26: 783–787.

120. Chen, W.-C., Lee, S.-J., Lee, L.-H., Lin, J.-L. (1999) Synthesis and characterization of trialkoxysilane-capped poly(methyl methacrylate)-titania hybrid optical thin films. *Journal of Materials Chemistry*, 9: 1999–2003.

121. Mazzocchetti, L., Cortecchia, E., Scandola, M. (2009) Organic-inorganic hybrids as transparent coatings for UV and X-ray shielding. *ACS Applied Materials & Interfaces*, 1(3): 726–734.

122. Que, W.X., Hu, X. (2003) Influence of titanium content and temperature on optical and mechanical properties of sol–gel derived TiO_2/γ-glycidoxypropyltrimethoxysilane and methyltrimethoxysilane hybrid organic–inorganic films. *Journal of Physics D: Applied Physics*, 36: 908–914.

123. Xiang, H.P., Ge, J.F., Cheng, S.H., Han, H.M., Cui, S.W. (2011) Synthesis and characterization of titania/MQ silicone resin hybrid nanocomposite via sol–gel process. *Journal of Sol-Gel Science and Technology*, 59: 635–639.

124. Osmond, M.J., McCall, M.J. (2010) Zinc oxide nanoparticles in modern sunscreens: an analysis of potential exposure and hazard. *Nanotoxicology*, 4(1): 15–41.

125. Wang, L.L., Zhang, X.T., Li, B., Sun, P.P., Yang, J.K., Xu, H.Y., Liu, Y.C. (2011) Superhydrophobic and ultraviolet-blocking cotton textiles. *ACS Applied Materials & Interfaces*, 3(4): 1277–1281.

126. Li, Y.-Q., Fu, S.-Y., Mai, Y.-W. (2006) Preparation and characterization of transparent ZnO/epoxy nanocomposites with high-UV shielding efficiency. *Polymer*, 47: 2127–2132.

127. Li, H.Y., Chen, Y.F., Ruan, C.X., Gao, W.M., and Xie Y.S. (2001) Preparation of organic–inorganic multifunctional nanocomposite coating via sol-gel routes. *Journal of Nanoparticle Research*, 3: 157–160.

128. Khrenov, V., Schwager, F., Klapper, M., Koch, M., Müllen, K. (2007) Compatibilization of inorganic particles for polymeric nanocomposites. Optimization of the size and the compatibility of ZnO particles. *Polymer Bulletin*, 58: 799–807.

129. Weichelt, F., Emmler, R., Flyunt, R., Beyer, E., Buchmeiser, M.R., Beyer, M. (2010) ZnO-Based UV nanocomposites for wood coatings in outdoor applications. *Macromolecular Materials and Engineering*, 295: 130–136.

130. Li, S.H., Toprak, M.S., Jo, Y.S., Dobson, J., Kim, D.K., Muhammed, M. (2007) Bulk synthesis of transparent and homogeneous polymeric hybrid materials with ZnO quantum dots and PMMA. *Advanced Materials*, 2007(19): 4347–4352.

131. Tu, Y., Zhou, L., Jin, Y.Z., Gao, C., Ye, Z.Z., Yang, Y.F., Wang, Q.L. (2010) Transparent and flexible thin films of ZnO-polystyrene nanocomposite for UV-shielding applications. *Journal of Materials Chemistry*, 20: 1594–1599.

132. Liu, H.T., Zeng, X.F., Zhao, H., Chen, J. F. (2012) Transparent and multifunctional polymer nanohybrid film with superhigh ZnO content synthesized by a bulk polymerization method. *Industrial & Engineering Chemistry Research*, 51: 6753–6759.

133. Jeeju, P.P., Sajimol, A.M., Sreevalsa, V.G., Varma, S.J., Jayalekshmi, S. (2011) Size-dependent optical properties of transparent, spin-coated polystyrene/ZnO nanocomposite films. *Polymer International*, 60: 1263–1268.

134. Zhang, L.Y., Yin, L.W., Wang, C.X., Lun, N., Qi, Y.X. (2010) Sol-gel growth of hexagonal faceted ZnO prism quantum dots with polar surfaces for enhanced photocatalytic activity. *ACS Applied Materials & Interfaces*, 2: 1769–1773.

135. Wang, X.L., Zhou, S.X., Wu, L.M. (2013) Facile encapsulation of SiO_2 on ZnO quantum dots and its application in waterborne UV-shielding polymer coatings. *Journal of Materials Chemistry C*, 1(45): 7547-7553.

136. Yabe, S., Yamashita, M., Momose, S., Tahira, K., Yoshida, S., Li, R.X., Yin, S., Sato, T. (2001) Synthesis and UV-shielding properties of metal oxide doped ceria via soft solution chemical processes. *International Journal of Inorganic Materials*, 3(7): 1003–1008.

137. Saadat-Monfared, A., Mohseni, M. (2014) Polyurethane nanocomposite films containing nano-cerium oxide as UV absorber; part 2: structural and mechanical studies upon UV exposure. *Colloids and Surfaces A: Physicochemical and Engineering Aspects*, 441: 752–757.

138. de Lima, J.F., Serra, O.A. (2013) Cerium phosphate nanoparticles with low photocatalytic activity for UV light absorption application in photoprotection. *Dyes and Pigments*, 97(2): 291–196.

139. Zimmennann, L., Weibel, M., Caseri, W., Suter, U.W. (1993) High refractive index films of polymer nanocomposites. *Journal of Materials Research*, 8(7): 1742–1748.

140. Antonello, A., Brusatin, G., Guglielmi, M., Martucci, A., Bello, V., Mattei, G., Mazzoldi, P., Pellegrini, G. (2010) Hybrid organic–inorganic ZnS-loaded nanocomposite films for stable optical coatings. *Thin Solid Films*, 518: 6781–6786.

141. Franc, J., Blanc, D., Zerroukhi, A., Chalamet, Y., Last, A., Destouches, N. (2006) Organo-silica–titania nanocomposite elaborated by sol–gel processing with tunable optical properties. *Materials Science and Engineering B*, 129: 180–185.

142. Chau, J.L.H., Tung, C.-T., Lin, Y.-M., Li, A.-K. (2008) Preparation and optical properties of titania/epoxy nanocomposite coatings. *Materials Letters*, 62: 3416–3418.

143. Antonello, A., Brusatin, G., Guglielmi, M., Bello, V., Mattei, G., Zacco, G., Martucci, A. (2011) Nanocomposites of titania and hybrid matrix with high refractive index. *Journal of Nanoparticle Research*, 13: 1697–1708.

144. Chang, C.-C., Cheng, L.-P., Huang, F.-H., Lin, C.-Y., Hsieh, C.-F., Wang, W.-H. (2010) Preparation and characterization of TiO_2 hybrid sol for UV-curable high-refractive-index organic–inorganic hybrid thin films. *Journal of Sol-Gel Science and Technology*, 55: 199–206.

145. Liu, B.-T., Tang, S.-J., Yub, Y.-Y., Lin, S.-H. (2011) High-refractive-index polymer/inorganic hybrid films containing high TiO_2 contents. *Colloids and Surfaces A: Physicochemical and Engineering Aspects*, 377: 138–143.

146. Takeda, M., Tanabe, E., Iwaki, T., Yabuki, A., Okuyama, K. (2008) High-concentration transparent TiO_2 nanocomposite films prepared from TiO_2 nanoslurry dispersed by using bead mill. *Polymer Journal*, 40(8): 694–699.

147. Nakayama, N., Hayashi, T. (2007) Preparation and characterization of TiO_2-ZrO_2 and thiol-acrylate resin nanocomposites with high refractive index via UV-induced crosslinking polymerization. *Composites: Part A*, 38: 1996–2004.

148. Nakayama, N., Hayashi, T. (2008) Synthesis of novel UV-curable difunctional thiourethane methacrylate and studies on organic-inorganic nanocomposite hard coatings for high refractive index plastic lenses. *Progress in Organic Coatings*, 62: 274–284.

149. Chang, C.-M., Chang, C.-L., Chang, C.-C. (2006) Synthesis and optical properties of soluble polyimide/titania hybrid thin films. *Macromolecular Materials and Engineering*, 291: 1521–1528.

150. Chang, W.-L., Su, H.-W., Chen, W.-C. (2009) Synthesis and properties of photosensitive polyimide–nanocrystalline titania optical thin films. *European Polymer Journal*, 45: 2749–2759.

151. Liou G.-S., Lin, P.-H., Yen, H.-J., Yu, Y.-Y., Tsai, T.-W., Chen, W.-C. (2010) Highly flexible and optical transparent 6F-PI/TiO_2 optical hybrid films with tunable refractive index and excellent thermal stability. *Journal of Materials Chemistry*, 20: 531–536.

152. Lee, S., Shin, H.-J., Yoon, S.-M., Yi, D.K., Choi, J.-Y., Paik, U. (2008) Refractive index engineering of transparent ZrO_2–polydimethylsiloxane nanocomposites. *Journal of Materials Chemistry*, 18: 1751–1755.

153. Tsuzuki, T. (2008) Abnormal transmittance of refractive-index-modified ZnO/organic hybrid films. *Macromolecular Materials and Engineering*, 293: 109–113.

154. Lü, C.-L., Yang, B. (2009) High refractive index organic–inorganic nanocomposites: design, synthesis and application. *Journal of Materials Chemistry*, 19: 2884–2901.

155. Tao P., Viswanath, A., Schadler, L.S., Benicewicz, B.C., Siegel, R.W. (2011) Preparation and optical properties of indium tin oxide/epoxy nanocomposites with polyglycidyl methacrylate grafted nanoparticles. *ACS Applied Materials & Interfaces*, 3: 3638–3645.

156. Liu, H.T., Zeng, X.F., Kong, X.R., Bian, S.G., Chen, J.F.. (2012) A simple two-step method to fabricate highly transparent ITO/polymer nanocomposite films. *Applied Surface Science*, 258: 8564–8569.

157. Feng, J., Huang, B.Y., Zhong, M.Q. (2009) Fabrication of superhydrophobic and heat-insulating antimony doped tin oxide/polyurethane films by cast replica micromolding. *Journal of Colloid and Interface Science*, 336: 268–272.

158. Wang, X., Hu, Y., Song, L., Xing, W.-Y., Lu, H.-D., Lv, P., Jie, G.-X. (2010) Effect of antimony doped tin oxide on behaviors of waterborne polyurethane acrylate nanocomposite coatings. *Surface & Coatings Technology*, 205(7): 1864–1869.

159. Zhang, G.-J., Chen, Z.-H., Zeng, X.-R., Yu, F., Wang, J. (2011) Preparation and characterization of poly(MMA–BA)/nano-ATO hybrid latex via miniemulsion polymerization. *Journal of Coatings Technology and Research*, 8(4): 505–511.

160. Li, Y.-Q., Kang, Y., Xiao, H.-M., Mei, S.-G., Zhang, G.-L., Fu, S.-Y. (2011) Preparation and characterization of transparent Al doped ZnO/epoxy composite as thermal-insulating coating. *Composites: Part B*, 42: 2176–2180.

161. Stefan, S., Smith, G.B. (2003) Dilute LaB6 nanoparticles in polymer as optimized clear solar control glazing. *Applied Physics Letters*, 82(24): 4346–4348.

162. Yuan, Y.-F., Zhang, L., Hu, L.-J., Wang, W., Min, G.-H. (2011) Size effect of added LaB6 particles on optical properties of LaB6/polymer composites. *Journal of Solid State Chemistry*, 184(12): 3364–3367.

163. Takeda, H., Adachi, K. (2007) Near infrared absorption of tungsten oxide nanoparticle dispersions. *Journal of American Ceramic Society*, 90(12): 4059–4061.

164. Lu, Y.F., Zhou, S.X., Gu, G.X., Wu, L.M. (2013) Preparation of transparent, hard thermochromic polysiloxane/tungsten-doped vanadium dioxide nanocomposite coatings at ambient temperature. *Thin Solid Films*, 534: 231–237.

165. Lange, J., Wyser, Y. (2003) Recent innovations in barrier technologies for plastic packaging—a review. *Packaging Technology and Science*, 16: 149–158.

166. Fredrickson, G.H., Bicerano, J. (1999) Barrier properties of oriented disk composites. *Journal of Chemical Physics*, 110(4): 2181–2188.

167. Gusev, A.A., Lusti, H.R. (2001) Rational design of nanocomposites for barrier applications. *Advanced Materials*, 13(21): 1641–1643.

168. Xiao, J., Huang, Y.L., Manke, C. W. (2010) Computational design of polymer nanocomposite coatings: a multiscale hierarchical approach for barrier property prediction. *Industrial & Engineering Chemistry Research*, 49: 7718–7727.

169. Sun, Q.H., Schork, F.J., Deng, Y.L. (2007) Water-based polymer/clay nanocomposite suspension for improving water and moisture barrier in coating. *Composites Science and Technology*, 67: 1823–1829.

170. Woo, R.S.C., Zhu, H.G., Chow, M.M.K., Leung, C.K.Y., Kim, J.K. (2008) Barrier performance of silane–clay nanocomposite coatings on concrete structure. *Composites Science and Technology*, 68(14): 2828–2836.

171. Pavlacky, E., Ravindran, N., Webster, D.C. (2012) Novel in situ synthesis in the preparation of ultraviolet-curable nanocomposite barrier coatings. *Journal of Applied Polymer Science*, 125: 3836–3848.

172. Möller, M.W., Kunz, D.A., Lunkenbein, T., Sommer, S., Nennemann, A., J, Breu (2012) UV-cured, flexible, and transparent nanocomposite coating with remarkable oxygen barrier. *Advanced Materials*, 24: 2142–2147.

173. Heidarian, M., Shishesaz, M.R., Kassiriha, S.M., Nematollahi, M. (2011) Study on the effect of ultrasonication time on transport properties of polyurethane/organoclay nanocomposite coatings. *Journal of Coatings Technology and Research*, 8(2): 265–274.

174. Priolo, M.A., Gamboa, D., Holder, K.M., Grunlan, J.C. (2010) Super gas barrier of transparent polymer-clay multilayer ultrathin films. *Nano Letters*, 10: 4970–4974.

175. Goebbert, C., Bisht, H., Al-Dahoudi, N., Nonninger, R., Aegerter, M.A., Schmidt, H. (2000) Wet chemical deposition of crystalline, redispersable ATO and ITO nanoparticles. *Journal of Sol-Gel Science and Technology*, 19: 201–204.

176. Al-Dahoudi, N., Bisht, H., Gobbert, C., Krajewski, T., Aegerter, M.A. (2001) Transparent conducting, anti-static and anti-static-anti-glare coatings on plastic substrates. *Thin Solid Films*, 392: 299–304.

177. Aegerter, M.A., Al-Dahoudi, N., Solieman, A., Kavak, H., Oliveira, P. (2004) Transparent conducting coatings made by chemical nanotechnology process. *Molecular Crystals and Liquid Crystals*, 417: 105–114.

178. Kèoniger, T., Mèunstedt, H. (2009) Influence of polyvinylpyrrolidone on properties of flexible electrically conducting indium tin oxide nanoparticle coatings. *Journal of Materials Science*, 44: 2736–2742.

179. Maksimenko, I., Gross, M., Königer, T., Münstedt, H., Wellmann, P.J. (2010) Conductivity and adhesion enhancement in low-temperature processed indium tin oxide/polymer nanocomposites. *Thin Solid Films*, 518(10): 2910–2915.

180. Soloukhin, V.A., Brokken-Zijp, J.C.M., Dewith, G. (2007) Conductive ATO-acrylate nanocomposite hybrid coatings: experimental results and modeling. *Journal of Polymer Science Part B: Polymer Physics*, 45: 2147–2160.

181. Sun, J.K., Gerberich, W.W., Francis, L.F. (2003) Electrical and optical properties of ceramic–polymer nanocomposite coatings. *Journal of Polymer Science Part B: Polymer Physics*, 41: 1744–1761.

182. Zhao, W.M., Li, M., Peng, H.-X. (2010) Functionalized MWNT-doped thermoplastic polyurethane nanocomposites for aerospace coating applications. *Macromolecular Materials and Engineering*, 295(9): 838–845.

183. De, S., Lyons, P.E., Sorel, S., Doherty, E.M., King, P.J., Blau, W.J., Nirmalraj, P.N., Boland, J.J., Scardaci, V., Joimel, J., Coleman, J.N. (2009) Transparent, flexible, and highly conductive thin films based on polymer-nanotube composites. *ACS Nano*, 3(3): 714–720.

184. Wang, W., Fernando, K.A.S., Lin, Y., Meziani, M.J., Veca, L.M., Cao, L., Zhang, P.Y., Kimani, M.M., Sun, Y.-P. (2008) Metallic single-walled carbon nanotubes for conductive nanocomposites. *The Journal of American Chemical Society*, 130: 1415–1419.

185. Allain, E., Besson, S., Durand, C., Moreau, M., Gacoin, T., Boilot, J.P. (2007) Transparent mesoporous nanocomposite films for self-cleaning applications. *Advanced Functional Materials*, 17: 549–554.

186. Habibi, M.H., Nasr-Esfahani, M., Egerton, T.A. (2007) Preparation, characterization and photocatalytic activity of TiO_2/methylcellulose nanocomposite films derived from nanopowder TiO_2 and modified sol–gel titania. *Journal of Materials Science*, 42: 6027–6035.

187. Ding, X.F., Zhou, S.X., Wu, L.M., Gu, G.X., Yang, J.T. (2010) Formation of supra-amphiphilic self-cleaning surfaces through sun-illumination of titania-based nanocomposite coatings. *Surface & Coatings Technology*, 205(7): 2554–2561.

188. Pinho, L., Mosquera, M.J. (2011) Titania-silica nanocomposite photocatalysts with application in stone self-cleaning. *The Journal of Physical Chemistry C*, 115: 22851–22862.

189. Dong, H.C., Ye, P.L., Zhong, M.J., Pietrasik, J., Drumright, R., Matyjaszewski, K. (2010) Superhydrophilic surfaces via polymer-SiO_2 nanocomposites. *Langmuir*, 26(19): 15567–15573.

190. Hwang J.-H., Lee, B.I., Klep, V., Luzinov, I. (2008) Transparent hydrophobic organic–inorganic nanocomposite films. *Materials Research Bulletin*, 43: 2652–2657.

191. Lin, C.-C., Hsu, S.-H., Chang, Y.-L. Su, W.-F. (2010) Transparent hydrophobic durable low moisture permeation poly(fluoroimide acrylate)/SiO_2 nanocomposite from solventless photocurable resin system. *Journal of Materials Chemistry*, 20: 3084–3091.

192. Liu, S.X., He, J.H., Xue, J.F., Ding, W.J. (2009) Efficient fabrication of transparent antimicrobial poly(vinyl alcohol) thin films. *Journal of Nanoparticle Research*, 11: 553–560.

193. Cioffi, N., Torsi, L., Ditaranto, N., Sabbatini, L., Zambonin, P.G., Tantillo, G., Ghibelli, L., D'Alessio, M.D., Bleve-Zacheo, T., Traversa, E. (2004) Antifungal activity of polymer-based copper nanocomposite coatings. *Applied Physics Letters*, 85(12): 2417–2419.

194. Xu, T., Xie, C.S. (2003) Tetrapod-like nano-particle ZnO/acrylic resin composite and its multi-function property. *Progress in Organic Coatings*, 46: 297–301.

CHAPTER 2

Superhydrophobic and Superoleophobic Polymeric Surfaces

Jie Zhao and W. (Marshall) Ming
Department of Chemistry, Georgia Southern University, Statesboro, GA, USA

2.1 INTRODUCTION

There are many examples of elegantly designed surface wettability in Mother Nature, including lotus leaf [1–3], legs of water strider [4], and gecko's feet [5]. The self-cleaning property of the lotus leaf can be attributed to the combination of a proper surface chemistry and a peculiar topographic feature based on dual-scale roughness: the coarse-scale rough structure is about 10–20 μm in size, whereas the finer structure on top of the coarse structure is in the range of a few hundred nanometers [1–3]. This unique surface roughness has shown to be very effective in obtaining superhydrophobicity and, in particular, low water roll-off angles. There have been many biomimetic approaches to obtain superhydrophobic surfaces, which have been extensively reviewed [6–16]. However, most of the reported superhydrophobic surfaces were not oil-repellent. It is highly desirable for superhydrophobic surfaces to be also oil repellent to maintain their superhydrophobicity; otherwise, a superhydrophobic surface with poor oil repellency can be easily contaminated by oily substances in an industrial or a household environment. An ideal super-repellent surface would be both superhydrophobic and superoleophobic [17, 18].

Superoleophobic surfaces have recently been obtained on the basis of reentrant surface structure [19, 20], as exemplified by low contact angle hysteresis (CAH) for probe liquids of low surface tension (<30 mN/m) like octane and ethanol. The reentrant surface structure ensures the entrapment of air beneath, which appears to be a key in obtaining superoleophobicity [21–25]. However, the fabrication of these superoleophobic surfaces involves lithography and etching steps. Developing a robust superoleophobic polymeric surface via facile and low-cost methods will enable wide practical applications of super-repellent surfaces.

Functional Polymer Coatings: Principles, Methods, and Applications, First Edition.
Edited by Limin Wu and Jamil Baghdachi.

In this chapter, we will focus primarily on superhydrophobic and superoleophobic *polymeric* surfaces. Polymer-based super-repellent surfaces hold great promise to be prepared in large scales via easily implementable techniques and may subsequently find more practical applications in the near future. We will give an overview on various approaches to obtain super-repellent polymeric surfaces and their potential applications in self-cleaning, anti-bioadhesion, anti-icing, oil/water separation, and so on; we will finish up with an outlook on further development of super-repellent surfaces.

2.2 SURFACE WETTABILITY

The contact angle (CA) of a liquid on an ideal, atomically smooth solid surface is indicative of the solid's surface wettability, as given by the Young's equation [26]:

$$\cos\theta = \frac{\gamma_{SV} - \gamma_{SL}}{\gamma_{LV}} \tag{2.1}$$

where γ_{SV}, γ_{SL}, and γ_{LV} refer to the interfacial surface tensions, with S, L, and V corresponding to the solid, liquid, and gas phase, respectively.

Surface wettability can be evaluated with two types of CAs of a probe liquid: static and dynamic CA. Static CA represents an approximation to equilibrium CA, but it can be easily affected by a variety of factors such as surface roughness, chemical homogeneity, and surface reorganization upon interrogation of a probe liquid. In contrast, dynamic CAs, namely, advancing (θ_A) and receding (θ_R) CAs, are much more reproducible in terms of measurement and may reveal much more valuable information about a solid surface than a static CA. The difference between θ_A and θ_R is often defined as contact angle hysteresis (CAH). CAH is an important parameter in determining how easily (or difficult) a droplet may move on a surface and directly impacts roll-off angle or sliding angle of the droplet on the solid surface. Roll-off angle is defined as the angle at which the droplet starts to roll off of an inclined surface, while sliding angle corresponds to the critical angle at which the droplet with a certain weight begins to slide down the inclined surface. An obvious difference between roll-off angle and sliding angle is that the roll-off angle is closely related to a super-repellent surface, whereas the sliding angle applies to any surface [27–29].

CA values are conventionally used to describe the surface wettability of a solid. When the probe liquid is water, a solid surface can be categorized as hydrophilic when the water contact angle (WCA) is less than 90° or as hydrophobic when WCA is >90°. On a given smooth surface, the maximum WCA is about 120° [30, 31] (on a self-assembled perfluoroalkyl monolayer). In the meantime, for a more hydrophobic liquid with a typical surface tension <30 mN/m, the maximum CA on any given smooth surface would be less than 90°. Both values are significantly smaller than 150°, which is arguably the lowest CA, in conjunction with a low roll-off angle, for a surface to be deemed "super-repellent."

To achieve super-repellency, the surface roughness of a solid plays a critical role. Two distinct models, the Wenzel [32] and Cassie–Baxter [33, 34] models, account for the roughness effect on the CAs of liquids on a roughened surface. When the

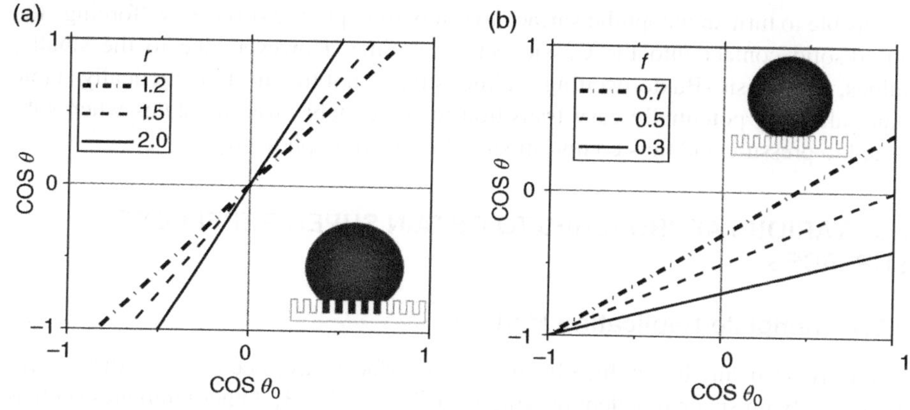

FIG. 2.1 Two different wetting regimes on a roughened surface: (a) Wenzel regime; (b) Cassie–Baxter regime.

surface roughness is low or moderate, there is always intimate contact between the liquid and the solid surface (Fig. 2.1a), and the wetting is in the Wenzel regime:

$$\cos\theta_{\mathrm{W}} = r\cos\theta_0 \tag{2.2}$$

where θ_0 and θ_{W} are the intrinsic CA (on the smooth surface) and the apparent CA on the roughened surface, respectively, and r is the roughness factor, defined as the ratio of the real surface area over the projected area.

It becomes obvious from Equation (2.2) that the roughness factor r, which is ≥ 1, would amplify the hydrophobicity/oleophobicity of an initially hydrophobic/oleophobic surface, as well as the hydrophilicity/oleophilicity of an initially hydrophilic/oleophilic surface, as shown in Fig. 2.1a. Due to the increased liquid/solid interfacial area, a roughened surface usually leads to high CAH, despite that θ_{A} may be greater than 150°. Oils normally have a relatively low surface tension and their intrinsic CAs on a smooth surface are normally below 90°, so it is impossible to achieve superoleophobicity when the wetting is in the Wenzel regime.

When the surface roughness exceeds a threshold, air becomes entrapped between the solid surface and liquid, and the wetting state transitions into the Cassie–Baxter regime [34, 35].

$$\cos\theta_{\mathrm{CB}} = f\cos\theta_0 + f - 1 \tag{2.3}$$

where θ_{CB} is the apparent CA in the Cassie–Baxter regime and f is the liquid/solid contact area fraction. Equation (2.3) and Fig. 2.1b depict a simplified scenario in which the liquid/solid contact area is flat. Obviously, the smaller the f value, the greater the θ_{CB} value and the more liquid repellent the surface becomes (Fig. 2.1b). Depending on the f value, the Cassie–Baxter wetting regime may lead to much reduced liquid/solid contact area, thus potentially allowing the liquid droplet to roll off easily. More importantly, if f is small enough, the θ_{CB} value may surpass 150° even with a $\theta_0 < 90°$; for instance, suppose $f = 0.1$ and $\theta_0 = 70°$, then $\theta_{\mathrm{CB}} = 150°$. This clearly implies that it is feasible to turn an oleophilic surface to a superoleophobic surface by "forcing" the liquid/solid contact into the Cassie–Baxter regime. However, due to the small f values, this Cassie–Baxter wetting regime is likely not the most energetically stable state, allowing potentially easy transition to more stable wetting states when subjected to pressure and thus compromising the superoleophobicity.

2.3 VARIOUS APPROACHES TO OBTAIN SUPER-REPELLENT SURFACES

2.3.1 Template-Replicating Methods

Naturally occurring hierarchically structured surfaces have been used as direct templates to obtain super-repellent polymeric surfaces [36–43]. Other templates such as inorganic/metallic membranes [44–51] and colloidal, lithographic, and woven materials [10, 52, 53] have been used for replication as well. Generally, the template-replicating process may be divided into three steps: (i) preparation of a master template with desired feature (it may be a naturally occurring object); (ii) replication of template by molding; and (iii) removal of the template. The method of replicating a template is quite straightforward to reproduce a structured surface with high fidelity.

Fürstner et al. used some plant leaves as templates to obtain superhydrophobic silicone (polydimethylsiloxane (PDMS)) surfaces without any surface chemical modification [36]. As shown in Fig. 2.2 (left), the microstructure of a replicate made from *Nelumbo nucifera* (lotus leaf) resembles that of a real leaf, and the replicate showed similar super-repellency (WCA of 158° and roll-off angle of 8°) to that of a lotus leaf as well.

As shown in Fig. 2.2 (right), Lee et al. [37] demonstrated a simple method for fabricating well-defined nanostructured polymeric surfaces of controllable aspect ratios with aligned nanoembosses or nanofibers, showing remarkable structural

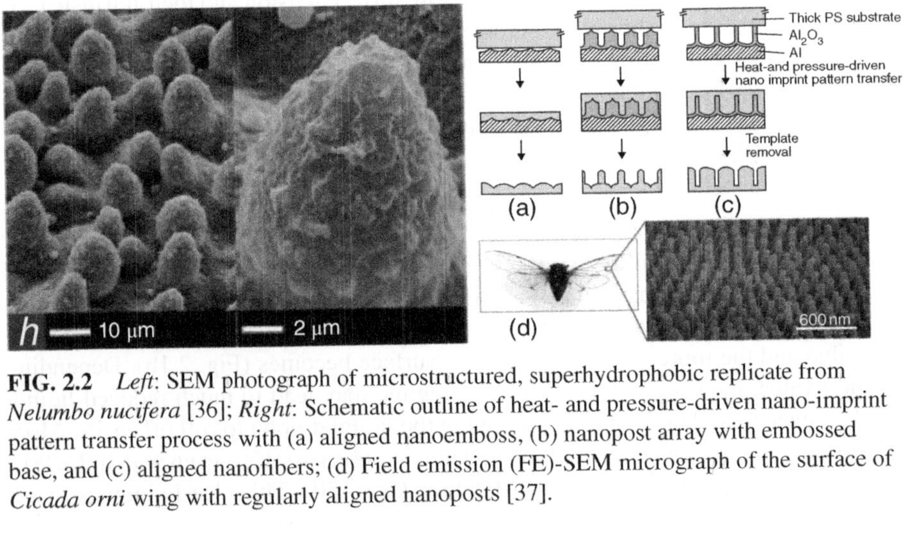

FIG. 2.2 *Left*: SEM photograph of microstructured, superhydrophobic replicate from *Nelumbo nucifera* [36]; *Right*: Schematic outline of heat- and pressure-driven nano-imprint pattern transfer process with (a) aligned nanoemboss, (b) nanopost array with embossed base, and (c) aligned nanofibers; (d) Field emission (FE)-SEM micrograph of the surface of *Cicada orni* wing with regularly aligned nanoposts [37].

similarity with naturally occurring biostructures such as the surface of the *Cicada orni* wing. The nanofabricating method can be readily extended to other thermoplastic polymeric materials.

Jiang et al. [44] chose an anodic aluminum oxide membrane as the template and prepared aligned polyacrylonitrile (PAN) nanofibers by extrusion of the PAN precursor solution into a solidifying solution under pressure. The surface of the as-made PAN nanofibers turned out to be superhydrophobic (CA = 173.8° ± 1.3°) without further modification.

Jung and Bhushan [51] reported a two-step molding process for the preparation of epoxy replicates, using shark skin and micropatterned Si as templates. Then *n*-perfluoroeicosane ($C_{20}F_{42}$) with low surface energy (6.7 mN/m) was deposited on the specimen surfaces by thermal evaporation to produce superhydrophobic and highly oleophobic surfaces.

2.3.2 Hierarchically Structured Particles

Inspired by the multilength-scale structured surface of the lotus leaf, there have been quite some efforts in mimicking this hierarchical structure to make super-repellent surfaces. Ming et al. [54] first prepared well-defined raspberry-like silica particles and then aligned them on an epoxy matrix, leading to a dual-scale structured surface (Fig. 2.3); further surface modification with a reactive PDMS rendered the structured surface superhydrophobic, as manifested by high-advancing WCA (165°) and low roll-off angle (3 ± 1°, 10 μl) [54]. The superhydrophobic surface demonstrated self-cleaning property, similar to the lotus leaf. On the other hand, if only single-sized particles (microsized or nanosized) were used, low-receding WCA was observed despite a relatively high advancing WCA, implying that the water droplet was in the Wenzel regime and highlighting the effectiveness of using dual-size particles to help transition from a Wenzel to a Cassie–Baxter wetting state.

Atomic force microscopy (AFM) revealed a dual-scale topographic structure of the raspberry-like superhydrophobic film (Fig. 2.4a): the micrometer-level structure

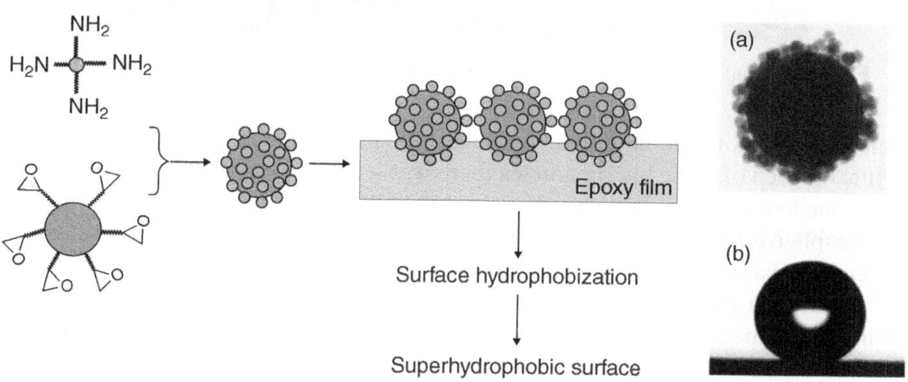

FIG. 2.3 Preparation of superhydrophobic films by using raspberry-like particles [54]. (a) TEM image of a raspberry-like particle; (b) a water droplet sitting on the superhydrophobic surface.

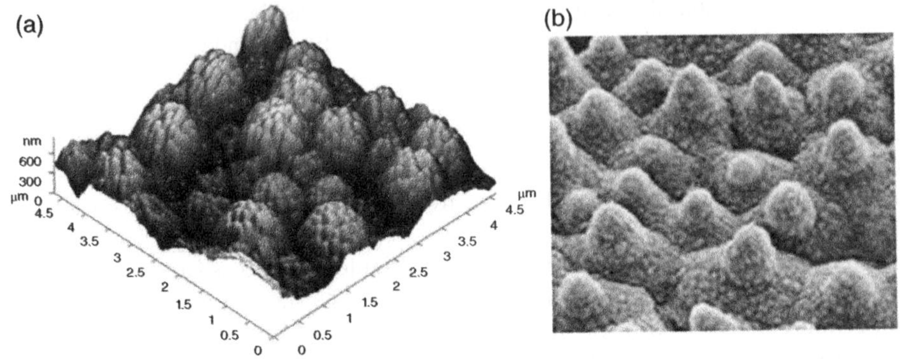

FIG. 2.4 (a) Surface topography (AFM) of a superhydrophobic film from raspberry-like particles [54], which to some extent resembles (b) the surface morphology (SEM) of the lotus leaf.

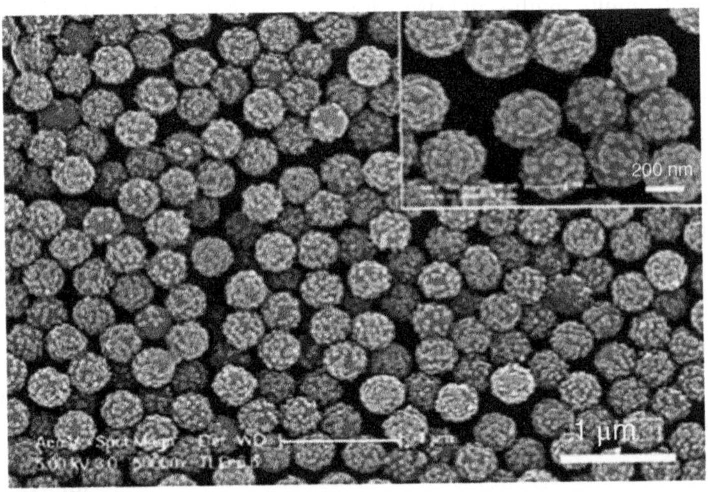

FIG. 2.5 Superhydrophobic surfaces based on PS-silica raspberry-like particles [55].

can be ascribed to the large core particles (700 nm) of the raspberry-like particles, while on each of these microparticles there was a finer structure at a submicrometer level due to the 70-nm shell nanoparticles [54]. Obviously, the topographic feature of the raspberry-like particles was preserved in the superhydrophobic film, which resembles that of a lotus leaf (Fig. 2.4b).

A different kind of raspberry-like particle was synthesized by Qian et al. [55], using preformed polystyrene (PS) particles as seeds, followed via self-condensation of tetraethyl orthosilicate to form a layer of smaller silica particles on the perimeter of the PS particles. These particles were assembled on a flat substrate to form a dual-scale structured surface (Fig. 2.5), which was then rendered superhydrophobic via surface hydrophobization [55].

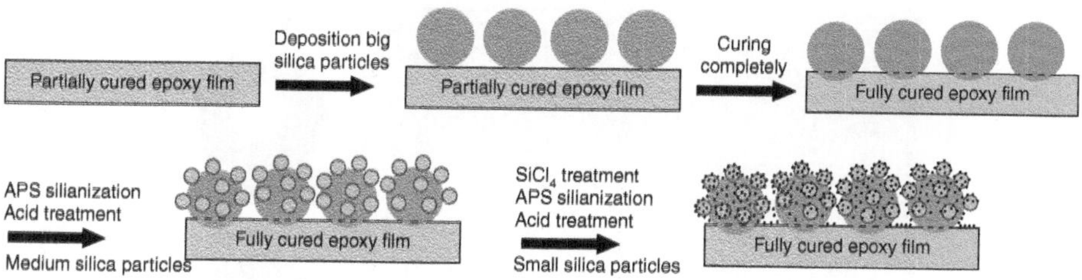

FIG. 2.6 Procedure to prepare triple-scale structured surfaces [56].

FIG. 2.7 SEM images of triple-scale structured surface [56]. Insets: photos of static water (10 μl) and $C_{16}H_{34}$ (5 μl) droplets on the sample.

For the structured surfaces from raspberry-like particles, an obvious disadvantage was the poor mechanical robustness since there was only point contact between the particles and the underlying substrate. To enhance mechanical robustness of the structured surface, a layer-by-layer (LbL) particle deposition approach [56] was developed to obtain multilength-scale structured films (Fig. 2.6) with particles partially embedded into the films. The mechanical robustness of the triple-scale structured coating was enhanced by $SiCl_4$-based cross-linking between silica particles. After chemical modification with a perfluoroalkyl silane, the triple-scale structured surface (Fig. 2.7) was turned superhydrophobic, on which the WCA and roll-off angle were, respectively, $167 \pm 3°$ and approximately 1° for a 10-μl water droplet [56]. The triple-scale structured surface was also highly oleophobic, as indicated by a hexadecane CA of 135°.

The presence of a multilength-scale surface roughness has proven to be important to the surface super-repellency, especially in terms of low water roll-off angles. The triple-scale structure appeared to allow tiny water droplets as small as 5 μl to reside in the Cassie–Baxter wetting regime, which was much more stable than on dual- or single-scale structured surfaces, as demonstrated by a water vapor condensation and a compression test [56].

Beside flat substrates, the raspberry-like topography was also superimposed onto woven structures (Fig. 2.8), such as cotton textiles, to render the normally hydrophilic textile super-repellent [57]. The modified textiles were completely nonwettable by

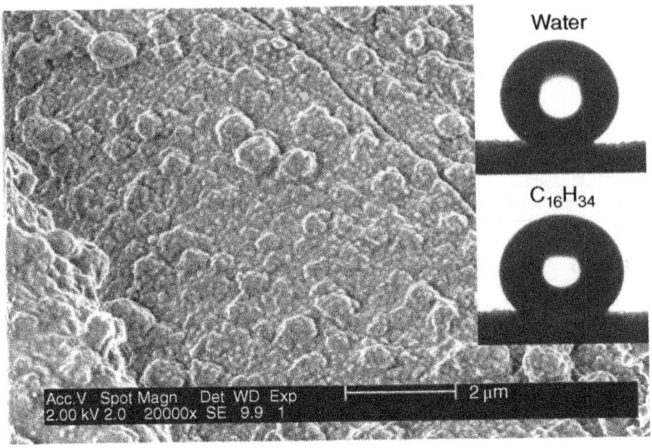

FIG. 2.8 Morphology of cotton fiber with raspberry-like particles by SEM [57]. (Insets: profiles of water and hexadecane on the super-repellent textile.)

both water and hexadecane, which both showed high CAs (>150°) and low roll-off angles (<10°). It appeared to be essential to add the nanoparticle layer in achieving superoleophobicity, especially in terms of low roll-off angles for hexadecane.

2.3.3 LbL Deposition

LbL deposition of polyelectrolyte multilayers (PEMs) has been often used to fabricate thin films with molecular-level control over film thickness and chemical composition. A unique feature for LbL deposition is that PEMs can be constructed onto both exterior and interior surfaces of a complex object. To render PEMs super-repellent, PEMs need to be roughened via, for instance, the introduction of porosity, deposition onto nanofibrous substrate, or involving micro-/nanoparticles, followed by surface hydrophobization [58–65]. The surface hydrophobization is often necessary because polyelectrolytes are hydrophilic.

Shiratori et al. [58] reported one of the first examples of PEM-based superhydrophobic coatings. To create the surface texture needed for superhydrophobic behavior, the PEM film was heated to 650°C to remove the polyelectrolytes. Using a combination of PEMs, silica nanoparticles, and a simple hydrophobic surface treatment, Zhai et al. [61] reported a straightforward procedure for creating honeycomb-like, stable superhydrophobic coatings (Fig. 2.9). The surface maintained its superhydrophobic character even after extended immersion in water.

Rutledge et al. [62] prepared hierarchically roughened superhydrophobic fabrics by coating electrospun fibers with silica nanoparticles using the LBL self-assembly technique and subsequently depositing a semifluorinated silane onto the surface to decrease the surface energy. This multistep process improved the unstable WCA of the electrospun nylon fiber from 135° to a stable WCA of 168° with almost zero hysteresis.

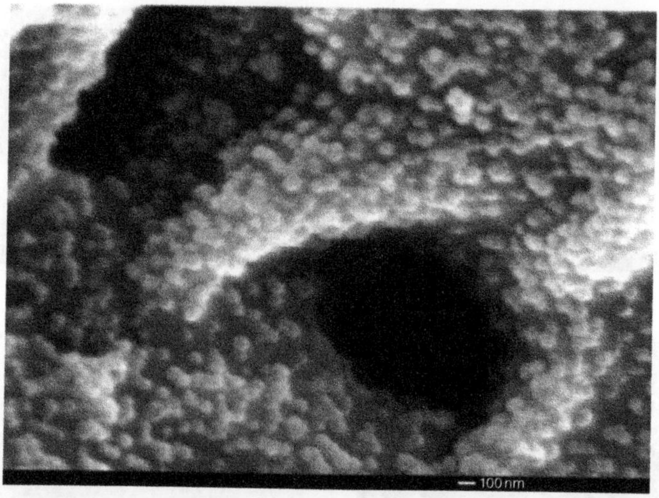

FIG. 2.9 SEM image of a superhydrophobic coating made from LbL PEM assembly, followed by deposition of silica nanoparticles [61].

Stability of LbL-deposited PEM coatings depends upon the nature of the multivalent interactions within the films, and it can be enhanced by postdeposition, chemical cross-linking between polyelectrolyte components. Lynn et al. [64, 65] reported an approach on the design of superhydrophobic polymer multilayers based on the covalent LbL assembly of poly(ethylene imine) and azlactone-containing polymer, poly(2-vinyl-4,4-dimethylazlactone). This approach led to covalent multilayers that can be further functionalized to impart new surface properties by reactions with amine-functionalized molecules.

2.3.4 Plasma Treatment

Plasma treatment has been widely applied in making super-repellent surfaces, including plasma etching and plasma polymerization. Plasma etching has been used to not only generate but also, to some extent, control surface roughness. The generated surface roughness may be used as a basis to obtain surface super-repellency.

Youngblood and McCarthy [66] used inductively coupled radio frequency argon plasma to prepare super-repellent polypropylene (PP) surface via simultaneous etching of PP and etching/sputtering of poly(tetrafluoroethylene) (PTFE). The semicrystalline PP surface was roughened due to the different ablation rates for the crystalline and amorphous regimes and was fluorinated by the fluorocarbon plasma from the ablation/depolymerization of PTFE. Changing the etching time could adjust the surface roughness (Fig. 2.10). The most hydrophobic surfaces exhibited advancing and receding WCAs of 172°/169°.

Minko et al. [67] reported a route to fabricate two-level structured self-adaptive surfaces (Fig. 2.11) of polymer materials via plasma treatment of PTFE. The first level of the structure was built by a rough polymer film that consisted of

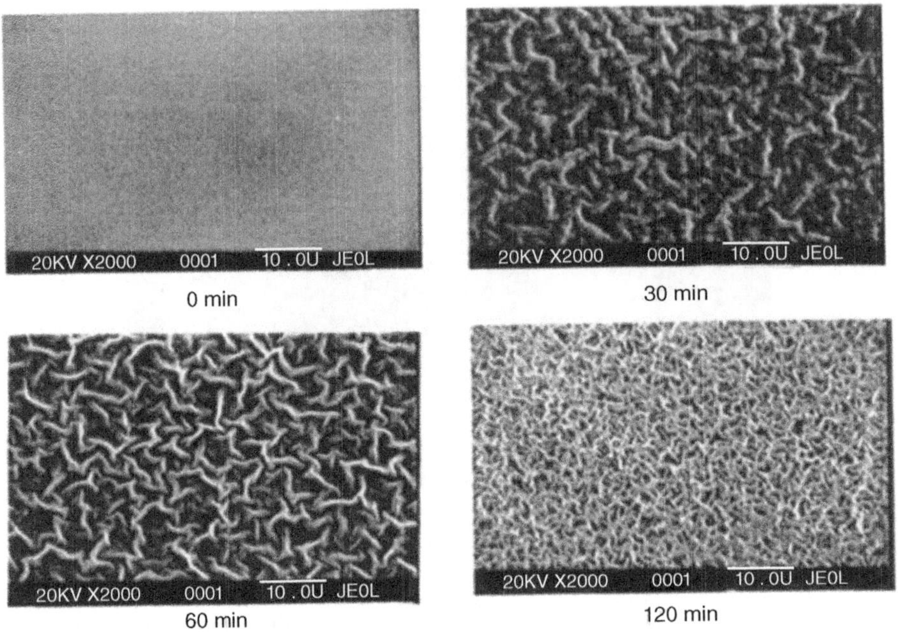

FIG. 2.10 Scanning electron micrographs of polypropylene as a function of argon/PTFE plasma reaction time [66].

micron-sized needle-like structures. The second level of the structure was formed by the nanoscopic self-assembled domains of a demixed polymer brush irreversibly grafted onto the needles. By exposing the surface to solvents that were selective to one of the components of the brush, the surface properties were reversibly tuned. The large-scale surface structure amplified the response and enabled to control wettability, adhesion, and chemical composition of the surface over a wide range.

Ellinas et al. [68] spin-coated PS microparticles on poly(methyl methacrylate) (PMMA) substrates, followed by anisotropic and isotropic oxygen plasma etching, to generate a hierarchical structure comprising PMMA pillar with roughened PS microsphere on the top (Fig. 2.12). The dimension of the hierarchical micropillars was controlled with the plasma etching process, leading to various roughened, super-repellent surfaces.

Zhang and Seeger [69] successfully fabricated superoleophobic surfaces with high CAs for various nonpolar liquids using the combination of organosilanes in a simple grow-from method. Silicone nanofilaments with different microstructures were first grown in toluene onto glass slides by simply regulating the water concentration during hydrolysis and condensation of trichloromethylsilane (TCMS). Subsequently, the nanofilaments were treated with oxygen plasma and modified with 1*H*,1*H*,2*H*,2*H*-perfluorodecyltrichlorosilane, leading to superoleophobic surfaces (Fig. 2.13) with high CAs and low roll-off angles for various nonpolar liquids.

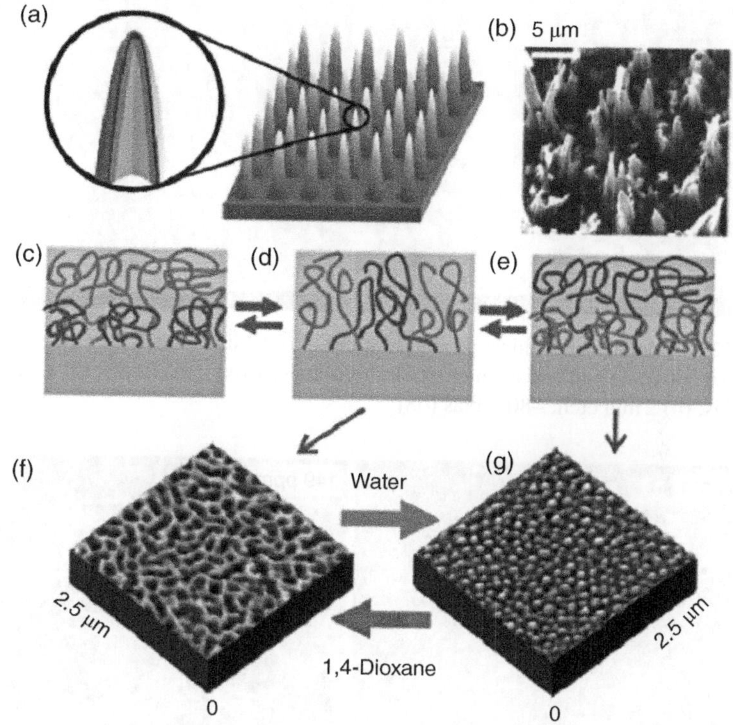

FIG. 2.11 Two-level structure of self-adaptive surface (SAS): schematic representation of needle-like surface morphology of PTFE surface (a) and SEM image of the PTFE film after 600 s of plasma etching (b). (c–e) Schematic illustration of each needle being covered by a covalently grafted mixed brush consisting of hydrophobic and hydrophilic polymers. (f, g) AFM images of different morphologies after exposure to different solvents [67].

2.3.5 Chemical Vapor Deposition

Chemical vapor deposition (CVD) is a chemical process involving gas-phase reaction of molecules inside a reactor and subsequent deposition onto a substrate to form a nonvolatile solid film. The conformal nature of the CVD process enables coating on complex substrates, which allows formation of films of nanoscale thicknesses on the substrates. The surface roughness depends mostly on the topography of the substrate used as a template. Through selection of gaseous reactants and adjustment of reaction conditions, tunable layer thickness can be realized on substrates with complex morphologies.

By combining electrospinning and initiated chemical vapor deposition, a hierarchically structured, superhydrophobic surface was reported by Seeger et al. [70]; poly(caprolactone) (PCL) was first electrospun and then coated with a thin layer of hydrophobic poly(perfluoroalkylethyl methacrylate) by CVD.

Plasma-enhanced chemical vapor deposition technique was also used to deposit vertically aligned carbon nanotube (CNT) forest in which the diameter and areal

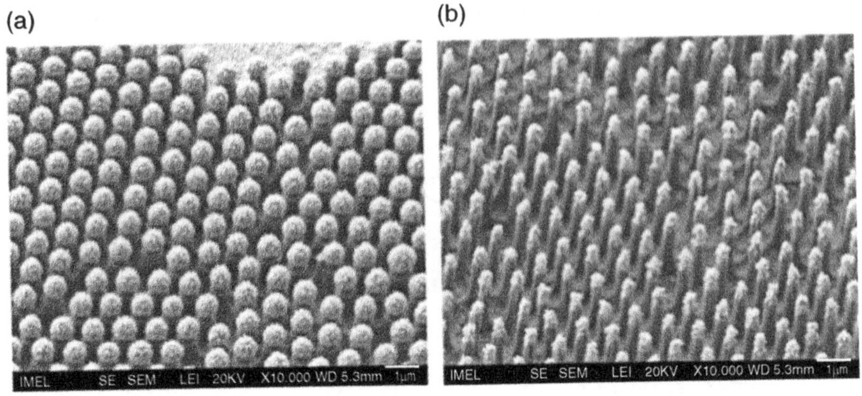

FIG. 2.12 Ordered micropillars produced after colloidal lithography of 1-µm PS particles on PMMA substrate and one etching step in high-density oxygen plasma, (a) 1 min etch, −80 V bias; (b) 2 min etch, −80 V bias [68].

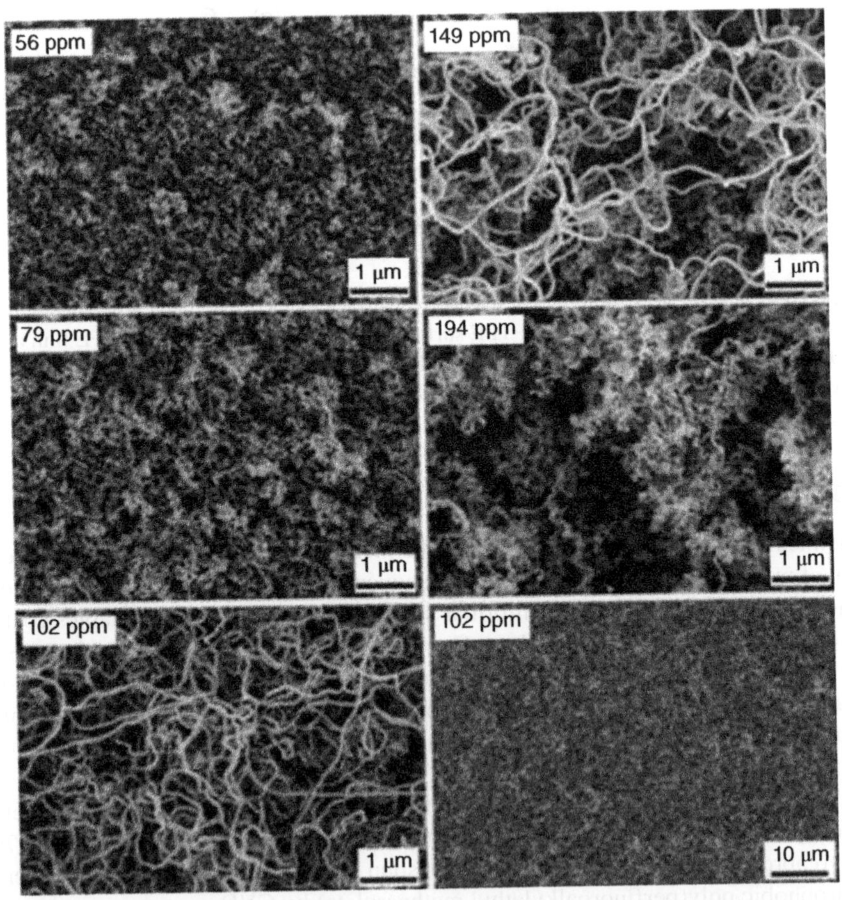

FIG. 2.13 Surface morphology of the TCMS-coated glass slides at different water concentrations in toluene during the TCMS-coating procedure [69]. Scale bars: 1 µm, except for the image at the bottom right corner (10 µm).

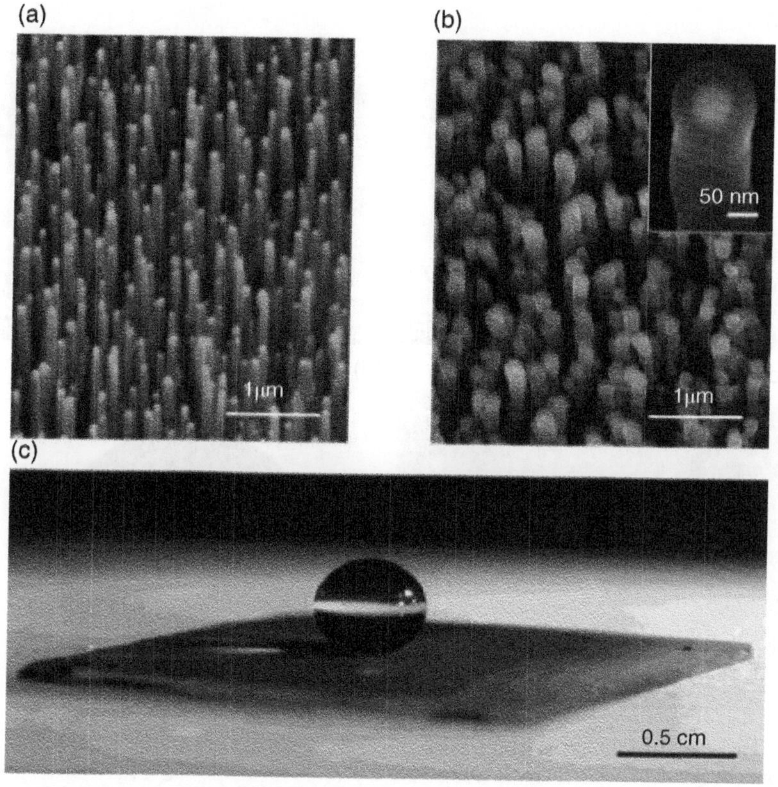

FIG. 2.14 SEM images of CNT forest: (a) as grown forest, (b) PTFE-coated forest, and (c) an essentially spherical water droplet suspended on the PTFE-coated CNT forest [71].

density of CNT can be conveniently controlled [71]. After that, PTFE coating was applied onto the CNT forest through a hot-filament chemical vapor deposition process to obtain stable, superhydrophobic surface (Fig. 2.14).

Jin et al. [72] described the preparation of super-repellent nanocellulose aerogels, consisting of fibrillar networks and aggregates with structures at different length scales, by fluorinating aerogel membranes via CVD method. The prepared aerogel can support considerable load on a water or an oil surface. In addition, aerogel-assisted chemical vapor deposition was shown to be a facile single-step strategy toward organic–inorganic nanocomposite coatings [73]; these coatings exhibited superhydrophobic behavior due to the combination of a low-surface-energy organic component and an adequate surface roughness.

2.3.6 Electrospinning

Electrospinning is a versatile technique for the fabrication of polymer fibers. Electrospinning offers many controllable parameters that can be used to tailor surface morphology and produce micro- and nanofibers from a wide range of synthetic

FIG. 2.15 (a) SEM image of film prepared from a 7 wt% PS/ dimethylformamide (DMF) solution; (b) 3D network structure; (c) surface nanostructure of a single porous microsphere; (d) water droplet on the film [76].

and natural polymer materials. Due to their very large surface area and high curvature, these fibers are good candidates to make super-repellent surfaces [74, 75].

A superhydrophobic PS film with a composite structure consisting of porous microspheres and nanofibers has been prepared by electrospinning [76]. By adjusting spinning parameters, the bead-fiber morphology could be readily controlled (Fig. 2.15). The final structured surfaces made of microspheres (3–7 μm) and nanofibers (60–140 nm) were superhydrophobic (WCA = $160° \pm 1°$) [76].

By one-step deposition of PS nanofibers onto a stainless steel mesh via electrospinning, a superhydrophobic–superoleophilic membrane was successfully prepared [77]. This membrane can selectively adsorb oil but is water-repellent, indicating potential application in oil–water separation area. Very recently, a superhydrophobic surface with strong adhesion to water droplets was fabricated by electrospinning fluorinated polyimide into dented nano- and/or micro-bowl-like particles linked with nanofibers [78].

As of superoleophobic surfaces, it is still challenging to prepare by electrospinning. It is necessary to use a fluoropolymer to attain a low-surface-energy top layer for superoleophobic surfaces, but normally fluoropolymers are not electrospinnable because their low dielectric constant prevents sufficient charging. Using a coaxial

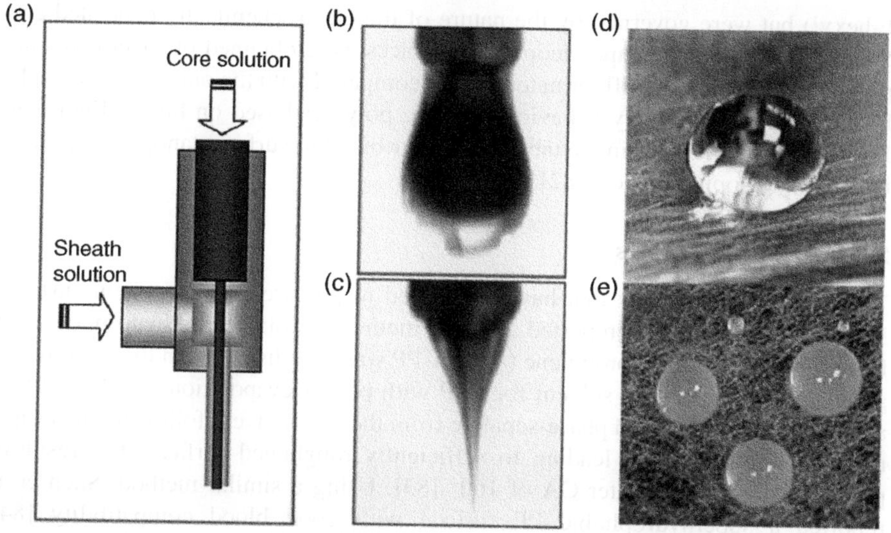

FIG. 2.16 Coaxial electrospinning operation: (a) diagram of the coaxial nozzle; (b) core-sheath droplet without bias; and (c) Taylor cone and coaxial jet formation at 12.5 kV; photographs of hydrophobic (d) and oleophobic (e) liquids of Teflon-coated gelatin membranes [79].

electrospinning method, superoleophobic fibers were produced with Teflon® AF (DuPont) sheath and PCL core materials [79]. The coaxially electrospun fiber membrane showed oleophobicity whereas the PCL-only surface was oleophilic (Fig. 2.16).

In addition, surfaces with reentrant texture were designed and developed by electrospinning blends of polyhedral oligomeric silsesquioxane (POSS) with PMMA. The surface morphology can be tuned systematically from a corpuscular or spherical microstructure to a beads-on-string structure and finally to bundled fibers by controlling the solution concentration and molecular weight of the sprayed polymer [80].

2.3.7 Electrochemical Polymerization

Electrochemical polymerization has been shown to be an effective method to prepare super-repellent surfaces. The possibility of incorporating hydrophobic groups consisting of fluorinated or hydrocarbon substituents into the monomer structure allowed a one-pot method to obtain superoleophobic films. Yan et al. developed superhydrophobic polypyrrole films on a conducting surface by electrochemical polymerization [81]. By simply adjusting electrical potential, switchable wettability between superhydrophobicity and superhydrophilicity was realized on the film surface, making it possible to construct multifunctional conducting polymer devices combining special mechanical, electrical, optical, and interfacial properties [81]. Guittard et al. reported electrodeposited, super-repellent polymers by using fluorinated 3,4-ethylene dioxypyrrole (EDOP) and 3,4-propylene dioxypyrrole (ProDOP) monomers [82]. Their nonwetting properties did not depend on the fluorocarbon chain length (F-octyl to

F-hexyl) but were governed by the nature of the electrochemically deposited core. Superhydrophobic and superoleophobic surfaces were obtained by electrochemical polymerization of highly fluorinated EDOP compared with the only superhydrophobic surfaces with a sticky behavior from the polymer based on highly fluorinated ProDOP. The difference in wettability was attributed to surface nanoporosity resulting from the doping process [82].

2.3.8 Other Methods

There are other methods that have been used to prepare super-repellent surfaces. Notably, Erbil et al. [83] reported a simple method to obtain superhydrophobic surface using isotactic polypropylene (*i*-PP): *i*-PP was first dissolved in the mixture of a good solvent and a poor solvent for *i*-PP with proper evaporation profile and then allowed to crystallize and phase-separate from the poor solvent following the evaporation of both solvents, leading to sufficiently roughened surface. The resulting porous surface had a water CA of 160° [83]. Using a similar method, Shen et al. reported a superhydrophobic PP surface with good blood compatibility [84]. However, this physical method would not lead to superoleophobic surfaces.

2.4 APPLICATIONS OF SUPER-REPELLENT POLYMERIC SURFACES

Due to the exceptional nonwetting characteristics of super-repellent surfaces, a myriad of applications can be envisaged. Obviously, super-repellent surfaces can be used as self-cleaning coatings; other potential applications for super-repellent polymeric surfaces including anti-bioadhesion, anti-icing, and water–oil separation will be briefly reviewed here.

2.4.1 Self-Cleaning

Self-cleaning phenomenon is ubiquitously present in nature and can be generalized as the so-called "lotus effect" [1, 2]. On a superhydrophobic surface, dirt particles can be easily picked up and taken away by rolling water droplets; minimized interfacial contact between the roughened surface and dirt particles also facilitates the self-cleaning effect. However, superhydrophobicity alone would not sustain the self-cleaning effect since oily substance can easily wet a superhydrophobic but oleophilic surface. To make a true self-cleaning surface, a superhydrophobic surface must also be superoleophobic (lotus leaf can maintain its self-cleaning character only when dealing with water, which is normally the case for a lotus growing in a pond). Many recently developed super-repellent surfaces (both superhydrophobic and superoleophobic) may potentially be used as long-term-effective, self-cleaning coatings; however, there are major challenges that need to be addressed prior to practical applications, including long-term durability and mechanical robustness.

An interesting recent development is a super-repellent surface inspired by the slippery pitcher plant surface [85], without relying on a hierarchically structured surface. Aizenberg et al. created a super-liquid-repellent surface by infiltrating a micro-/nanoporous substrate with a lubricating liquid to produce a thin, ultrasmooth lubricating layer [85]. Due to the lubricating layer, this surface outperforms its natural counterparts and many other state-of-the-art synthetic liquid-repellent surfaces in its capability to repel various liquids from water, hydrocarbons, crude oil, to blood. Another elegant feature of this surface is that it could quickly restore its liquid-repellency even after physical damage (within 0.1–1 s) [85].

2.4.2 Anti-bioadhesion

Undesired protein, platelet, and bacterial adsorptions are challenging in various biomedical applications and have caused a lot of issues. When a water droplet containing biospecies sits on a super-repellent surface, air becomes entrapped between water and the structured surface in the Cassie–Baxter wetting regime, and the liquid/solid interfacial area will be significantly reduced, thus potentially minimizing bioadhesion. Recently, the feasibility of using super-repellent surfaces to reduce bioadhesion has been investigated [86–91].

Schoenfisch et al. [86] demonstrated that the combination of micro- and nanostructured roughness from silica colloids, and fluorinated silane xerogel produced a highly bacteria-repellent surface that reduced the adhesion of highly pathogenic *Staphylococcus aureus* and *Pseudomonas aeruginosa* by approximately two orders of magnitude versus controls, making these surfaces good candidates as medical device coatings. Recently, a potential-induced reversibly wettable polythiophene film was developed by Advincula et al. [87], on which hydrophilicity and superhydrophobicity can be easily varied by simply changing its redox property via a potential switching. As a result, the wettability change in return can control the protein adsorption and bacterial adhesion. The superhydrophobic surface inhibited the adhesion of *fibrinogen* and *Escherichia coli*, whereas the hydrophilic surface led to increased attachment of both protein and bacteria (Fig. 2.17) [87].

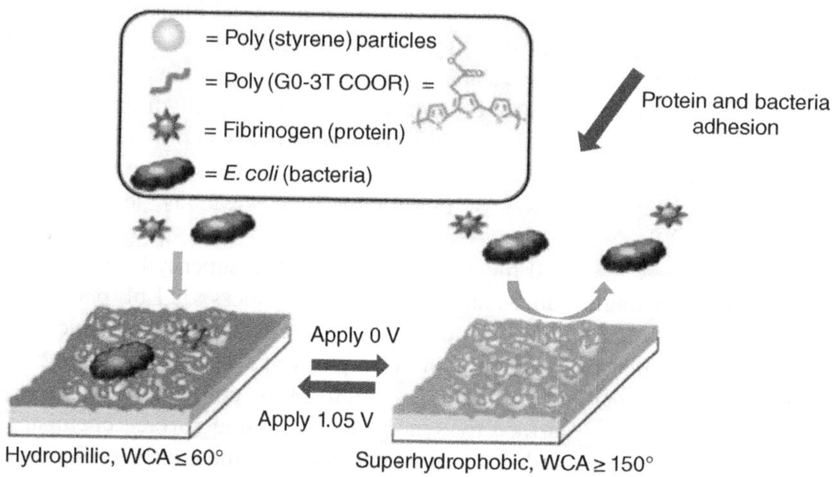

FIG. 2.17 Protein (fibrinogen) and bacterial (*Escherichia coli*) adhesion onto undoped (*right*) and doped (*left*) colloidally templated polythiophene (poly(ethyl 2-(2,5-di(thiophen-2-yl) thiophen-3-yl)acetate)/PS) surfaces [87].

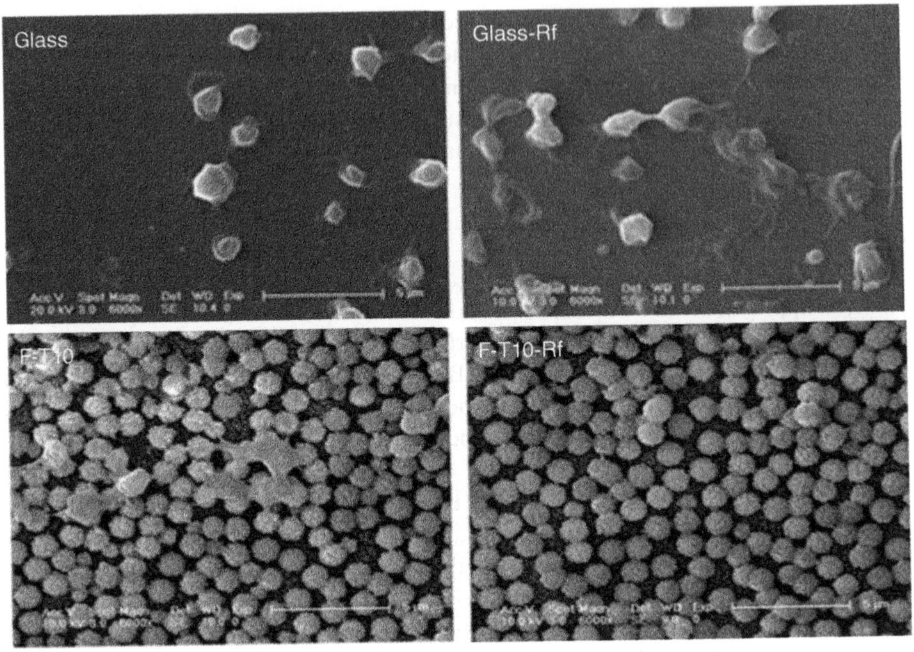

FIG. 2.18 SEM images of platelets adhered to control glass and triple-scale structured surfaces before (*left*) and after (*right*) perfluorination [91].

The prevention of *E. coli* growth on the shrinkage-induced superhydrophobic PS, polycarbonate, and polyethylene surfaces was also investigated [88]. Only 2% of the initial bacterial cells adhered onto the superhydrophobic surface and the following rinsing can remove nearly all the cells (<0.1% of the initial bacterial cells) on the surface. Sun et al. prepared nanostructured superhydrophobic surfaces with excellent platelet adhesion resistance and relatively low platelet activation by using aligned CNT as structured substratum [89]. Protein adsorption on superhydrophobic surfaces is inevitable due to the fact that there is always contact between protein and the structured surface; however, binding strength of protein toward the surface may be greatly reduced, allowing easy removal of protein by flow shear, as demonstrated by Shirtcliffe et al. [90].

More recently, we designed hierarchically structured, superhydrophobic epoxy-based surfaces with single-, dual-, and triple-scale roughness via LbL particle deposition (Fig. 2.18) [56]. The dual-/triple-scale structured, superhydrophobic surfaces exhibited significantly reduced protein adsorption (up to 90% decrease) [91]. Furthermore, the platelet adhesion and activation can be completely suppressed on the triple-scale structured surface by synergistically tuning surface chemistry and topography (Fig. 2.18). The reduction of bioadhesion was the most significant on the triple-scale structured surface, obviously due to the smallest possible interfacial area allowed for platelets or protein to physically adsorb.

2.4.3 Anti-Icing

Ice formation on material surfaces has significantly impacted on safety and energy consumption in many infrastructures, transportation, and heating systems such as aircrafts, ships, offshore oil platforms, telecommunications, and solar panels. One common way to minimize ice formation is to use anti-icing chemicals, which are only temporary and often toxic. An alternative way to solve this problem is the application of superhydrophobic surface. Superhydrophobic surfaces can minimize or even eliminate ice formation by repelling impinging water drops before they can freeze [92].

Gao et al. [93] reported that superhydrophobic surfaces consisting of nanoparticle-polymer composites exhibited anti-icing capability to prevent ice formation upon impact of supercooled water. They found that the anti-icing capability of these composites depended not only on their superhydrophobicity but also on the size of the particles exposed on the surface. The critical particle sizes that determined the superhydrophobicity and the anti-icing property were in two different length scales.

Through analysis of the temperature-dependent droplet/surface interaction, Aizenberg et al. [94] revealed that highly ordered superhydrophobic materials may remain entirely ice-free down to −25 to −30°C, due to their ability to repel impacting water before ice nucleation occurred. Furthermore, ice accumulated below these temperatures can be easily removed. However, not all superhydrophobic surfaces were suitable for anti-icing applications under different conditions, and some surfaces failed in high-humidity conditions due to water condensation and frost formation and even led to increased ice adhesion.

2.4.4 Oil–Water Separation

Oil–water separation has recently become a global challenge because of frequent occurrence of oil spill accidents, and there is an increasing demand for the development of effective and inexpensive approaches for cleanup of oil pollution in water system. Moreover, oil–water separation also has other applications, including wastewater treatment, fuel purification, and separation of commercially relevant emulsions. Although traditional techniques such as oil skimmers, centrifuges, settling tanks, depth filters, and flotation technologies are useful for separation of immiscible oil/water mixtures, they are not always effective for emulsified oil/water mixtures, especially not for surfactant-stabilized emulsions. Filtration membranes have been successfully applied for the separation of various emulsions; however, the low flux and quick decline of the permeation rate due to surfactant adsorption and/or pore plugging by oil droplets led to severe fouling problem [95]. Recently, a membrane-based single-unit operation was developed to achieve the gravity-driven separation of various oil–water mixtures under an electric field. Tuteja et al. obtained a continuous oil–water emulsion separation apparatus that was triggered on demand and could remove more than 99% of the emulsified droplets [96]. Furthermore, they also developed a novel membrane with responsive surfaces [97]; a mesh membrane coated with a blend of perfluorodecyl POSS and cross-linked poly (ethylene glycol) diacrylate exhibited both superhydrophilic and superoleophobic properties in air and under water.

Uniquely, this reconfiguration was reversible. Compared with the traditional membrane-based separation technologies, this membrane exhibited higher separation efficiency and can separate, for the first time, a range of different oil–water mixtures in a single-unit operation, with >99.9% separation efficiency, by using the difference in capillary forces acting on the two phases [97].

2.5 SUMMARY AND OUTLOOK

Significant progress has been made over the past two decades regarding the preparation of super-repellent (i.e., both superhydrophobic and superoleophobic) surfaces. In this chapter, we briefly reviewed major approaches in designing and making super-repellent polymeric surfaces, together with their potential applications. Polymer-based super-repellent surfaces may hold great promise for future practical applications due to their intrinsic advantages including (large-scale) processability, flexibility, variation in material properties from molecular design, and low cost. However, there are significant hurdles that need to be overcome. Long-term durability is the first Achilles' heel: the hierarchical structure that proves to be effective in generating super-liquid-repellency normally has poor mechanical stability and may be easily compromised over time, leading to reduced or even total loss of liquid-repellency. It is therefore of critical importance to construct mechanically robust, super-repellent surfaces. Second, to obtain superoleophobic surfaces it is necessary to modify the outer surface with perfluoroalkyl moiety, which usually is significantly enriched at the surface. However, the thin fluorine-rich layer is unlikely to sustain against abrasion and scratch, and the super-repellency would gradually diminish due to the slow erosion of low-surface-energy species. A promising strategy is to make low-surface-energy species self-replenishing, toward self-healing super-repellent surfaces [98–101]. However, it is still very challenging to autonomously restore the damaged hierarchical structure when that does occur. For a superoleophobic surface, another major issue is its metastable Cassie–Baxter wetting regime, which under various circumstances may easily transition into the Wenzel wetting regime, which is energetically more stable, and lose oil-repellency. It is desirable to design and construct a hierarchical structure that offers energy barriers so high that the Cassie–Baxter to Wenzel regime transition becomes difficult, thus possibly preserving superoleophobicity.

ACKNOWLEDGMENTS

Financial support from USDA/NIFA is gratefully acknowledged.

REFERENCES

1. W. Barthlott, C. Neinhuis, Purity of the sacred lotus, or escape from contamination in biological surfaces. *Planta* 1997, 202, 1.

2. C. Neinhuis, W. Barthlott, Characterization and distribution of water-repellent, self-cleaning plant surfaces. *Ann. Bot.* 1997, 79, 667.
3. S. Herminghaus, Roughness-induced non-wetting. *Europhys. Lett.* 2000, 52, 165.
4. X. Gao, L. Jiang, Biophysics: water-repellent legs of water striders. *Nature* 2004, 36, 432.
5. K. Autumn, Y.A. Liang, S.T. Hsieh, W. Zesch, W.P. Chan, T.W. Kenny, R. Fearing, R.J. Full, Adhesive force of a single gecko foot-hair. *Nature* 2000, 405, 681.
6. T.L. Sun, L. Feng, X.F. Gao, L. Jiang, Bioinspired surfaces with special wettability. *Acc. Chem. Res.* 2005, 38, 644.
7. M. Callies, D. Quere, On water repellence. *Soft Matter* 2005, 1, 55.
8. X.J. Feng, L. Jiang, Design and creation of superwetting/antiwetting surfaces. *Adv. Mater.* 2006, 18, 3063.
9. J. Genzer, K. Efimenko, Recent developments in superhydrophobic surfaces and their relevance to marine fouling: a review. *Biofouling* 2006, 22, 339.
10. P. Roach, N.J. Shirtcliffe, M.I. Newton, Progress in superhydrophobic surface development. *Soft Matter* 2008, 4, 224.
11. K. Koch, H.F. Bohn, W. Barthlott, Hierarchical sculpturing of plant surfaces and superhydrophobicity. *Langmuir* 2009, 25, 14116.
12. M. Liu, Y. Zheng, J. Zhai, L. Jiang, Bioinspired super-antiwetting interfaces with special liquid-solid adhesion. *Acc. Chem. Res.* 2010, 43, 368.
13. K. Liu, X. Yao, L. Jiang, Recent developments in bio-inspired special wettability. *Chem. Soc. Rev.* 2010, 39, 3240.
14. K. Liu, Y. Tian, L. Jiang, Bio-inspired superoleophobic and smart materials: design, fabrication, and application. *Prog. Mater. Sci.* 2013, 58, 503.
15. M. Wolfs, T. Darmanin, F. Guittard, Superhydrophobic fibrous polymers. *Polym. Rev.* 2013, 53, 460.
16. E. Celia, T. Darmanin, E. Taffin de Givenchy, S. Amigoni, F. Guittard, Recent advances in designing superhydrophobic surfaces. *J. Colloid Interface Sci.* 2013, 402, 1.
17. Z. Xue, M. Liu, L. Jiang, Recent developments in polymeric superoleophobic surfaces. *J. Polym. Sci. B: Polym. Phys.* 2012, 50, 1209.
18. X.M. Li, D. Reinhoudt, M. Crego-Calama, What do we need for a superhydrophobic surface. *Chem. Soc. Rev.* 2007, 36, 1350.
19. A. Tuteja, W. Choi, M.L. Ma, J.M. Mabry, S.A. Mazzella, G.C. Rutledge, G.H. McKinley, R.E. Cohen, Designing superoleophobic surfaces. *Science* 2007, 318, 1618.
20. A. Ahuja, J.A. Taylor, V. Lifton, A.A. Sidorenko, T.R. Salamon, E.J. Lobaton, P. Kolodner, T.N. Krupenkin, Approach to superlyophobic surfaces. *Langmuir* 2008, 24, 9.
21. W. Choi, A. Tuteja, J.M. Mabry, R.E. Cohen, G.H. McKinley, A modified Cassie–Baxter relationship to explain contact angle hysteresis and anisotropy on non-wetting textured surfaces. *J. Colloid Interface Sci.* 2009, 339, 208.
22. A. Tuteja, W. Choi, G.H. McKinley, R.E. Cohen, M.F. Rubner, Design parameters for superhydrophobicity and superoleophobicity. *MRS Bull.* 2008, 33, 757.
23. S.S. Chhatre, A. Tuteja, W. Choi, A. Revaux, D. Smith, J. M. Mabry, G.H. McKinley, R.E. Cohen, Thermal annealing treatment to achieve switchable and reversible oleophobicity on fabrics. *Langmuir* 2009, 25, 13625.
24. B.W. Choi, A. Tuteja, S. Chhatre, J.M. Mabry, R. Cohen, G.H. McKinley, Fabrics with tunable oleophobicity. *Adv. Mater.* 2009, 21, 2190.

25. A. Steele, I. Bayer, E. Loth, Inherently superoleophobic nanocomposite coatings by spray atomization. *Nano Lett.* 2009, 1, 501.
26. T. Young, An essay on the cohesion of fluids. *Philos. Trans, R. Soc. Lond.* 1805, 95, 65.
27. M. Miwa, A. Nakajima, A. Fujishima, K. Hashimoto, T. Watanabe, Effects of the surface roughness on sliding angles of water droplets on superhydrophobic surfaces. *Langmuir* 2000, 16, 5754.
28. A. Marmur, The lotus effect: superhydrophobicity and metastability. *Langmuir* 2004, 20, 3517.
29. B. Balu, V. Breedveld, D.W. Hess, Fabrication of "rolloff" and "sticky" superhydrophobic cellulose surfaces via plasma processing. *Langmuir* 2008, 24, 4785.
30. S.R. Coulson, I. Woodward, J.P.S. Badyal, S.A. Brewer, C. Willis, Super-repellent composite fluoropolymer surfaces. *J. Phys. Chem. B* 2000, 104, 8836.
31. T. Nishino, M. Meguro, K. Nakamae, M. Matsushita, Y. Ueda, The lowest surface free energy based on-CF_3 alignment. *Langmuir* 1999, 15, 4321.
32. R.N. Wenzel, Resistance of solid surfaces to wetting by water. *Ind. Eng. Chem.* 1936, 28, 988.
33. A.B.D. Cassie, S. Baxter, Wettability of porous surfaces. *Trans. Faraday Soc.* 1944, 40, 546.
34. A.B.D. Cassie, Contact angles. *Discuss. Faraday Soc.* 1948, 3, 11.
35. J. Jopp, H. Grull, R. Yerushalmi-Rozen, Wetting behavior of water droplets on hydrophobic microtextures of comparable size. *Langmuir* 2004, 20, 10015.
36. R. Fürstner, W. Barthlott, C. Neinhuis, P. Walzel, Wetting and self-cleaning properties of artificial superhydrophobic surfaces. *Langmuir* 2005, 21, 956.
37. W. Lee, M.K. Jin, W.C. Yoo, J. K. Lee, Nanostructuring of a polymeric substrate with well-defined nanometer-scale topography and tailored surface wettability. *Langmuir* 2004, 20, 7665.
38. L. Feng, Y. Zhang, J. Xi, Y. Zhu, N. Wang, F. Xia, L. Jiang, Petal effect: a superhydrophobic state with high adhesive force. *Langmuir* 2008, 24, 4114.
39. B. Bhushan, Y.C. Jung, Micro- and nanoscale characterization of hydrophobic and hydrophilic leaf surfaces. *Nanotechnology* 2006, 17, 2758.
40. Y. Lee, S.H. Park, K.B. Kim, J.K. Lee, Fabrication of hierarchical structures on a polymer surface to mimic natural superhydrophobic surfaces. *Adv. Mater.* 2007, 19, 2330.
41. K. Koch, B. Bhushan, Y.C. Jung, W. Barthlott, Fabrication of artificial Lotus leaves and significance of hierarchical structure for superhydrophobicity and low adhesion. *Soft Matter* 2009, 5, 1386.
42. E. Mele, S. Girardo, D. Pisignano, Strelitzia reginae leaf as a natural template for anisotropic wetting and superhydrophobicity. *Langmuir* 2012, 28, 5312.
43. P. Peng, Q. Ke, G. Zhou, T. Tang, Fabrication of microcavity-array superhydrophobic surfaces using an improved template method. *J. Colloid Interface Sci.* 2013, 395, 326.
44. L. Feng, S. Li, H.H. Li, J. Zhai, Y. Song, L. Jiang, D. Zhu, Super-hydrophobic surface of aligned polyacrylonitrile nanofibers. *Angew. Chem. Int. Ed.* 2002, 41, 1221.
45. B. He, N.A. Patankar, J. Lee, Multiple equilibrium droplet shapes and design criterion for rough hydrophobic surfaces. *Langmuir* 2003, 19, 4999.
46. L. Feng, Y. Song, J. Zhai, B. Liu, J. Xu, L. Jiang, D. Zhu, Creation of a superhydrophobic surface from an amphiphilic polymer. *Angew. Chem. Int. Ed.* 2003, 42, 7, 800.

47. C. Guo, L. Feng, J. Zhai, G. Wang, Y. Song, L. Jiang, D. Zhu, Large-area fabrication of a nanostructure-induced hydrophobic surface from a hydrophilic polymer. *ChemPhysChem* 2004, 5, 750.

48. Q.F. Xu, B. Mondal, A.M. Lyons, Fabricating superhydrophobic polymer surfaces with excellent abrasion resistance by a simple lamination templating method. *ACS Appl. Mater. Interfaces* 2011, 3, 3508.

49. C. Su, Y. Xu, F. Gong, F. Wang, C. Li, The abrasion resistance of a superhydrophobic surface comprised of polyurethane elastomer. *Soft Matter* 2010, 6, 6068.

50. L. Zhang, Z. Zhou, B. Cheng, J.M. DeSimone, E.T. Samulski, Superhydrophobic behavior of a perfluoropolyether lotus-leaf-like topography. *Langmuir* 2006, 22, 8576.

51. Y.C. Jung, B. Bhushan, Wetting behavior of water and oil droplets in three-phase interfaces for hydrophobicity/philicity and oleophobicity/philicity. *Langmuir* 2009, 25, 14165.

52. J. Li, J. Fu, Y. Cong, Y. Wu, L.J. Xue, Y.C. Han, Macroporous fluoropolymeric films templated by silica colloidal assembly: a possible route to super-hydrophobic surfaces. *Appl. Surf. Sci.* 2006, 252, 2229,

53. S.H. Li, S.B. Zhang, X.H. Wang, Fabrication of superhydrophobic cellulose-based materials through a solution-immersion process. *Langmuir* 2008, 24, 5585.

54. W. Ming, D. Wu, R. van Benthem, G. de With, Superhydrophobic films from raspberry-like particles. *Nano Lett.* 2005, 5, 2298.

55. Z. Qian, Z.C. Zhang, L.Y. Song, H.R. Liu, A novel approach to raspberry-like particles for superhydrophobic materials. *J. Mater. Chem.* 2009, 19, 1297.

56. J. Zhao, B. Leng, Z. Shao, G. de Withc, W. Ming, Triple-scale structured superhydrophobic and highly oleophobic surfaces. *RSC Adv.* 2013, 3, 22332.

57. B. Leng, Z. Shao, G. de With, W. Ming, Superoleophobic cotton textiles. *Langmuir* 2009, 25, 2456.

58. T. Soeno, K. Inokuchi, S. Shiratori, Ultra water-repellant surface resulting from complicated microstructure of SiO_2 nanoparticles. *Trans. Mater. Res. Soc. Jpn.* 2003, 28, 1207.

59. X. Zhang, F. Shi, X. Yu, H. Liu, Y. Fu, Z. Wang, L. Jiang, X. Li, Polyelectrolyte multilayer as matrix for electrochemical deposition of gold clusters: toward super-hydrophobic surface. *J. Am. Chem. Soc.* 2004, 126, 3064.

60. R.M. Jisr, H.H. Rmaile, J.B. Schlenoff, Hydrophobic and ultrahydrophobic multilayer thin films from perfluorinated polyelectrolytes. *Angew. Chem. Int. Ed.* 2005, 44, 782.

61. L. Zhai, F.C. Cebeci, R.E. Cohen, M.F. Rubner, Stable superhydrophobic coatings from polyelectrolyte multilayers. *Nano Lett.* 2004, 4, 1349.

62. M.L. Ma, M. Gupta, Z. Li, L. Zhai, K.K. Gleason, R. Cohen, M.F. Rubner, G.C. Rutledge, Decorated electrospun fibers exhibiting superhydrophobicity. *Adv. Mater.* 2007, 19, 255.

63. J. Yang, Z. Zhang, X. Men, X. Xu, X. Zhu, X. Zhou, Q. Xue, Rapid and reversible switching between superoleophobicity and superoleophilicity in response to counterion exchange. *J. Colloid Interface Sci.* 2012, 366, 191.

64. M.E. Buck, J. Zhang, D.M. Lynn, Layer-by-layer assembly of reactive ultrathin films mediated by click-type reactions of poly(2-alkenyl azlactone)s. *Adv. Mater.* 2007, 19, 3951.

65. U. Manna, M.J. Kratochvil, D.M. Lynn, Superhydrophobic polymer multilayers that promote the extended, long-term release of embedded water-soluble agents. *Adv. Mater.* 2013, 25, 6405.

66. J.P. Youngblood, T.J. McCarthy, Ultrahydrophobic polymer surfaces prepared by simultaneous ablation of polypropylene and sputtering of poly(tetrafluoroethylene) using radio frequency plasma. *Macromolecules* 1999, 32, 6800.
67. S. Minko, M. Muller, M. Motornov, M. Nitschke, K. Grundke, M. Stamm, Two-level structured self-adaptive surfaces with reversibly tunable properties. *J. Am. Chem. Soc.* 2003, 125, 3896.
68. K. Ellinas, A. Tserepi, E. Gogolides, From superamphiphobic to amphiphilic polymeric surfaces with ordered hierarchical roughness fabricated with colloidal lithography and plasma nanotexturing. *Langmuir* 2011, 27, 3960.
69. J.P. Zhang, S. Seeger, Superoleophobic coatings with ultralow sliding angles based on silicone nanofilaments. *Angew. Chem. Int. Ed.* 2011, 50, 6652.
70. J. Zimmermann, F.A. Reifler, G. Fortunato, L.C. Gerhardt, S. Seeger, A simple, one-step approach to durable and robust superhydrophobic textiles. *Adv. Mater.* 2008, 18, 3662.
71. K.K.S. Lau, J. Bico, K.B.K. Teo, M. Chhowalla, G.A.J. Amaratunga, W.I. Milne, G.H. McKinley, K.K. Gleason, Superhydrophobic carbon nanotube forests. *Nano Lett.* 2003, 3, 1701.
72. H. Jin, M. Kettunen, A. Laiho, H. Pynnonen, J. Paltakari, A. Marmur, O. Ikkala, R.H.A. Ras, Superhydrophobic and superoleophobic nanocellulose aerogel membranes as bioinspired cargo carriers on water and oil. *Langmuir* 2011, 27, 1930.
73. R.G. Palgrave, I.P. Parkin, Aerosol assisted chemical vapor deposition using nanoparticle precursors: a route to nanocomposite thin films. *J. Am. Chem. Soc.* 2006, 128, 1587.
74. I. Sas, R.E. Gorga, J.A. Joines, K.A. Thoney, Literature review on superhydrophobic self-cleaning surfaces produced by electrospinning. *J. Polym. Sci. B: Polym. Phys.* 2012, 50, 824.
75. N. Nuraje, W.S. Khan, Y. Lei, M. Ceylan, R. Asmatulu, Superhydrophobic electrospun nanofibers. *J. Mater. Chem. A* 2013, 1, 1929.
76. L. Jiang, Y. Zhao, J. Zhai, A lotus-leaf-like superhydrophobic surface: a porous microsphere/nanofiber composite film prepared by electrohydrodynamics. *Angew. Chem. Int. Ed.* 2004, 116, 4438.
77. M.W. Lee, S. An, S.S. Latthe, C. Lee, S.K. Hong, S.S. Yoon, Electrospun polystyrene nanofiber membrane with superhydrophobicity and superoleophilicity for selective separation of water and low viscous oil. *ACS Appl. Mater. Interfaces* 2013, 5, 10597.
78. G. Gong, J.T. Wu, Y. Zhao, J.G. Liu, X. Jin, L. Jiang, A novel fluorinated polyimide surface with petal effect produced by electrospinning. *Soft Matter* 2014, 10, 549.
79. D. Han, A.J. Steckl, Superhydrophobic and oleophobic fibers by coaxial electrospinning. *Langmuir* 2009, 25, 9454.
80. S. Srinivasan, S.S. Chhatre, J.M. Mabry, R.E. Cohen, G.H. McKinley, Solution spraying of poly(methyl methacrylate) blends to fabricate microtextured, superoleophobic surfaces. *Polymer* 2011, 52, 3209.
81. L. Xu, W. Chen, A. Mulchandani, Y. Yan, Reversible conversion of conducting polymer films from superhydrophobic to superhydrophilic. *Angew. Chem. Int. Ed.* 2005, 44, 6009.
82. T. Darmanin, F. Guittard, Molecular design of conductive polymers to modulate superoleophobic properties. *J. Am. Chem. Soc.* 2009, 131, 7928.
83. H.Y. Erbil, A.L. Demirel, Y. Avci, O. Mert, Transformation of a simple plastic into a superhydrophobic surface. *Science* 2003, 299, 1377.

84. X. Hou, X. Wang, Q. Zhu, J. Bao, C. Mao, L. Jiang, J. Shen, Preparation of polypropylene superhydrophobic surface and its blood compatibility. *Colloids Surf., B* 2010, 80, 247.

85. T.S. Wong, S.H. Kang, S.K.Y. Tang, E.J. Smythe, B.D. Hatton, A. Grinthal, J. Aizenberg, Bioinspired self-repairing slippery surfaces with pressure-stable omniphobicity. *Nature* 2011, 477, 443.

86. B.J. Privett, J. Youn, S.A. Hong, J. Lee, J. Han, J.H. Shin, M.H. Schoenfisch, Antibacterial fluorinated silica colloid superhydrophobic surfaces. *Langmuir* 2011, 27, 9597.

87. R. Pernites, C. Santos, M. Maldonado, R. Ponnapati, D. Rodrigues, R. Advincula, Tunable protein and bacterial cell adsorption on colloidally templated superhydrophobic polythiophene films. *Chem. Mater.* 2012, 24, 870.

88. L.R. Freschauf, J. McLane, H. Sharma, M. Khine, Shrink-induced superhydrophobic and antibacterial surfaces in consumer plastics. *PLoS One* 2012, 7, e40987.

89. T.L. Sun, H. Tan, D. Han, Q. Fu, L. Jiang, No platelet can adhere-largely improved blood compatibility on nanostructured superhydrophobic surfaces. *Small* 2005, 1, 959.

90. Y. Koc, A.J. de Mello, G. McHale, M.I. Newton, P. Roach, N.J. Shirtcliffe, Nano-scale superhydrophobicity: suppression of protein adsorption and promotion of flow-induced detachment. *Lab Chip* 2008, 8, 582.

91. J. Zhao, L. Song, Y. Jin, W. Ming, Anti-bioadhesion on hierarchically structured, superhydrophobic surfaces. *Chem. Commun.* 2013, 49, 9191.

92. A.J. Meuler, G.H. McKinley, R.E. Cohen, Exploiting topographical texture to impart icephobicity. *ACS Nano* 2010, 4, 7048.

93. L. Cao, A.K. Jones, V.K. Sikka, J. Wu, D. Gao, Anti-icing superhydrophobic coatings. *Langmuir* 2009, 28, 12444.

94. P. Kim, T.S. Wong, J. Alvarenga, M.J. Kreder, W.E. Adorno-Martinez, J. Aizenberg, Design of ice-free nanostructured surfaces based on repulsion of impacting water droplets. *ACS Nano* 2010, 4, 7699.

95. D. Rana, T. Matsuura, Surface modifications for antifouling membranes. *Chem. Rev.* 2010, 110, 2448.

96. G. Kwon, A.K. Kota, Y. Li, A. Sohani, J.M. Mabry, A. Tuteja, On-demand separation of oil-water mixtures. *Adv. Mater.* 2012, 24, 3666.

97. A.K. Kota, G. Kwon, W. Choi, J.M. Mabry, A. Tuteja, Hygro-responsive membranes for effective oil-water separation. *Nat. Commun.* 2012, 3, 1025.

98. Y. Li, L. Li, J.Q. Sun, Bioinspired self-healing superhydrophobic coatings. *Angew. Chem. Int. Ed.* 2010, 122, 6265.

99. H.X. Wang, Y.H. Xue, J. Ding, L.F. Feng, X.G. Wang, T. Lin, Durable, self-healing superhydrophobic and superoleophobic surfaces from fluorinated-decyl polyhedral oligomeric silsesquioxane and hydrolyzed fluorinated alkyl silane. *Angew. Chem. Int. Ed.* 2011, 50, 11433.

100. H. Zhou, H. Wang, A. Gestos, T. Lin, Robust, self-healing superamphiphobic fabrics prepared by two-step coating of fluoro-containing polymer, fluoroalkyl silane, and modified silica nanoparticles. *Adv. Funct. Mater.* 2013, 23, 1664.

101. A.C.C. Esteves, Y. Luo, M.W.P. van de Put, C.C.M. Carcouët, G. de With, Self-replenishing dual structured superhydrophobic coatings prepared by drop-casting of an all-in-one dispersion. *Adv. Funct. Mater.* 2014, 24, 986.

CHAPTER 3

Superhydrophilic and Superamphiphilic Coatings

Sandro Olveira, Ana Stojanovic, and Stefan Seeger
Department of Chemistry, University of Zurich, Zurich, Switzerland

3.1 INTRODUCTION

The concept of *superhydrophilicity* was introduced following intensive research on superhydrophobic surfaces and in response to a request for coatings and surfaces that display strong affinity to water [1]. When water droplets are in contact with a superhydrophilic surface, they completely spread over the surface. When such a surface additionally enables total spreading of oily droplets (superoleophilic property), it is defined as superamphiphilic. Over the past decade, surfaces exhibiting superhydrophilicity or superamphiphilicity have been the subjects of immense interest because of their potential applications in various fields, including the development of microfluidic devices, liquid–liquid separation membranes, antifogging, antireflective, self-cleaning, and antifouling coatings [2–7].

In 2000, the term *superhydrophilicity* was used for the first time in papers published by three different research groups from Japan [1, 8, 9]. Earlier, in 1997, Wang et al. published a seminal paper on a superamphiphilic coating that consisted of a TiO_2 polycrystalline film [10].

This chapter reviews the current state of research on superhydrophilic and superamphiphilic coatings, and it is organized as follows: Section 3.2 summarizes important fundamentals and definitions that apply to artificial superhydrophilic and superamphiphilic surfaces. The next section (Section 3.3) presents (i) examples of naturally occurring superhydrophilic and/or superamphiphilic surfaces and (ii) the most prominent examples of artificial superwetting coatings are illustrated in Section 3.4. Since high-quality coatings exhibiting superhydrophilicity or superamphiphilicity cannot be produced without manipulation and control of surface

Functional Polymer Coatings: Principles, Methods, and Applications, First Edition.
Edited by Limin Wu and Jamil Baghdachi.

chemistry and surface structure, Section 3.5 contains an overview of the most common techniques used for manufacturing such coatings. Then, the next section (Section 3.6) describes the most explored applications of superhydrophilic and superamphiphilic coatings, which include antifogging films, antireflective coatings, enhanced boiling heat transfer, separation membranes, and smart surfaces with reversible switching abilities. Section 3.7 provides an overview of commercially available superhydrophilic and superamphiphilic coatings. In the last Section 3.8 conclusions about the current state of research and commercial applications of these superwetting coatings are drawn.

3.2 BASIC CONCEPTS OF SUPERHYDROPHILICITY

Surface wettability is generally characterized by the value of the contact angle. A surface is called superhydrophilic (or alternatively, superwetting) if the apparent contact angle of water on the surface is less than 5°. In addition to either superhydrophilicity or superhydrophobicity, a surface can also exhibit either superoleophilicity or superoleophobicity. On a superoleophilic surface, the contact angle of a polar droplet exhibits a value lower than 10°. A surface that is both superhydrophilic and superoleophilic is called superamphiphilic.

The wettability of a solid surface is controlled by the surface free energy and the geometric structure of the surface. Commonly, the presence of polar groups on the surface decreases the contact angle of liquids. Water can completely spread over a few smooth surfaces, including those of glass [11], quartz [12], amorphous silica [13], gold [14], selected oxides (carrying OH groups on the surface) [15], and selected self-assembled monolayers with OH-based functionalities (OH, COOH, POOH). Such strong affinity for water is typically short term, and in the presence of any contamination, the contact angle increases to a few tens of degrees [16, 17]. According to Young's equation, the contact angle of a liquid drop on a solid surface results from equilibrium between cohesive forces in the liquid and adhesive forces between the solid and the liquid [18]. For a certain liquid, the predominant contribution to the contact angle originates from the interfacial character of the solid material, which is related to its surface structure [19]. Wettability is determined by the surface free energy of a solid surface, which is commonly expressed by Young's equation,

$$\cos\theta = \frac{\gamma_{sv} - \gamma_{sl}}{\gamma_{lv}} \tag{3.1}$$

Here, θ is the contact angle in Young's mode, and γ_{lv}, γ_{sv}, and γ_{lv} are the three different types of surface tension (liquid/vapor, solid/vapor, and solid/liquid) involved in the system.

In recent studies, Drelich et al. suggested that surfaces are truly superhydrophilic only if the surface is textured and/or structured [20, 21]. Rough or porous surfaces possess a roughness factor r (where r is the ratio of the real surface area to the projected surface area) that is defined by the Wenzel equation [22]. The Wenzel theory predicts that, for any rough surface, the actual surface area will be greater than the geometric surface area. This surface ratio is called the roughness factor and is defined by

$$r = \frac{\cos\theta}{\cos\theta^*} \tag{3.2}$$

where θ and θ^* are the actual and geometric contact angles, respectively. In other words, an increase in the surface area (due to the presence of texture) amplifies the natural hydrophilicity of the material.

Hence, according to the definition by Drelich et al., truly superhydrophilic surfaces possess roughness factors greater than one, and water spreads completely over them. Figure 3.1 shows the relationship between the contact angle on a smooth surface (Young's contact angle, θ) and the minimum value of the roughness factor (r) that is required for the same surface to promote complete spreading of the liquid. The figure shows that with moderate roughening of the surface in which r is between 1.2 and 2, superhydrophilicity should be conceivable on any material having an intrinsic contact angle less than 60°. For materials with θ>65–70°, roughening might not be practical, since extremely high values for r are needed. Theoretically, on any rough material, a liquid should spread to zero (or nearly zero) apparent contact angle; however, in practice, liquid penetration into the rough structure of a substrate might be difficult. Such a system is essentially a three-phase system trapped in a metastable state, and the surface should be treated more like a porous or solid–air composite material [23, 24].

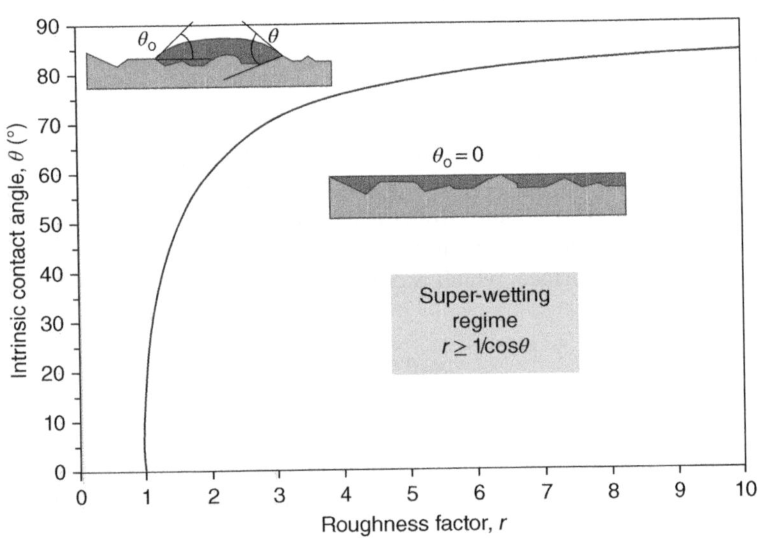

FIG. 3.1 Minimum values of roughness factor necessary to promote the complete spreading of liquid on a surface with varying Young's (intrinsic) contact angle. Used with permission from Ref. 21. © Royal Society of Chemistry.

Besides being rough, porous surfaces also can exhibit superhydrophilic properties via wicking. Wicking or spontaneous imbibition is the suction of a liquid into a porous medium due to negative capillary pressure created at the liquid–air interface [25]. Wetting on a three-dimensional (3D) porous surface that exhibits a 3D capillary effect was investigated theoretically by Quéré et al. [24, 26, 27] According to the authors, a "hemi-wicking" behavior is likely on superhydrophilic 3D porous media; this behavior is between droplet spreading and penetration. The critical contact angle, θ_c, below which the penetration of the porous surface by a liquid will take place, is given by

$$\cos\theta_c = \frac{1-\varphi_s}{r-\varphi_s} \tag{3.3}$$

where φ_s is the solid fraction remaining dry during the wicking process and r (≥ 1) is the surface roughness. For a porous surface, r goes to infinity, and (3.3) predicts that the microstructure will be fully invaded by any liquid having a contact angle (as measured on a flat surface) of less than 90°. For rough surfaces ($r > 1$ but not infinity), the critical angle can vary between 0 and 90°. In the case of 3D porous materials, it is possible to switch from a superhydrophobic to a superhydrophilic state by slight changes in the surface chemistry [28]. This extreme transition in wetting behavior is enabled by the fact that only filled or empty pores are energetically favorable. Thus, compared to roughness-induced superhydrophilicity, porosity-induced superhydrophilicity offers some unusual possibilities for designing functional surfaces. In this regard, nanoporous thin films are particularly attractive since, in contrast to microporous films, they do not scatter light due to their small pore size [29].

Besides 3D capillary effects, a surface can be completely wetted by a 2D capillary effect. The 2D capillary effect has been observed on metal oxide semiconductors [30]. In general, ultraviolet (UV) light induces the formation of hydrophilic and oleophilic nanodomains on the surface. These different domains lead to nanostructured and microstructured flow channels for both aqueous and oily liquids. Channels for water flow are formed by oleophilic walls, while those for oily liquids are formed by hydrophilic walls; these structures may exhibit behavior that resembles the 2D capillary effect [31]. The hydrophilic areas are higher in position than oleophilic areas; this increases the formation of flow channels [30]. When a liquid droplet comes in contact with such a surface, it will flow along these nano- and microchannels and form a very thin film on the surface. This effect usually leads to superamphiphilicity; it was first observed in TiO_2 coatings. The existence of surface superamphiphilicity has been explained as follows: after UV illumination, oxygen vacancies are created in the surface, and these vacancies induce the translation of the corresponding Ti^{4+} sites to Ti^{3+}. The as-formed Ti^{3+} sites are favorable to dissociate water molecules and further monolayers or multilayers of water form by molecular adsorption [32]. This results in the formation of surface hydrophilic domains, while leaving the rest of the surface oleophilic. In many cases, 2D and 3D capillary effects can coexist and work cooperatively on the same surface [30].

3.3 NATURALLY OCCURRING SUPERHYDROPHILIC AND SUPERAMPHIPHILIC SURFACES

During evolution, plants and animals developed complex strategies to handle liquids, particularly for water management. With their sophisticated surface structures, living organisms can react to the environmental changes and deal with the extreme living conditions. There is enormous diversity in biological, multifunctional, and protective surfaces formed in different environments [33]. One of the most prominent examples is the surface of the lotus leaf, which exhibits extraordinary purity because of its extreme water-repellent properties. The reason for this superhydrophobic behavior, discovered by Barthlott in the 1990s, [34] is a surface structure consisting of an epicuticular wax layer and a microscopic papillae structure. The wax causes low surface energy and the papillae, which consist of further branch-like nanostructures, provide high surface roughness. These two features, the chemical composition and the topographic micro- and nanostructure of the surface, are most important for determining the wettability of a surface [35, 36]. In contrast to the lotus leaf, some plants show superhydrophilic properties. If a drop of water comes in contact with the surface of such plants, it will spread within seconds or even milliseconds. With regard to wetting behavior and environmental conditions, superhydrophilic plants can be divided into three groups: those that are permanently wet, those on which water is absorbed at surfaces, and those on which water spreads over surfaces. [37] The leaves of submerged water plants, for example, *Anubias barteri*, have permanently wet surfaces that consist of smooth cell surfaces without any waxes, papilla, or hairs. Plants that absorb water have porous surfaces or pores or multicellular hairs. For example, *Sphagnum mosses* consist of pores (10–20 μm in diameter) that form a sponge-like surface structure; this enables the plant to absorb amount of water that is up to 20 times the weight of the plant itself. This strategy of absorbing water is particularly important as a way for rootless plants to take up nutrients and as a way for plants to retain water in dry areas where dew is almost the only water source [38]. *Ruellia devosiana* belongs to the group of plants on which water spreads on surfaces. The leaves of this plant contain a complex surface with different cell types, including hair papilla, papilla cells, and glands; such a complex surface can spread a 5-μl drop of water within 0.2 s [38]. This is the fastest spreading behavior known for a plant species. Such a superhydrophilic surface is very important for plants in areas having high densities of precipitation, such as rainforests. Fast spreading creates a larger water–air interface, leading to an increase in evaporation, compared to hydrophilic, hydrophobic, or superhydrophobic surfaces. Through fast evaporation, a smooth gas exchange at the surface is ensured, and the growth of microorganisms at the surface is hindered. [37] The leaves of *R. devosiana* show another interesting property. If a 10-μl droplet of oil is placed on its leaf, it also spreads to a flat film within 0.6 s. Surprisingly, water on a vertically oriented leaf stripe can move against gravity. Within 31 s, water on a leaf can move a vertical distance of 5 cm. This phenomenon of water transport without any pressure against gravity might be exploited in applications of liquid-transport devices [39].

3.4 ARTIFICIAL SUPERHYDROPHILIC COATINGS

3.4.1 TiO_2 Coatings

The first superamphiphilic coating that consisted of a thin TiO_2 polycrystalline film from anatase sol on a glass substrate was reported by Wang et al. [10]. When the thin TiO_2 film is exposed to UV radiation, the surface turns from slightly hydrophilic to superhydrophilic. The film exhibits a water contact angle of $72° \pm 1°$ before UV irradiation (Fig. 3.2a). After irradiation, water droplets spread on the film, resulting in a contact angle of $(0 \pm 1)°$ (Fig. 3.2b). This change in wettability is more obvious when a TiO_2-coated glass is exposed to water vapor. Without UV irradiation, the glass fogs (Fig. 3.2c), but upon irradiation, the glass becomes transparent (Fig. 3.2d), displaying an outstanding antifogging effect.

All nonpolar liquids (e.g., glycerol trioleate and hexadecane) spread across the surface upon UV irradiation, exhibiting contact angles of (0 ± 1) [10]. The same wettability change is observed on both anatase and rutile TiO_2 surfaces, independent of their photocatalytic activities. Even after storage in dark conditions for a few days, the high amphiphilicity of the TiO_2 surface was maintained. A longer storage period induced a gradual increase in the water contact angle, revealing the surface wettability trend toward hydrophobicity. However, high amphiphilicity was repeatedly regenerated by

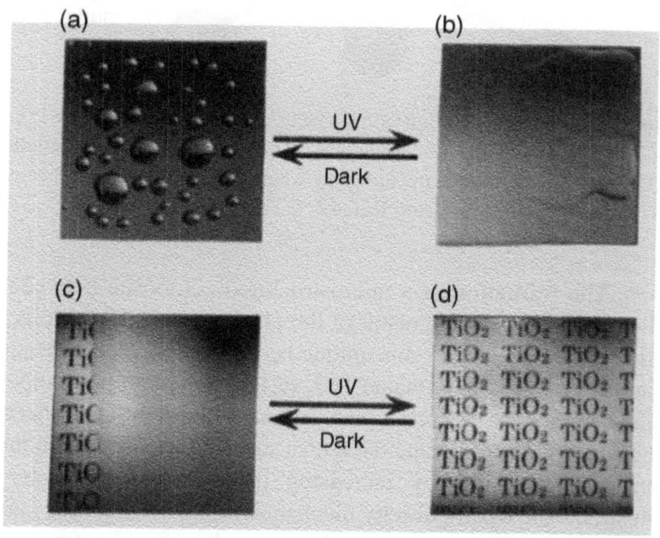

FIG. 3.2 (a) A hydrophobic surface before ultraviolet irradiation. (b) A highly hydrophilic surface after ultraviolet irradiation. (c) Exposure of a hydrophobic TiO_2-coated glass to water vapor. The formation of fog (small water droplets) hinders the view of the text on paper placed behind the glass. (d) Result of ultraviolet irradiation, creating an antifogging surface. The high hydrophilicity prevents the formation of water droplets, making the text clearly visible. Used with permission from Ref. 10. © Nature Publishing Group.

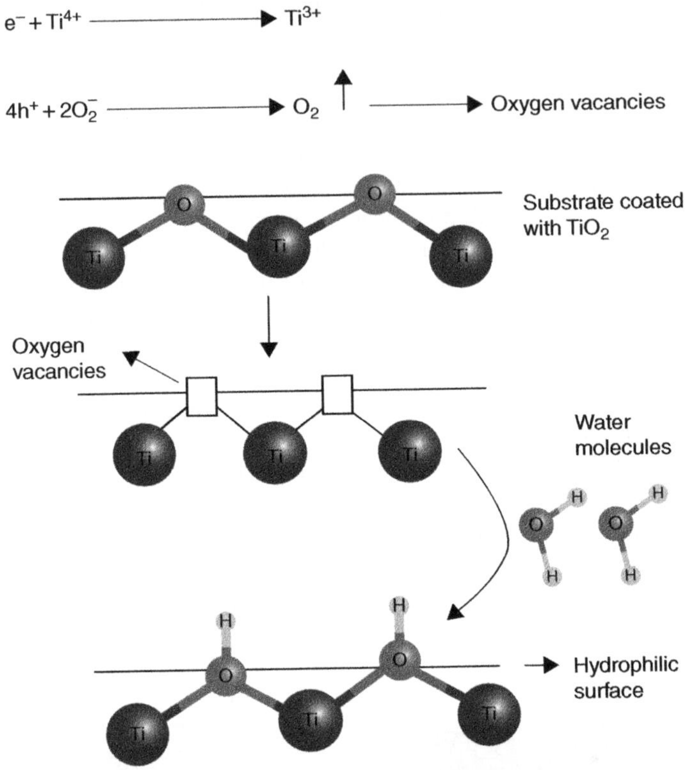

FIG. 3.3 Mechanism of hydrophilicity on surfaces coated by TiO_2. Under UV radiation, the valence of Ti^{4+} changes to Ti^{3+}, accompanied by the release of O_2. This creates oxygen vacancies on the surface that can be occupied by water; hence, the surface becomes more hydrophilic. Used with permission from Ref. 42. © Elsevier.

UV irradiation. The formation of a microstructured composite between hydrophilic and oleophilic phases, which is a result of the photogenerated Ti^{3+} defects at definite sites, is considered to account for this unique behavior [40]. As a consequence, water can spread rapidly on a UV-illuminated TiO_2 surface, which imparts superhydrophilic properties. ZnO possesses the same photoinduced hydrophilicity mechanism [41]. Increasing the concentration of nano-TiO_2 on a surface increases the number of accessible oxygen vacancies, thereby increasing the capacity for water absorption. Figure 3.3 illustrates the hydrophilicity mechanism for a TiO_2-coated surface.

When placed in a dark environment, TiO_2-based surface coatings typically lose their superhydrophilic properties within minutes to hours; this limits their practical applications. In order to improve that Machida et al. found that by adding 30 mol% SiO_2 to a TiO_2 coating, the contact angle of water is low immediately after the production, and hydrophilicity is preserved in a dark place [43]. At this point, extensive research has focused on chemical modifications to solid surfaces, such as ion doping, [44] metal deposition, [45] semiconductor coupling, [11, 20, 46], and further compositing with SiO_2. [47] Moreover, superhydrophilicity might also be strengthened by increasing surface acidity or the number of hydroxyl groups at the surface.

Today, there are numerous studies dealing with superhydrophilic self-cleaning surfaces containing titanium, but the majority of these publications still use glass as a standard substrate. On the other hand, one of excellent potential target for photoinduced cleaning, UV protection, and antimicrobial effects is the textile industry [48]. Special clothes, especially those that are endangered by staining with heavy contaminants such as soot, oils, or lubricants, are good candidates for applying superhydrophilic self-cleaning surfaces. This is particularly important in the case of textile products that are either utilized outdoors or cannot be washed (because of their size or water sensitivity). Recently, a TiO_2 coating was reported to be deposited on textile material using radiofrequency plasma-enhanced chemical vapor deposition (RF PECVD) technique [49]. In this procedure, titanium (IV) chloride was used as the titanium source, oxygen was supplied as O_2 gas, and a cotton fabric served as the substrate. The stability of the coating remains unchanged after washing the fabric in a detergent solution, even after subsequent storage for 18 months. The number of nanoparticles absorbed on surfaces of fabrics and subsequent superhydrophilicity are affected by the efficiency of pretreatments [50]. For instance, studies have been performed on the effectiveness of some pretreatment methods, such as activating textile surfaces using plasma and vacuum UV irradiation [50, 51]. In these methods, fabric surfaces were modified by introducing negatively charged groups, thereby increasing the hydrophilicity of the fabrics. Alternatively, cross-linking agents can be used to immobilize TiO_2 nanoparticles on surfaces of the wool [52].

Besides glass and textile substrates, Ti-containing mesoporous silica thin films (Ti-MSTFs) have been prepared on Al and Al–Mg alloy substrates via a sol–gel/spin-coating method [53]. This coating method is applicable to various materials that have low thermal resistance. The resulting Al and Al–Mg alloy substrates coated with Ti-MSTFs had highly hydrophilic properties, even under dark conditions, and showed photo-induced superhydrophilicity under UV irradiation.

3.4.2 SiO_2 Coatings

Previous examples illustrate that addition of SiO_2 to titanium enhances the durability of superhydrophilic coatings. However, very recently, superhydrophilic coatings have been reported to be produced from mesoporous SiO_2 without the addition of titanium [54]. The early theory established by Quéré et al. suggests that it is possible to enhance the wetting property of a surface by introducing roughness at the right scale [23, 55]. As a result, a mesoporous SiO_2 superhydrophilic thin film was produced by the sol–gel method. When one coating layer of the film was applied, the contact angle for water decreased below 5° in 4 s; on films coated 6–12 times, the contact angle decreased below 5° in less than 1 s. Thus, superhydrophilicity increases as the number of coatings increases. The mechanism for such behavior can be understood from the simple relation derived by Wenzel et al. (see Section 3.2).

In addition to mesoporous substances, nanostructured materials having structural elements between 1 and 100 nm have the potential to improve surface functionalities of thin films. Materials with well-ordered pores have been intensely investigated in many fields, including catalytic chemistry, adsorption chemistry, electrochemistry, and materials science. The most important families of silica-based porous materials are zeolites, which have microporous structures with pore sizes less than 2 nm, and mesoporous silicas, which have mesoporous structures with pore sizes between 2 and 50 nm. Mesoporous structures have been formed via evaporation-induced self-assembly methods. The significant hydrophilic behavior can be achieved when transparent mesoporous silica thin films containing single-site photocatalysts are used. The single-site photocatalysts include moieties of Ti-, V-, Cr-, Mo-, and W-oxide; [56] these photocatalysts show exclusive and remarkable catalytic properties that are not demonstrated by bulk catalysts [57, 58]. Because of electron localization, substitution sites for heteroatoms would attract water molecules; therefore, these materials become hydrophilic. After coating, the materials exhibit significant hydrophilic properties under dark conditions and photoinduced superhydrophilicity under UV irradiation. Among them, the W-containing mesoporous silica thin film shows the best hydrophilic properties. This coating can be applied to various materials, including Al, Al–Mg alloys, and polycarbonate.

3.5 METHODS FOR FABRICATING SUPERHYDROPHILIC AND SUPERAMPHIPHILIC SURFACES

Most solids are naturally rough, but their roughness is usually insufficient to reinforce a superhydrophilic state on a material surface. Inspired by naturally occurring examples of plants with superwetting properties for both water and oily liquids, such as *R. devosiana*, scientists have started creating artificial surfaces that exhibit similar properties [37]. Within the past decade, many different methods and materials have been implemented for the production of superhydrophilic and superamphiphilic surfaces on various substrates. Here, we review the most common techniques used for producing these coatings.

3.5.1 Sol–Gel Method

In general, the sol–gel method is used to synthesize porous network structures. It is a low-temperature technique that is simple, affordable, and easy to control. By adjusting the composition of the precursor solution along with the hydrolysis and polycondensation processes, the resulting films can exhibit different morphologies and can contain different chemical components at the surface. The sol–gel method is widely used and is a convenient process for coating surfaces to obtain superhydrophilic and/or superamphiphilic properties.

TiO_2–polydimethylsiloxane (TiO_2–PDMS) composite films can be prepared by the sol–gel method from a $Ti(OBu)_4$-benzoylacetone solution containing PDMS [59]. Contact angles measured for the TiO_2–PDMS thin films show a wettability transition from hydrophobic to superhydrophilic states after treatment with oxygen plasma for 1 s. Ma et al. [60] used this conventional technique to produce transparent mesoporous silica coatings that showed permanent superamphiphilicity with very fast spreading rates of within a few microseconds. To form the sol, they mixed tetraethyl orthosilicate (TEOS) and a poloxamer (Pluronic F-127) with nitric acid; the solution was stirred at room temperature for 2 h. Next, glass substrates were spin-coated with the sol, and further drying steps were performed. The approach aimed to create a superamphiphilic surface by increasing the roughness of an amphiphilic surface. Therefore, they introduced mesopores on the surface and obtained a superamphiphilic coating without using UV illumination. The ability of water and oily liquids, such as hexadecane, to spread with a contact angle of 0° on this surface is caused by high surface energy and high surface roughness. The high surface energy results from Si–O–Si and Si–OH groups present on the surface. The high capillary pressure inside the mesopores leads to the fast spreading mentioned above. Furthermore, the coating is transparent, because the pore diameter is in the range of 10 nm, and therefore much less than the wavelength of visible light. This is an important feature with regard to possible industrial applications.

Kako and Ye used a similar sol–gel method to produce a complex oxide ($InNbO_4$) coating with UV-induced superamphiphilic properties [61]. A powder of $In(NO_3)_3$ and niobium ethoxide was mixed in ethanol and spin-coated on a quartz substrate. After 30 s of UV exposure, the contact angle of a water droplet changes from 55° to less than 5°, indicating superhydrophilicity. Moreover, the contact angles of CH_2I_2 and dodecane decrease to 15° and 0°, respectively, on such a surface. $InNbO_4$ is the first complex oxide to show superamphiphilicity after UV illumination.

3.5.2 Layer-By-Layer Assembly

In 1966, Iler published a method for creating multilayers of inorganic "colloidal particles" without using any organic molecules; the method is now known as layer-by-layer (LbL) deposition [62]. He reported that multilayers of oppositely charged nanoparticles can be assembled by the sequential adsorption of oppositely charged nanoparticles onto substrates from aqueous suspensions. Although Iler's work did not attract attention at the time, 25 years later, Decher et al. developed the LbL process to fabricate multilayer thin films from oppositely charged polyelectrolytes. [63] Over the past 15 years, the LbL technique has received enormous attention. To enhance the wettability of the films, micro- or nanoparticles are commonly integrated to increase surface roughness. The LbL assembly has the advantage of precisely controlling the film thickness; thus, it is a desirable technique for fabricating transparent coatings. Liu and He [64] reported an LbL method for obtaining superhydrophilic coatings, in which raspberry-like silica nanospheres were prepared by the electrostatic self-assembly of polyelectrolytes and monodispersed silica nanoparticles of two different sizes.

Alternatively, the production of superhydrophilic coatings composed completely of nanoparticles has been proposed [65]. In addition, thin films consisting of TiO_2 and SiO_2 nanoparticles have been prepared via LbL deposition. The presence of nanopores in TiO_2/SiO_2 nanoparticle coatings leads to useful functionalities, including antireflective and antifogging properties. The last step of this method includes calcination at high temperatures. This well-known calcination process burns out the polymer component of the film and fuses the silica nanoparticles together via the formation of stable siloxane bridges [66]. Unfortunately, the high temperatures employed during this curing process also limit the substrate materials that can be coated; plastics with low melting points are not suitable for this type of coating.

3.5.3 Electrochemical Methods

Electrochemical methods include electrochemical deposition, anodization, galvanic cell reactions, and electrochemical polymerization. These are facile methods for constructing rough surfaces, regardless of the size and shape of the substrate.

Shibuichi et al. have shown that the treatment of the surface of an aluminum plate with anode oxidation leads to superamphiphilic properties [67]. The plate was dipped into an acidic solution, and oxidation was performed at a current density of $10\,mA/cm^2$. This method increases surface roughness; in fact, the analysis of the aluminum surface showed it to be fractal. Such a rough surface showed superwettable properties for both polar and nonpolar solvents.

Nanostructured conducting polymers generally show superhydrophilic properties. Among these, the surface of polyaniline exhibits amphiphilic behavior. Therefore, Zhang et al. used an electrochemical-template-free method for the direct deposition of nanostructured polyaniline (PANI) on a substrate such as stainless steel [68]. This substrate was chosen, because it is widely used in industrial equipment. The PANI nanofibers were prepared using sulfuric acid and aniline at 0.85 V on a stainless steel electrode. The structure and size of the nanofibers can be controlled by polymerization time. A water droplet on this modified PANI surface spreads very fast and reaches a contact angle of 5° within a few milliseconds. Droplets of organic solvents, such as acetone and hexane, spread even faster than a water droplet. Such a superamphiphilic surface is of high interest from both scientific and industrial viewpoints, as PANI shows simple nonredox doping/dedoping chemistry, is environmentally stable, and the superwetting property is permanently stable. This method exemplifies a convenient way for producing polymer-functionalized surfaces that exhibit superamphiphilic properties.

3.5.4 Electrospinning

Electrospinning is a technique used to produce fibers with diameters ranging from micrometers to nanometers [69, 70]. In this technique, the sample solution is pumped through a nozzle to which a high electric voltage is applied. Owing to the evaporation of the solvent, the solution jet solidifies and forms fibers that are deposited on a collector. The final morphology strongly depends on the starting solution concentration.

Superhydrophilic surfaces can be generated by α-Fe_2O_3 nanofibers with contact angles of 0° for water [71]. The α-Fe_2O_3 nanofibers are produced by electrospun poly(vinyl alcohol)/ferrous acetate composite nanofiber precursors and high-temperature calcination in air. The experimental results show that the morphology and crystalline phase of α-Fe_2O_3 nanofibers are influenced by the content of ferrous acetate in composite nanofibers and the calcination temperature. By controlling the calcination temperature, the magnetic property of α-Fe_2O_3 nanofibers can be tuned from superparamagnetic to ferromagnetic.

Superamphiphilic coating applicable to textiles can be produced by electrospinning. Lim et al. [72] fabricated Janus fabrics with superwetting properties by using polyacrylonitrile (PAN) as the starting material. The method is very useful, because it enables the synthesis of micro- and nanofibers with definite diameters that can be coated on different substrates. To produce superamphiphilic Janus fabrics, a polymer solution of PAN, dimethylformamide, and TEOS was electrospun into nanofibers. Next, the fibrous mats were heated at 200°C, and the polymer solution mentioned earlier was electrospun onto the treated mats. Then, the mats were peeled from the substrate. This simple method provides a new way for producing functionally smart materials, which are of high interest for possible industrial applications.

3.5.5 Etching

Etching is often used as an additional step in many procedures to improve the roughness of substrates on different scales. For example, Kim et al. demonstrated a facile chemical etching method for fabricating a superamphiphilic surface on a silicon wafer [73]. On a clean silicon wafer, gold nanoclusters were deposited by thermal evaporation and etched in an HF/H_2O_2/DI water solution. The gold clusters catalyzed the etching process at room temperature; therefore, the wafer was selectively etched. Next, the surface was coated with a self-assembled monolayer material. The surface of the silicon wafer showed superamphiphilic behavior, upon exposure of a coated wafer to deep UV light ($\lambda \sim 254$ nm).

A silicon wafer exhibiting superhydrophilicity can be produced by an electroless (EE) silicon etching method [74]. The EE method is a top–down technique that can be used to modify the morphology of the surface with nanoscale structures over a large area at room temperature.

In industry, titanium alloys are often coated with commercially pure titanium (Ti) via physical vapor deposition (PVD), especially if they are used in the clinical sector, such as in the hospital. The coated surfaces have high roughness and show superoleophilicity. A droplet of mineral oil spreads with a critical contact angle of about 0°, while the contact angle of water droplet was 145°, indicating hydrophobicity. Jennissen and Lüers developed a simple chemical treatment to improve the wetting properties of such commercially available Ti-PVD surfaces to superamphiphilicity [75]. After cleaning, the samples are etched with chromosulfuric acid at 240°C for 30 min. A water droplet spread on the treated Ti-PVD surface to a critical contact angle of 0°, indicating additional superhydrophilicity.

3.5.6 Plasma Treatment

Because of its ease in handling and very effective results, plasma treatment is one of the most used techniques for producing superhydrophilic or superamphiphilic surfaces. Oxygen plasma treatment modifies the surface properties of polymers, resulting in the hydrophilization of their surfaces [76]. The roughness and morphology of a treated polymer surface depend on the time of exposure [77].

Zimmermann et al. used this method to prepare superamphiphilic surfaces based on a silicone nanofilament coating [78]. First, a desired substrate, such as a glass slide, was coated with silicone nanofilaments via a simple CVD coating at room temperature. The resulting surface was completely wetted by a droplet of hexadecane, which indicates superoleophilic and superhydrophobic properties. After exposing the surface to oxygen plasma, droplets of both hexadecane and water spread completely on the surface, thus showing superamphiphilic properties.

Liu et al. used a plasma method to introduce superamphiphilic properties to a surface consisting of carbon nanotubes (CNT) that were decorated with silver nanoparticles (Ag) [79]. First, a silicon-wafer substrate was coated with a paste of Ag@CNT. By treating the modified substrate with Ar plasma, the contact angle of a water droplet decreased from 85° to almost 0° within 5 min. The treated surface also showed superoleophilic properties with contact angles of 0° for diiodomethane and ethylene glycol. The plasma treatment increases the number of hydrophilic functional groups, such as hydroxyl and carboxyl groups, on the surface. In addition, proton donor components are increased, which leads to an increase in oleophilicity due to the capillary effect. This approach can be used, for instance, for the production of superamphiphilic Ag-CNT electrodes.

Polymeric superamphiphilic surfaces were produced by Ellinas et al. using a plasma etching method [80]. Polystyrene colloid particles were deposited on a poly(methylmethacrylate) (PMMA) surface via spin coating. This colloidal lithography was followed by oxygen plasma etching, leading to a surface with highly ordered arrays of micropillars of PMMA. The height and diameters of the pillars can be adjusted very accurately by controlling the etching time and voltage. This method led to high surface roughness due to the introduced micro/nanoscale topography. The superwetting properties of such a nanotextured surface are stable over a long time.

3.5.7 Hydrothermal Method

The hydrothermal method is a wet chemical procedure in which single crystals are synthesized in hot water under high pressure. This method allows precise control over crystal morphology and composition.

A microwave-assisted hydrothermal method for the fabrication of superamphiphilic titanate network (STN) films was developed by Li et al. [81]. The SNT films consisted of twisted multiwalled titanate nanotubes grown on a Ti foil. Compared to other hydrothermal methods, this method is rapid (10 min) and simple. Droplets of both polar and nonpolar solvents spread immediately with contact angles below 1°. On untreated Ti samples, the contact angles for water and CCl_4 were 73° and 15°, respectively. This facile production method also allows the integration of other atoms, such as PbS or CdS. This is of high interest for possible indoor applications, because Ti-containing coatings demand UV light. By modifying the surface with materials sensitive to visible light, such as PbS or CdS, the capability for light absorption can be shifted to longer wavelengths. Films of PbS-STN and CdS-STN were produced via the same microwave-assisted hydrothermal method and showed superamphiphilicity similar to the STN films. Compared to TiO_2 surfaces, the superamphiphilicity remained permanently stable, even after 6 months, without the need for light illumination.

3.5.8 Dip Coating

Dip coating is a very simple and popular method for creating thin films. Uniform films can be applied to flat or cylindrical substrates. A related technique often used in industrial applications is spin coating.

Chen et al. developed a superhydrophilic, superamphiphilic, scratch-resistant coating via a dip-coating process [82]. The coating consisted of aggregated zeolite nanoparticles, which resulted in the useful properties mentioned earlier. The prepared surfaces had higher roughness and strength compared to amorphous SiO_2 surfaces. In addition, the coatings showed a high antireflective and superamphiphilic properties that are of high interest in, for instance, the production of solar panels.

In addition, surfaces with tunable wettability for guiding water droplets can be produced by dip coating [83]. For example, silicon nanowires can be dip-coated in dodecyltrichlorosilane to attain a superhydrophobic state, and then the wettability can be converted to hydrophilic via UV-enhanced photodecomposition.

3.5.9 Phase Separation

The phase-separation method is commonly used for the fabrication of porous polymer coatings. In this method, the starting material is typically a polymer solution or a polymer blend, and phase separation is induced by changing temperature or pressure or both. For example, porous polymers with switchable wettability can be produced by the condensation of organo-triethoxysilane in a mixture of an organic solvent and water [28].

In another approach, Zhang et al. reported a one-step production method for a superhydrophilic polymer surface. A nylon 6,6 plate was swelled by formic acid and then immersed in a coagulate bath to induce precipitation. Microparticles with nanoprotrusions were generated and linked together, covering the surface. After drying, the as-formed surface showed superhydrophilic abilities due to the hydrophilic nature of nylon and the network of micro/nano flower-like particles [84]. Poly(L-lactic acid) substrates can be prepared using a phase-separation-based method. Using argon-plasma posttreatment, the wettability of the surfaces can be controlled in the range from superhydrophobic to superhydrophilic [85].

3.5.10 Templating Method

Templating is an effective method for constructing surfaces with highly controlled morphology. The inverse of a template can be formed and replicated again for producing positive replicas and offering the possibility to template natural biosurfaces.

Nanoporous anodic aluminum oxide (AAO) has been commonly used for the pressure-driven imprint process. By choosing AAO replications with different pore diameters and channel lengths, the diameter and height of surface-projecting nanostructures can be controlled [86–88].

Another widely used template material is polystyrene. Li et al. used a polystyrene colloidal monolayer as a template to produce a hierarchically ordered TiO_2 hemispherical array with hexagonal not-close-packed tops [89]. The obtained coating exhibited excellent superhydrophilicity with a contact angle of 0° without UV radiation. A very novel approach for making superhydrophobic/superhydrophilic patterns is based on printing an "ink" (an ethanol solution of a phospholipid) on a porous superhydrophobic surface [90].

3.6 APPLICATIONS

3.6.1 Self-Cleaning

One of the most attractive applications, which is already being commercialized, is the production of self-cleaning coatings. Self-cleaning coatings are broadly classified into two major categories: superhydrophilic (water contact angles close to 0°) and superhydrophobic (water contact angles >150°). Both clean themselves by the action of water. On a superhydrophilic surface, water spreads, and pollutants can be removed by a stream of water [91]. Water spreads because adhesive forces between the liquid and the substrate play a more important role than internal cohesive forces within the liquid [92]. In addition, these strong adhesive forces between water and surface prevent interactions between impurities and the surface, enabling their easy removal [93]. Water penetrates between impurities and the surface so that the impurities can be washed away [37]. On superhydrophilic surfaces, traces of water will evaporate much faster than on superhydrophobic surfaces, thereby contributing to a cleaner surface. On superhydrophilic coatings consisting of TiO_2 or other semiconductor materials, two self-cleaning mechanisms can occur [94]. One is the photocatalytic effect that is induced by sunlight to chemically breakdown organic pollutants. This effect is a consequence of its semiconductor nature [95]. Photocatalytic properties of TiO_2 are beyond the scope of this chapter and can be found elsewhere [96]. The decomposed impurities can be easily washed away by water that immediately spreads to a film on superhydrophilic surfaces [97]. The other self-cleaning mechanism involved in TiO_2 coatings is the superoleophilic effect. Superamphiphilic surfaces enhance self-cleaning, because oily liquids spread completely, thereby increasing the contact area with the TiO_2 coating and promoting the faster decomposition of contaminants.

The self-cleaning features of superhydrophilic and/or superamphiphilic surfaces are useful for a broad range of possible applications. For example, a soiled kitchen exhaust fan consisting of plastic cannot be cleaned only with water. However, if the exhaust fan is coated with a superhydrophilic and/or superamphiphilic surface, oily contaminants can easily be removed by a stream of water [97]. Traffic signs soiled by exhaust gases from automobiles could be cleaned in the same way or even by rainfall

FIG. 3.4 Comparison of a conventional tent material (left) and a self-cleaning material coated with TiO_2 (right). Used with permission from Ref. 98. © The Japan Society of Applied Physics.

[94]. Self-cleaning is of particular interest for building walls and windows where cleaning is very difficult. A building wall (such as in Fig 3.4) can be cleaned just by rainwater, saving cleaning costs and time [97].

3.6.2 Antifogging and Antireflective Coatings

Antifogging surfaces are needed to maintain visibility through transparent surfaces in high-humidity environments. Since relative humidity is a strong function of temperature, a vapor can easily reach its saturation limit in response to changes in temperature. In addition, condensation often occurs when a cold surface rapidly comes into contact with warm moist air. The resulting condensation appears as tiny droplets. These droplets randomly scatter light, causing the surfaces to be translucent or foggy. A superhydrophilic coating can prevent fogging, because water spreads on the rough hydrophilic surface to form a thin film instead of droplets. In addition to antifogging, superhydrophilic coatings commonly exhibit antireflective properties.

Generally speaking, surface roughness and transparency are competitive properties. Hydrophilicity increases as surface roughness increases, while the transparency of rough surfaces often decreases owing to the Mie scattering effect [99]. Antireflective properties reduce glare and maximize the amount of light passing through, an effect that shows promise for improving materials used in greenhouses and solar-cell panels. As a consequence, it is of great importance to develop a simple method for fabricating transparent and superhydrophilic coatings with specific functional properties. To facilitate commercialization, such coatings also need to be mechanically stable and cost-effective.

Antireflective properties are very important for the production of solar modules, because solar conversion efficiency depends on optical transmission [82]. Son et al. showed that the efficiency of solar panels coated with a superhydrophilic surface is reduced by only 1.4% after 12 weeks in an outdoor environment. In comparison, the efficiency of uncoated solar panels is reduced by 7.8% [3].

Rubner's Research Group at the Massachusetts Institute of Technology (MIT) fabricated a superhydrophilic polyelectrolyte multilayer film with exceptional antifogging and antireflective properties [29]. This film was created from LbL assembled SiO_2 nanoparticles and a polycation. With suitable control over the processing conditions (pH, concentration, etc.) and proper choice of nanoparticle size, multifunctional nanoporous thin-film coatings were created. The resulting film was superhydrophilic, antifogging, and significantly suppressed the reflection of light (antireflective). To improve mechanical durability, the final deposited film was heated to about 500°C for 4 h. After this process, the resulting thin-film coating could withstand aggressive rubbing and easily passed a standard scotch-tape peel test. A simple experiment was carried out to demonstrate the antifogging properties of the obtained superhydrophilic coating. Two glass slides, one coated and the other uncoated, were placed in a refrigerator at −18°C for some time. After removal from the refrigerator, the coated slide remained transparent and not fogged, while water condensate readily formed on the uncoated slide. On a coated surface, water remains on a wet surface as a continuous sheet instead of dewetting to form droplets. This property of coated films provides the opportunity for maintaining good visibility even when the nanopores in the film are fully saturated with water. Since these coatings are nanoporous structures composed of SiO_2 and air-filled pores, their refractive index lies between those of silica ($n \approx 1.45$) and air ($n = 1$); this makes them good candidates for antireflective applications.

As one of the first and most prominent superhydrophilic coatings, TiO_2 (discussed earlier) also possesses impressive antifogging properties. However, the need for UV radiation to produce desirable surface properties remains a major drawback to using this coating material [10]. To eliminate this drawback, porous TiO_2/SiO_2 (TS) composite thin-film coatings with superhydrophilic performance have been produced by a sol–gel process under template-free conditions [47]. When the SiO_2 content was set to 20%, the TS composite coatings exhibited a water contact angle of 2.5° without UV radiation, and the time needed for a water droplet to completely spread over the surface was less than 1 s. The resulting coatings showed excellent antifogging properties, which are attributed to the instantaneous spreading of droplets absorbed on the coated glass surface to form sheets similar to a water membrane. The water therefore evaporates immediately, providing quicker drying of the surface and keeping it clear for a long time (Fig. 3.5).

There are approaches other than TiO_2 technology for obtaining antifogging and antireflective coatings [100, 101]. You et al. produced a thin-film coating of $La(OH)_3$ nanorods on a glass substrate by simple self-stacking methods [102]. These single-layer coatings significantly reduced reflective losses for visible light. To improve the roughness of these coatings, silica nanoparticles were deposited. The resulting

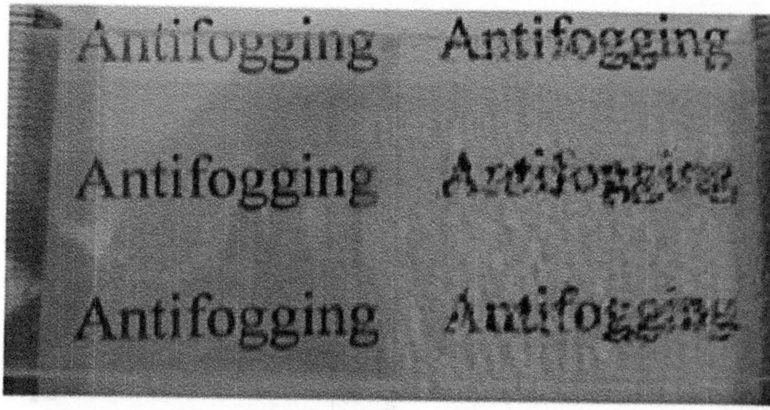

FIG. 3.5 Comparison of antifogging behavior of a bare glass slide (right) with a slide partially coated (left) with superhydrophilic porous TS film. Used with permission from Ref. 47. © Elsevier.

$La(OH)_3/SiO_2$ film showed nanoporosity-driven superhydrophilicity and the antifogging property with no significant loss in the antireflective property.

Recently, a novel antifogging coating has been developed for plastic substrates; the coating consists of a hydrophilic/hydrophobic bilayer structure [103]. The bottom layer is hydrophobic colloidal silica, which acts as a mechanical support and a hydrophobic barrier against water penetration. Atop this layer, an antifogging coating was applied; this top layer incorporates a superhydrophilic species synthesized from Tween-20 (surfactant), isophorone diisocyanate (coupling agent), and 2-hydroxyethyl methacrylate (monomer). The resulting coating was transparent, wearable, and could be soaked in water for 7 days at 25°C without downgrading its antifogging capability.

Besides superhydrophilic and highly hydrophilic surfaces [104–107], the so-called zwitter-wettable surfaces also exhibit excellent antifogging and antifrost properties [108]. Zwitter-wettable coatings have the ability to rapidly absorb molecular water from the environment while simultaneously appearing hydrophobic when probed with water droplets (Fig. 3.6). They are prepared by using hydrogen-bonding-assisted LbL assembly of poly(vinyl alcohol) (PVA) and poly(acrylic acid) (PAA). In an additional step, functionalizing the nanoblended PVA/PAA multilayer with poly(ethylene glycol methyl ether) (PEG) segments produced significantly enhanced antifogging and frost-resistant properties.

Antifogging coatings undoubtedly impact diverse applications such as sports and sanitary equipment, lenses for optical devices, automobile windshields, windows, eyeglasses, camera lenses, or any other transparent glass or plastic surface. For example, when a food item is packaged and displayed in a refrigerated cabinet, the relative humidity inside the package increases because of the decrease in temperature. Consequently, water tends to condense on the inner surfaces of packages, which, if treated to be antifogging, can enhance the visual display of the packaged items. Commercially available antifogging and antireflective coatings will be discussed in the following sections.

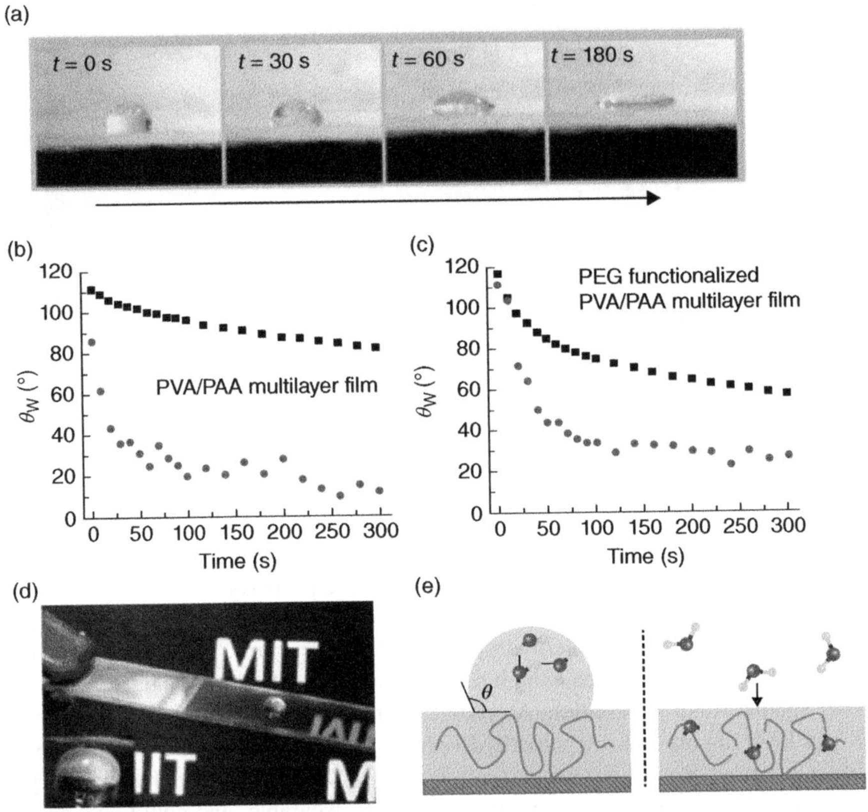

FIG. 3.6 (a) Temporal evolution of water-drop profiles on a PEG-functionalized PVA/PAA multilayer film at 37°C and 80% RH. (b) Changes in water contact angle over time for a PVA/PAA multilayer film at 37°C and 80% RH. (c) Changes in water contact angle over time for a PEG-functionalized PVA/PAA multilayer film at 22 ± 1°C and 40 ± 10% RH and at 37°C and 80% RH. (d) Photograph of a water drop placed on a PEG-functionalized PVA/PAA multilayer film after being transferred from −20°C to 22 ± 1°C, 40 ± 10% RH. Inset photograph shows the magnified image of the water drop with a contact angle above 90°. Only the glass coated with PEG-functionalized PVA/PAA multilayer resists formation of frost. (e) Schematic of zwitter-wettability. MIT logo in the figure is used with permission from the Massachusetts Institute of Technology. Used with permission from Ref. 108. © American Chemical Society.

3.6.3 Antifouling Properties

Fouling is the deposition of an unwanted material on solid surfaces to the detriment of function. In marine engineering, fouling refers to the growth of microorganisms, algae, plants, etc., on a surface immersed in seawater.

In membrane technologies, fouling is the deposition of retained particles, macromolecules, and salts at the membrane surface or inside the pores. This fouling is caused by interactions between the membrane surface and the foulants in many different forms. The foulants not only physically interact with the membrane surface but also chemically degrade the membrane material [109]. It is generally assumed that fouling decreases with an increase in the hydrophilicity of the polymeric material. This assumption seems reasonable, since with an increase in membrane surface hydrophobicity, hydrophobic organic molecules are driven toward the surface, enhancing surface contamination. Water separation membranes should be designed to maximize their surface affinity with water so as to increase their resistance to fouling [110]. Antifouling properties arise because of the strong hydration layer of the hydrophilic surface, which opposes the adsorption of molecules and particles on the membrane surface [111]. Elimelech et al. experimentally confirmed these hypotheses by producing superhydrophilic thin-film composite membranes on which increased resistance to fouling was observed [112, 113].

Biomedical devices can be subjected to fouling via the deposition of surplus cells, proteins, and biomolecules. Patel et al. recently examined two types of superhydrophilic surfaces as potential surfaces in microfluidic devices: [6] polyester films treated by oxygen plasma and indium–tin-oxide-coated glasses treated by an electrochemical method. Fluorescence microscopy studies confirmed the significantly reduced adhesion of fluorescein and fluorescent proteins after the surfaces were treated to be superhydrophilic, thereby indicating their potential for antifouling applications.

Many types of hydrophilic polymer surfaces with suitable wettability and antifouling properties, especially bio-antifouling, have been proposed and prepared by surface-initiated controlled radical polymerization of vinyl monomers with specially designed hydrophilic functional groups; this process leads to densely grafted polymers on the solid surface, which are the so-called polymer brushes [114, 115]. Kobayashi et al. investigated the behavior of different polyelectrolyte polymer brushes [116]. They observed that polyelectrolyte brushes repel both air bubbles and hexadecane in water. Even when silicone oil was spread on the polyelectrolyte brush surfaces in air, once they were immersed in water, the oil quickly rolls up and detaches from the brush surface. Figure 3.7 shows the contact angle of silicone oil on the surface of a poly[2-(methacryloyloxy)ethyl phosphorylcholin] (PMCP) brush and on an unmodified silicon wafer in air and in water. The oil detachment observed on the superhydrophilic polyelectrolyte brush in water was caused by low adhesive forces between the brush and the oil; this could contribute to its excellent antifouling and self-cleaning properties.

3.6.4 Enhanced Boiling Heat Transfer

Over the past 80 years heat transfer under boiling conditions has been investigated by numerous scientists worldwide [117, 118]. One of the most important parameters in the performance of pool boiling is the critical heat flux (CHF). The CHF is the maximum heat flux at which boiling heat transfer sustains its high cooling efficiency. When a surface reaches CHF, it is coated with a vapor film, which then interferes with contact between the surface and the liquid and decreases the heat-transfer efficiency. Subsequently, the system temperature increases, and if it exceeds the limits of its containment materials, system breakdown occurs. Because of this, every

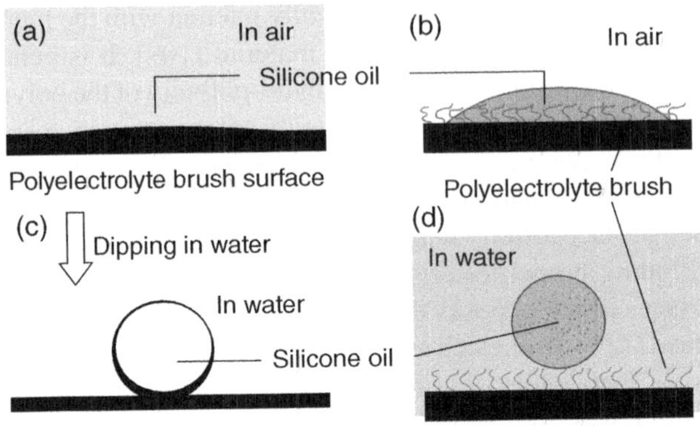

FIG. 3.7 Wettability-reversion phenomena of a silicone oil droplet (5.0 μl, Shin-Etsu Chemical Co. KF-96-100CS) on a PMPC brush (a, b) in air and (c,d) in water. Photograph (c) (side view) displays an oil droplet on a PMPC brush substrate in water, showing superoleophobicity, with a contact angle of 173°. Used with permission from Ref. 116. © American Chemical Society.

system integrates a safety margin by operating at a heat flux much lower than CHF; of course, this safety measure limits the efficiency of the system.

In 1993, Wang and Dhir showed that CHF can be increased by enhancing surface wettability [119]. Since then, a number of studies have reported that surface wettability is an important factor affecting boiling heat transfer [120, 121]. In 1995, Choi introduced the concept of nanofluids [122]. Nanofluids are a new class of nanotechnology-based heat-transfer fluids, engineered by dispersing and stably suspending nanoparticles (with dimensions on the order of 1–50 nm) in traditional heat-transfer fluids. The base fluids include water, ethylene, oil, biofluids, and polymer solutions. Various materials are commonly used as nanoparticles, including chemically stable metals (e.g., copper, gold, silver), metal oxides (e.g., alumina, bismuth oxide, silica, titania, zirconia), several allotropes of carbon (e.g., diamond, CNTs, fullerenes), and functionalized nanoparticles [123]. Vertically aligned nanoforests of hydrophilic/superhydrophilic nanorods [124], nanowires [125], and water-based alumina nanofluids [126] have shown the potential for considerably improving boiling heat transfer. The increase in CHF is attributed to roughness, high surface-tension forces of superhydrophilic nanostructures for pumping in fresh liquid, and capillary wicking phenomena [127].

Phan et al. investigated the influence of surface wettability on nucleate boiling heat transfer by varying the water contact angle [120]. It was found that increasing surface wettability increases the vapor-bubble departure radius and reduces the bubble emission frequency. Hsu et al. coated a plain copper surface with silica nanoparticles and found that the superhydrophilic surface with a contact angle less than 10° has a larger CHF than the hydrophilic one with a contact angle of 16° [121]. The superhydrophilic surface exhibits an increase in CHF of approximately 100%

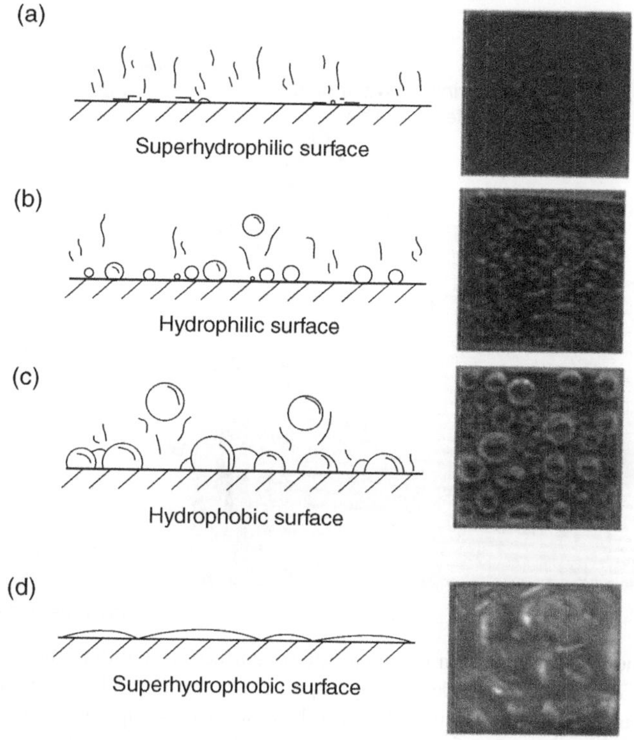

FIG. 3.8 Effects of surface wettability on the growth of bubbles during boiling on (a) superhydrophilic surface, (b) hydrophilic surface, (c) hydrophobic surface, and (d) superhydrophobic surface. In (d), the bubbles coalesce into a thin film that impedes heat transfer from the surface. Used with permission from Ref. 120. © Elsevier.

compared to a plain copper surface. For superhydrophilic surfaces, small bubbles form on the surface when the wall temperature is 100°C (illustration in Fig. 3.8). These small growth bubbles move and merge with other bubbles and then depart the surface. However, the size and number of growth bubbles on the heating surface both increase when the surface is more hydrophobic, as shown in Fig. 3.8b and c. Fig. 3.8d shows the effects of growth bubbles during boiling on a superhydrophobic surface; bubbles spread over the surface and coalesce with bubbles formed at other sites, causing a large area of the surface to be covered with a vapor film [13]. This vapor film interferes with contact between the surface and the surrounding liquid and decreases heat-transfer efficiency.

Since CHF is the upper limit for nucleate boiling, the enhancement of CHF offers the potential for major improvements in the performance of many practical applications. For example, the use of superhydrophilic coatings with higher CHFs could enable the effective thermal management of even smaller and more powerful electronic devices, improve power-up rates in commercial nuclear plants, allow the design of more compact heat exchangers for the chemical industry, among others.

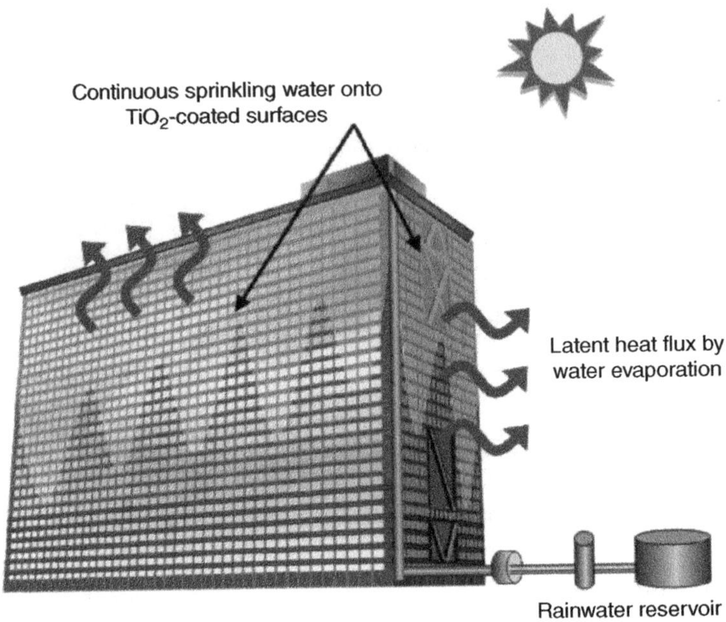

FIG. 3.9 Energy-saving system in which exterior surfaces of buildings are covered with a TiO_2 coating. The coating is made superhydrophilic by exposure to UV radiation in sunlight and then by pumping of stored rainwater over those surfaces. Evaporation of water from the surfaces helps cool the building and reduces load on air conditioning equipment. Used with permission from Ref. 98. © The Japan Society of Applied Physics.

3.6.5 Efficient Water Evaporation

A current problem in big cities around the world is the so-called "heat island phenomenon." Increasing amounts of exhaust gases from traffic and decreasing areas of lakes and green land significantly increase temperatures in cities. Hashimoto et al. suggested the use of superhydrophilic surfaces to prevent the heat island phenomenon. The facades of a building equipped with a superhydrophilic surface can be completely covered by a thin water film. Through quick and efficient water evaporation, the building can be cooled by the flux in latent heat (Fig. 3.9). Therefore, if small amounts of collected rainwater are continuously sprinkled onto the superhydrophilic surface, the temperature increase in cities could be reduced. Thinner water layers (in the range of 0.1 mm) improve the efficiency of cooling buildings and the surrounding air. The decrease in temperature brings another positive impact: less air conditioning is needed, leading to a total decrease in energy consumption of more than 10%.

3.6.6 Switchable and Patterned Wettability Coatings

Combining the two extremes, superhydrophilicity and superhydrophobicity on the same surface in precise 2D patterns opens the possibilities for exciting new functionalities in a wide variety of applications. Generally, superhydrophilic–superhydrophobic patterned surfaces are used to control bioadhesive and nonbioadhesive regions. The "switch" between the two regimes can be triggered by heat, light, or a solvent.

Switching between superhydrophobicity and superhydrophilicity in porous materials was predicted theoretically and demonstrated experimentally using a thermally induced change in contact angle [28]. The porous materials used in that study were produced by a phase-separation method. Reaction occurs through the hydrolysis of the ethoxy groups and the polymerization of the silanol groups thus formed. Polymerization causes a decrease in the dipole moment, leading to hydrophobic phase separation. The dried material had organic groups on its surface, causing the foam to be superhydrophobic, but after thermal treatment to 400°C, the surface was superhydrophilic. [48] This sudden hydrophilic–hydrophobic transition is due to the cross-linking of the silica backbone, which causes the redistribution of organic groups from the surface into the bulk of the material. In porous materials, the transition is very sharp, and because the pores can only be empty or filled, partial states are not energetically favored. Beside porous surfaces, rough surfaces also can exhibit wetting transitions [74].

Han et al. showed that superhydrophilic channels photo-patterned in a superhydrophobic porous polymer layer can separate peptides of different hydrophobicities and isoelectric points by 2D thin-layer chromatography [128]. Recently, another facile and versatile method has been presented for creating superhydrophilic patterns in superhydrophobic porous polymer films by UV-initiated photografting [129]. The extreme difference in wettability between superhydrophilic and superhydrophobic areas permits the use of superhydrophilic patterns as microfluidic channels. The method allows precise control of the size and geometry of photo-grafted superhydrophilic patterns. For mixed polymer brushes that consist of incompatible hydrophobic and hydrophilic components attached to a substrate, the top morphology and composition of the films can be switched by exposure to different solvents, which in turn results in changes in the surface energy and water contact angle [30, 130, 131].

There are many emerging technologies that can benefit from the combination of extreme wettabilities on one substrate. Some of those advantages are patterning of complex geometries with liquids, production of cell microarrays, offset printing, and control of the adhesion of proteins, cells, or bacteria.

3.6.7 Other Applications

Besides the main applications of self-cleaning and antifogging, superhydrophilic surfaces are also very suitable for filtration processes. For example, Sun et al. found that a superhydrophilic polypropylene filter shows remarkable filtration efficiency and may serve as an ultrafilter. Opposite, uncoated polypropylenes filter presents poor filtration properties because of its hydrophobic properties [132]. The superhydrophilic property of substrates, particularly of metals, can also be used for the attachment of biocoats as BMP-2. These coatings exhibit bioactive properties in bone and are therefore suited for implants [133].

The examples of applications mentioned above illustrate the immense range of potential applications of superhydrophilic surfaces. There is continuing development of novel and improved superhydrophilic surfaces for new applications. However, the remaining question is how many of possible applications are already being implemented in real products. This question is addressed in the following section.

3.7 COMMERCIAL COATINGS

Some of the aforementioned technologies are already commercialized products available in the market. This section gives a sampling of companies that offer superhydrophilic products in different industrial fields. On the one hand, there are finished products for end users, and on the other hand, there are chemical solutions that can be easily sprayed onto a surface to obtain superhydrophilic properties.

In particular, coatings containing TiO_2 nanoparticles are widely offered for different uses. One of the first products in the market with a self-cleaning property was a glass cover for lights in tunnels [98]. These covers are mainly used in tunnels in Japan; they remain clean due to a photocatalytic process induced by sodium lamps that emit UV light. In contrast, conventional glass covers are covered by exhaust compounds (Fig. 3.10), leading to dark tunnels.

The German company "GXC Coatings GmbH" offers a group of GXC NuGlass PK coatings that modify different surface substrates, including glass, polycarbonate, polymethylmethacrylate, and various metals. The coating includes TiO_2 nanoparticles and exhibits a self-cleaning effect. It is mainly used in cover panels for exterior lighting systems in the automotive industry, for covers of measuring instruments, and in architectural elements [134].

Self-cleaning windows constitute a major part of the market of superhydrophilic products. The first glass with self-cleaning properties, "Pilkington Active," was introduced by the British company Pilkington in 2001 [135]. The technology is based on TiO_2 particles; the surface exhibits superhydrophilicity and can decompose dirt with the help of sunlight [136, 137]. Pilkington Active™ windows are easily cleaned by

FIG. 3.10 Glass covers on lighting fixtures in a highway tunnel. Without a TiO_2 coating, the covers are darkened by automobile exhaust (left). With a TiO_2 coating, the covers remain clean (right). Used with permission from Ref. 98. © The Japan Society of Applied Physics.

rain and are therefore used in many public and private buildings, including the Town Hall in France, the Britomart Transport Centre in New Zealand, and the Hilton Hotel in Finland. Besides the construction industry, the technology is also used in the automotive industry for the production of side mirrors for cars [98, 138].

In 2001, Saint Gobain, a competitor of Pilkington, presented a product called SGG Aquaclean at the international building exhibition Batimat [139, 140]. In contrast to most other self-cleaning surfaces, the self-cleaning effect is based only on the superhydrophilic property of the surface without any catalytic effect; it can be applied to all building elements that are exposed to rain. Only 1 year later, PPG Industries introduced a self-cleaning glass called Sun Clean [141]. Similar to almost all self-cleaning windows, the Sun Clean coating is based on TiO_2 technology and is mainly used in commercial building applications [142]. There are many other competitors in the market offering self-cleaning windows, including, for example, Neat Glass from Cardinal Glass Industries [143]. This company claims that Neat Glass shows the same self-cleaning effects as Pilkington Active and Sun Clean, but it also transmits more visible light and reflects less. This coating consists of a mixture of titanium dioxide and silicon dioxide [144].

Besides self-cleaning windows, there are other superhydrophilic materials developed for use in the construction industry. In 2002, the Japanese company Toto Ltd. introduced a superhydrophilic and photocatalytic paint for walls called Hydrotect [145, 146]. The system consists of three-layer painting that can be easily applied on outside walls to produce a self-cleaning surface. The technology is based on TiO_2 nanoparticles; besides self-cleaning, it also leads to surfaces that can purify the atmosphere by a catalytic process. Toto Ltd. claims that up to now, more than 1000 public and private buildings are equipped with the Hydrotect technology. Compared to uncoated buildings, the walls stay much cleaner (see Fig. 3.11). For example, the facade of a Toyota factory in Aichi is coated with Hydrotect to make a contribution

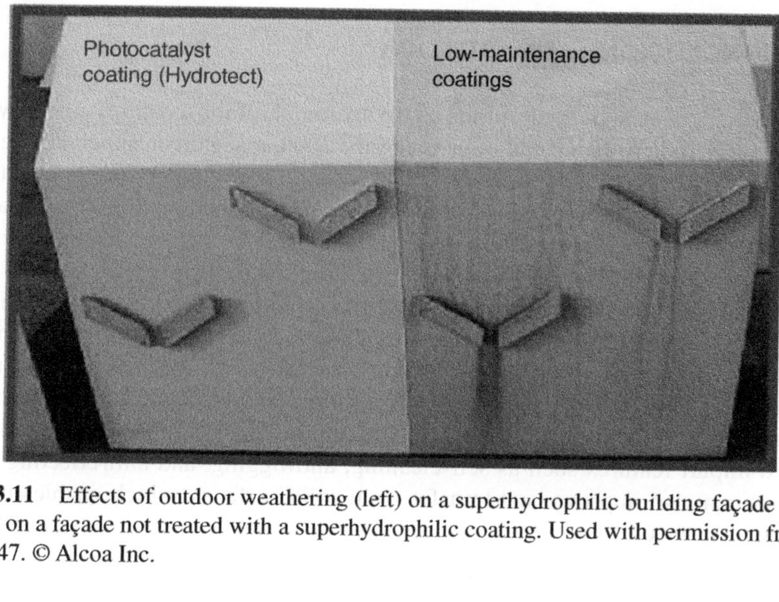

FIG. 3.11 Effects of outdoor weathering (left) on a superhydrophilic building façade and (right) on a façade not treated with a superhydrophilic coating. Used with permission from Ref. 147. © Alcoa Inc.

toward sustainable production [148]. One of the leading manufacturers of ceramic tiles, Casalgrande Padana, developed, in cooperation with Toto Ltd Bios Self Cleaning Ceramics®, a product line of porcelain tiles with self-cleaning properties. The products are based on the Hydrotect technology from Toto and can be used in both interior and exterior architectural applications [149]. The tiles also exhibit antibacterial and smell-reducing properties, making them very suited for indoor applications [150, 151]. Alcoa Architectural Products used the Hydrotect technology to offer a self-cleaning aluminum panel suited for facades; the product is called Reynobond with EcoClean [152]. The panels also clean the surrounding air by a photocatalytic effect [153].

Self-cleaning is not the only important property of superhydrophilic surfaces that has been commercialized. There are already products in the market with antifogging properties. The company Akzente Oberflächen- und Vertriebs GmbH in Germany developed an antifogging coating based on plastics [154]. Acryl groups are a main component used during the synthesis of the plastics. The antifogging coating is mainly used in the production of motorcycle helmets and goggles.

The products mentioned above are mainly suited for specific applications. In addition, some superhydrophilic products in the market can be used in our daily lives. Superhydrophilic coating solutions that can be easily sprayed on various substrates to get self-cleaning glasses or antifogging contact lenses are offered by several companies, including Laiyang Zixilai Environmental Protection Technology Co., Ltd. and Tomorrow Nano Science and Technology Co. in China and iCoat Company in America [155–157]. For example, the product IC No-Fog from iCoat can be applied to various lens materials, both plastic and glass.

This selection of superhydrophilic products shows that such novel technologies will improve living standards and make contributions to solve important everyday problems.

3.8 CONCLUSIONS AND OUTLOOK

The production of superhydrophilic and superamphiphilic coatings is a new and interesting field in research and industry. Since 2000, research on superwetting coatings has expanded, resulting in a significant increase in the number of publications.

Generally, artificial superhydrophilicity and/or superamphiphilicity can be achieved by two ways: a texture-induced method or a photoinduced strategy. Texture-induced superhydrophilicity is attained by creating rough surface structures on materials having high surface energies. Alternatively, photoinduced superhydrophilicity using TiO_2 and ZnO has attracted significant attention as an intriguing phenomenon that can provide antifogging and self-cleaning properties. Regardless of which mechanism is used, it is clear that the incorporation of superhydrophilicity into commercial products can have significant benefits. Superhydrophilic coatings are already being used to impart features such as self-cleaning, antifogging, and antireflecting; these coatings can also improve heat transfer from solid surfaces, which enables the fast cooling of surfaces.

Significant studies have focused on the production of superhydrophilic surfaces with additional features, including wettability switching, patterning, and gradient wetting. These surfaces are expected to have exciting applications in microfluidics, microarrays, offset printing, etc.

However, there are still many challenges in the production of artificial superhydrophilic coatings. For instance, the production of photoinduced coatings having superhydrophobicity faces serious obstacles in terms of cost and materials. When texture-induced superhydrophilicity is used as a production method, another challenge is poor mechanical stability, which is currently addressed through self-curing and self-healing mechanisms. The investigation of these mechanisms should provide further future opportunities. In particular, coatings containing TiO_2 have to be improved, because they are easily wiped off or damaged due to the lack of hardness. Up to now, most photoinduced coatings can only be used for outdoor applications, because superhydrophilicity is induced by UV light. Therefore, there is a need to develop superhydrophilic coatings with high hardness and optical transparency but without any need for external stimuli. Some research groups have already started to develop such coatings, but they are in the minority [158–160]. Although some superhydrophilic coatings are in the market, there is a demand for new technologies that provide superhydrophilic coatings with stable properties and that can be scaled-up for commercial production.

REFERENCES

1. Fujishima, A., Rao, T. N., Tryk, D. A. (2000) TiO_2 photocatalysts and diamond electrodes, *Electrochimica Acta*, 45: 4683–4690.
2. Tahk, D. et al. (2010) Fabrication of antireflection and antifogging polymer sheet by partial photopolymerization and dry etching, *Langmuir*, 26: 2240–2243.
3. Son, J. et al. (2012) A practical superhydrophilic self cleaning and antireflective surface for outdoor photovoltaic applications, *Solar Energy Materials and Solar Cells*, 98:46–51.
4. Anandan, S. et al. (2013) Super-hydrophilic graphene loaded TiO_2 thin-film for self cleaning applications, *ACS Applied Materials & Interfaces*, 5: 207–212.
5. Yang, Y.-F., Li, Y., Li, Q.-L., Wan, L.-S., Xu, Z.-K. (2010) Surface hydrophilization of microporous polypropylene membrane by grafting zwitterionic polymer for anti-biofouling, *Journal of Membrane Science*, 362: 255–264.
6. Patel, P., Choi, C. K., Meng, D. D. (2010) Superhydrophilic surfaces for antifogging and antifouling microfluidic devices, *Journal of the Association for Laboratory Automation*, 15: 114–119.
7. Yang, J. et al. (2012) Superhydrophilic–superoleophobic coatings, *Journal of Materials Chemistry*, 22: 2834–2837.
8. Tadanaga, K., Morinaga, J., Minami, T. (2000) Formation of superhydrophobic-superhydrophilic pattern on flowerlike alumina thin film by the sol-gel method, *Journal of Sol-Gel Science and Technology*, 19: 211–214.
9. Hattori, A. et al. (2000) Ultrathin SiOx film coating effect on the wettability change of TiO_2 surfaces in the presence and absence of UV light illumination, *Journal of Colloid and Interface Science*, 232: 410–413.

10. Wang, R. et al. (1997) Light-induced amphiphilic surfaces, *Nature*, 388: 431–432.
11. Quéré, D. (2005) Non-sticking drops, *Reports on Progress in Physics*, 68: 2495.
12. Lamb, R. N., Furlong, D. N. (1982) Controlled wettability of quartz surfaces, *Journal of the Chemical Society, Faraday Transactions*, 178: 61–73.
13. Asay, D. B., Barnette, A. L., Kim, S. H. (2009) Effects of surface chemistry on structure and thermodynamics of water layers at solid–vapor interfaces, *The Journal of Physical Chemistry C*, 113: 2128–2133.
14. Smith, T. (1980) The hydrophilic nature of a clean gold surface, *Journal of Colloid and Interface Science*, 75: 51–55.
15. Harju, M., Levänen, E., Mäntylä, T. (2006) Wetting behavior of plasma sprayed oxide coatings, *Applied Surface Science*, 252: 8514–8520.
16. Takeda, S., Fukawa, M., Hayashi, Y., Matsumoto, K. (1999) Surface OH group governing adsorption properties of metal oxide films, *Thin Solid Films*, 339: 220–224.
17. Takeda, S., Yamamoto, K., Hayasaka, Y., Matsumoto, K. (1999) Surface OH group governing wettability of commercial glasses, *Journal of Non-Crystalline Solids*, 249: 41–46.
18. Fox, H. W., Zisman, W. A. (1950) The spreading of liquids on low-energy surfaces. I. polytetrafluoroethylene, *Journal of Colloid Science*, 5: 514–531.
19. Carré, A., Gastel, J.-C., Shanahan, M. E. R. (1996) Viscoelastic effects in the spreading of liquids, *Nature*, 379: 432–434.
20. Drelich, J., Chibowski, E. (2010) Superhydrophilic and superwetting surfaces: definition and mechanisms of control, *Langmuir*, 26: 18621–18623.
21. Drelich, J., Chibowski, E., Meng, D. D., Terpilowski, K. (2011) Hydrophilic and superhydrophilic surfaces and materials, *Soft Matter*, 7: 9804–9828.
22. Wenzel, R. N. (1936) Resistance of solid surfaces to wetting by water, *Industrial & Engineering Chemistry*, 28: 988–994.
23. Bico, J., Marzolin, C., Quéré, D. (1999) Pearl drops, *EPL (Europhysics Letters)*, 47: 220.
24. Ishino, C., Okumura, K., Quéré, D. (2004) Wetting transitions on rough surfaces, *EPL (Europhysics Letters)*, 68: 419.
25. Masoodi, R. (2013) *Wicking in porous materials. Traditional and modern modeling approaches* (CRC Press, Boca Raton, FL).
26. Bico, J., Tordeux, C., Quéré, D. (2001) Rough wetting, *EPL (Europhysics Letters)*, 55: 214.
27. Quéré, D. (2008) Wetting and roughness, *Annual Reviews Materials Research*, 38: 71–99.
28. Shirtcliffe, N. J., McHale, G., Newton, M. I., Perry, C. C., Roach, P. (2005) Porous materials show superhydrophobic to superhydrophilic switching, *Chemical Communications*, 25, 3135–3137.
29. Cebeci, F. Ç., Wu, Z., Zhai, L., Cohen, R. E., Rubner, M. F. (2006) Nanoporosity-driven superhydrophilicity: a means to create multifunctional antifogging coatings, *Langmuir*, 22: 2856–2862.
30. Feng, X. J., Jiang, L. (2006) Design and creation of superwetting/antiwetting surfaces, *Advanced Materials*, 18: 3063–3078.

31. Wang, R. et al. (1998) Photogeneration of highly amphiphilic TiO_2 surfaces, *Advanced Materials*, 10: 135–138.

32. Sun, R.-D., Nakajima, A., Fujishima, A., Watanabe, T., Hashimoto, K. (2001) Photoinduced surface wettability conversion of ZnO and TiO_2 thin films, *The Journal of Physical Chemistry B*, 105: 1984–1990.

33. Ferrari, M. (2012) Switching surface wettability properties, *Journal of Adhesion Science and Technology*, 28: 791–814.

34. Barthlott, W., Neinhuis, C. (1997) Purity of the sacred lotus, or escape from contamination in biological surfaces, *Planta*, 202: 1–8.

35. Gennes, P. G. de. (1985) Wetting: statics and dynamics, *Reviews of Modern Physics*, 57: 827–863.

36. Tsujii, K., Yamamoto, T., Onda, T., Shibuichi, S. (1997) Super oil-repellent surfaces, *Angewandte Chemie International Edition in English*, 36: 1011–1012.

37. Zhang, J., Severtson, S. J. (2014) Fabrication and use of artificial superhydrophilic surfaces, *Journal of Adhesion Science and Technology*, 28: 751–768.

38. Koch, K., Barthlott, W. (2009) Superhydrophobic and superhydrophilic plant surfaces: an inspiration for biomimetic materials, *Philosophical Transactions of the Royal Society A: Mathematical, Physical and Engineering Sciences*, 367: 1487–1509.

39. Koch, K., Blecher, I. C., König, G., Kehraus, S., Barthlott, W. (2009) The superhydrophilic and superoleophilic leaf surface of *Ruellia devosiana (Acanthaceae)*: a biological model for spreading of water and oil on surfaces, *Functional Plant Biology*, 36: 339–350.

40. Wang, R. et al. (1998) Photogeneration of highly amphiphilic TiO_2 surfaces, *Advanced Materials*, 10: 135–138.

41. Miyauchi, M., Shimai, A., Tsuru, Y. (2055) Photoinduced hydrophilicity of heteroepitaxially grown ZnO thin films, *The Journal of Physical Chemistry B*, 109: 13307–13311.

42. Montazer, M., Pakdel, E. (2011) Functionality of nano titanium dioxide on textiles with future aspects: focus on wool, *Journal of Photochemistry and Photobiology C: Photochemistry Reviews*, 12: 293–303.

43. Machida, M., Norimoto, K., Watanabe, T., Hashimoto, K., Fujishima, A. (1999) The effect of SiO_2 addition in super-hydrophilic property of TiO_2 photocatalyst, *Journal of Materials Science*, 34: 2569–2574.

44. Kontos, A. G. et al. (2011) Visible light induced wetting of nanostructured N–F co-doped titania films, *Photochemical & Photobiological Sciences*, 10: 350–354.

45. Limage, H. et al. (2011) Study of the effect of a silver nanoparticle seeding layer on the crystallisation temperature, photoinduced hydrophylic and catalytic properties of TiO_2 thin films deposited on glass by magnetron sputtering, *Surface and Coatings Technology*, 205: 3774–3778.

46. Miyauchi, M., Nakajima, A., Watanabe, T., Hashimoto, K. (2002) Photoinduced hydrophilic conversion of TiO_2/WO_3 layered thin films, *Chemistry of Materials*, 14: 4714–4720.

47. Huang, T., Huang, W., Zhou, C., Situ, Y., Huang, H. (2012) Superhydrophilicity of TiO_2/SiO_2 thin films: synergistic effect of SiO_2 and phase-separation-induced porous structure, *Surface and Coatings Technology*, 213: 126–132.

48. Montazer, M., Pakdel, E. (2011) Functionality of nano titanium dioxide on textiles with future aspects: focus on wool, *Journal of Photochemistry and Photobiology C: Photochemistry Reviews*, 12: 293–303.
49. Sobczyk-Guzenda, A. et al. (2013) Morphology, photocleaning and water wetting properties of cotton fabrics, modified with titanium dioxide coatings synthesized with plasma enhanced chemical vapor deposition technique, *Surface and Coatings Technology*, 217: 51–57.
50. Daoud, W. A. et al. (2008) Self-cleaning keratins, *Chemistry of Materials*, 20: 1242–1244.
51. Bozzi, A., Yuranova, T., Kiwi, J. (2005) Self-cleaning of wool-polyamide and polyester textiles by TiO_2-rutile modification under daylight irradiation at ambient temperature, *Journal of Photochemistry and Photobiology A: Chemistry*, 172: 27–34.
52. Montazer, M., Pakdel, E. (2011) Self-cleaning and color reduction in wool fabric by nano titanium dioxide, *The Journal of the Textile Institute*, 102: 343–352.
53. Horiuchi, Y., Ura, H., Kamegawa, T., Mori, K., Yamashita, H. (2010) Design of superhydrophilic surfaces on metallic substrates by the fabrication of Ti-containing mesoporous silica thin film, *Applied Catalysis A: General*, 387: 95–99.
54. Chen, P., Hu, Y., Wei, C. (2012) Preparation of superhydrophilic mesoporous SiO_2 thin films, *Applied Surface Science*, 258: 4334–4338.
55. Bico, J., Thiele, U., Quéré, D. (2002) Wetting of textured surfaces, *Colloids and Surfaces A: Physicochemical and Engineering Aspects*, 206: 41–46.
56. Horiuchi, Y., Yamashita, H. (2011) Design of mesoporous silica thin films containing single-site photocatalysts and their applications to superhydrophilic materials, *Applied Catalysis A: General*, 400: 1–8.
57. Yamashita, H., Nishio, S., Katayama, I., Nishiyama, N., Fujii, H. (2006) Photoinduced super-hydrophilic property and photocatalysis on transparent Ti-containing mesoporous silica thin films, *Catalysis Today*, 111: 254–258.
58. Yamashita, H., Mori, K. Applications of single-site photocatalysts implanted within the silica matrixes of zeolite and mesoporous silica, *Chemistry Letters*, 36: 348–353.
59. Nakata, K. et al. (2011) Rapid erasing of wettability patterns based on TiO_2-PDMS composite films, *Materials Chemistry and Physics*, 126: 484–487.
60. Ma, J., Yang, Z., Qu, Z., Yang, Z. (2006) Supra-amphiphilic transparent mesoporous silica coating, *Chinese Science Bulletin*, 51: 2572–2575.
61. Kako, T., Ye, J. (2007) Photoinduced amphiphilic property of InNbO4 thin film, *Langmuir*, 23: 1924–1927.
62. Iler, R. K. (1966) Multilayers of colloidal particles, *Journal of Colloid and Interface Science*, 21: 569–594.
63. Decher, G., Hong, J.-D. (1991) Buildup of ultrathin multilayer films by a self-assembly process, 1 consecutive adsorption of anionic and cationic bipolar amphiphiles on charged surfaces (Wiley Online Library), *Makromolekulare Chemie, Macromolecular Symposia*, 46: 321–327.
64. Liu, X., He, J. (2007) Hierarchically structured superhydrophilic coatings fabricated by self-assembling raspberry-like silica nanospheres, *Journal of Colloid and Interface Science*, 314: 341–345.

65. Lee, D., Rubner, M. F., Cohen, R. E. (2006) All-nanoparticle thin-film coatings, *Nano Letters*, 6: 2305–2312.

66. Unger, K. K. (1979) *Porous silica, its properties and use as support in column liquid chromatography* (Elsevier Scientific Pub. Co.; Distributors for the U.S.A. and Canada, Elsevier, North-Holland, Amsterdam, New York, NY).

67. Shibuichi, S., Yamamoto, T., Onda, T., Tsujii, K. (1998) Super water-and oil-repellent surfaces resulting from fractal structure, *Journal of Colloid and Interface Science*, 208: 287–294.

68. Zhang, H. et al. (2008) Preparation of Nanostructured Polyaniline and its Super-Amphiphilic Behavior, *Macromolecular Rapid Communications*, 29: 68–73.

69. Huang, Z.-M., Zhang, Y.-Z., Kotaki, M., Ramakrishna, S. (2003) A review on polymer nanofibers by electrospinning and their applications in nanocomposites, *Composites Science and Technology*, 63: 2223–2253.

70. Jiang, L., Zhao, Y., Zhai, J. (2004) A lotus-leaf-like superhydrophobic surface: a porous microsphere/nanofiber composite film prepared by electrohydrodynamics, *Angewandte Chemie*, 116: 4438–4441.

71. Zhu, Y., Zhang, J. C., Zhai, J., Jiang, L. (2006) Preparation of superhydrophilic α-Fe_2O_3 nanofibers with tunable magnetic properties, *Thin Solid Films*, 510: 271–274.

72. Lim, H. S. et al. (2010) Superamphiphilic Janus Fabric, *Langmuir*, 26: 19159–19162.

73. Kim, T.-I., Tahk, D., Lee, H. H. (2009) Wettability-controllable super water-and moderately oil-repellent surface fabricated by wet chemical etching, *Langmuir*, 25: 6576–6579.

74. Kim, B. S., Shin, S., Shin, S. J., Kim, K. M., Cho, H. H. (2011) Control of superhydrophilicity/superhydrophobicity using silicon nanowires via electroless etching method and fluorine carbon coatings, *Langmuir*, 27: 10148–10156.

75. Jennissen, H. P., Lüers, S. (2011) Lotus-effect and inverse lotus-effect in connection with extremely rough titanium surfaces, *Materialwissenschaft und Werkstofftechnik*, 42: a331.

76. Jokinen, V., Suvanto, P., Franssila, S. (2012) Oxygen and nitrogen plasma hydrophilization and hydrophobic recovery of polymers, *Biomicrofluidics*, 6: 16501.

77. Skarmoutsou, A., Charitidis, C. A., Gnanappa, A. K., Tserepi, A., Gogolides, E. (2012) Nanomechanical and nanotribological properties of plasma nanotextured superhydrophilic and superhydrophobic polymeric surfaces, *Nanotechnology*, 23: 505711.

78. Zimmermann, J., Rabe, M., Artus, G. R. J., Seeger, S. (2008) Patterned superfunctional surfaces based on a silicone nanofilament coating, *Soft Matter*, 4: 450–452.

79. Liu, X., Pan, D., Choi, H.-S., Lee, J. K. (2013) Superamphiphilic Ag–CNTs electrode by atmosphere plasma treatment, *Current Applied Physics*, 13: 122–126.

80. Ellinas, K., Tserepi, A., Gogolides, E. (2011) From superamphiphobic to amphiphilic polymeric surfaces with ordered hierarchical roughness fabricated with colloidal lithography and plasma nanotexturing, *Langmuir*, 27: 3960–3969.

81. Li, Q., Kako, T., Ye, J. (2010) PbS/CdS nanocrystal-sensitized titanate network films: enhanced photocatalytic activities and super-amphiphilicity, *Journal of Materials Chemistry*, 20: 10187–10192.

82. Chen, C.-H., Li, S.-Y., Chiang, A. S. T., Wu, A. T. , Sun, Y. S. (2011) Scratch-resistant zeolite anti-reflective coating on glass for solar applications, *Solar Energy Materials and Solar Cells*, 95: 1694–1700.

83. Seo, J., Lee, S., Lee, J., Lee, T. (2011) Guided transport of water droplets on superhydrophobic–hydrophilic patterned Si nanowires, *ACS Applied Materials & Interfaces*, 3: 4722–4729.
84. Zhang, L. et al. (2010) Micro–nano hierarchically structured nylon 6, 6 surfaces with unique wettability, *Journal of Colloid and Interface Science*, 345: 116–119.
85. Song, W., Veiga, D. D., Custódio, C. A., Mano, J. F. (2009) Bioinspired degradable substrates with extreme wettability properties, *Advanced Materials*, 21: 1830–1834.
86. Abdelsalam, M. E., Bartlett, P. N., Kelf, T., Baumberg, J. (2005) Wetting of regularly structured gold surfaces, *Langmuir*, 21: 1753–1757.
87. Guo, C. et al. (2004) Large-area fabrication of a nanostructure-induced hydrophobic surface from a hydrophilic polymer, *ChemPhysChem*, 5: 750–753.
88. Buijnsters, J. G., Zhong, R., Tsyntsaru, N., Celis, J.-P. (2013) Surface wettability of macroporous anodized aluminum oxide, *ACS Applied Materials & Interfaces*, 5: 3224–3233.
89. Li, Y., Sasaki, T., Shimizu, Y., Koshizaki, N. (2008) A hierarchically ordered TiO_2 hemispherical particle array with hexagonal-non-close-packed tops: synthesis and stable superhydrophilicity without UV irradiation, *Small*, 4: 2286–2291.
90. Li, J. S., Ueda, E., Nallapaneni, A., Li, L. X., Levkin, P. A. (2012) Printable superhydrophilic–superhydrophobic micropatterns based on supported lipid layers, *Langmuir*, 28: 8286–8291.
91. Xue, C.-H., Jia, S.-T., Zhang, J., Ma, J.-Z. (2010) Topical review: large-area fabrication of superhydrophobic surfaces for practical applications: an overview, *Science and Technology of Advanced Materials*, 11: 3002.
92. Yuranova, T., Laub, D., Kiwi, J. (2007) Synthesis, activity and characterization of textiles showing self-cleaning activity under daylight irradiation, *Catalysis Today*, 122: 109–117.
93. Gould, P. (2003) Smart, clean surfaces, *Materials Today*, 6: 44–48.
94. Carp, O., Huisman, C. L., Reller, A. (2004) Photoinduced reactivity of titanium dioxide, *Progress in Solid State Chemistry*, 32: 33–177.
95. Reyes-Coronado, D. et al. (2008) Phase-pure TiO_2 nanoparticles: anatase, brookite and rutile, *Nanotechnology*, 19: 145605.
96. Fujishima, A., Zhang, X., Tryk, D. A. (2008) TiO_2 photocatalysis and related surface phenomena, *Surface Science Reports*, 63: 515–582.
97. Fujishima, A., Rao, T. N., Tryk, D. A. (2000) Titanium dioxide photocatalysis, *Journal of Photochemistry and Photobiology C: Photochemistry Reviews*, 1: 1–21.
98. Hashimoto, K., Irie, H., Fujishima, A. (2005) TiO_2 photocatalysis: a historical overview and future prospects, *Japanese Journal Of Applied Physics Part 1 Regular Papers Short Notes And Review Papers*, 44: 8269.
99. Bobbert, P. A., Vlieger, J., Greef, R. (1988) Theory of light reflection from a substrate sparsely seeded with spheres: comparison with an ellipsometric experiment, *Thin Solid Films*, 164: 63–67.
100. Chang, C.-C. et al. (2012) Preparation of water-resistant antifog hard coatings on plastic substrate, *Langmuir*, 28: 17193–17201.
101. Chen, Y. et al. (2012) Transparent superhydrophobic/superhydrophilic coatings for self-cleaning and antifogging, *Applied Physics Letters*, 101: 033701–033704.

102. You, J.-H., Lee, B.-I., Lee, J., Kim, H., Byeon, S.-H. (2011) Superhydrophilic and antireflective $La(OH)_3/SiO_2$-nanorod/nanosphere films, *Journal of Colloid and Interface Science*, 354: 373–379.

103. Chang, C.-C. et al. (2012) Preparation of water-resistant antifog hard coatings on plastic substrate, *Langmuir*, 28: 17193–17201.

104. Tricoli, A., Righettoni, M., Pratsinis, S. E. (2009) Anti-fogging nanofibrous SiO_2 and nanostructured SiO_2–TiO_2 films made by rapid flame deposition and in situ annealing, *Langmuir*, 25: 12578–12584.

105. Nuraje, N., Asmatulu, R., Cohen, R. E., Rubner, M. F. (2011) Durable antifog films from layer-by-layer molecularly blended hydrophilic polysaccharides, *Langmuir*, 27: 782–791.

106. Chevallier, P., Turgeon, S., Sarra-Bournet, C., Turcotte, R., Laroche, G. (2011) Characterization of multilayer anti-fog coatings, *ACS Applied Materials & Interfaces*, 3: 750–758.

107. Introzzi, L. et al. (2012) "Wetting enhancer" pullulan coating for antifog packaging applications, *ACS Applied Materials & Interfaces*, 4: 3692–3700.

108. Lee, H., Alcaraz, M. L., Rubner, M. F., Cohen, R. E. (2013) Zwitter-wettability and antifogging coatings with frost resisting capabilities, *ACS Nano*, 7 (3): 2172–2185.

109. Rana, D., Matsuura, T. (2010) Surface modifications for antifouling membranes, *Chemical Reviews*, 110: 2448.

110. Elimelech, M., Phillip, W. A. (2011) The future of seawater desalination: energy, technology, and the environment, *Science*, 333: 712–717.

111. Morra, M. (2000) On the molecular basis of fouling resistance, *Journal of Biomaterials Science, Polymer Edition*, 11: 547–569.

112. Tiraferri, A., Kang, Y., Giannelis, E. P., Elimelech, M. (2012) Superhydrophilic thin-film composite forward osmosis membranes for organic fouling control: fouling behavior and antifouling mechanisms, *Environmental Science & Technology*, 46: 11135–11144.

113. Tiraferri, A., Kang, Y., Giannelis, E. P., Elimelech, M. (2012) Highly hydrophilic thin-film composite forward osmosis membranes functionalized with surface-tailored nanoparticles, *ACS Applied Materials & Interfaces*, 4: 5044–5053.

114. Chen, Y. et al. (2007) Controlled grafting from poly (vinylidene fluoride) microfiltration membranes via reverse atom transfer radical polymerization and antifouling properties, *Polymer*, 48: 7604–7613.

115. Kobayashi, M., Takahara, A. (2010) Tribological properties of hydrophilic polymer brushes under wet conditions, *The Chemical Record*, 10: 208–216.

116. Kobayashi, M. et al. (2012) Wettability and antifouling behavior on the surfaces of superhydrophilic polymer brushes, *Langmuir*, 28: 7212–7222.

117. van Carey, P. (1992) *Liquid-vapor phase-change phenomena* (Hemisphere, New York, NY).

118. Nukiyama, S. (1966) The maximum and minimum values of the heat Q transmitted from metal to boiling water under atmospheric pressure, *International Journal of Heat and Mass Transfer*, 9: 1419–1433.

119. Wang, C. H., Dhir, V. K. (1993) Effect of surface wettability on active nucleation site density during pool boiling of water on a vertical surface, *Journal of Heat Transfer*, 115: 659–669.

120. Phan, H. T., Caney, N., Marty, P., Colasson, S., Gavillet, J. (2009) Surface wettability control by nanocoating: the effects on pool boiling heat transfer and nucleation mechanism, *International Journal of Heat and Mass Transfer*, 52: 5459–5471.
121. Hsu, C.-C., Chen, P.-H. (2012) Surface wettability effects on critical heat flux of boiling heat transfer using nanoparticle coatings, *International Journal of Heat and Mass Transfer*, 55:3713–3719.
122. Choi, S. U. S., Eastman, J. A. (1995) *Enhancing thermal conductivity of fluids with nanoparticles* (Argonne National Lab., Illinois).
123. Kim, H. (2011) Enhancement of critical heat flux in nucleate boiling of nanofluids: a state-of-art review, *Nanoscale Research Letters*, 6: 1–18.
124. Li, C. et al. (2008) Nanostructured copper interfaces for enhanced boiling, *Small*, 4: 1084–1088.
125. Chen, R. et al. (2009) Nanowires for enhanced boiling heat transfer, *Nano Letters*, 9: 548–553.
126. Jung, J.-Y., Kim, E. S., Kang, Y. T. (2012) Stabilizer effect on CHF and boiling heat transfer coefficient of alumina/water nanofluids, *International Journal of Heat and Mass Transfer*, 55: 1941–1946.
127. Ye, J., Yin, Q., Zhou, Y. (2009) Superhydrophilicity of anodic aluminum oxide films: from "honeycomb" to "bird's nest," *Thin Solid Films*, 517: 6012–6015.
128. Han, Y. et al. (2010) Monolithic superhydrophobic polymer layer with photopatterned virtual channel for the separation of peptides using two-dimensional thin layer chromatography-desorption electrospray ionization mass spectrometry, *Analytical Chemistry*, 82: 2520–2528.
129. Zahner, D., Abagat, J., Svec, F., Fréchet, J. M. J., Levkin, P. A. (2011) A facile approach to superhydrophilic–superhydrophobic patterns in porous polymer films, *Advanced Materials*, 23: 3030–3034.
130. Zhao, B., Brittain, W. J. (2000) Synthesis, characterization, and properties of tethered polystyrene-b-polyacrylate brushes on flat silicate substrates, *Macromolecules*, 33: 8813–8820.
131. Julthongpiput, D., Lin, Y.-H., Teng, J., Zubarev, E. R., Tsukruk, V. V. (2003) Y-shaped polymer brushes: nanoscale switchable surfaces, *Langmuir*, 19: 7832–7836.
132. Sun, T., Feng, L., Gao, X., Jiang, L. (2005) Bioinspired surfaces with special wettability, *Accounts of Chemical Research*, 38: 644–652.
133. Jennissen, H. P. (2005) Ultra-hydrophilic transition metals as histophilic biomaterials, *Macromolecular Symposia*, 225: 43–70.
134. GXC Coatings GmbH. Clean with Nature's Forces—by Photocatalysis. Available at http://www.gxc-coatings.com/products/self-cleaning.html. (accessed December 6, 2014.)
135. Nippon Sheet Glass Co. L. Selbstreinigendes Glas. Available at http://www.glas-hanelt.de/service.html?file=files/Glas_Hanelt/PDF/PilkingtonActive _b2bbrosch.pdf. (accessed February 11, 2015.)
136. GKT Unternehmensgruppe. Pilkington Activ. Available at http://www.gktgruppe.de/produkte/isolierglas/pilkington-activ/index.html. (accessed December 6, 2014.)
137. Nippon Sheet Glass Co. L. Pilkington Activ™. Available at http://www.gktgruppe.de/produkte/isolierglas/pilkington-activ/index.html. (accessed February 11, 2015.).
138. Pilkington UK Ltd. Pilkington Activ™ Global Brochure. Available at http://www.pilkington.com/products/bp/bybenefit/selfcleaning/literature.htm. (accessed December 6, 2014.)

139. KREA Ltd. Saint-Gobain Glass Introduces SGG Aquaclean, A New Self-Cleaning Glass. Available at http://www.glassonweb.com/news/index/127/. (accessed December 6, 2014.)

140. The United States Patent and Trademark Office. Trademark Search. Available at http://tmsearch.uspto.gov/bin/gate.exe?f=searchss&state=4807:rgnnt6.1.1. (accessed February 11, 2015.)

141. PPG Industries. PPG Introduces SUNCLEAN Self-Cleaning Glass for Commercial Applications. Available at http://www.ppg.com/en/newsroom/news/Pages/20121022B.aspx. (accessed December 6, 2014.)

142. PPG Industries. How It Works: Self-Cleaning Window Glass Technology. Available at http://www.ppg.com/en/newsroom/news/pages/20121022b.aspx. (accessed February 11, 2015.)

143. Cardinal Glass Industries. Let Nature Service Your Windows. Available at http://www.cardinalcorp.com/products/neat-glass/. (accessed December 6, 2014.)

144. Cardinal Glass Industries. Architectural Glass Guide. Available at http://www.cardinalcorp.com/wp-content/uploads/pdf/cardinal-architectural-brochure.pdf. (accessed December 6, 2014.)

145. TOTO. HYDROTECT. Available at http://de.toto.com/technologie/technologie-einzelansicht/Technology/show/HYDROTECT/. (accessed December 6, 2014.)

146. Japan Echo Inc. Self-Cleaning Walls and Windows. Available at http://web-japan.org/trends/08_sci-tech/sci090116.html. (accessed December 6, 2014.)

147. Alcoa Architectural Products. The EcoClean The EcoClean™ Self Cleaning Process. Available at http://www.alcoa.com/bcs/aap_eastman/ecoclean/en/pdfs/Web_support_slides_self_cleaning.pdf. (accessed December 6, 2014.)

148. ARCHmatic. Superhydrophilie verspricht selbst- und luftreinigende Fassaden. Available at http://www.baulinks.de/webplugin/2010/1944.php4. (accessed December 6, 2014.)

149. Casalgrande Padana S.p.A. A New Step Forward in the Area of Bioactive Ceramics. Available at http://www.casalgrandepadana.com/index.cfm/1,869,2432,0,html/A-NEW-STEP-FORWARD-IN-THE-AREA-OF-BIOACTIVE-CERAMICS#.UWGe-1ccfmg. (accessed December 6, 2014.)

150. TOTO. Casalgrande Padana Launch Bios Self Cleaning Ceramics® with HYDROTECT®. Available at http://gb.toto.com/news/press-releases/press-releases/?tx_ttnews[tt_news]=284&c. (accessed February 11, 2015.)

151. Mauro Manfredini. Percorsi in Ceramica. Available at http://www.casalgrandepadana.com/index.cfm/1,869,2432,0,html/A-NEW-STEP-FORWARD-IN-THE-AREA-OF-BIOACTIVE-CERAMICS. (accessed February 11, 2015.)

152. Alcoa Architectural Products. Smog-Eating Buildings? Available at http://www.alcoa.com/aap/north_america/pdf/ecoclean/EcoClean_Newsletter1.pdf. (accessed December 6, 2014.)

153. ARCHmatic. Aluminiumfassaden als Schmutz- und Smogvernichter. Available at http://www.baulinks.de/webplugin/2011/1287.php4. (accessed December 6, 2014.)

154. Akzente Oberflächen- und Vertriebs GmbH. Antifog Beschichtung. Available at http://www.akzente-hc.de/antifog-beschichtung.html. (accessed December 6, 2014.)

155. Focus Technology Co. L. Super Hydrophilic Nano Self Cleaning Coating for Car Glass Building Walls. Available at http://www.made-in-china.com/showroom/laiyangzixilai/

product-detailkMSxVTRGHKpN/China-Super-Hydrophilic-Nano-Self-Cleaning-Coating-for-Car-Glass-Building-Walls.html. (accessed December 6, 2014.)

156. weiku. Nano Self-Cleaning and AntiBacteria Coating. Available at http://en.zixilai.com/products_list/&pmcId=26.html. (accessed February 11, 2015.)

157. iCoat Company. IC No-Fog. Available at http://www.icoatcompany.com/ICNoFog.aspx. (accessed December 6, 2014.)

158. Forbes, P. (2008) Self-cleaning materials: lotus leaf-inspired nanotechnology, *Scientific American Magazine*, 8: 1–5.

159. Yamasaki, S. et al. (2010) Development of the organic-inorganic hybrid super-hydrophilic layer, *Fuji film Research & Development*, 55: 29–33.

160. Zorba, V., Chen, X., Mao, S. S. (2010) Superhydrophilic TiO_2 surface without photocatalytic activation, *Applied Physics Letters*, 96: 093702–093703.

CHAPTER 4

Self-Healing Polymeric Coatings

A.C.C. Esteves[1] and S.J. García[2]

[1] Materials and Interface Chemistry Group, Chemical Engineering and Chemistry Department, Eindhoven University of Technology, Eindhoven, The Netherlands

[2] Novel Aerospace Materials Group, Faculty of Aerospace Engineering, Delft University of Technology, Delft, The Netherlands

4.1 INTRODUCTION

Coatings have become one of the most ubiquitous end applications for polymeric materials. A typical industrial coating formulation is composed of binders (polymers), fillers, solvents, pigments, and additives. The term "polymeric coatings" is used here to refer to coatings in which "organic polymers" act as binders and are the main components of the formulation.

Over the years, materials have been evolving to meet the demands of our high-tech society. Coatings are no longer protective and decorative only, but are also required to deliver additional functions. Advanced functional polymeric coatings with self-cleaning, low-adherence, antifouling (AF), antimicrobial, or anticorrosive properties have been extensively investigated over the past decades [1, 2]. One of the greatest advantages of using functional coatings is that with a rather small amount of material, thin layer(s) of a few tens to hundreds of microns, it is possible to functionalize the whole substrate with little or no impact on other properties, such as mechanical strength.

Functional coatings are generally active at the different interfaces involved, that is, polymer–air, polymer–substrate (metal, wood, plastic, or glass), polymer–liquid (water or oil) and eventually polymer–polymer, in multilayer systems [3]. Hence, coatings consist of periphery structures that are permanently in contact with the environment and users, being constantly subjected to the aggression by different factors, for example, wear, rain, dust, oils, chemicals, or bacteria, which gradually decrease the materials performance leading to shorter life times. In this context,

Functional Polymer Coatings: Principles, Methods, and Applications, First Edition.
Edited by Limin Wu and Jamil Baghdachi.

functional coatings have very much in common with the boundary structures of living organisms such as our skin, plants cuticles, mussel's shells, or cellular membranes. However, synthetic polymeric coatings of nowadays are still far from biological surfaces, which are rough and flexible, chemically heterogeneous, selectively permeable and stimuli-reactive. Moreover, these biological surfaces have an extremely important characteristic on which the survival of the species strongly depends: the ability to self-recover functionalities and maintain the boundary characteristics, that is, self-healing.

Damage on coatings will always remain unavoidable. Introducing self-repairing mechanisms into coatings is one way to ensure a high level of performance and an extended service life-time, with reduced energy and cost efforts with maintenance and repair [4]. This is particularly important for coatings applications on high-added value products, for example, solar cells, electronic devices, or other applications where regular maintenance and repair is expensive or not easy to perform, such as aircrafts, marine vessels, sky scrapers, or green-houses. For these reasons, polymeric coatings that can respond to damages inflicted by the environment or users have been intensively investigated. Although a significant literature foundation has been set through scientific publications [5–8] and a few book chapters [2, 9, 10], self-healing polymeric coatings remains a challenging topic.

In this chapter, we present a revision of the self-healing approaches and mechanisms that we believe have demonstrated potential to find applications in the broad and challenging smart coatings field.

4.1.1 Self-Healing Materials

From its conception, advanced materials are initially designed or formulated to work at a desired performance level during a certain expected service-time through which, ideally, this performance should be kept at a constant level. However, the continuous or arbitrary but severe use of the material inevitably leads to wearing, aging, or other accidental damages, which lower the performance of the material. These damages may occur abruptly or over time, reaching unacceptable performance levels determining the ultimate failure of the material, structure, or device crossing the limit of reliability (Fig. 4.1a) [11]. A traditional way of extending the life time of the materials is the improvement of its resistance to damages (Fig. 4.1b). In this case, the material may start-off with an improved performance so that it can be used for a longer time but, if the material is not maintained or repaired, the performance will irreversibly decrease and soon reach the critical limit of reliability (Fig. 4.1b). Self-healing materials, however, are based on the recognition and management of the damage. This concept implies that the material will be able to react, and partially or totally recover the lost performance, before it reaches a critical level of performance which could rapidly lead to failure (Fig. 4.1c). Ideally, the self-healing material should be able to self-repair back to its original properties multiple times, before a critical value of performance is reached (Fig. 4.1d). When analyzing Fig. 4.1, it thus becomes obvious that the implementation of the self-healing concept aims at extending the service lifetime of materials, structures, and

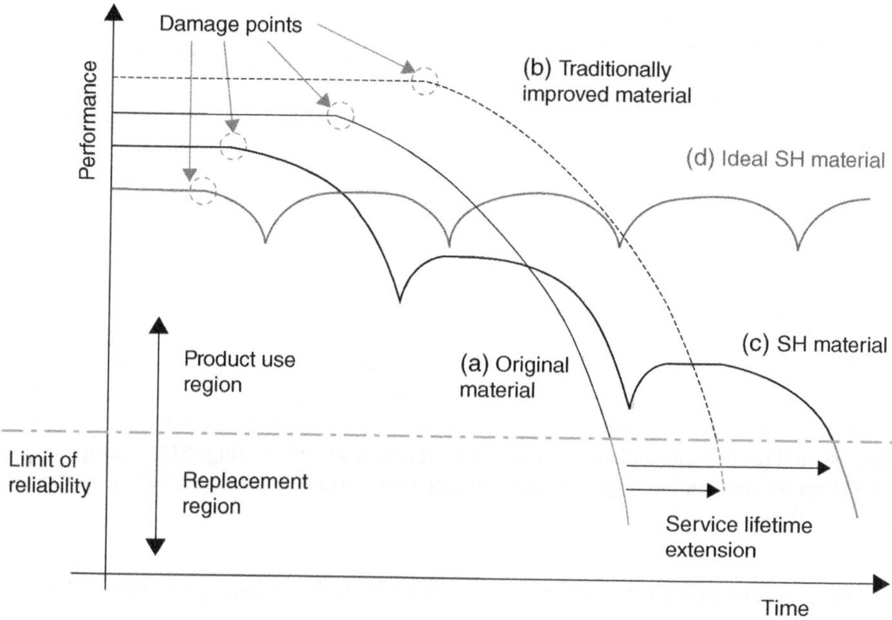

FIG. 4.1 Performance of a material, structure, or device along its service life time: (a) original material; (b) traditional improved material; (c) self-healing material, and (d) "ideal" self-healing material. Reproduced with permission from Ref. 11. © Elsevier.

devices, not by increasing their initial performance but by implementing the concept of *autonomous* or *induced* self-repair.

Since the landmark publication by White et al. [12] on self-healing polymer composites, the topic has been ripening across different materials areas [4, 7]. Researchers have now realized that to implement a successful healing process on synthetic materials and structures, including polymeric coatings, a number of requirements need to be fulfilled: (i) *detection and activation mechanisms*—incorporation of mechanisms to recognize a local damage and trigger the healing process; (ii) *healing agents*—add extrinsic or intrinsically designed components that performs the healing process; (iii) *transport of the healing agents to the damage loci*—implement a driving force and mobility within the material or of a phase or component of it, and (iv) *local reconstruction or repair*—incorporate a healing mechanism that may be autonomous or require an additional energy trigger, for example, UV, local temperature, high kinetic energy, or resting time.

Furthermore, to design healing strategies for polymeric coatings and implement the aforementioned requirements, the scale and extent of the damage—defined here as an immediate or imminent loss of a property or function of the material—should be taken in consideration. Three damage *scales* can be considered, as initially proposed by van Benthem et al. [13] and later on adapted by Garcia et al. [8]: (i) *macro*, related to the scratches or delamination, (ii) *micro*, for example, formation

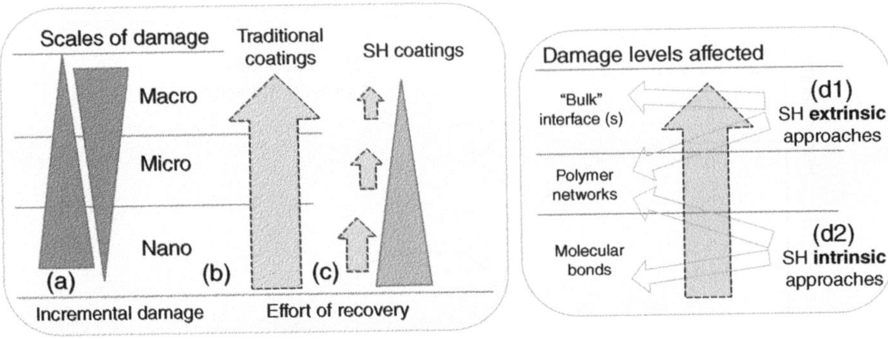

FIG. 4.2 Damage scales classification and incremental damage *versus* efforts involved in damage recovery of traditional and self-healing (SH) coatings: (a) damage increment across scales and from different directions, (b) efforts required for recovery accordingly to scales of damage, and (c) damage and effort of recovery upon use of self-healing (SH) coatings. (d) SH approaches for coatings and levels of action according to damage scales: (1) extrinsic and (2) intrinsic.

of micro-cracks or crazing, and (iii) *nano*, related to the breaking of chemical bonds (Fig. 4.2).

For the material recovery, the damage scale hierarchy has to be considered. A small damage due to breaking of chemical bonds (*nanoscale*) will inevitably affect the chemical structure or network properties. This initial damage will—in time (e.g., by fatigue damage such as aging) or at a sudden event (impact or accidental scratch)—increment the damage (across the *microscale*) and lead to the partial or total loss of the original properties/functionalities of the material (finally reaching the *macroscale*) (Fig. 4.2a). Accordingly, the recovery efforts demanded are proportional to the scale and extent of the damage (Fig. 4.2b).

Note also that in coatings, damage can start either from the *nano-up* or *macro-down* side and go in any direction (Fig. 4.2a). From a *macro-down* direction, the exposure to UV-solar radiation or a superficial scratch may destroy the surface chemical groups and rupture network segments, leading to changes on the wettability properties, for example, loss of the hydrophobic character. If it penetrates further throughout the coating bulk, it will allow direct contact with the environment and lead to extensive damage at all scales. Other damages can occur from *nano-up* direction. Cracks can originate from internal or interfacial stresses build up during the chemical network formation or from stresses developed upon severe usage (e.g., repeated mechanical strain).

All these levels of complexity make damage management, aimed for self-healing coatings, a rather challenging matter and calls for different approaches. The decision to tackle a specific damage scale will most likely determine the choice of the best self-healing approach. If the healing action should tackle the *macroscale*, an *extrinsic* approach may be more suitable (Fig. 4.2d1). In this case, a localized, large proportion (volume) damage that already affects the bulk and interface levels can be more efficiently repaired or prevented from further propagation. On the other hand, to mitigate damages started at the *nano-* or *microscales*, randomly distributed along a smaller volume of the material, an *intrinsic* healing approach is probably preferred. Broken molecular bonds and ruptured networks can be steadily repaired avoiding increment of damage (Fig. 4.2d2). Noteworthy, once again the different scales are intertwined and a straightforward choice of strategy is not always trouble-free.

4.1.2 Self-Healing Polymeric Coatings

Coatings are often applied over a large surface area and are therefore the most likely setting for the first damages to occur. Furthermore, for polymeric coatings in particular, the nature of the materials used does not allow to build-up a large margin of protection against routines damages. Typically, organic polymers exhibit by nature a limited mechanical, thermal, and UV resistance, and when used in coatings, they undergo through natural weathering and aging processes under the influence of moisture, aggressive chemicals, UV radiation, and oxygen from air, via chemical processes such as photolysis, photooxidation, hydrolysis, and thermooxidative degradation. The ideal of a self-healing polymer coating which will continuously and infinitely auto-repair is therefore, not feasible on the basis of organic polymers only (excluding the use of additives), but the mitigation of the damages to slow down these damaging processes can still be reached.

Several self-healing approaches have been reported for polymers and/or polymeric coatings, which can be classified in different ways, according to the healing mechanism: autonomous *versus* non-autonomous (triggered) [14], reversible *versus* irreversible [15], and extrinsic *versus* intrinsic healing [8]. Noteworthy, the borders between the different classifications may overlap in a number of new systems. The third classification seems to be collecting much consensus in the coatings field and will also be followed in this chapter.

An extrinsic approach involves the addition of external components to the system or formulation, which are not a part of the main polymer network (e.g., filled capsules [16–19] or microvascular networks [20, 21]). An intrinsic approach uses healing agents inherent to the material which are typically a part of the network [22–25]. Both approaches have advantages and disadvantages. The external healing concept allows a localized response and gives certain flexibility in choosing different external components, but it is rather limited for multiple healing events. The intrinsic approach may allow multiple healing and a restoration of the materials to a level which is very close to the original state (e.g., almost full restoration of broken bonds in polymer networks). On the other hand, it typically requires a certain mobility of the system—or at least of a large part of it—and in the majority of the cases, an external energy trigger (e.g., light and temperature) as well.

Some requirements desired for both approaches are a good miscibility (or distribution) and compatibility of the *healing agents* with(in) the other components, and a minimal effect on the chemical/physical properties of the coatings, upon incorporation in the formulation or during film formation.

4.2 SELF-HEALING APPROACHES FOR FUNCTIONAL POLYMERIC COATINGS

4.2.1 Intrinsic Healing

In self-healing intrinsic approaches, damage is repaired via a temporary and local mobility increase of a phase or component of the system, which results in reflow of the material toward the damaged area. Such behavior is highly dependent on the use of reversible physical or chemical bonds as well as chain mobility. The combined effect of the polymer architecture on mechanical and healing behavior is expected to gain a major interest in the coming years due to the need of direct industrial applications of self-healing polymers [11].

In this section, we revise several *intrinsic* healing approaches for polymeric coatings. An overview—comprehensive, but not necessarily absolute—of intrinsic healing systems and mechanisms potentially applicable to polymeric coatings is provided in Table 4.1, for easy guidance and quick access to the latest results and further references. In the following subsections, we discuss in detail some of the systems/mechanisms, in light of main molecular interactions that can be used to introduce intrinsic healing mechanisms on polymeric coatings: (i) molecular chains inter-diffusion and segregation, (ii) reversible bonds, and (iii) noncovalent bonds. For other examples, we refer the reader to the additional references provided in Table 4.1.

4.2.1.1 *Molecular Inter-Diffusion and Chain Segregation* One of the self-healing intrinsic mechanisms with high potential for polymeric coatings relies on molecular chains diffusion and/or chains segregation, to create or reestablish physical linkages. This mechanism is based on the molecular mobility of polymer chains and their ability to heal the material by establishing chain entanglements across polymer–polymer interfaces formed when the two broken parts come together. Hence, a concentration gradient or surface tension differences leading to chain diffusion are seen here as the *driving forces* for the healing process. These mechanisms have a strong potential for self-repairing of polymeric coatings damaged at different scales (from *nano* to *macro*, Fig. 4.2a), being particularly interesting for functional coatings in which the functionality relies on presence of specific chemical groups at surfaces or interfaces, as will be discussed later in Section 4.3.

The diffusion of polymer chains in concentrated solutions and melts has been described already in the 1970s, namely by the well-known reptation model of chain dynamics described by de Gennes [86]. Wool and coworkers [26, 87, 88] applied such reptation model to explain thermal-induced crack healing at polymer–polymer interfaces. According to this model, the healing of a crack is achieved once the polymer chains near the interface completely disentangle from their "tubes" and continuously rearrange until a new equilibrium is reached, hereby eliminating the polymer–polymer interfaces. The overall healing process evolves in five stages: (i) surface rearrangement, (ii) surface approach, (iii) wetting of surfaces, (iv) low level of diffusion across interfaces, and (v) high level of diffusion, equilibration, and randomization (Fig. 4.3).

TABLE 4.1 Selection of Intrinsic Self-Healing Polymer Systems Which Have Been (or Have the Potential to be) Implemented in Polymeric Coatings

	Healing Interaction	Polymeric System	Healing Mechanism/ Chemistry	Healing Conditions (Temp., Time, Medium, etc)	Reference(s)
Molecular inter-diffusion and chains segregation (Section 4.2.1.1)	Dangling chains diffusion (*bulk healing*)	Poly(styrene) (PS)	—	70–120°C ($T_g \approx 100$°C)	[26]
		Poly(methyl methacrylate) (PMMA)	Locally effective lowering of T_g	$T > T_g$	[27–31]
		Poly(carbonate)	Locally effective lowering of T_g	$T > T_g$	[32]
		Poly(styrene)-*b*-poly(methyl methacrylate)	—	30–125°C ($T_g \approx 110$°C)	[33]
	Self-replenishing (*surface healing*)	Poly(urethane polyesters)/fluorinated dangling chains	Surface segregation	RT, a few min to hr	[34–36]
		Nanocomposite-porous films/fluorinated compounds/poly(ethylene oxide chains) (PEO) chains	Surface segregation	RT, a few min to hr	[37–40]
Reversible bonds (Section 4.2.1.2)	Thermo-reversible bonds	Furan-maleimide based thermosets	Diels–Alder (DA)	145°C, 25 min	[41, 42]
		Macrocyclic derivative of DCPD	DA	DA at RT, *r*DA at 90–120°C	[43]
		Furan-maleimide functionalized epoxy, polyketones, PS, PE, PMMA, polyamides and polyesters	DA	DA at RT, rDA at 90–120°C	[44–57]
	Sulfide bonds	Disulfide (S–S), tetrasulfide, (S–S–S–S), thiol (S–H) exchange	Sulfur bonds	Various Temp and UV	[24, 58, 59]
		Trithiocarbonate containing polymer—poly(*n*-butyl acrylate)	Sulfur bonds	UV	[60, 61]
	Other interactions	Chitosan in cross-linked PU network. oxetane groups	Urea bonds cleavage	UV	[37, 38, 40]
		Alkoxyamine functionalized poly(methacrylic esters)	Radical recombination	RT, wetting with DMF, 24 h	[62, 63]
		Cinnamoyl functionalized polymers	Cycloaddition	UV	[64, 65]

(Continued)

TABLE 4.1 *(Continued)*

	Healing Interaction	Polymeric System	Healing Mechanism/ Chemistry	Healing Conditions (Temp., Time, Medium, etc)	Reference(s)
Noncovalent bonds (Section 4.2.1.3)	Supramolecular (H-bonds)	Fatty acid derivatives with diethylenetriamine reacted with urea; poly(urethanes); epoxy-amine	H-bond networks Quadrupole H-bonds	Pressure	[23, 66–70]
		Poly(urethane) (PU)	H-bond networks	Electro-magnetic radiation	[71, 72]
	Ionomers	Thin films/Poly(ethylene-*co*-methacrylic acid)	Ionic groups re-arrangements	Various	[73–77]
	π-π Donor-Acceptor stacking	Poly(diimide/pyrenyl) end capped polyamide or poly(siloxane)	π-π stacking networks	50–80°C	[78, 79] [80]
	Other interactions	Layers of polyelectrolytes (layers of anionic microgels and poly cationic amines) on PDMS	H-bond ion-networks	Water or humid environment	[81]
		Linear polymers linked with pyridine groups cross-linked with metal complexes	Metal-ligand cleavage	Immersion in complexes solution, heating	[82–85]

DA, Diels-Alder; DCPD, dicyclopentadiene; h, hours; min, minutes; rDA, Reverse Diels-Alder; RT, room temperature/ambient conditions.

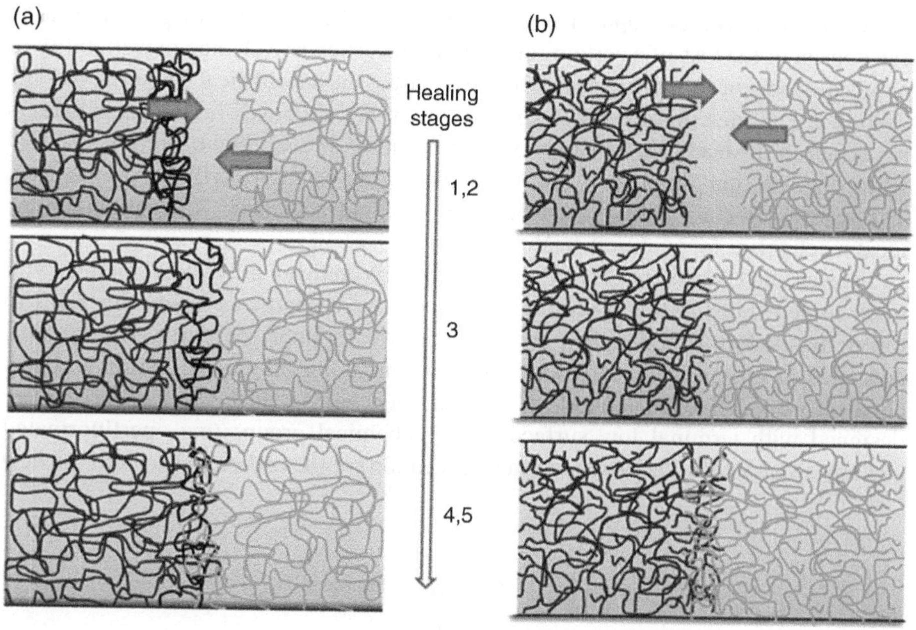

FIG. 4.3 Schemes of the (a) molecular inter-diffusion and (b) chain-end segregation models showing the stages of healing occurring at polymer–polymer interfaces: (1) surface rearrangement, (2) surface approach, (3) wetting of surfaces, (4) low level of diffusion across interfaces, and (5) high level of diffusion, equilibration, and randomization. Adapted from Ref. 7.

Thermal-induced healing of cracks has been largely studied since a couple of decades, by bringing two pieces of the same polymer into contact at a temperature above the glass transition temperature (T_g), thus allowing higher mobility of the molecular chains and leading to the gradual disappearance of the interface, with an increase in the mechanical strength due to chain entanglements (Fig. 4.3a). One example is the thermal-induced healing of cracks in poly(methyl methacrylate) (PMMA) and PMMA-poly(methoxy ethylacrylate) (PMEA) copolymer films which has been demonstrated to occur at temperature of 5°C above T_g, with a healing time slightly larger than 1 min. Solvent-induced healing has also been investigated, and in contrast to thermal-induced, it can be undertaken at temperatures below T_g. Typically, the solvent(s) are introduced into the polymers to assist the wetting and diffusion during the healing process and removed thereafter. Methanol and/or ethanol have been used to assist the healing of cracks in PMMA films [27–31], while carbon tetrachloride was used for polycarbonate [32]. In many cases, however, the use of solvents for self-healing purposes causes excessive plasticization and swelling of the polymeric films, limiting the final healing efficiency and leading to incomplete recovery of the mechanical strength. Note that in this case, the use of solvents can also be seen as an extrinsic healing approach.

As mentioned earlier, apart from the formation of cracks, a damage of a coating can consist of the loss of a certain function in time, such as a change in wettability or reduced adhesion to the substrate. In these cases, the mechanical properties of the coating are not deteriorated; hence, rebonding of broken bonds (at the *nano*scale) is not required. Instead, the transport or the relocation of a *healing agent* into the loci where the function diminishes, in a self-replenishing way [34], will just do the trick. The rearrangement of chain-ends occurring at interfaces (Fig. 4.3b) provides the ideal mechanism for such self-healing purposes.

Esteves and coworkers [34, 35] developed poly(urethane polyester) coatings with a self-replenishing ability, by exploring differences of energy as the *driving force* to repair the surface functionality, for example, hydrophobic property. The systems consisted of cross-linked polymer networks containing polymeric dangling chains covalently bonded to the network. These dangling chains—the *healing agents*—were designed with terminal low surface energy chemical groups (e.g., perfluorinated chains) which guide the reorientation (segregation) of the end-chains toward the new

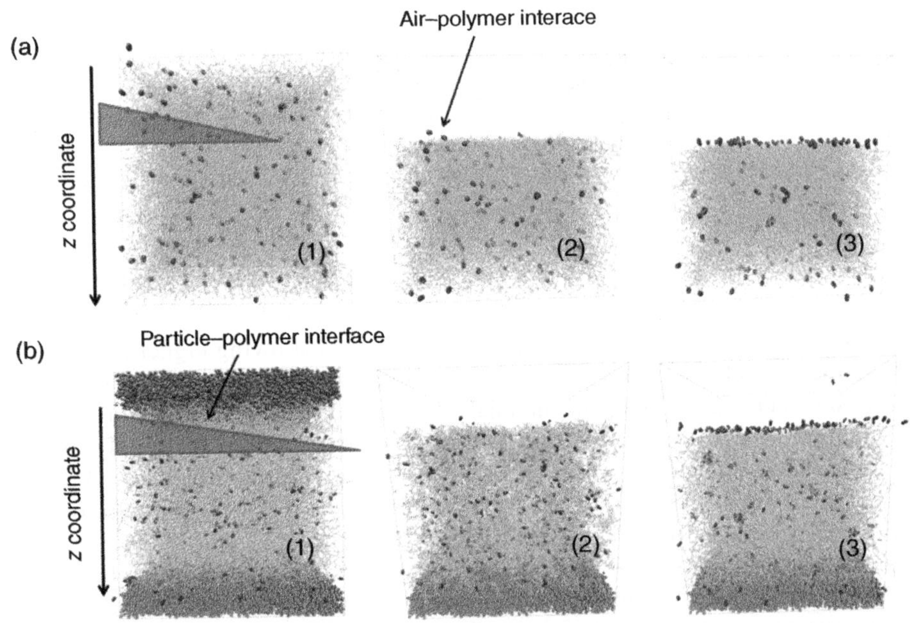

FIG. 4.4 Simulation of the self-replenishing of cross-linked polymeric coatings containing: (a) polymers only and (b) polymers and (inorganic) particles (surfaces only, represented). Snapshots of the polymer layer: (1) before submitted to damage; (2) after damage, with a new air-polymer interface, and (3) after self-replenishing (equilibration), that is, recovery of surface properties. Low surface energy components (e.g., fluorinated dangling chains) are represented by the darker and larger beads which were increased in size (4×) for visualization purposes. (a) Reproduced with permission from Ref. 35. © 2014, AIP Publishing LLC and (b) Adapted from Ref. 39. © Wiley.

air interfaces created by the damage, for example, upon removal of top polymer layers (wear) ([34], Fig. 4.4a).

One of the advantages of this approach is that the healing process can be autonomous and spontaneous, since the trigger and driving force are intrinsically available. The recreation of new air-interfaces resets an energy difference which further promotes the mobility of the dangling chains toward the damaged areas. A dual simulation-experimental study on these systems has proved multiple self-replenishing events with self-healing efficiency of approximately 98% [34, 35].

It should be noticed, however, that the intrinsic mobility of the end chains, as well as of the bulk material (i.e., polymers T_g), is critical for efficient self-replenishing and the time-scales involved in the recovery. Phase-separated polymer systems made of PDMS-*grafted*-(polystyrene-*block*-poly(maleic-anhydride) (PSMA)) (co)polymers, have been reported to fully recover their initial hydrophobicity in a few tens of minutes, while the poly(urethane polyester) systems containing the perfluorinated dangling chains took about 2 h to recover [36], both at room temperature. This difference has been mainly attributed to the presence of the PDMS chain grafts ($T_g \sim -128$°C) in the first system, which contribute for surface rearrangement with a very "soft" and mobile component in a rather "stiff" bulk PSMA phase ($T_g \sim +162$°C) [36].

4.2.1.2 Reversible Bonds Another self-healing intrinsic mechanism presenting a high potential for developing polymeric coatings makes use of reversible (dynamic) chemical bonds, which can associate/dissociate under specific conditions, and allow for rapid conformational changes that ensure a healing action in response to damage. Such behavior depends on the chemical bonds association/dissociation characteristics, the overall mobility of the molecules, and the system environment. This intrinsic mechanism is particularly relevant for recovering damages at the *micro-* and *nano*scales (Fig. 4.2a), acting at the polymer networks and molecular bonds levels (Fig. 4.2b).

A wide range of well-known reversible chemistries is available for designing intrinsic self-healing mechanisms, such as Diels–Alder (DA) reactions [41–43] (namely with furan-maleimide derivative functionalized epoxies, polyketones, poly(styrene), poly(methyl methacrylates), polyamides, and polyesters [44–57]), disulfide bonds [24, 58, 59], trithiocarbonate exchange reactions [60, 61], radical recombination [62, 63], and others (see Table 4.1).

Some of these chemistries have already been specifically investigated for polymeric coatings, namely by the incorporation of reversible cross-linking functionalities into the polymer backbone, such as DA adducts. Wouters *et al.* [54] reported thermally reversible polymeric networks based on *block*-(co)polymers containing furan moieties which were further cross-linked with bismaleimides, according to the DA reactions (see Fig. 4.5a and b). These systems could be implemented in acrylic-based powder coatings. For alternative (epoxy-amine)-based coatings formulations, the authors made a diamine to react with a bifunctional epoxy containing reversible DA adducts [54]. Another example was reported by Scheltjens et al. [45] with furan and maleimide functionalized epoxy resins, which were reacted with diamines (e.g., jeffamine) and further cross-linked with a bismaleimide (Fig. 4.5c–e).

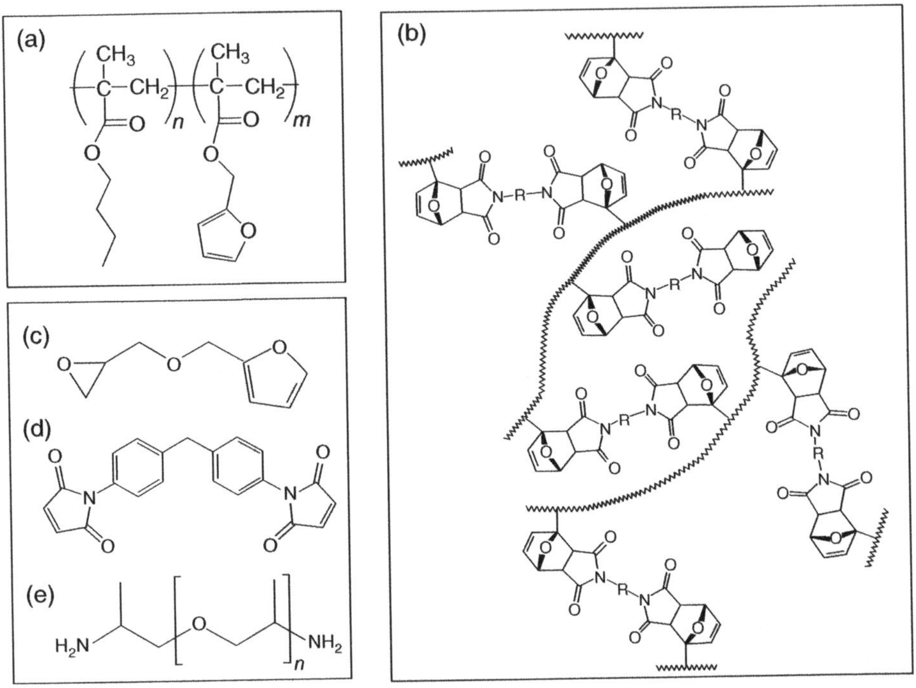

FIG. 4.5 (a) Furfuryl methacrylate–*block*-butyl methacrylate (FMA-*b*-BA) (co)polymer and (b) schematic representation of the cross-linked network made by reacting (a) with a bismaleimide according to DA (reversible acrylic-based polymer networks for powder coatings); (c) furan-modified and (d) maleimide-modified epoxies and (e) jeffamine as components for reversible epoxy-amine networks containing Diels-Alder (DA) adducts. Reprinted with permission from Ref. 54. © Elsevier.

Reversible redox reactions of sulfur bonds, which can be triggered by light or the presence of oxidants/reducers, have also been extensively investigated for self-healing materials, specially thiol–disulfide reactions, which in the oxidized state form disulfide bonds (S–S) that can be ruptured by reduction to form thiol groups (S–H) [25, 58]. Matyjaszewski et al. [60, 61] reported photo-induced healing of polymers based on reshuffling of trithiocarbonate bonds, and Canadell et al. [24] designed promising polymer systems based on disulfide interchange reactions (S–S to S–S) with very interesting applicability to polymeric coatings, as they use moderate temperatures and times to heal (~1 h, 60°C).

These approaches are potentially attractive for polymeric coatings especially due to the different existing chemical routes for promoting healing, the possibility to induce multiple self-repairing events, and availability of a range of different triggers.

One general disadvantage of using reversible bonds for intrinsic self-healing mechanisms may be the fact that the dynamic reactions depend on energy triggers, such as thermal treatment, light (UV) irradiation [60, 61, 64, 65, 89], pH changes [90], or addition of catalytic additives, and hence healing is not autonomous.

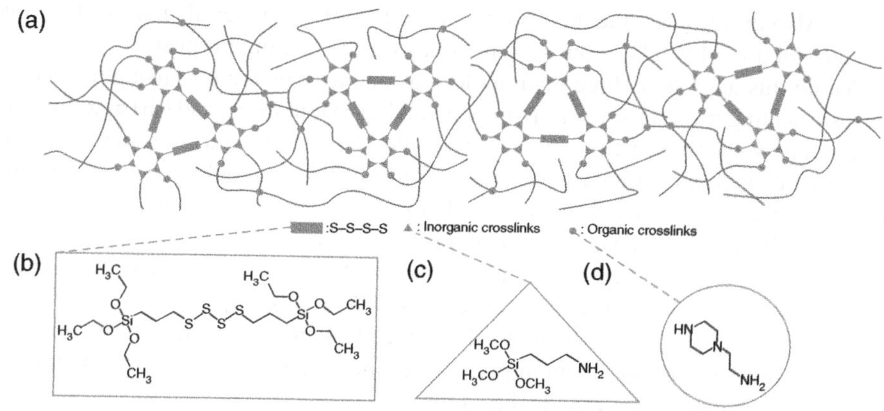

FIG. 4.6 (a) Idealized structure of the hybrid sol–gel intrinsic self-healing polymer network. Insets show some of the key chemical structures used on the hybrid systems: (b) reversible tetrasulfide (S–S–S–S) bonds; (c) inorganic cross-links, and (d) organic epoxy-amine cross-links. Adapted from Ref. 59. © Wiley.

Moreover, the dynamic bonds are typically distributed throughout the whole material and when the external trigger is applied, the parts that have not been damaged will also be affected, leading to possible changes of shape or mechanical properties. Noteworthy, for coatings, this may not necessarily consist of a major concern since the shape of the thin polymer film is mostly confined by the substrate to which it is adhered. In relation to this aspect, an important consideration for an efficient healing of such dynamic systems is also that the local reversibility (i.e., bond forming-breaking reactions) must be significantly faster than the overall changing processes (e.g., polymer flow and macroscopic deformation).

An appealing thought to solve problems related to a weakening of the self-healing polymer during the healing step is the design of intrinsic self-healing polymeric systems with hybrid architectures containing both reversible and irreversible bonds allowing high versatility by varying the polymer architecture [11]. A particularly interesting hybrid system was recently reported by AbdohlaZadeh et al. [59] consisting of dual organic–inorganic cross-linked networks made by a sol–gel process. These systems contain nonreversible (covalent) cross-links as well as reversible tetrasulfide (S–S–S–S) groups. The hybrid sol–gel architecture exhibits an attractive combination of significant recovery of bulk damage as a result of mesoscopic flow at modest temperatures (e.g., the maximum healing kinetics was observed at 70°C) while preserving adequate mechanical properties during the healing stage due to the presence of a stable cross-linked network (Fig. 4.6).

4.2.1.3 Noncovalent Bonds

The principle behind intrinsic healing mechanisms using noncovalent bonds is similar to what has been described in the previous case of reversible (covalent) bonds. High mobility can be gained by breaking bonds and allowing material reflow, and mechanical strength can be restored by reforming bonds. Also in this case, the mechanical properties of the materials may suffer from the introduction of weaker noncovalent bonds for the healing purposes.

Within this approach, three main molecular interactions are promising to develop intrinsic healing mechanisms: (i) hydrogen bonds, (ii) ionomers, and (iii) π–π stacking interactions. Other interesting interactions could be metal–ligand coordination bonds (Table 4.1) which are still at an initial stage of research, but could be very promising for instance, for the recovery of adhesion between coatings and metallic surfaces.

Generally, most of these molecular interactions will require energy triggers to work, for example, temperature or/and pressure. Noteworthy, the typical temperatures required here are in a more accessible range as compared to the previous systems (e.g., Diels–Alder), making these mechanisms very interesting for polymeric coatings. The requirement of pressure application is, however, more restrictive for polymeric coatings for practical reasons.

The use of hydrogen bonds is among the most successful healing routes so far, as demonstrated by Cordier et al. [23, 66] for supramolecular rubbery materials. After a cut-damage into two pieces, the rubber could be healed by the application of a small pressure between the loose ends combined with the heat to increase the healing ability. In spite of a clear interest in these systems, their application on polymeric coatings is probably not very realistic due to the requirement of application of a controlled pressure.

Sijbesma and coworkers [67, 68] reported a special type of supramolecular polymer networks, based on quadruple hydrogen bonding ureido-pyrimidinone moieties, which show more promising features for self-healing polymeric coatings. In these materials, the healing mechanism acts at the *nano*scale, in response to local stresses, avoiding the growth of damages into the *micro*scale, in a preemptive mode. Recently, a similar concept of delayed elasticity and preemptive healing was proposed for epoxy-amine coatings, using single H-bond moieties introduced in the polymer network via an amide (acetamide), which was coreacted with a diamine (jeffamine) and an epoxy resin, in different ratios of the (co)amines in relation to the epoxy [69]. The introduction of H-bonded cross-links allowed enough flow to fully recover approximately 70 μm deep cut on a coating, in about 10 min at 45°C, with very minor effects on the polymer mechanical properties or adhesion to the substrate. The introduction of larger content or multiple H-bonds in the system could excel the healing performance of these traditional epoxy-amine coatings, however, that would be at the cost of their already good adhesion, barrier and protective properties. Nevertheless, such self-healing polymer materials could be very interesting for instance for avoiding fatigue-related damages in polymeric coatings.

Ionomers are a second type of molecular interactions with high potential for self-healing materials. The systems based on ionomers contain species like acid groups in the form of metal salts (ionic species) that are bonded to the polymer structure creating electrostatic interactions or aggregates. These ionic clusters can undergo transitions and relaxations at certain temperature ranges, leading to an increase in the mobility of the polymers which can have a significant impact on the mechanical and physical properties of the materials [73]. Ionomers have been applied for ballistic healing [74, 75], and recent studies show their applicability as coatings as well [91].

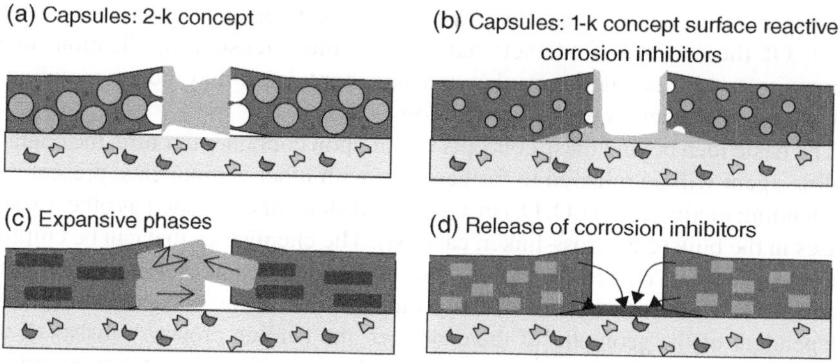

FIG. 4.7 Extrinsic healing approaches in polymeric coatings: (a) 2-k capsules with liquid healing agent [92], (b) 1-k capsules with liquid healing agent [93], (c) expansive phases [94], and (d) release of corrosion inhibitors from inorganic carriers [95].

The third noncovalent type of interactions worth mentioning for self-healing materials are based donor–acceptor π–π stacking. So far, only a few works [78, 79] have reported the healing abilities of these systems, although with promising results. The concept proposes a mechanically robust healable polymer system based on the complexation of a chain-folding (co)polyimide with a pyrenyl end-capped polyamide. The polyimide chains form noncovalent bonds by multiple intercalations of π-electron-rich pyrenyl end groups, into the designed polyimide chain folds. These systems reached a healing efficiency of 100% in about 5 min at 50°C, and raise high expectations for polymeric coatings, although the current research is still insufficient to evaluate possible limitations and strengths of the systems.

4.2.2 Extrinsic Healing

An extrinsic self-healing approach refers to all systems in which the healing agent (monomer, oligomer, solvent, passivating agent, etc.) is introduced in the polymer matrix as a secondary phase. All container-based approaches such as capsules, fibers, nanocarriers, and other type of particles fall under this concept. The main advantage of this approach relies on an almost instantaneous localized response to damage without any external trigger, other than the damaging event itself, for example, a scratch, or the immediate result of the damaging event, such as ions being released from a metal surface upon exposure to the corrosive environment. Figure 4.7 shows a scheme of the most successful concepts used in extrinsic self-healing polymeric coatings.

4.2.2.1 Containers Filled with Liquid Healing Agents Four routes have been reported so far using the idea of encapsulating liquid healing agents: (i) spherical [12, 92] or elongated [96] polymeric capsules containing a liquid healing agent, (ii) single elongated hollow fibers filled with the healing agent [97–99], (iii) compartmented fibers [100] where the healing agent is contained in high aspect ratio compartments distributed along the fiber, and (iv) vascular networks [101]. Of these, fibers find their natural and most feasible application area in composites while isolated capsules and nanocontainers may be more suited for self-healing polymeric coating applications.

The basic idea behind these concepts is that upon container fracture, the contained healing agent will be released to further react with other components, present in the surrounding media (e.g., H_2O, O_2) in a dispersed state, or embedded as other extrinsic phases in the bulk (e.g., cross-linker, catalyst). The chemistries that can be employed in this concept are numerous and depend on the application (healing function they intend to heal, environment of use, etc.) as well as on the encapsulation method. Independent of the geometry of the container, the healing process consists of three steps with three distinct time constants: (i) damage of carrier at $t \ll$ seconds, (ii) release of healing agent at seconds $< t <$ minutes, and (iii) reaction of healing agent minutes $< t <$ days. The main disadvantages of this approach are the impossibility of multiple healing events at one damage zone and the likely formation of a gap where the healing agent was contained.

Since the establishment of this route to implement healing, research has focused on several technological and scientific improvements: (i) improvement on encapsulation techniques (i.e., higher efficiency, stronger shell walls, capsule homogeneity, capsule stability, capsule geometries—from spherical to ellipsoidal, capsule size, etc. [96, 102]), (ii) selection of healing agent–cross-linker/catalyst pairs suitable for different matrices and encapsulating shell materials (i.e., use of efficient less expensive catalysts, use of healing agents adapted to the media of use, and use of new encapsulated chemistries), (iii) development of healing agents not requiring cross-linker or catalyst (i.e., solvents [103, 104], water and surface reactive agents such as silyl esters [93] and oils [105, 106]), and (iv) implementation of the encapsulation concept to more application-oriented research. In polymeric coatings, the main application of this concept has focused on barrier restoration and corrosion protection, as will be discussed in Section 4.3, while almost no studies exist on the use of such concept to restore or heal other coating functions such as color or hydrophobicity [8].

4.2.2.2 *Solid Inorganic Nano- and Microcarriers* A different approach to the encapsulation of liquids is the use of solid or porous inorganic micro- and nanoparticles. The most common examples are (i) expansive phases such as hydraulic inorganic grains as calcium silicate and calcium aluminate [107] or expansive clay layers [94, 108] and (ii) the use of inorganic nanocarriers such as montmorillonite [109], hydrotalcites [95, 110], halloysite nanotubes [111], and zeolites [112, 113] containing active agents such as corrosion inhibitors. In expansive phases, the main healing principle relies on the reaction of the embedded particle with the environment (e.g., H_2O) leading to an expansion able to close damages. In the nanocarriers approach, the healing mechanism comes after the release of molecules or ions from the carrier followed by the reaction of these with the surface of the surrounding substrate and/or coating. Opposite to the liquid-container approach, no gaps are created here, when the healing agent is released or has reacted. On the other hand, the scale of the damage that can be restored is significantly different; while in liquid encapsulation, damages in the *micro* up to *macro*scale are in principle feasible, in the solid nanocontainers, the type of damage that can be healed remains at the *nano* and *micro*scale (Fig. 4.2).

4.3 FUNCTIONALITIES RECOVERY AND POSSIBLE APPLICATIONS

4.3.1 Surface Properties: Wettability and Anti-(bio)adhesion

Many of the current functional polymeric coatings rely on their surface chemical groups for a special performance or functionalities, such as extreme water repellence, self-cleaning behavior (via superhydrophobic or superhydrophilic mechanisms), or anti-(bio)adhesion (i.e., low adherence of proteins and other biofoulants).

Since surface functionalities are strongly related to the molecular characteristics of the surface, self-repairing mechanisms acting at the *nano* and *micro*scale seem to be more suitable to repair surfaces; hence, intrinsic self-healing approaches have been mainly investigated for this purpose. In particular, promising systems have been reported for polymeric coatings using molecular interdiffusion—of chain ends or impregnated additives—and self-replenishing mechanisms.

Special wettability states, ranging from (super)hydrophobicity to (super)hydrophilicity, are utmost desired for functional coatings with self-cleaning, anti-fouling (AF), or low-friction properties. Accordingly, self-healing mechanisms that aim retrieving the surface chemical composition upon surface damage, to maintain the wettability characteristics, have been intensively investigated.

In this context, superhydrophobic surfaces raise an additional problem, not only the surface composition has to be repaired but also a surface topography needs to be reconstructed. A coarse but rather efficient approach to achieve this is to use low-surface-energy polymer layers (e.g., PDMS or fluorinated polymers) heavily loaded with inorganic nanoparticles (even largely aggregated) [114, 115]. When the surface is damaged, for example by subtle abrasion or wear, new topographic surfaces are recreated which still contain aggregates and the polymer, and thus the superhydrophobic character is retrieved. One of the drawbacks of these approaches is the very high load of aggregates, which results in very rough and opaque coatings, and in some cases rather stiff polymeric layers which are susceptible to crack and other mechanical failures.

Li et al. [37] prepared porous polymeric coatings via a layer-by-layer method which are composed of poly(allylamine hydrochloride), a sulfonated poly-ether-ketone, and poly(acrylic acid), impregnated with a reactive low-surface energy component, 1*H*,1*H*,2*H*,2*H*-perfluoro-octyltriethoxysilane (POTS). Upon removal of top layers by O_2 plasma treatment and recreation of new air-interfaces, the POTS molecules segregate to the new surfaces recovering the initial low surface energy. Due to the porosity of the coatings, the damaged surfaces are rough, and thus the initial superhydrophobic character is restored. Note that superhydrophobic surfaces rely on the combination of low surface energy combined with a rough topography. In this case, the *driving force* for the relocation of the *healing agents* was the energy difference between air-interfaces and the coatings bulk, but a certain humidity level was necessary to trigger the healing agents (POTS molecules) mobility.

A self-replenishing mechanism was applied to superhydrophobic surface-structured polyurethane-based coatings by Esteves et al. [38]. After harsh abrasion with sand paper (500 cycles), the composite coatings were able to maintain a rather high (advancing) water contact angle of approximately 145°, spontaneously and autonomously. In this case, the surface topography was replicated upon damage, through the presence of dual-size SiO_2 nanoparticles, chemically bonded to a cross-linked poly(urethane polyester) network with low-surface energy dangling ends, and homogenously distributed throughout the polymer layer (Fig. 4.4b). With a combined simulation-experimental approach, the authors showed that as long as a thin (in this case ~30 nm) polymer layer is present, covering the dual-size nanoparticles surfaces, the coatings were able to self-replenish multiple times, with healing efficiencies of about 95–98% [38, 39].

Using a similar self-replenishing principle, Minko *et al.* [40] reported recently self-healing AF coatings. A polymer system consisting of a pH-responsive poly(2-vinylpyridine) (PVP) network was grafted with poly(ethylene oxide) (PEO) chains (brushes), both to the surface and inside the host network material (3D grafting). Note that PEO-surface grafted polymer films have been extensively investigated for their AF behavior, that is, low-protein adhesion properties. To mimic a surface damage, a substantial fraction of the top polymer layers was removed on the PVP-PEO 3D-grafted polymers and sequentially flushed with a solution at physiological conditions (pH=7.4 and 37°C). The 3D-grafted PVP–PEO polymer networks showed a fourfold increase in longevity of the AF behavior, as compared to polymer films in which only the surface was grafted with PEO brushes [40]. Also in this case, the surface properties healing was attributed to the spontaneous surface replenishment with PEO chain segments stored inside the film, and in proximity of the interface, driven by an emerging gradient in a chemical potential.

The self-healing principles applied so far to hydrophobic [34, 35], superhydrophobic [38, 39], and AF coatings [40], can be expected to expand to coatings with other functionalities, such as antibacteria, antifog, anti-icing, or even drag reduction, which will find application in new technological areas, for example, biomedical, optical devices, microfluidics, and microelectronics. Although hydrophilic and superhydrophilic surfaces are equally relevant for many functional coatings, no studies seem to have been reported so far for self-healing (super)hydrophilic polymeric coatings.

Another interesting possibility for repairing surface functionalities is using reactive chain ends for reestablishing molecular interactions across surfaces/interfaces. This principle has been reported to repair broken bonds in the main chain of *bulk* networks, for example, by Takeda et al. [116] on poly(phenylene-ether) composite materials which recovered spontaneously by a reaction of the chain ends with a copper/amine complex catalyst added to the system; and by Urban et al. [89] on polyurethane matrices with additional nonreacted oxetane groups which were further reacted on damaged sites upon UV irradiation.

In functional polymeric coatings, such mechanisms could assist, for instance, the readhesion of films at the polymer-substrate interfaces, the rebonding of layers on multilayer coatings systems, or the firm reattachment of opposite faces of a crack once they get into contact. In this case, the reactive groups should only be "consumed" once they reach an interface, and sufficient chain mobility should be granted by tailoring the polymer chain reactive segments.

Extrinsic self-healing approaches have understandably been less investigated for the recovery of surface properties. There could be, however, interesting possibilities using extrinsic self-healing agents, such as using the controlled release of biocides encapsulated on *healing reservoirs*, which could reduce the adhesion of (bio)foulants to the surface, and hence extending the AF performance of the polymeric coatings [117]. In any case, a step of surface-segregation of the healing components would always have to be included or stimulated by additional means, for example, increasing mobility by heating, swelling with solvents, or washing with solutions.

4.3.2 Barrier and Corrosion Protection

Both extrinsic and intrinsic approaches have been employed to restore barrier and corrosion protection. For corrosion protection, the extrinsic route is by far the most explored one, the main reason being the relative ease of incorporation of both capsules with liquid healing agent and doped nanoparticles in existing coatings.

Two main approaches have been reported so far using liquid healing agents for corrosion protection (as schematized in Fig. 4.7a and b): (i) 2-k concept using encapsulated agent with dispersed catalyst [92] and (ii) 1-k concept using water reactive agents such as tung oil [105] and water and surface reactive systems such as silyl esters [93]. Following this last idea, a recent work has successfully shown the possibility of encapsulating an oil-corrosion inhibitor double agent (triazole derivative) for active corrosion protection [118]. The mechanism of capsule opening has also been object of research where the opening mechanism can be mechanical as in the original concept by White et al. [12], triggered by the corrosion process itself using pH changes [119] or redox variations [120, 121] or due to environmental factors such as UV irradiation [122].

Inorganic nanoparticles (Fig. 4.7d) have been used mainly to restore the protective layer on top of a metallic surface without necessarily filling the gap created during the damaging effect (i.e., in case of mechanical damage). The so-far reported concepts focus on the use of the corrosion processes themselves to trigger the release of the functional agents, namely corrosion inhibitors or passivating agents. Several carriers have been reported so far to be efficient, although the ones that have attracted more attention till date are hydrotalcites [95, 110] and more recently zeolites [112, 113]. Independently of the carrier, the traditional and most extended approach consists on the introduction of one corrosion inhibitor in the carrier. The inhibitor will be released by ion exchange due to the presence of metal cations in the surrounding or anions from the corrosive environment (e.g., Cl^-) [95, 110], pH changes as in particles coated with polyelectrolyte shells [123], or by desorption control from adsorbed corrosion inhibitors on nanoparticles [124]. Recently, a new route using two different inhibitors doped in one single carrier has been successfully proposed [112]. The concept has been demonstrated using a NaY zeolite carrier and Ce^{3+} and diethyldithiocarbamate as the corrosion inhibitors for AA2024. These concepts propose the fast release of one inhibitor by desorption and the slowest and sustained release of the second inhibitor by ion exchange thus leading to a very beneficial fast and sustained response [112] and hence keeping the corrosion process under control for longer times.

Intrinsic healing concepts have also been proposed to restore barrier and hence potentially corrosion protection with the advantage that the healing event can be repeated multiple times. Most of the reported concepts for intrinsic healing polymeric coatings employ temperature as the trigger leading to the necessary mobility to close a damage. DA-based polymers were successfully used in powder coating systems [54] as well as liquid-based polymers [125]. Although these systems show good healing behavior, the temperatures needed to reach the necessary flow (around 120–150°C) limit their applications. Disulfide-based systems on the other hand offer sufficient flow at moderate temperatures (60–70°C). Canadell et al. [24] presented an epoxy coating with disulfide bonds capable of restoring small damages. More recently, AbdohlaZadeh et al. [59] introduced a hybrid organic–inorganic coating capable of restoring relatively big damages (500 μm) upon the application of moderate temperatures and pressures while maintaining good mechanical properties. Supramolecular approaches may be also applied in the restoration of damaged coatings [8]. Coatings using triggers other than temperature have also been reported such as moisture-promoted healing copolymer proposed by Zhang et al. [126]. The initial reported results show gap closure by means of a zipping mechanism promoted by water, although the question remains if these systems will be capable of maintaining good properties after long water immersion.

The development of self-healing coatings also requires the development of quantitative and reliable techniques [127]. Since its proposition as technique to quantify the level of corrosion protection using doped nanoparticles [95] and barrier restoration with encapsulated healing agents [93], electrochemical impedance spectroscopy (EIS) has become one of the most commonly used techniques due to its ease of application. An alternative to traditional EIS is odd random phase multisine electrochemical impedance spectroscopy (ORP-EIS) which accelerates the measurement reducing the influence of instabilities [128]. Despite the potential of EIS, attention must be put to the correct interpretation of the results obtained with this technique and even more when dealing with the selection of electrical equivalent circuits for more quantitative analysis [129].

Local electrochemical techniques have also been proposed to quantify the level of restoration at the damage itself (e.g., scratch, hole). Scanning vibrating electrode technique (SVET) allows *in situ* monitoring of the distribution of anodic and cathodic areas at the surface. This technique has been used for self-healing coatings using corrosion inhibitors [130, 131] where the suppression of the corrosion activity is shown as disappearance of the local activity. SVET has been proven to be very effective also in systems aiming at gap closure or surface coverage using microencapsulated liquid healing agents [93]. Although a bit more complex in use, scanning electrochemical microscopy has been proven as very powerful technique to follow the healing process of corrosion inhibitors, encapsulated agents, and intrinsic polymers both by

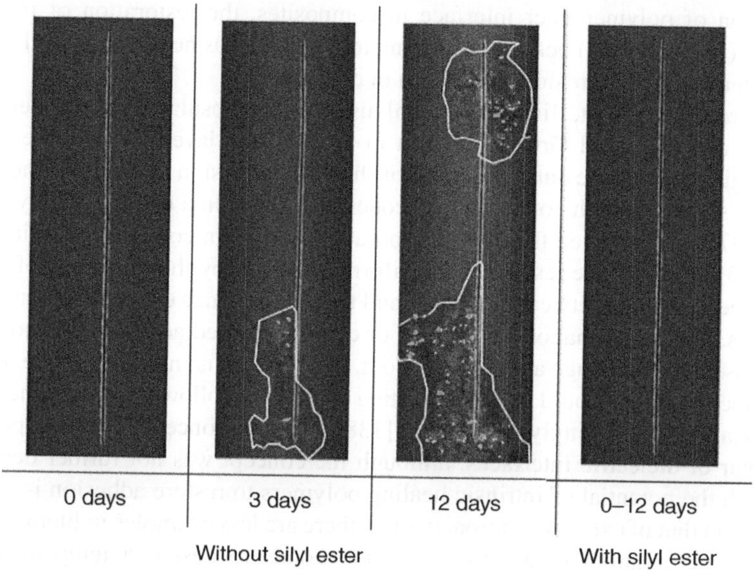

FIG. 4.8 Delamination (white irregular lines) and pitting (white-gray dots) around a scribe (vertical line) of a coated AA7050 panel exposed to 0.05 M NaCl in distilled water. Figure 4.8 shows a significant time delay in the appearance of both delamination and pitting using a silyl ester as healing agent (from 3 days to more than 12 days). Adapted from Ref. 134. © NACE.

following the redox activity at the metallic surface (redox mode) [132] or the redox mode combined with the feedback mode (to follow topography changes) [133].

In order to obtain direct visual information about the level and type of protection offered by a self-healing coating, Garcia et al. recently proposed the use of x-ray tomography coupled to EIS [134]. Figure 4.8 shows how a silyl ester is capable of reducing both the under-film delamination (white lines) and pitting (white dots) of a coated AA7050 panel immersed in a corrosive solution. A quantification of these parameters led to a major understanding of the effect of the silyl ester otherwise not fully understood by electrochemical methods alone.

4.3.3 Interfacial Bonding Between Dissimilar Materials

Adhesion between similar materials (e.g., polymer A–polymer A) as well as between dissimilar materials (e.g., metal–polymer, polymer A–polymer B) is a critical aspect in many fields such as coatings technology and composite materials. As highlighted in one of our previous works, loss of adhesion is a critical factor in coatings technology leading to accelerated localized corrosion [8]. In other application fields such as packaging and electronic devices such as LEDs, loss of adhesion is one of the most critical reasons for device malfunction [135]. It is clear that many application fields could significantly benefit of the implementation of self-healing concepts aiming at interfacial bonding restoration. Despite there are several works focusing on the restoration of polymer-fiber interface in composites, the restoration of interfacial bonding (i.e., adhesion) between a coating and a metal has not yet attracted as much attention as barrier/corrosion restoration in coatings.

In a recent attempt, Jin et al. [136] used an encapsulated dicyclopentadiene (DCPD) and dispersed Grubbs catalyst in an adhesive layer. Despite the initially encouraging results, the authors could not show full adhesion recovery as the studied adhesive showed a high component of cohesive failure. In a different study, Mardel et al. [137] demonstrated that the incorporation of certain corrosion inhibitors (i.e., Ce(dbp)3) improves the resistance to filiform corrosion by the formation of a somewhat adhesive oxide between the metal and the coating once exposed to humid environments. The use of nanoencapsulated or compartmented adhesion promoters that are released upon damage and that wet both, the coating and metal interface, reacting with either materials could be an interesting route to be followed to implement interfacial healing. Interestingly, Lane et al. [138] proved a concept for this approach of the repair of dielectric interfaces, although the concept was not further developed. Although the potential of intrinsic healing polymers to restore adhesion is probably higher than that of extrinsic approaches [8], there are less examples in literature using this approach. Lafont et al. used disulfide bonds to design a temperature- and pressure-triggered self-healing adhesive [139]. It is expected that in the near future, more works addressing the interfacial restoration will be published.

4.4 CONCLUDING REMARKS AND CHALLENGES

In this chapter, several approaches with potential for self-healing polymeric coatings have been presented. It should be noted that the selection of a self-healing approach and its final level of success fully depend on appropriateness of the approach for a given final application and functionality to be healed.

Certain extrinsic approaches, such as the use of vascular networks or fibers do not seem very likely to make it to real polymeric coatings applications, mainly due to the detrimental impact on other properties of the coatings and inherent fabrication limitations. Other extrinsic concepts such as liquid encapsulation and the use of doped inorganic nanoparticles have already reached promising results so as to reach the industrial sector. Recently, intrinsic approaches appear to be the most promising route, namely using thermally reversible interactions or self-segregated molecular chains, in the particular case of functional polymeric coatings. With these approaches, multiple healing events can be reached; however, limitations regarding the temperatures or other stimuli required for the healing and possible side effects on other properties (e.g., lower mechanical stability) by the implementation of the healing mechanisms should been taken into consideration.

In spite of the many steps given toward self-healing polymeric coatings, optimal solutions have not been found for the following questions: how could the coating material distinguish accurately between nondamaged and damaged areas? How to control precisely the transport or the mobility of desired healing agent to damaged areas only, without affecting the neighboring nondamaged, coated and uncoated areas? How can we visualize the self-repairing state of a coating at a specific time of use? How can we introduce multiple healing mechanisms that will simultaneously address damages across several scales (*nano* to *macro*)?

The development of the self-healing materials field in the upcoming years may offer alternative solutions that will bring us closer to industrial applications. A possible next step for a successful implementation of self-healing mechanisms in polymeric coatings is the use of systems based on detection-actuation-healing, for example, by the implementation of new visual damage–detecting mechanisms, such as color change, or specific and location-targeted driving forces, such as inclusion of charge gradients.

Finally, there is also a need to dedicate more attention to the in-depth understanding of the reported self-healing mechanisms, which may require combined experiments, simulations, and mathematical modeling approaches, so that we can gain a better control over their implementation and optimization on (functional) polymeric coatings.

ACKNOWLEDGMENTS

P.T.M Albers is acknowledged for is contribution in selecting literature references about the self-healing intrinsic approaches. The authors gratefully thank the financial support throughout several years from the Dutch Ministry of Economic Affairs, Agriculture and Innovation via the Innovative Research Program on Self-Healing Materials (IOP-SHM) in the Netherlands, which allowed the consolidation of their research on self-healing materials.

REFERENCES

1. Feng, W., Patel, S. H., Young, M. Y., Zunino, J. L., Xanthos, M., Smart polymeric coatings—recent advances, *Adv Polym Technol* 2007, 26, 1.
2. Baghdachi, J., Smart coatings, in: *Smart Coatings II* (Ed. T. Provder, J. Baghdachi), Vol. 1002, American Chemical Society, Washington, DC 2009.
3. Ghosh, S. K., *Functional Coatings and Microencapsulation: A General Perspective*, Wiley-VCH Verlag GmH & Co KGaA, Weinheim 2006.
4. van der Zwaag, S., *Self-Healing Materials: An Alternative Approach to 20 Centuries of Materials Science*, Vol. 100, Springer, Dordrecht 2007.
5. Yang, Y., Urban, M. W., Self-healing polymeric materials, *Chem Soc Rev* 2013, 42, 7446.
6. Murphy, E. B., Wudl, F., The world of smart healable materials, *Prog Polym Sci* 2010, 35, 223.
7. Wu, D. Y., Meure, S., Solomon, D., Self-healing polymeric materials: a review of recent developments, *Prog Polym Sci* 2008, 33, 479.
8. Garcia, S. J., Fischer, H. R., van der Zwaag, S., A critical appraisal of the potential of self healing polymeric coatings, *Prog Org Coat* 2011, 72, 211.

9. van Benthem, R. A. T. M., Ming, W., de With, G., Self-Healing Materials: An approach to 20 Centuries of Materials Science, Vol 100, Springer Series in Materials Science, Dordrecht 2007.
10. Baghdachi, J., Perez, H., Shah, A., Design and development of self-healing polymers and coatings, in: *Smart Coatings III* (Ed. J. Baghdachi, T. Provder), Vol. 1050, American Chemical Society, Washington, DC 2010.
11. Garcia, S. J., Effect of polymer architecture on the intrinsic self-healing character of polymers, *Eur Polym J* 2014, 53, 118.
12. White, S. R., Sottos, N. R., Geubelle, P. H., Moore, J. S., Kessler, M. R., Sriram, S. R., Brown, E. N., Viswanathan, S., Autonomic healing of polymer composites, *Nature* 2001, 409, 794.
13. van Benthem, R. A. T. M., Ming, W., De With, G., *Self-Healing Materials: An Approach to 20 Centuries of Materials Science*, Vol. 100, Springer Series in Materials Science, Dordrecht 2007, 142–145.
14. Williams, K. A., Dreyer, D. R., Bielawski, C. W., The underlying chemistry of self-healing materials, *Mrs Bull* 2008, 33, 759.
15. Syrett, J. A., Becer, C. R., Haddleton, D. M., Self-healing and self-mendable polymers, *Polym Chem* 2010, 1, 978.
16. Yang, J. L., Keller, M. W., Moore, J. S., White, S. R., Sottos, N. R., Microencapsulation of isocyanates for self-healing polymers, *Macromolecules* 2008, 41, 9650.
17. Huang, M. X., Yang, J. L., Facile microencapsulation of HDI for self-healing anticorrosion coatings, *J Mater Chem* 2011, 21, 11123.
18. Jackson, A. C., Bartelt, J. A., Marczewski, K., Sottos, N. R., Braun, P. V., Silica-protected micron and sub-micron capsules and particles for self-healing at the microscale, *Macromol Rapid Commun* 2011, 32, 82.
19. Blaiszik, B. J., Caruso, M. M., McIlroy, D. A., Moore, J. S., White, S. R., Sottos, N. R., Microcapsules filled with reactive solutions for self-healing materials, *Polymer* 2009, 50, 990.
20. Toohey, K. S., Sottos, N. R., Lewis, J. A., Moore, J. S., White, S. R., Self-healing materials with microvascular networks, *Nat Mater* 2007, 6, 581.
21. Hamilton, A. R., Sottos, N. R., White, S. R., Self-healing of internal damage in synthetic vascular materials, *Adv Mater* 2010, 22, 5159.
22. Tournilhac, F., Cordier, P., Montarnal, D., Soulie-Ziakovic, C., Leibler, L., Self-healing supramolecular networks, polymer networks: synthesis, properties, theory and applications, *Macromolecular Symp* 2010, 291–292, 84.
23. Cordier, P., Tournilhac, F., Soulie-Ziakovic, C., Leibler, L., Self-healing and thermoreversible rubber from supramolecular assembly, *Nature* 2008, 451, 977.
24. Canadell, J., Goossens, H., Klumperman, B., Self-healing materials based on disulfide links, *Macromolecules* 2011, 44, 2536.
25. Pepels, M., Filot, I., Klumperman, B., Goossens, H., Self-healing systems based on disulfide-thiol exchange reactions, *Polym Chem-UK* 2013, 4, 4955.
26. Wool, R. P., Oconnor, K. M., Craze healing in polymer glasses, *Polym Eng Sci* 1981, 21, 970.
27. Lin, C. B., Lee, S. B., Liu, K. S., Methanol-induced crack healing in poly(methyl methacrylate), *Polym Eng Sci* 1990, 30, 1399.
28. Wang, P. P., Lee, S., Harmon, J. P., Ethanol-induced crack healing in poly(methyl methacrylate), *J Polym Sci Pol Phys* 1994, 32, 1217.

29. Kawagoe, M., Nakanishi, M., Qiu, J., Morita, M., Growth and healing of a surface crack in poly(methyl methacrylate) under case II diffusion of methanol, *Polymer* 1997, 38, 5969.

30. Hsieh, H. C., Yang, T. J., Lee, S., Crack healing in poly(methyl methacrylate) induced by co-solvent of methanol and ethanol, *Polymer* 2001, 42, 1227.

31. Liu, C. K., Yang, T. J., Shen, J. S., Lee, S., Some recent results on crack healing of poly(methyl methacrylate), *Eng Fract Mech* 2008, 75, 4876.

32. Wu, T., Lee, S., Carbon tetrachloride-induced crack healing in polycarbonate, *J Polym Sci Pol Phys* 1994, 32, 2055.

33. Yufa, N. A., Li, J., Sibener, S. J., Diblock copolymer healing, *Polymer* 2009, 50, 2630.

34. Dikic, T., Ming, W., van Benthem, R. A. T. M., Esteves, A. C. C., de With, G., Self-replenishing surfaces, *Adv Mater* 2012, 24, 3701.

35. Esteves, A. C. C., Lyakhova, K., van Riel, J. M., van der Ven, L. G. J., van Benthem, R. A. T. M., de With, G., Self-replenishing ability of cross-linked low surface energy polymer films investigated by a complementary experimental-simulation approach, *J Chem Phys* 2014, 140, 124902.

36. Esteves, A. C. C., Gunbas, I. D., van Riel, J. M., Noordover, B. A. J., de With, G., van Benthem, R. A. T. M., Time-resolved surface rearrangements of polymer films monitored by dynamic recovery contact angle (DRCA), *Rsc Adv* 2014, 4, 20094.

37. Li, Y., Li, L., Sun, J. G., Bioinspired self-healing superhydrophobic coatings, *Angew Chem Int Edit* 2010, 49, 6129.

38. Esteves, A. C. C., Luo, Y., van de Put, M. W. P., Carcouët, C. C. M., de With, G., Self-replenishing dual structured superhydrophobic coatings prepared by drop-casting of an all-in-one dispersion, *Adv Funct Mater* 2014, 7, 986.

39. Lyakhova, K., Esteves, A. C. C., van de Put, M. W. P., van der Ven, L. G. J., van Benthem, R. A. T. M., de With, G., Simulation-experimental approach to investigate the role of interfaces in self-replenishing composite coatings, *Adv Mater Inter* 2014, 1, 1400053.

40. Kuroki, H., Tokarev, I., Nykypanchuk, D., Zhulina, E., Minko, S., Stimuli-responsive materials with self-healing antifouling surface via 3D polymer grafting, *Adv Funct Mater* 2013, 23, 4593.

41. Chen, X. X., Dam, M. A., Ono, K., Mal, A., Shen, H. B., Nutt, S. R., Sheran, K., Wudl, F., A thermally re-mendable cross-linked polymeric material, *Science* 2002, 295, 1698.

42. Chen, X. X., Wudl, F., Mal, A. K., Shen, H. B., Nutt, S. R., New thermally remendable highly cross-linked polymeric materials, *Macromolecules* 2003, 36, 1802.

43. Murphy, E. B., Bolanos, E., Schaffner-Hamann, C., Wudl, F., Nutt, S. R., Auad, M. L., Synthesis and characterization of a single-component thermally remendable polymer network: Staudinger and Stille revisited, *Macromolecules* 2008, 41, 5203.

44. Tian, Q., Yuan, Y. C., Rong, M. Z., Zhang, M. Q., A thermally remendable epoxy resin, *J Mater Chem* 2009, 19, 1289.

45. Scheltjens, G., Brancart, J., De Graeve, I., Van Mele, B., Terryn, H., Van Assche, G., Self-healing property characterization of reversible thermoset coatings, *J Therm Anal Calorim* 2011, 105, 805.

46. Tian, Q. A., Rong, M. Z., Zhang, M. Q., Yuan, Y. C., Synthesis and characterization of epoxy with improved thermal remendability based on Diels-Alder reaction, *Polym Int* 2010, 59, 1339.

47. Zhang, Y., Broekhuis, A. A., Picchioni, F., Thermally self-healing polymeric materials: the next step to recycling thermoset polymers? *Macromolecules* 2009, 42, 1906.

48. Canary, S. A., Stevens, M. P., Thermally reversible cross-linking of polystyrene via the furan-maleimide Diels-Alder reaction, *J Polym Sci Pol Chem* 1992, 30, 1755.

49. Kavitha, A. A., Singha, N. K., A tailor-made polymethacrylate bearing a reactive diene in reversible Diels-Alder reaction, *J Polym Sci Pol Chem* 2007, 45, 4441.

50. Liu, Y. L., Chen, Y. W., Thermally reversible cross-linked polyamides with high toughness and self-repairing ability from maleimide- and furan-functionalized aromatic polyamides, *Macromol Chem Phys* 2007, 208, 224.

51. Liu, Y. L., Hsieh, C. Y., Chen, Y. W., Thermally reversible cross-linked polyamides and thermo-responsive gels by means of Diels-Alder reaction, *Polymer* 2006, 47, 2581.

52. Magana, S., Zerroukhi, A., Jegat, C., Mignard, N., Thermally reversible crosslinked polyethylene using Diels-Alder reaction in molten state, *React Funct Polym* 2010, 70, 442.

53. Toncelli, C., De Reus, D. C., Picchioni, F., Broekhuis, A. A., Properties of reversible Diels-Alder furan/maleimide polymer networks as function of crosslink density, *Macromol Chem Phys* 2012, 213, 157.

54. Wouters, M., Craenmehr, E., Tempelaars, K., Fischer, H., Stroeks, N., van Zanten, J., Preparation and properties of a novel remendable coating concept, *Prog Org Coat* 2009, 64, 156.

55. Yoshie, N., Watanabe, M., Araki, H., Ishida, K., Thermo-responsive mending of polymers crosslinked by thermally reversible covalent bond: Polymers from bisfuranic terminated poly(ethylene adipate) and tris-maleimide, *Polym Degrad Stabil* 2010, 95, 826.

56. Peterson, A. M., Jensen, R. E., Palmese, G. R., Reversibly cross-linked polymer gels as healing agents for epoxy-amine thermosets, *Acs Appl Mater Inter* 2009, 1, 992.

57. Kavitha, A. A., Singha, N. K., "Click chemistry" in tailor-made polymethacrylates bearing reactive furfuryl functionality: a new class of self-healing polymeric material, *Acs Appl Mater Inter* 2009, 1, 1427.

58. Yoon, J. A., Kamada, J., Koynov, K., Mohin, J., Nicolay, R., Zhang, Y. Z., Balazs, A. C., Kowalewski, T., Matyjaszewski, K., Self-healing polymer films based on thiol-disulfide exchange reactions and self-healing kinetics measured using atomic force microscopy, *Macromolecules* 2012, 45, 142.

59. AbdohlaZadeh, M., Esteves, A. C. C., van der Zwaag, S., Garcia, S. J., Healable dual organic–inorganic crosslinked sol–gel based polymers: crosslinking density and tetrasulfide content effect, *J Polym Sci A Polym Chem* 2014, 52, 1953.

60. Amamoto, Y., Kamada, J., Otsuka, H., Takahara, A., Matyjaszewski, K., Polymers through reshuffling of trithiocarbonate units, *Angew Chem Int Edit* 2011, 50, 1660.

61. Nicolay, R., Kamada, J., Van Wassen, A., Matyjaszewski, K., Responsive gels based on a dynamic covalent trithiocarbonate cross-linker, *Macromolecules* 2010, 43, 4355.

62. Imato, K., Nishihara, M., Kanehara, T., Amamoto, Y., Takahara, A., Otsuka, H., Self-healing of chemical gels cross-linked by diarylbibenzofuranone-based trigger-free dynamic covalent bonds at room temperature, *Angew Chem Int Edit* 2012, 51, 1138.

63. Yuan, C., Rong, M. Z., Zhang, M. Q., Zhang, Z. P., Yuan, Y. C., Self-healing of polymers via synchronous covalent bond fission/radical recombination, *Chem Mater* 2011, 23, 5076.

64. Cho, S. Y., Kim, J. G., Chung, C. M., A fluorescent crack sensor based on cyclobutane-containing crosslinked polymers of tricinnamates, *Sensor Actuat B-Chem* 2008, 134, 822.

65. Chung, C. M., Roh, Y. S., Cho, S. Y., Kim, J. G., Crack healing in polymeric materials via photochemical [2+2] cycloaddition, *Chem Mater* 2004, 16, 3982.

66. Montarnal, D., Cordier, P., Soulie-Ziakovic, C., Tournilhac, F., Leibler, L., Synthesis of self-healing supramolecular rubbers from fatty acid derivatives, diethylene triamine, and urea, *J Polym Sci Pol Chem* 2008, 46, 7925.

67. Wietor, J. L., Dimopoulos, A., Govaert, L. E., van Benthem, R. A. T. M., de With, G., Sijbesma, R. P., Preemptive healing through supramolecular cross-links, *Macromolecules* 2009, 42, 6640.

68. Nieuwenhuizen, M. M. L., de Greef, T. F. A., van der Bruggen, R. L. J., Paulusse, J. M. J., Appel, W. P. J., Smulders, M. M. J., Sijbesma, R. P., Meijer, E. W., Self-assembly of ureido-pyrimidinone dimers into one-dimensional stacks by lateral hydrogen bonding, *Chem-Eur J* 2010, 16, 1601.

69. Villani, M., Camlibel, C., Deshmuk, Y., Esteves, A. C. C., de With, G., *CoSi 2014—Coatings Science International Conference-Book of Abstracts*, Noordwijk 2014.

70. Montarnal, D., Tournilhac, F., Hidalgo, M., Couturier, J. L., Leibler, L., Versatile one-pot synthesis of supramolecular plastics and self-healing rubbers, *J Am Chem Soc* 2009, 131, 7966.

71. Huang, L., Yi, N., Wu, Y., Zhang, Y., Zhang, Q., Huang, Y., Ma, Y., Chen, Y., Multichannel and repeatable self-healing of mechanical enhanced graphene-thermoplastic polyurethane composites, *Adv Mater* 2013, 25, 2224.

72. Skorb, E. V., Skirtach, A. G., Sviridov, D. V., Shchukin, D. G., Mohwald, H., Laser-controllable coatings for corrosion protection, *ACS Nano* 2009, 3, 1753.

73. Tachino, H., Hara, H., Hirasawa, E., Kutsumizu, S., Tadano, K., Yano, S., Dynamic mechanical relaxations of ethylene ionomers, *Macromolecules* 1993, 26, 752.

74. Varley, R. J., van der Zwaag, S., Towards an understanding of thermally activated self-healing of an ionomer system during ballistic penetration, *Acta Mater* 2008, 56, 5737.

75. Kalista, S. J., Ward, T. C., Thermal characteristics of the self-healing response in poly (ethylene-co-methacrylic acid) copolymers, *J R Soc Interface* 2007, 4, 405.

76. Kalista, S. J., Self-healing of poly(ethylene-co-methacrylic acid) copolymers following projectile puncture, *Mech Adv Mater Struct* 2007, 14, 391.

77. Kalista, S. J., Pflug, J. R., Varley, R. J., Effect of ionic content on ballistic self-healing in EMAA copolymers and ionomers, *Polym Chem* 2013, 4, 4910.

78. Burattini, S., Colquhoun, H. M., Fox, J. D., Friedmann, D., Greenland, B. W., Harris, P. J. F., Hayes, W., Mackay, M. E., Rowan, S. J., A self-repairing, supramolecular polymer system: healability as a consequence of donor-acceptor pi-pi stacking interactions, *Chem Commun* 2009, 44 6717.

79. Burattini, S., Colquhoun, H. M., Greenland, B. W., Hayes, W., A novel self-healing supramolecular polymer system, *Faraday Discuss* 2009, 143, 251.

80. Corten, C. C., Urban, M. W., Repairing polymers using an oscillating magnetic field, *Adv Mater* 2009, 21, 5011.

81. South, A. B., Lyon, L. A., Autonomic self-healing of hydrogel thin films, *Angew Chem Int Edit* 2010, 49, 767.

82. Herrmann, W. A., Weskamp, T., Bohm, V. P. W., Metal complexes of stable carbenes, *Adv Organomet Chem* 2001, 48, 1.

83. Karthikeyan, S., Potisek, S. L., Piermattei, A., Sijbesma, R. P., Highly efficient mechanochemical scission of silver-carbene coordination polymers, *J Am Chem Soc* 2008, 130, 14968.

84. Kersey, F. R., Yount, W. C., Craig, S. L., Single-molecule force spectroscopy of bimolecular reactions: System homology in the mechanical activation of ligand substitution reactions, *J Am Chem Soc* 2006, 128, 3886.

85. Piermattei, A., Karthikeyan, S., Sijbesma, R. P., Activating catalysts with mechanical force, *Nat Chem* 2009, 1, 133.
86. de Gennes, P. G., Reptation of a polymer chain in presence of fixed obstacles, *J Chem Phys* 1971, 55, 572.
87. Kim, Y. H., Wool, R. P., A theory of healing at a polymer polymer interface, *Macromolecules* 1983, 16, 1115.
88. Wool, R. P., Self-healing materials: a review, *Soft Matter* 2008, 4, 400.
89. Ghosh, B., Urban, M. W., Self-repairing oxetane-substituted chitosan polyurethane networks, *Science* 2009, 323, 1458.
90. Deng, G. H., Tang, C. M., Li, F. Y., Jiang, H. F., Chen, Y. M., Covalent cross-linked polymer gels with reversible sol-gel transition and self-healing properties, *Macromolecules* 2010, 43, 1191.
91. Vega, J. M., Grande, A. M., Van der Zwaag, S., Garcia, S. J., On the role of clusters and free carboxylic group content on the surface scratch healing of ionomers, *Eur Polym J* 2014, 57, 121.
92. Cho, S. H., White, S. R., Braun, P. V., Self-healing polymer coatings, *Adv Mater* 2009, 21, 645.
93. Garcia, S. J., Fischer, H. R., White, P. A., Mardel, J., Gonzalez-Garcia, Y., Mol, J. M. C., Hughes, A. E., Self-healing anticorrosive organic coating based on an encapsulated water reactive silyl ester: Synthesis and proof of concept, *Prog Org Coat* 2011, 70, 142.
94. Micciche, F., Fischer, H., Varley, R., van der Zwaag, S., Moisture induced crack filling in barrier coatings containing montmorillonite as an expandable phase, *Surf Coat Technol* 2008, 202, 3346.
95. Buchheit, R. G., Guan, H., Mahajanam, S., Wong, F., Active corrosion protection and corrosion sensing in chromate-free organic coatings, *Prog Org Coat* 2003, 47, 174.
96. Mookhoek, S. D., Fischer, H. R., van der Zwaag, S., A numerical study into the effects of elongated capsules on the healing efficiency of liquid-based systems, *Comp Mater Sci* 2009, 47, 506.
97. Pang, J. W. C., Bond, I. P., A hollow fibre reinforced polymer composite encompassing self-healing and enhanced damage visibility, *Compos Sci Technol* 2005, 65, 1791.
98. Liu, H. A., Gnade, B. E., Balkus, K. J., A delivery system for self-healing inorganic films, *Adv Funct Mater* 2008, 18, 3620.
99. Dry, C. M., Sottos, N. R., Passive smart self-repair in polymer matrix composite materials, in *Proc SPIE 1916, Smart Structures and Materials 1993: Smart Materials*, 1993.
100. Mookhoek, S. D., Fischer, H. R., van der Zwaag, S., Alginate fibres containing discrete liquid filled vacuoles for controlled delivery of healing agents in fibre reinforced composites, *Compos Part A Appl S* 2012, 43, 2176.
101. Toohey, K. S., Sottos, N. R., White, S. R., Characterization of microvascular-based self-healing coatings, *Exp Mech* 2009, 49, 707.
102. Nesterova, T., Self-healing anticorrosive coatings, PhD dissertation, Technical University of Denmark, Kongens Lyngby 2012.
103. Caruso, M. M., Delafuente, D. A., Ho, V., Sottos, N. R., Moore, J. S., White, S. R., Solvent-promoted self-healing epoxy materials, *Macromolecules* 2007, 40, 8830.
104. Mookhoek, S. D., Mayo, S. C., Hughes, A. E., Furman, S. A., Fischer, H. R., van der Zwaag, S., Applying SEM-based X-ray microtomography to observe self-healing in solvent encapsulated thermoplastic materials, *Adv Eng Mater* 2010, 12, 228.

105. Kumar, A., Stephenson, L. D., Murray, J. N., Self-healing coatings for steel, *Prog Org Coat* 2006, 55, 244.

106. Samadzadeh, M., Boura, S. H., Peikari, M., Ashrafi, A., Kasiriha, M., Tung oil: an autonomous repairing agent for self-healing epoxy coatings, *Prog Org Coat* 2011, 70, 383.

107. Sugama, T., Gawlik, K., Self-repairing poly(phenylenesulfide) coatings in hydrothermal environments at 200 degrees C, *Mater Lett* 2003, 57, 4282.

108. Rey, R., Javierre, E., Garcia, S. J., van der Zwaag, S., Garcia-Aznar, J. M., Numerical study of the scratch-closing behavior of coatings containing an expansive layer, *Surf Coat Technol* 2012, 206, 2220.

109. Bohm, S., McMurray, H. N., Powell, S. M., Worsley, D. A., Novel environment friendly corrosion inhibitor pigments based on naturally occurring clay minerals, *Mater Corros* 2001, 52, 896.

110. Zheludkevich, M. L., Poznyak, S. K., Rodrigues, L. M., Raps, D., Hack, T., Dick, L. F., Nunes, T., Ferreira, M. G. S., Active protection coatings with layered double hydroxide nanocontainers of corrosion inhibitor, *Corros Sci* 2010, 52, 602.

111. Shchukin, D. G., Lamaka, S. V., Yasakau, K. A., Zheludkevich, M. L., Ferreira, M. G. S., Mohwald, H., *J Phys Chem C* 2008, 112, 958.

112. Ferrer, E. L., Rollon, A. P., Mendoza, H. D., Lafont, U., Garcia, S. J., Double-doped zeolites for corrosion protection of aluminium alloys, *Micropor Mesopor Mater* 2014, 188, 8.

113. Dias, S. A. S., Lamaka, S. V., Nogueira, C. A., Diamantino, T. C., Ferreira, M. G. S., Sol-gel coatings modified with zeolite fillers for active corrosion protection of AA2024, *Corros Sci* 2012, 62, 153.

114. Wu, L., Zhang, J. P., Li, B. C., Fan, L., Li, L. X., Wang, A. Q., Facile preparation of super durable superhydrophobic materials, *J Colloid Interface Sci* 2014, 432, 31.

115. Yilgor, I., Bilgin, S., Isik, M., Yilgor, E., Facile preparation of superhydrophobic polymer surfaces, *Polymer* 2012, 53, 1180.

116. Imaizumi, K., Ohba, T., Ikeda, Y., Takeda, K., Self-repairing mechanism of polymer composite, *Mater Sci Res Int* 2001, 7, 249.

117. Jamsa, S., Mahlberg, R., Holopainen, U., Ropponen, J., Savolainen, A., Ritschkoff, A. C., Slow release of a biocidal agent from polymeric microcapsules for preventing biodeterioration, *Prog Org Coat* 2013, 76, 269.

118. Koh, E., Baek, S. Y., Kim, N. K., Lee, S., Shin, J., Kim, Y. W., Microencapsulation of the triazole derivative for self-healing anticorrosion coatings, *New J Chem* 2014, 38, 4409.

119. Fu, J. J., Chen, T., Wang, M. D., Yang, N. W., Li, S. N., Wang, Y., Liu, X. D., Acid and alkaline dual stimuli-responsive mechanized hollow mesoporous silica nanoparticles as smart nanocontainers for intelligent anticorrosion coatings, *ACS Nano* 2013, 7, 11397.

120. Lv, L. P., Zhao, Y., Vilbrandt, N., Gallei, M., Vimalanandan, A., Rohwerder, M., Landfester, K., Crespy, D., Redox responsive release of hydrophobic self-healing agents from polyaniline capsules, *J Am Chem Soc* 2013, 135, 14198.

121. Vimalanandan, A., Lv, L. P., Tran, T. H., Landfester, K., Crespy, D., Rohwerder, M., Redox-responsive self-healing for corrosion protection, *Adv Mater* 2013, 25, 6980.

122. Dispinar, T., Colard, C. A. L., Du Prez, F. E., Polyurea microcapsules with a photocleavable shell: UV-triggered release, *Polym Chem* 2013, 4, 763.

123. Zheludkevich, M. L., Shchukin, D. G., Yasakau, K. A., Mohwald, H., Ferreira, M. G. S., Anticorrosion coatings with self-healing effect based on nanocontainers impregnated with corrosion inhibitor, *Chem Mater* 2007, 19, 402.

124. Zheludkevich, M. L., Serra, R., Montemor, M. F., Ferreira, M. G. S., Oxide nanoparticle reservoirs for storage and prolonged release of the corrosion inhibitors, *Electrochem Commun* 2005, 7, 836.

125. Scheltjens, G., Diaz, M. M., Brancart, J., Van Assche, G., Van Mele, B., A self-healing polymer network based on reversible covalent bonding, *React Funct Polym* 2013, 73, 413.

126. Zhang, Z. H., Hu, Y. L., Liu, Z. S., Guo, T. Y., Synthesis and evaluation of a moisture-promoted healing copolymer, *Polymer* 2012, 53, 2979.

127. Bose, R., Lafont, U., Vega, J. M., Garcia, S. J., Van der Zwaag, S., *Methods to Monitor and Quantify (Self-)Healing in Polymers and Polymer Systems*, Wiley-VCH verlag GmbH, Weinheim 2013.

128. Jorcin, J. B., Scheltjens, G., Van Ingelgem, Y., Tourwé, E., Van Assche, G., De Graeve, I., Van Mele, B., Terryn, H., Hubin, A., Investigation of the self-healing properties of shape memory polyurethane coatings with the "odd random phase multisine" electrochemical impedance spectroscopy, *Electrochim Acta* 2010, 55, 6195.

129. Garcia, S. J., Markley, T. A., Mol, J. M. C., Hughes, A. E., Unravelling the corrosion inhibition mechanisms of bi-functional inhibitors by EIS and SEM-EDS, *Corros Sci* 2013, 69, 346.

130. Lamaka, S. V., Zheludkevich, M. L., Yasakau, K. A., Montemor, M. F., Cecilio, P., Ferreira, M. G. S., TiOx self-assembled networks prepared by templating approach as nanostructured reservoirs for self-healing anticorrosion pre-treatments, *Electrochem Commun* 2006, 8, 421.

131. Lamaka, S. V., Zheludkevich, M. L., Yasakau, K. A., Montemor, M. F., Ferreira, M. G. S., High effective organic corrosion inhibitors for 2024 aluminium alloy, *Electrochim Acta* 2007, 52, 7231.

132. Gonzalez-Garcia, Y., Mol, J. M. C., Muselle, T., De Graeve, I., Van Assche, G., Scheltjens, G., Van Mele, B., Terryn, H., SECM study of defect repair in self-healing polymer coatings on metals, *Electrochem Commun* 2011, 13, 169.

133. Gonzalez-Garcia, Y., Garcia, S. J., Hughes, A. E., Mol, J. M. C., A combined redox-competition and negative-feedback SECM study of self-healing anticorrosive coatings, *Electrochem Commun* 2011, 13, 1094.

134. Garcia, S. J., Wu, X. M., van der Zwaag, S., Combined electrochemical impedance spectroscopy and X-ray-computed tomography study of the effect of a silyl ester on delamination and underfilm pit formation in a coated AA7050 sample, *Corrosion* 2014, 70, 475.

135. Lafont, U., van Zeijl, H., van der Zwaag, S., Increasing the reliability of solid state lighting systems via self-healing approaches: a review, *Microelectron Reliab* 2012, 52, 71.

136. Jin, H. H., Mangun, C. L., Stradley, D. S., Moore, J. S., Sottos, N. R., White, S. R., Self-healing thermoset using encapsulated epoxy-amine healing chemistry, *Polymer* 2012, 53, 581.

137. Mardel, J., Garcia, S. J., Corrigan, P. A., Markley, T., Hughes, A. E., Muster, T. H., Lau, D., Harvey, T. G., Glenn, A. M., White, P. A., Hardin, S. G., Luo, C., Zhou, X., Thompson, G. E., Mol, J. M. C., The characterisation and performance of Ce(dbp)(3)-inhibited epoxy coatings, *Prog Org Coat* 2011, 70, 91.

138. Lane, M. W., Roush, A., Callahan, S. E., Repair of dielectric interfaces with chemistry specific coupling agents, *AIP Conf Proc* 2009, 1143, 71.

139. Lafont, U., Moreno-Belle, C., van Zeijl, H., van der Zwaag, S., Self-healing thermally conductive adhesives, *J Intel Mater Syst Struct* 2014, 25, 67.

CHAPTER 5

Stimuli-Responsive Polymers as Active Layers for Sensors

Sergio Granados-Focil

Gustaf Carlson School of Chemistry and Biochemistry, Clark University, Worcester, MA, USA

5.1 INTRODUCTION

The importance of sensors in life is unequivocal as they allow the perception of outside stimuli. All living organisms have specialized senses that have evolved to help them identify the world around them by obtaining both chemical and physical information about their surrounding environment. The same principles have been implemented during the design of "physical" sensors. This gave rise to a large variety of devises that can be used to measure the physical properties of an object such as the temperature, pressure or electrical conductivity as well as their chemical properties [1]. According to the International Union of Pure and Applied Chemistry, a chemical sensor is "a device that transforms chemical information, ranging from the concentration of a specific sample component to total composition analysis, into a useful analytical signal" [1]. The process of chemical sensing can be divided into the three distinctive stages of selective recognition, reversible signal transduction, and transmission of the generated signal in the form of an electrical signal. This process is illustrated in Fig. 5.1.

Chemical sensors can be categorized into five major groups (depending on the signal transduction): thermal, mechanical, magnetic, optical, and electrical sensors. The latter are by far the most studied type of chemical sensors. However, traditional sensors are considered to be inherently "passive." The next step in the development of sensing technologies is to create sensors that can automatically adapt to the measuring tasks and exhibit self-diagnostic and self-calibrating properties. These "smart" sensors would allow the user to actively change the sensor response mechanism by the application of external stimuli such as light, temperature, ionic strength, and electrical current [2]. The development of these stimuli-responsive devices

Functional Polymer Coatings: Principles, Methods, and Applications, First Edition.
Edited by Limin Wu and Jamil Baghdachi.

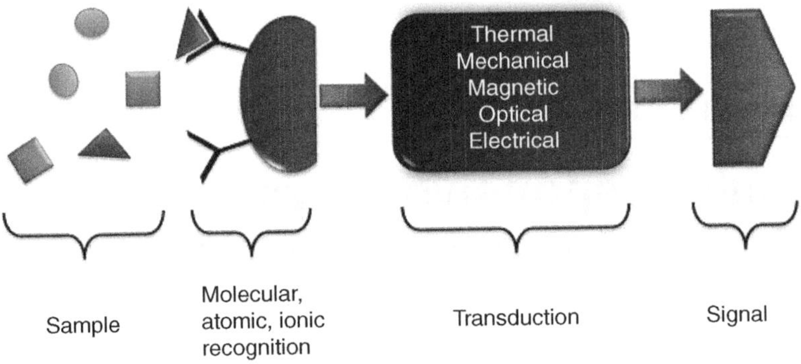

FIG. 5.1 Chemical sensor operation.

requires new materials that are capable of transforming an external signal into either a chemical or a conformational change.

This chapter will describe the latest progress in the field of stimuli-responsive soft materials with special emphasis on stimuli that could be used to design dynamic sensors, such as temperature changes, electromagnetic radiation exposure, magnetic fields, electrical fields, and selective binding of biochemically relevant molecules. A second section will describe the progress in the growing area of multi-stimuli-responsive polymers. The development of models describing the effect of simultaneous or sequential stimuli on the size, chain stiffness, and solvent compatibility of these macromolecules will lead to the development of complex multi-signal processing sensors. These devices will rely on the stimuli-driven changes in polymer structure to control the diffusion and binding of analytes into the sensing layers. The latest trends in the use of stimuli-responsive polymers, as active layers in a wide variety of sensors, will then be summarized. Finally, the remaining challenges and future directions for the development of stimuli-responsive polymeric materials for "smart" sensors will be discussed.

Several essential fields for the development of the next generation of smart sensing layers, such as the development of multistimuli-responsive materials, and stimuli-responsive electrochemical sensors, have reached a crucial state of advancement where there may have produced "solutions looking for a problem." The purpose of this review is to group in the same document the results from disciplines that may not regularly exchange ideas, to help identify promising solutions and approaches that could be useful across fields.

5.2 STIMULI-RESPONSIVE SOFT MATERIALS

The vast majority of the stimuli-responsive soft materials either are polymeric in nature or involve the use of self-assembled amphiphilic structures. They can be used to produce solid structures, gels, or solutions that share the ability to respond to chemical, physical, or biochemical stimuli [3]. In solution or the swollen gel state, the stimuli–response can be divided into changes in the macromolecular size, secondary structure, solubility, and degree of supramolecular interaction. The trigger of these responses generally involves reversible formation and dissociation of non-covalent bonds (hydrogen bonding, hydrophobic effects, electrostatic interactions, etc.), simple reactions (e.g., acid–base reactions) of polymeric side groups, and/or osmotic pressure gradients arising from such interactions. The most common chemical stimuli that have been used to develop "smart" materials include changes in the pH [4], ionic strength [5], selective absorption of solvents or gases [6], and redox reactions [7]. In terms of physical stimuli, the most promising systems involve the use of changes in electrical fields [8], temperature [9], electromagnetic radiation [10], or mechanical stress [11].

A very promising, relatively new, category is the bioinspired use of responses to ligands, enzymes, antigens, and other biochemical agents [3]. Many biological processes use feedback-controlled communication via stimuli-responsive polymers (i.e., proteins, nucleic acids, polysaccharides, etc.) that change conformation due to changes in the environment. The same functionality can be added to their synthetic counterparts, producing materials capable of acting, not only as passive structural supports but also as dynamic stimuli transducers. Macromolecules bearing functional groups that are susceptible to stimuli-induced changes in chemical nature (polarity, solubility, shape, etc.) can amplify the relatively small changes in chemical structure into large changes in the macroscopic properties of the materials.

5.2.1 Thermally Responsive Polymers

The use of temperature changes as an external stimulus is particularly attractive because it can be provided noninvasively to a wide variety of systems. Several mechanisms can be used to achieve a temperature-driven response, depending on the physical state of the polymeric material, for solid matrices; most examples fall into one of two categories, shape memory materials and thermotropic liquid crystalline systems [12].

Shape memory systems typically consist of thermoplastic elastomers with two domains, one with high glass transition temperature (T_g1) and a second with either a lower T_g (T_g2) or a melting temperature, T_m, that enables the temperature-responsive behavior. These materials can thus be deformed above the highest T_g1 to produce a "permanent" shape. Deformation at a temperature between T_g1 and T_g2, or T_m, allows the "programming" of a second shape that can be fixed as the polymer is cooled below the lowest transition temperature (T_g2 or T_m), known as the "switching" temperature. The permanent shape can be recovered by heating the material above the switching temperature. Many examples of this approach have been reported and are extensively reviewed [13].

Liquid crystalline polymers can act as temperature-responsive switches, if the LC-isotropic transition results in a significant change in dimensions or refractive index. These behaviors have been exploited to develop materials with potential applications as artificial muscles [14] and the development of thermochromic coatings [15]. In solution, or as gels, the most widely studied stimuli-responsive polymeric systems are those that undergo a temperature-triggered change in solubility. The most intensely explored category of materials in recent years is the one exhibiting a lower critical solution temperature (LCST). The overwhelming majority of examples involve N-substituted polyacrylamides with different pendant chains [16, 17] and other classes of polymers such as poly(oligoethyleneoxide-(meth)acrylate)s [18] or poly(2-oxazoline)s [19]: the most commonly used material is poly(*N*-isopropylacrylamide) (PNIPAM). PNIPAM is an attractive choice because its LCST is close to the human body temperature at approximately 32°C [9]. A recent review describes the use of other materials beyond PNIPAM [20]. Some examples of potential applications of thermoresponsive polymers exhibiting LCST include biomedical devices, sensors, and water purification. Their use to generate multistimuli-responsive systems has received significant attention in recent years and will allow the creation of complex stimuli-responsive nanostructures and sensing layers [21]. Multiresponsive polymers will be described in more detail in a separate section of this chapter.

5.2.2 Field-Responsive Polymers

A large fraction of the stimuli-responsive polymers described in the literature exhibit relatively slow responses due to transport phenomena limitations of the relevant stimulus, such as ion diffusion for pH-responsive systems or heat transfer in temperature-sensitive systems [22]. An attractive alternative to generate materials with faster stimuli responses consists on using electric, magnetic, or electromagnetic fields. Their relatively fast response simplifies device fabrication by facilitating signal processing and control [23].

5.2.2.1 Electro-Responsive Polymers Polymeric responses to electrical stimuli occur through the collapse of a gel within an electric field, an electrochemical reaction, an electrically induced binding–debinding event, metal-charged polymer interactions, electrorheological phenomena, or changes in the polymer's electrophoretic mobility [24]. The majority of the electro-responsive polymeric matrices reported in the literature are polyelectrolyte hydrogels. However, there are a few examples of neutral polymers exhibiting electric field responses in nonconductive media. These systems tend to require a polarizable or charged component able to respond to electric field. One of these example involves the use of surface-charged colloidal titania particles embedded within a lightly cross-linked polydimethylsiloxane matrix [24].

5.2.2.2 Magneto-Responsive Polymers Traditionally, response to a magnetic field has been achieved by combining magnetic particles with polymeric materials and stabilizing the mixture through supramolecular interactions [25]. In more recent years, the development of more sophisticated polymer synthesis methods have allowed the covalent grafting of polymeric chains to magnetic particles, generating more stable particle/polymer composites. Pyun and coworkers have reported the use of nitroxide-mediated radical polymerization to prepare polymeric surfactants able to stabilize magnetic colloidal assemblies. The morphology of these nanoparticles can be tuned by varying the solvent and casting conditions [26]. In another example, magneto-responsive gels were prepared by surface-initiated ATRP. The resulting polymer-coated nanoparticles produced a gel with the magnetically active iron nanoparticles acting as cross-linking points [27].

5.2.2.3 *Photoresponsive Polymers* Photoresponsive polymers can be generated by synthesizing macromolecules whose properties change upon irradiation with light of a specific wavelength [28]. In general, the changes result from light-induced isomerizations of functional groups within the polymer backbone or side groups [29]. A wide variety of applications may be developed from these systems, including polymer rheological control, photomechanical actuation, tissue engineering, drug delivery, and sensors [30].

By far, the most widely studied examples use azobenzene as their light-responsive motifs [28]. This chromophore is attractive due to its well-documented light-induced *cis*-to-*trans* isomerization, which leads to a fast change in electronic structure, shape, and polarity (see Fig. 5.2 [31]. The change in polarity associated with the light-induced isomerization process has been used to produce photoresponsive micelles and vesicles. Azobenzene motifs were introduced into amphiphilic block copolymer structures to control the aggregation behavior of the resulting micelles upon irradiation [32]. Azobenzene-derived chromophores have been introduced into many polymeric backbones, such as poly(*N*-hydroxypropylmethacrylamide) (PHPMA) [33], poly(acrylic acid) (PAA) [34], poly (2-dimethylaminoethyl-methacrylate) [35], and PNIPAM [36]. Examples of photoresponsive dendritic structures bearing azobenzene derivatives have also been reported [37].

Besides azobenzene motifs, other chromophores have been introduced into polymeric structures to generate photoresponsive materials. Many examples have been reported using spiropyran as the light-responsive motif [38]. The spiropyran groups undergo a drastic polarity increase upon irradiation-induced isomerization from the spirostructure to the merocyanine zwitterionic structure (see Fig. 5.2b). This isomerization is reversible upon exposure of the merocyanine to visible light or

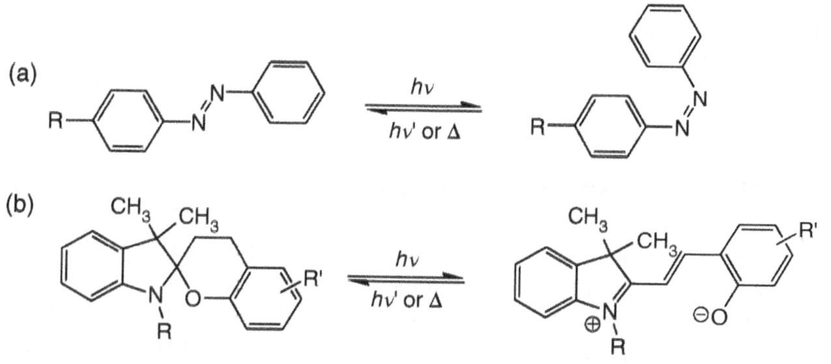

FIG. 5.2 Reversible photoinduced transformations of (a) azobenzene and (b) spiropyran derivatives. Reproduced with permission from Ref. 22. © Elsevier.

temperatures above 65°C. Many photoresponsive, spiropyran-bearing polymers have been reported using backbones such as PAA [39], PHPMA [38], and PNIPAM [40]. Light-responsive polymeric micelles have been produced using diblock copolymers containing spiropyran-containing moieties [41]. The micellar structure was completely disrupted upon UV irradiation and recovered when the solution was exposed to visible light. In another example, PHPMA was copolymerized with a cinnamate-containing monomer. Photoisomerization of the cinnamate motifs produced polymers with increased polarity and higher solubility in water. In contrast with other photoactive groups, the cinnamate motifs are chemically inert and exhibit thermal stability in both the *trans* and the *cis* isomeric forms [42].

Most reports describing photoresponsive polymeric materials require the use of UV and visible wavelengths; since that wavelength range is readily absorbed by living tissue, the noninvasive biomedical applications are limited. In contrast, infrared (IR) wavelengths can penetrate the skin and could be used as a photoactivation mechanism. Frechet and coworkers have reported the synthesis of diblock copolymer micelles containing 2-diazo-1,2-naphthoquinone in their core as an IR-active motif. Upon irradiation, the Wolff rearrangement of the hydrophobic 1,2-naphthoquinone units to the hydrophilic 3-indene carboxylate units increases the polarity of the micelle cores destabilizing their structures and leading to payload release [43]. In another example, diblock copolymer micelles containing the IR photocleavable 2-nitrobenzylester motifs in their cores were also shown to dissociate upon exposure to near-infrared radiation [44].

5.2.3 Biologically Responsive Polymer Systems

Interest in stimuli-responsive polymers has grown significantly within the biomedical research community. The wide range of potential applications, such as controlled drug delivery, smart surfaces for tissue engineering, sensing/diagnostic layers, or the in situ generation of porous networks, makes the use of low-cost, highly tunable, biocompatible polymeric materials very attractive [45].

Driven by these potential, many efforts have focused on developing synthetic polymers able to change their properties upon exposure to a biologically relevant small molecule or macromolecule. This has been achieved by either incorporating functional groups known to strongly interact with the species of interest or taking advantage of a chemical change resulting from conjugation of a biological substrate with the synthetic polymer.

5.2.3.1 Glucose-Responsive Polymers Glucose is a challenging yet potentially very rewarding target for stimuli-responsive materials. Efficient response to glucose *in vivo* coupled with fast glucose-driven actuation could produce systems able to test for diabetes in a passive, noninvasive way.

So far, the most common approach to achieve glucose response is to utilize a pH-responsive polymer that contains the glucose oxidase (GOx) enzyme. The by-product of the enzyme-catalyzed glucose oxidation (gluconic acid) will change the pH of the surrounding solution, which can be sensed by the pH-responsive polymer backbone. For applications intended to treat diabetes, the pH response produces a change in hydrodynamic volume of an insulin-containing matrix. Expansion or contraction of this matrix will result in glucose-driven insulin release. One example is the work of Imanishi and coworkers who modified a cellulose film with GOx-conjugated PAA [46]. At pH above 7, the PAA matrix remains extended due to the repulsive interaction between the carboxylate groups. This expansion maintained the pores of the cellulose membrane closed. As the gluconic acid was produced from the GOx-catalyzed glucose oxidation, the protonation of the neighboring carboxylates resulted in the collapse of the PAA chains unblocking the cellulose membrane pores and releasing the insulin trapped within the pores. Similar approaches using other pH-responsive materials, such as PAA-PVDF [47], PNIPAM-*co*-PAA [48], and a copolymer resulting from the copolymerization of dimethylacrylamide (DMA) with a sulfadimethoxine-bearing monomer [49], have been reported. The opposite approach where the gluconic acid production leads to a polymer chain expansion has also been explored using macromolecules containing basic motifs [50]. An interesting extension of this concept consists in the use of polymers that respond to the increased concentration of hydrogen peroxide, also a by-product of GOx-catalyzed glucose oxidation [51].

A different approach consists on taking advantage of binding between lectins and glucose. The lectin most heavily used to generate glucose-responsive matrices is concanavalin A (ConA). One example describes a ConA–glycogen gel that undergoes a gel–sol transition when the free glucose displaces the matrix-bound glycogen due to the more favorable glucose–ConA binding compared with the ConA–glycogen binding [52]. In a complementary approach, Hoffman and coworkers synthesized a hydrogel by copolymerizing vinyl-functionalized ConA macromonomers with glucose-bearing monomers. The swelling of the resulting material was dependent in the concentration of free glucose that will compete for the ConA-binding sites forming the hydrogel cross-links [53].

There have also been examples where the glucose response is generated by a polymer with nonnaturally occurring boronic acid motifs. Boronic acid can reversibly bind to sugars containing alpha-diols and beta-diols, which has made them a popular ligand for chromatography applications [3] (see Fig. 5.3). The cyclic boronic esters resulting from the condensation of the boronic acid (1) and the diol are hydrolytically

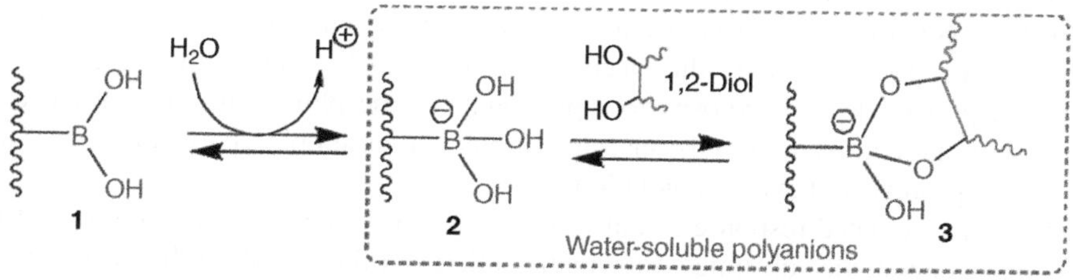

FIG. 5.3 Aqueous ionization equilibria of boronic acids. As the concentration of diol increases, the equilibria shift toward the anionic boronate forms of the boronic acid. Reproduced with permission from Ref. 22. © Elsevier.

unstable, but the esters produced from condensation of the anionic form (2) and the diol to yield a boronate ester (3), shifting the equilibrium toward the formation of the negatively charged forms (2) and (3) [54]. As the concentration of glucose rises, the relative amount of negatively charged boronates increases, making the system more hydrophilic. Thus, the charge density of the boronic acid–bearing polymers will depend on pH and on the concentration of diols present in solution [55]. Most of the examples using boronic acids as the glucose-sensing motifs consisted of hydrogels that changed volume upon exposure to glucose. A glucose-responsive hydrogel was synthesized by using a terpolymer of 3-acrylamidophenylboronic acid (APBA), (*N,N*-dimethylamino)propylacrylamide (DMAPA), and DMA [56]. At physiological pH, the boronic acid groups were complexed with poly(vinyl alcohol) (PVA), which acted as a cross-linker. Displacement of PVA motifs due to competitive binding of added glucose led to an increase in the swelling of the hydrogel due to a decrease in the cross-link density. Similar examples used PVA complexed with poly(*N*-vinyl-2-pyrrolidone-*co*-APBA) [57] or poly(DMA-*co*-3-methacrylamido-phenylboronic acid-*co*-DMAPA-*co*-butyl methacrylate) [58].

As illustrated in Fig. 5.3, solubility modulation of polymeric boronic acid requires the initial deprotonation of the boronic acid at physiological pH since most boronic acids have pK_a higher than 8. Potential *in vivo* applications would benefit from using more acidic boronic acids, with pK_a closer to 7.4. In one example, Karaoke and coworkers reported the synthesis of a glucose-responsive copolymer containing 4-(1,6-dioxo-2,5-diaza-7-oxamyl) phenylboronic acid (p$K_a \approx 7.8$) and *N*-isopropylacrylamide (NIPAM) [59].

The majority of the glucose-responsive polymers using boronic acid binding have been synthesized via traditional radical initiated polymerization, producing random copolymers [60] and gels [61]. Recently, there have been reports of better-defined block copolymer structures using ATRP [62] and RAFT [63]. The complex structures accessible from the diblock copolymer architectures could potentially lead to the development of glucose-responsive, insulin-delivering nanocapsules.

5.2.3.2 *Enzyme-Responsive Polymers*

Enzyme response can be achieved by combining an enzyme-sensitive substrate with an additional component that transforms the enzyme-catalyzed reaction products into a macroscopic transition [64]. Enzymatic activity can lead to hydrogel volume changes, changes in surface properties, and changes in the supramolecular architecture. Enzyme-induced response can be used to generate gels *in situ* in a noninvasive fashion. Xu and coworkers used enzymatic dephosphorylation to induce a gel–sol transition. Fluorenylmethyloxycarbonyl-tyrosine phosphate was exposed to a phosphatase; the phosphate removal led to a reduction in electrostatic repulsions, supramolecular assembly via pi-stacking of the fluorenyl groups, and eventual gelation.

Enzyme-mediated response can also be achieved by adding functional groups that can act as enzymatic substrates. Enzymatic action can then result in the formation of new covalent bonds. Ulijn and coworkers used a blood-clotting enzyme, transglutaminase to promote the formation of a hydrogel via the cross-linking of lysine and glutamine residues [64]. In another example, Hubbell and coworkers produced hydrogels in the presence of cells via a Michael reaction between a vinyl sulfone-functionalized telechelic PEG and cysteine-functionalized proteases [65]. The hydrogels produced by this process could be locally degraded in response to the cell-surface proteases illustrating the possibility of producing gels with templated paths for cell migration. Chymotrypsin-responsive materials were produced by introducing a labile CYKC tetrapeptide sequence as the cross-linker in polyacrylamide hydrogels [66]. When exposed to chymotrypsin solutions, the gels dissolved quickly dissolved due to the degradation of CYKC by chymotrypsin.

5.2.3.3 Antigen-Responsive Polymers Highly specific antigen–antibody interactions mediate complex immune responses that recognize and neutralize foreign objects in the body. Antigen–antibody interactions have been used to produce stimuli-responsive hydrogels containing either trapped or covalently attached antigen and antibody motifs within the hydrogel matrix [67]. In one report, antigen-responsive hydrogels were produced by covalently attaching antibody Fab′ fragments to a PNIPAM-*co*-PMBA hydrogel network. Binding of the antigen to the Fab′ fragment produced reversible changes in volume (see Fig. 5.4).

5.2.3.4 Redox-/Thiol-Responsive Polymers Polymers containing disulfide linkages can be used to generate both redox- and thiol-responsive matrices [68]. By taking advantage of the large difference in glutathione (GSH) concentration between the inside the cell and its surrounding environment [69], it is possible to design delivery systems that release their payload only after having penetrated the cell membranes.

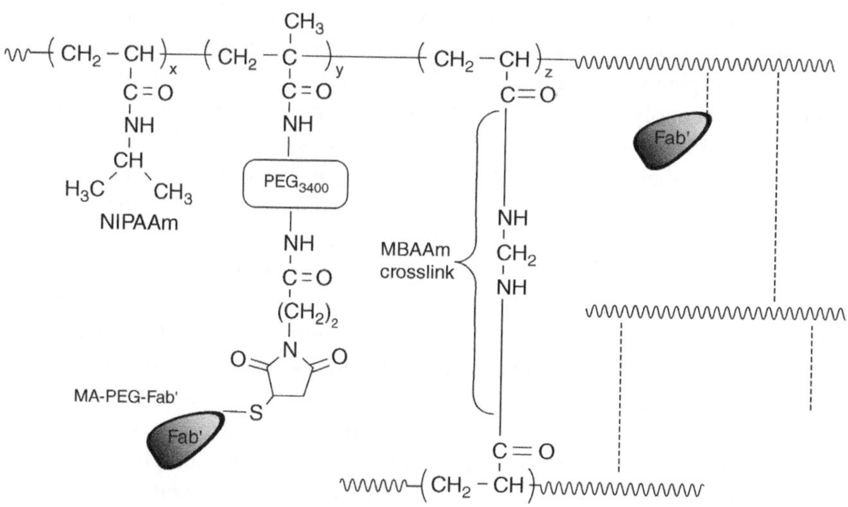

FIG. 5.4 Chemical structure of Fab′-containing antigen-responsive hydrogels. Addition of free sodium fluorescein led to competitive binding and loss of antigen–antibody cross-links. Reproduced with permission from Ref. 67. © Wiley

This significant variation in concentration has been utilized to design thiol-/redox-responsive drug delivery systems that specifically release biologically active molecules upon entry into cells. There are many examples of shell–cross-linked diblock copolymer micelles that utilize the S–S bond cleavage to trigger payload release [70]. A slightly different approach illustrates that redox response can also be induced in surfactant-based aggregates. For example, Thayumanavan and coworkers reported that supramolecular polymer–surfactant complexes can form micelles susceptible to thiol-induced dissociation [71].

5.2.4 Multistimuli-Responsive Materials

Recently, there has been an increased interest in producing materials that can respond to two or more types of stimuli, such as light and temperature, temperature and pH, or light and pH. Many other combinations, including redox reactions, and magnetic or electric fields are also conceivable. The addition of sensitivity to one or more additional stimuli will result in a richer landscape for the fine tuning of functionality and properties. The majority of the examples currently in the literature have been published in the last 10 years and focus primarily on light- and temperature-responsive systems, but considering that the field is still in its infancy, there is plenty of room for expansion [21].

5.2.4.1 *Temperature- and Light-Responsive Polymers* In the majority of the reports describing dual sensitivity to light and temperature changes, these properties are achieved by adding photochromic motifs to structures that exhibit LCST or upper critical solution temperature (UCST) behavior. The solubility of the resulting materials is affected by the changes in polarity originating from the isomerization of the photoresponsive motifs. The final transition temperatures can be tuned by varying the amount of photochromic groups added to the polymer chains. The first report of combined light and temperature response achieved this by using a copolymer of NIPAM with *N*-(4-phenylazo-phenyl)acrylamide [72]. LCST shifts of 6°C were observed after UV irradiation for a copolymer containing 2.7 mol% of the photochromic monomer.

Inspired by this work, many other examples have been reported in the last 20 years, including the synthesis of telechelic stimuli-responsive polymers via RAFT [73]; polymer brushes via ATRP [74]; and many other examples of copolymers containing azobenzene-bearing monomers and NIPAM, *N*-cyclopropylacrylamide [16], *N*-ethyl-*N*-methylacrylamide [75], and *N,N*-dimethylacrylamide [76]. Focused on the goal of producing self-assembled structures exhibiting multistimuli responses, many amphiphilic diblock and triblock copolymers have been synthesized [77]. One example illustrating the effect of multistimuli response on micelle formation is illustrated in Fig. 5.5.

Besides azobenzene, other photochromic motifs have been used to produce light- and temperature-responsive polymers. The most common examples are salicylideneaniline [79] and spiropyran [80]. In the case of the salicylideneaniline, the photochromic behavior stems from a photoinduced tautomerization process [81]. Copolymers of monomers bearing the photochromic groups as a side chain and NIPAM were synthesized and showed an aqueous LCST that depended on the amount of salicylideneaniline and its isomerization state.

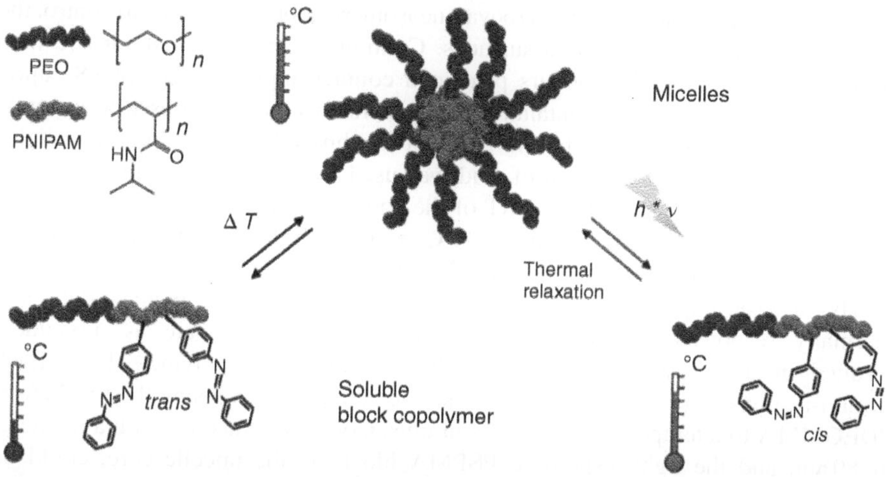

FIG. 5.5 Temperature- and light-controlled formation/disruption of micelles derived from PEO-b-PNIPAM block copolymers having azobenzene moieties in the thermoresponsive segment. Reproduced with permission from Ref. 78. © Royal Society of Chemistry.

As discussed before, spiropyran's photoinduced isomerization produces a large polarity change in the molecule. Spiropyran containing temperature- and light-responsive materials have been obtained by synthesizing copolymers of NIPAM and a photochromophore monomer. This copolymer's LCST ranged between 30 and 50°C, depending on spiropyran content [40]. Counterintuitively, the cloud point of the irradiated polymer did not increase as it would be expected for the resulting structure containing the more hydrophilic, zwitterionic, merocyanine motifs; instead it decreased after UV exposure. Furthermore, the ring-closing reverse reaction to the spiropyran form took more than 20 days to complete, suggesting that the merocyanine form is stabilized by interaction with the aqueous environment within the polymer backbone and that the presence of the zwitterionic motifs destabilizes the water NIPAM interactions, resulting in a decrease of the LCST.

A similar example of merocyanine stabilization by water was reported in 2004 [82]. In that case the authors used a copolymer of NIPAM and a spiropyran-containing acrylate to induce the photocomplexation of Pb^{2+} in aqueous solutions. PNIPAM hydrogels containing spiropyran groups have been synthesized [83]. The gels shrunk upon UV irradiation; the photoisomerization was accompanied by proton dissociation at neutral pH. The linear, thermally induced, bathochromic/ipsochromic shift of spiropyran's absorption maxima can be calibrated to use spiropyran-containing films as thermometers [84]. The temperature-induced bathochromic shifts can be ascribed to the formation of less polar domains due to polymer aggregation as the polymer backbone changes from a coiled to a globular structure above the LCST. The use of similar copolymers has been proposed for the fluorometric detection of cyanide anions at temperatures close to the LCST [85].

The light-induced spiropyran/merocyanine isomerization can be used to control the wettability of copolymer-coated surfaces. Coating a nanostructured surface with spiropyran-containing copolymers produced contact angle changes of 98° upon irradiation [86]. Spiropyran-containing copolymers have been also used to generate multiresponsive materials combining pH-, solvatochromic-, temperature-, and light-responsive ability [87]. Addition of acid was used to produce the isomerization of spiropyran to merocyanine. The LCST of the copolymer increased due to the higher hydrophilicity of the merocyanine domains. Conversely, the light-induced isomerization from merocyanine to spiropyran decreased the LCST.

The use of copolymers from light-responsive spiropyran methacrylate and thermally responsive di(ethylene glycol)methylethermethacrylate as monomers resulted in doubly responsive "schizophrenic" diblock copolymers [88]. Each stimuli-responsive block could be addressed selectively by exposing the PSPMA-b-PDEGMMA to a temperature or light stimulus. Core-shell micelles with a diameter of 80 nm, and the light-responsive PSPMA block on the micelle core, could be produced at 15°C under visible light. Upon UV exposure at room temperature, the micelles completely dissolved due to the increase in polarity of the PSPMA block. This transition was fully reversible, recovering the 80-nm micelles after exposure to visible light at 15°C. In a second experiment, the micelles were exposed to UV light, followed by a heating step producing 24-nm diameter inverse micelles with the thermally responsive PDEGMMA collapsed at the core and the light-responsive PSPMA, now in the merocyanine form, forming the shell (see Fig. 5.6). Triply responsive copolymers containing azobenzene as the light-responsive motif and PNIPAM as

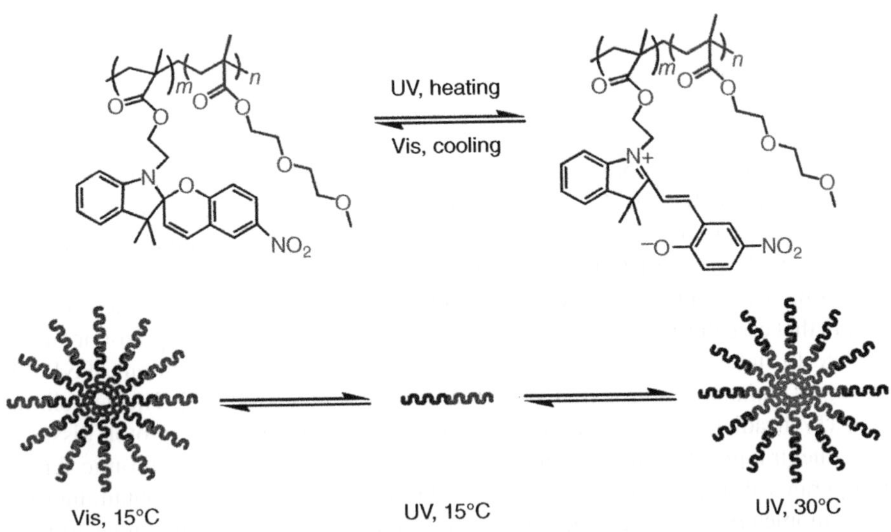

FIG. 5.6 Schematic representation of double-responsive micellization of "schizophrenic" PSPMA-b-PDEGMMA in aqueous solution. Depending on temperature and light, the block copolymer formed micellar structures with various segments as core and corona, respectively. Reproduced with permission from Ref. 88. © Wiley.

the temperature-responsive moieties have been prepared. These copolymer hydrogels respond to temperature, pH, and light [89].

5.2.4.2 *Thermoresponsive Materials Containing Cross-Linkable Motifs*

The light-induced dimerization of coumarin motifs has been used to produce thermo-responsive, photo–cross-linked gels with tunable swelling [90]. The authors describe temperature- and light-induced changes in the size of gels and nanoaggregates. Their system consists of a diblock copolymer with a PEO block and a poly(2-(2-methoxyethoxy)ethylmethacrylate-*co*-4-methyl-(7-(methacryloyl)oxy-ethyloxy]coumarin) (P(MEOMA-*co*-CMA)) block. Nanogels were produced by light-induced isomerization of the coumarin motifs, resulting in cross-linking of the micellar aggregates at temperatures below the LCST of the P(MEOMA-*co*-CMA) block. Exposure of the nanogels to light with a shorter wavelength (260nm) cleaved the coumarin dimers and reduced the cross-linking of the nanogels, resulting in a much higher swelling of the final material.

5.2.5 Stimuli-Responsive Hydrogels

An excellent review on the basics of stimuli-responsive behavior for different geometries and dimensions of stimuli-responsive cross-linked systems as well as their applications has been recently published [91]. The kinetics of the diffusion-controlled swelling processes in hydrogels have been described as being inversely proportional to the square of the gel radius. Therefore, the swelling/deswelling transitions occur faster in smaller gel particles. The first report describing the study of size dependence on solvent absorption and desorption phenomena in gels was done by Fillmore and coworkers [92].

One very attractive application of stimuli-responsive hydrogels is their use as catalyst supports. Immobilization of catalyst motifs within a polymeric scaffold improves catalyst stability and solubility while allowing the straightforward separation of the catalyst-gel particles by centrifugation and filtration [93]. The use of polymer supports allows the utilization of the catalysts under heterogeneous and homogeneous conditions [94]. Stimuli-responsive catalyst supports allow the control of substrate accessibility to the catalyst sites via the application of an external signal. This provides an efficient avenue to control the reaction progress [95]. Nanocomposite hydrogels produced by combination of inorganic nanoparticles and stimuli-responsive polymeric matrices have shown remarkable mechanical stability and swelling behavior [96]. There are several examples illustrating this approach using PNIPAM as the polymeric matrix to prepare silver [97] and palladium [98] nanoparticles as catalysts for oxidation and hydrogenation reactions. An interesting material was reported by Hamamoto and coworkers [99], who synthesized a PNIPAM-*co*-acrylamide copolymer containing quaternary ammonium moieties cross-linked with phosphotungstate anions. The resulting network can act as a hydrophobic component sponge at temperatures above the LCST (~60°C), where the hydrophobic component can be released from this matrix by lowering the temperature and increasing the hydrophilicity of the matrix. The phosphotungstate groups catalyzed the oxidative

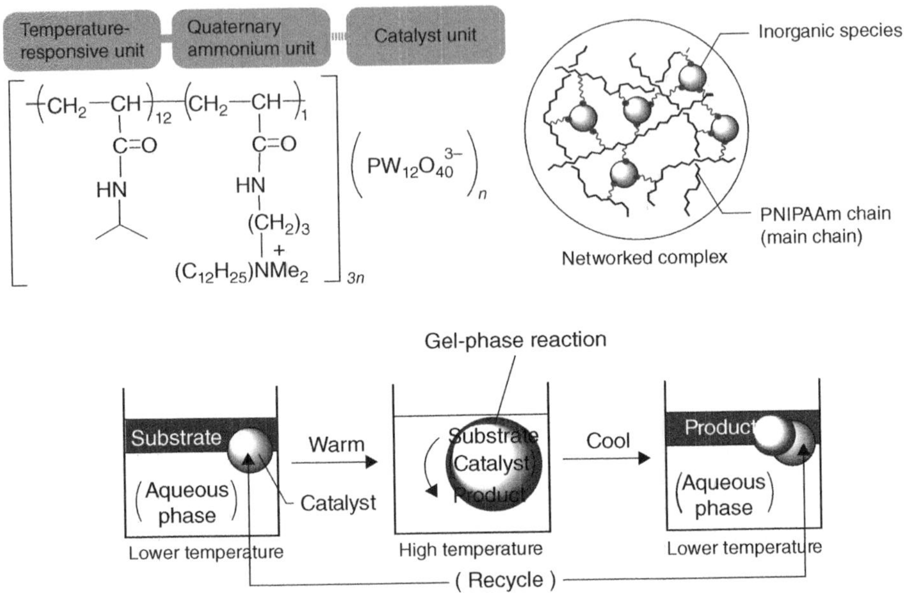

FIG. 5.7 PNIPAAm–$PW_{12}O_{40}^{3-}$ complex (upper left part); simplified structure of the PNIPAAm–inorganic complex network (upper right part); schematic representation of thermomorphic catalysis (lower part). Reproduced with permission from Ref. 99. © Wiley.

cyclization of pent-4-en-1-ol. This thermoresponsive catalyst system promoted substrate conversions higher than 80% in aqueous solution and allowed easy recovery of the catalytic matrix (see Fig. 5.7). A review illustrating the potential of these systems as catalysts and product separation matrices has been recently published by Ikegami and Hamamoto [100].

5.3 SENSORS FROM STIMULI-RESPONSIVE HYDROGEL LAYERS

Color-changing materials upon exposure to a particular solvent can be prepared by deposition of a PNIPAM microgel layer over a gold-coated glass substrate. This is then followed by deposition of a second gold layer on top of the microgel (see Fig. 5.8; [102]). The color produced by these devices depends on the composition and swelling response of the migrogels, which can be tuned to respond to temperature, pH, and analyte concentration. The etalons, also known as Fabry–Perot interferometers, can be used as sensors for pH, temperature, and glucose [103]. Several studies on the swelling kinetics of these systems have shown that faster responses can be achieved with thinner layers of the microgel and the coating gold layer [102].

Stimuli-responsive hydrogels can be potentially utilized to detect many types of analytes [104]. For example, DNA [105], humidity [106], and solvent-responsive layers [107] have been reported recently. Other areas of interest include the development of miniaturized pH sensors [108–112], sensors for toxic metallic ions

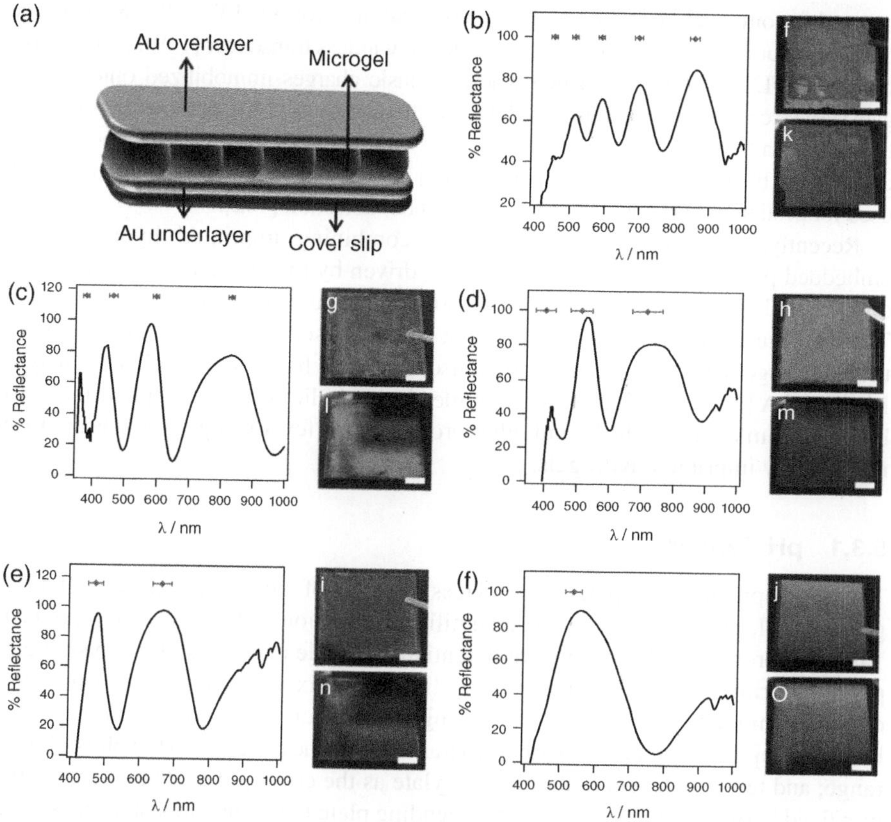

FIG. 5.8 (a) Schematic illustrating the structure of a poly (*N*-isopropylacrylamide) microgel-based etalon (b–f). Representative spectra for etalons composed of (b) large-diameter poly(*N*-isopropylacrylamide)-*co*-acrylic acid (PNIPAm-*co*-AAc) microgels (hydrodynamic diameter (DH) = 1548 ± 69 nm) (c) poly (*N*-isopropylacrylamide)-*co*-acrylamide (pNIPAm-*co*-AAm) (DH = 653 ± 10 nm) (d) medium-diameter poly (*N*-isopropylacrylamide)-*co*-acrylic acid (pNIPAm-*co*-AAc) microgels (DH = 653 ± 10 nm) (e) poly (*N*-isopropylacrylamide)-*co*-vinyl acrylic acid (pNIPAm-*co*-VAA) (DH = 603 ± 17 nm), and (f) small-diameter poly (*N*-isopropylacrylamide)-*co*-acrylic acid (pNIPAm-*co*-AAc) microgels (DH = 229 ± 10 nm), all obtained in pH 3 formate buffer. The average peak positions (± 1 standard deviation) from six spectra are shown in red above each spectrum. (f–j) Photographs for the etalons (b–f) in the dry state, respectively. (k–o) Photographs of the etalons (b–f) hydrated in pH 3 formate buffer. Scale bars in the photographs are 5 mm. Reproduced with permission from Ref. 101. © Elsevier.

such as Hg^{2+} and Pb^{2+} [113–119], and glucose sensors to monitor blood glucose levels in diabetics [120–135].

Other detection methods using hydrogel as the stimuli-responsive layer have been developed. For example, piezoresistive sensors where swelling of the hydrogel produces deflection of a silicon membrane have been developed [136]. These devices can be used for online environmental and biomedical monitoring [137, 138]. Miniaturized devices include the use of a flex-circuit inductive wireless transducer [139], an field effect transistor (FET)-based transformed where intrinsic charges immobilized onto the FET gate surface can be directly transduced into electrical signals [140], optical fibers coated at the tip with a hydrogel layer to produce a Fabry–Perot interferometer for high-resolution detection of the hydrogel's optical path length [129, 141], and magnetoelastic biosensors, used for pesticide (organophosporous compounds) detection [142].

Recently, the development of sensors combining imprinted hydrogels with embedded photonic crystals (PCs) has been driven by the interest in producing low-cost, visually readable, colorimetric sensors that require no external power source. The wide variety of functional groups that could be used to modulate the swelling of these hydrogel layers could enable their use as sensors for a large number of analytes. Bisphenol A [143], 3-pyridinecarboxamide [144], cholic acid [145], ephedrine [146], bovine serum albumin [147], and other proteins are a few examples of sensors using molecularly imprinted hydrogels.

5.3.1 pH Sensors

The development of implantable devices capable of online monitoring of blood glucose, pH, and CO_2 has received significant attention in the last few years [104, 148]. Piezoresistive pH sensors are potentially suitable candidates for physiological pH monitoring. In one example, a cross-linked matrix was synthesized using *N,N*-dimethylaminoethyl methacrylate, bearing the pH-sensitive tertiary amine group; hydroxypropyl methacrylate, to obtain a transition of the pH close to the physiological range; and tetraethyleneglycol dimethacrylate as the cross-linker. The hydrogel film was fixed between an Si substrate and a bending plate in the piezoresistive sensor, see Fig. 5.9a. Two other possible configurations are illustrated in Figs. 5.9b and c [137]. The hydrogel composition in this case was tuned to generate a close-to-linear response within the pH range of 6.8–7.8. The influence of the ionic strength can be quantified separately using a second sensor with a neutral pH-unresponsive hydrogel.

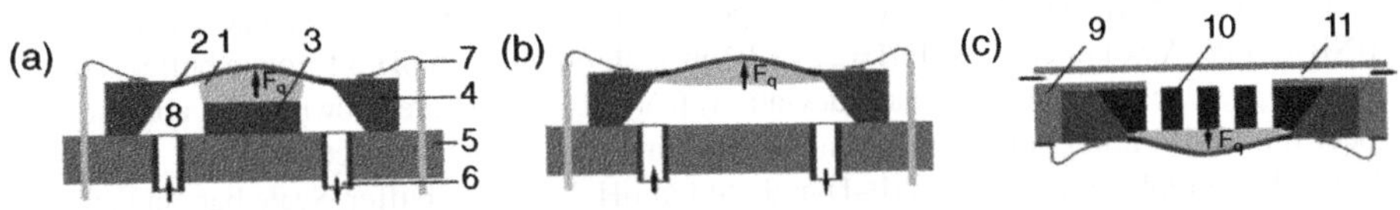

FIG. 5.9 Schematic cross-sectional view of the hydrogel-based piezoresistive pH micro-sensor in three different configurations: (a) a relatively thick hydrogel layer (100-mm range) is mechanically constrained at the bottom of the holder, (b) bimorph configuration with a thin hydrogel film directly bound to the bending plate, (c) a hydrogel sheet is clamped between a rigid grate and the chip. 1, hydrogel; 2, piezoresistor; 3, Si substrate; 4, Si pressure sensor chip; 5, socket; 6, fluidic channel for solution inlet–outlet; 7, electrical interconnect; 8, measuring solution; 9, backside clamp; 10, rigid grate; 11, fluidic channel (optional). Reproduced with permission from Ref. 108. © Elsevier.

Swelling or shrinkage of the responsive hydrogel due to a change of pH produces a certain pressure that deforms the bending plate. The bending is registered by 4 piezoresistors connected to a wheatstone bridge circuit. The deformation-induced mechanical strain in the piezoresistors changes the resistivity on the bridge circuit, proportionally affecting the output voltage. The mechanical work produced by the hydrogel swelling is transduced into an electrical signal. A different type of pH sensor using the swelling-induced change in the surface plasmon resonance (SPR) of noble metal nanoparticles embedded within a matrix of pH-responsive hydrogel has been reported [109]. In this particular example, thin gold layer (50 nm) was used as the SPR-active layer. The metallic layer was coated with the pH-responsive hydrogel, producing a top coating that changed refractive index upon swelling due to pH changes. This change in the refractive index led to shifts in the SPR maximum and extinction amplitude. The defection range of this pH sensor was 4.5–6.4 showing reproducible results after 30 days of use.

Griffete and coworkers reported a sensing layer built using an inverse opal hydrogel (IOH) structure with a planar defect layer [111]. This defect layer was introduced to increase the response and the decrease the response time. The defect-containing IOH showed a larger redshift upon swelling compared with the defect-free layers. Response times were measured in the highest sensitivity pH range between pH 7.5 and 8.5. The defect-containing IOH required approximately half the time to respond compared with the defect-free IOH sensor. Quantum dot (QD)–modified responsive hydrogels have been used to produce sensing layers that change photoluminescence intensity upon changing the pH [112]. The photoluminescence of the multilayer system was suppressed by excess negative charges generated upon the deprotonation of the pH-sensitive component, poly(methacrylic acid) PAA. Switching pH values from 8 to 3 resulted in a full recovery of the photoluminescence intensity. Measurements over several switching cycles showed that changes in the pH-induced photoluminescence were reversible and reproducible.

5.3.2 Metal Ion Sensors

The availability of drinking water is expected to be one of the biggest challenges facing humanity in the next 100 years. This makes the ability to reliably and cheaply monitor water quality a pressing problem. Visual color-changing sensors for trace heavy metal detection have been developed by using an Hg^{2+}-mediated thymine–thymine base pairing in DNA [119]. The hairpin structure produced by DNA in the presence of Hg^{2+} can be detected by binding of the fluorescent dye SYBR green I, which produces a strong green fluorescence. In the absence of Hg^{2+} binding of the DNA to SYBR green I, now in the random-coil conformation, produced a weak yellow fluorescence.

Almost no or yellow fluorescence was also observed when other metal ions like Zn^{2+}, Co^{2+}, Cu^{2+}, Mn^{2+}, Pb^{2+}, Fe^{3+}, Mg^{2+}, or Ca^{2+} were used. The best detection limits, 1.1 nM, were achieved using cationic hydrogels instead of their neutral counterparts. This same sensor design concept was used to develop Pb^{2+} sensors [117]. It was possible to detect 20 nM Pb^{2+} with the naked eye using a sample volume of 50 ml. The

US Environmental Protection Agency has defined a maximum acceptable limit for lead contamination in drinking water as 72 nM. Therefore, this sensor is suitable to analyze Pb^{2+} in drinking water. By using shape-encoded hydrogels, for example, a spherical for Pb^{2+} hydrogel and triangular for Hg^{2+} hydrogel, simultaneous analyte detection is possible.

A sensing layer was developed based on PC-containing interpenetrating hydrogel networks of PVA and PAA with poly(styrene) (PS) colloid crystals as the PC elements. This system can act as a pH sensor or as a metal ion sensor after functionalization of the PAA with a certain ligand that is selective for Ni^{2+}, Cu^{2+}, or Zn^{2+} [118]. All the photonic hydrogel films were mechanically robust, stable for at least 3 months, and exhibited responses that are completely reversible.

5.3.3 Humidity Sensors

Low-cost, no-power-consuming humidity sensors are a desirable alternative to the electronically based ones widely available today. These devices can be prepared from PC-bearing hydrogels [106, 149]. Xian and coworkers reported a system using 150-nm, monodisperse, latex spheres with a PS core and a PMMA-*co*-PAA shell. These polymeric PCs are then immobilized within poly(acrylamide) (PAAm) matrix by photochemically polymerizing a nanosphere/acrylamide suspension. When exposed to moisture, the hydrogel swells and the refractive index contrast between PC and the hydrogel matrix (PAAm) increases producing visible redshift in the hydrogel film: the color changed from transparent at 20% RH to violet, blue, cyan, green, and red as the RH increased to 50, 70, 80, 90, and 100%, respectively.

Xuan and coworkers produced a humidity sensor using magnetite colloids suspended in a poly(ethylene glycol) acrylate (PEGA) hydrogel [149]. The photonic structures were made by the magnetic self-assembly of the Fe_3O_4–SiO_2 colloids and fixed in the PEGA hydrogel by UV irradiation. The response time depended on film thickness and whether the hydrogel was swelling or shrinking. The best samples showed response times of 9 min for swelling induced by increases in RH and 3 min for the deswelling induced by decreases in RH.

One-dimensional PC lattices have been prepared to be used as solvent sensors for oil quality control, food industry quality assurance, and biotechnology [107]. The sensing membrane was produced by casting 10 alternating layers of PMMA-*co*-poly(hydroxyethyl methacrylate) (PHEMA)-*co*- poly(ethyleneglycol) dimethacrylate (PEGDMA) and titania. The high refractive index contrast between the titanium oxide and the polymer layers led to the expected photonics stopbands in several layers and the concomitant color changes. The shifts of the stopband upon exposure to carbon disulphide, ethyl acetate, ethanol, acetonitrile, dichloromethane, trichloromethane, butanone, tetrahydrofuran, 1,4-dioxane, acetone, dimethylformamide, and dimethylsulfoxide are 29, 37, 62, 113, 127, 171, 202, 253, 267, 285, 388, and 420 nm, respectively. Fig. 5.10 shows the color change of the PCs. Alcohol sensors were prepared using a similar approach with PAAm IOHs [150]. The blueshifts were ascribed to the change in refractive index and swelling ratio. Isopropanol showed a greater blueshift than *tert*-butanol.

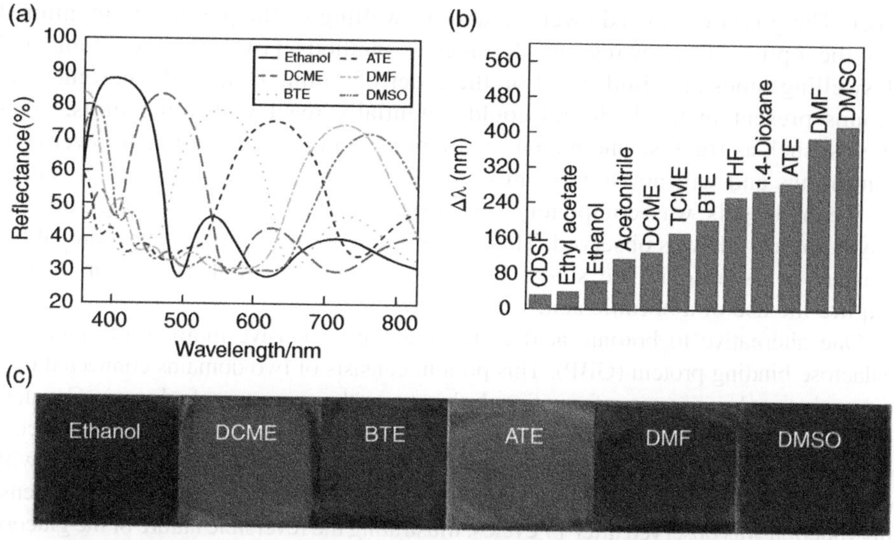

FIG. 5.10 (a) Reflective spectra of an 1D PC in different solvents. (b) Shift of the photonic stopband of an 1D PC in different solvents. (c) Photographs of an 1D PC in different solvents. Reproduced with permission from Ref. 107. © Royal Society of Chemistry.

5.3.4 DNA Sensors

DNA detection at low concentration plays a key role in fields such as genomics, disease diagnostics, forensic science, and viral or bacterial detection. This makes a simple, label-free, DNA detection method quite desirable. A promising approach was reported by Zhao and coworkers [151]; this group made an single stranded DNA (ssDNA) cross-linked hydrogel matrix capable of recognizing sequences complementary to those within the cross-linking segments. In order to produce a visible response, QDs were embedded within the hydrogel matrix to generate a photonics hydrogel. Upon complementary sequence binding, the conformation of the ssDNA originally forming the cross-links will change leading to hydrogel shrinkage. The dimensional change induced by the DNA binding produced a blueshift in the Bragg diffraction peaks from the PCs. DNA concentrations as low as 1 nM were detectable. Control experiments using single-mismatch target DNA and double-mismatch DNA showed that this approach is sequence specific. Simultaneous detection of several DNA sequences is possible by making hydrogels with different ssDNA sequences and different QDs. Different ratios of QD pairs with different emission wavelengths can be used to prepare an array of easily distinguishable hydrogels.

5.3.5 Glucose Sensors

A promising approach to develop sensors capable of *in vivo* glucose monitoring has been reported by Tierney and coworkers [129]. The sensing layer is a half-spherical glucose-responsive hydrogel covalently linked to the end of an optical fiber. The glucose-induced swelling and deswelling of the gel can be monitored via the optical fiber using interferometric methods [141]. The swelling and deswelling times are similar and on the order of 2 min. Though the boronic acid groups present in the hydrogel could potentially react with other diols, other sugars, such as fructose and mannose, are present in the body at much lower concentrations and will not significantly interfere with the measurements. Swelling of the hydrogels was completely reversible, but low sensitivity to glucose at physiological pH was observed. This is likely due to the relatively high pK_a of the boronic acids used in the study. Further optimization of these sensors would require the use of a boronic acid with a lower pK_a.

One alternative to boronic acid as the glucose-responsive motif is the glucose-/galactose-binding protein (GBP). This protein consists of two domains connected by a hinge that undergoes a conformational change in the presence of glucose. Covalent insertion of a modified GBP molecule produced hydrogels that shrink upon glucose exposure. The dimensional change was measured and found to positively correlate with changes in glucose concentration ranging from 0.01 to 1 mM. No loss of glucose-sensing function was observed after 17 cycles, illustrating the reversible nature of the glucose binding and the conformational change [125]. Nanogels capable of integrated glucose detection and self-regulated insulin delivery at physiological pH have been reported by Wu and coworkers [134]. The material consists of a gold nanoparticle core with a shell of poly(4-vinylboronic acid-*co*-2-(dimethylamino)-ethyl acrylate-*co*-*N,N*-methylenebisacrylamide). The polymeric shell swells in the presence of glucose, and the Au cores report the event as a change in the fluorescence signal. The nanogels were able to trap insulin and release it upon glucose-driven swelling of the shell molecules. The payload release occurred within 30 min, which is compatible with clinical needs.

5.4 IONOPHORE-BASED SENSORS

Ionophore-based sensors are a well-studied class of chemical sensors that can exhibit either electrochemical or optical transduction mechanisms. The former group is often referred to as ion-selective electrodes (ISEs), while the latter is typically called "optodes."

5.4.1 Ion-Selective Electrodes

ISEs are a type of sensors widely used in the field of chemical sensing because they offer a simple and convenient way of measuring changes in the concentration of one or several ions in a sample. Any changes in the activity of ions are followed by a simultaneous change in the electrical potential that can be later recorded by a voltmeter. A schematic representation of an ISE measuring-cell assembly is shown in Fig. 5.11.

The history of ISEs can be traced back to the pioneering work carried out by Max Cremer in 1906 [152]. He determined that the difference in the activity of hydrogen ions (H^+) between two solutions separated by a glass membrane can give rise to a membrane potential that is also known as an electromotive force (EMF). Walther H Nernst later explained this relationship. His findings indicated that at room

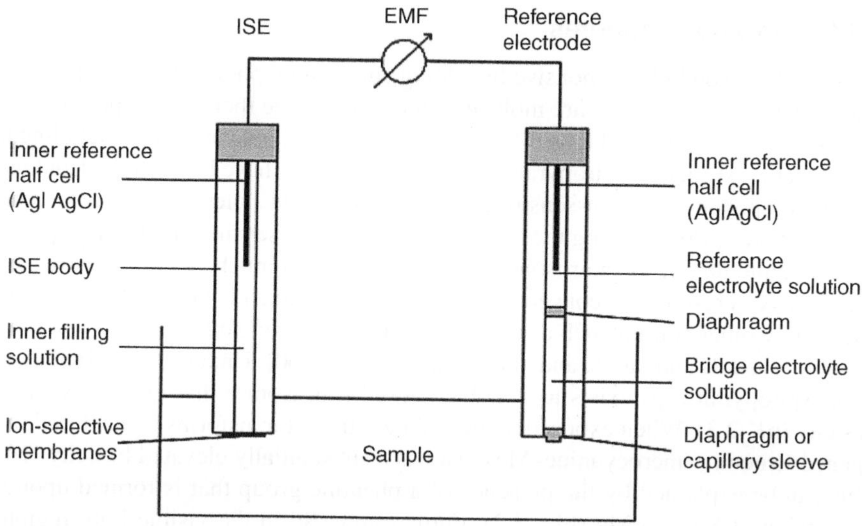

FIG. 5.11 Schematic representation of the experimental setup in a conventional ion-selective electrode measurement.

temperature a 10-fold increase in the activity of measured ions is followed by an increase in the EMF of 59.2 mV/Z_i [153]. This is illustrated by the following equation:

$$\mathrm{EMF} = E^0 + \frac{2.303\,RT}{Z_i F} \log a_i \qquad (5.1)$$

where EMF corresponds to the electrical potential developed between the working and reference electrodes; E^0 is the cell potential under standard conditions (273.15 K and 100 kP_a); a_i is the activity of primary ions; z_i is the ionic charge of the ion; F, T, and R are the Faraday constant, the absolute temperature, and the universal gas constant, respectively.

The real breakthrough in the field of ISEs came with the discovery of the ionophore-based membranes [154]. An ionophore is better known as a chemical compound that binds and mediates transport of ions across the biological membranes [155]. Early days' research illustrated that certain antibiotics, when used as doping agents, can significantly enhance the transport of monovalent cations between the water-immiscible organic liquids [156]. These properties were later exploited during the design of ISEs containing the antibiotic valinomycin as an ion carrier. These electrodes exhibited high selectivity toward the potassium ion (K^+), therefore allowing its measurements in various biological samples including blood and urine samples [157]. The selectivity of ISEs is mainly governed by the internal composition of a sensing membrane. Due to their excellent tunability, polymer-based membrane electrodes constitute one of the most successful types of ISEs. Main components include a polymer, a plasticizer, an ion exchanger, and an ion carrier [154].

5.4.2 Chromoionophores

ISEs can be made light-responsive by adding chromoionophores within the polymer matrix. Chromoionophores are molecules that can change their color upon the complexation with the targeted ions [158]. Many photoswitchable materials including the azobenzenes, stilbenes, spiropyrans, fulgides, and others have been studied and used as the components of light-sensitive systems and could potentially serve as chromoionophores [159]. Among the different kinds of light-sensitive materials, spiropyrans are quite unique as their isomers exhibit very different chemical properties (see Fig. 5.2). As a result, interconversion between the closed and open forms can also be induced by other stimuli such as the temperature, different solvents, the presence of metal ions, acids and bases, and a mechanical force [160]. Under visible light conditions, spiropyran (Sp) exists as a stable, closed-ring isomer that exhibits very low basicity (pK_a 2.3). When exposed to the UV radiation, it transforms to a colored ring-opened structure (merocyanine–Mc) that has a substantially elevated basicity [161]. This can be explained by the presence of a phenolic group that is formed upon the conversion of Sp into Mc (Fig. 5.2). Furthermore, Sp in the visible light region is optically transparent, while an Mc isomer displays a large absorption band in the blue–red region (λ 550–600 nm) [162]. Merocyanine also displays a much higher dipole moment compared with spiropyran, which may lead to chromoionophore leaching from the sensing membrane into the aqueous solution. It has been recently suggested that the covalent attachment of spiropyran to the support may significantly reduce the rate of leaching. Previous research has also identified that the immobilization of Sp moieties can (i) prevent Mc molecules from forming aggregates by reducing the intermolecular interactions; (ii) improve an emission signal by eliminating collisions of a fluorophore with the solution molecules; and (iii) reduce photodegradation [159]. The predicted response and lifetime of the spiropyran-based sensors are mainly dictated by their resistance to photofatigue. The effects of photodegradation are more pronounced with an increasing number of switching cycles. This can result in the irreversible changes in a response of a sensing material. Although isomerization of Sp to Mc is a nondestructive process, the side reactions that occur between the triplet oxygen and triplet excited state of Sp generate a singlet oxygen that causes an irreversible oxidation of a photoswitch [163]. Some researchers observed that the switching process, if performed in an airtight environment in the presence of antioxidants, yields better resistance to the photodegradation [164]. Another study demonstrated that the immobilization of Sp molecules to a solid support can significantly reduce the photodegradation rate [165].

The Mc isomer also exhibits ion-binding properties and on several occasions it was introduced as a metal ion chelator during the development of sensing membranes. Although a binding strength of the individual Mc moiety is relatively weak, the stability of ion–ligand complexes can be improved by attaching two or more Mc groups onto the support (e.g., polymer backbone) [166]. Alternatively, extra functional groups can be introduced to the molecule not only to improve the stability of the formed complexes but also to increase the selectivity toward targeted ions. This has resulted in the development of several sensors for the detection of transition and alkali metal ions [167].

5.4.3 Optodes

In addition to the traditional components in ISEs, optode sensors contain a chromoionophore that exhibits a high selectivity toward the hydrogen ions (H^+) [154]. The optode membranes, when introduced into a solution containing only a very small concentration of targeted ions, exhibit a coloration that is characteristic to the protonated form of a chromoionophore. A color change is observed with an increasing concentration of an analyte ion. This can be attributed to the formation of ion–ligand complexes in the membrane bulk and a concurrent transfer of protons to the aqueous solution [160]. Therefore, the membrane develops the color of the deprotonated chromoionophore, as demonstrated in the following equilibrium:

$$CH^+R^- + L + I^+_{aq} \rightleftarrows C + LI^+R^- + H^+_{aq} \tag{5.2}$$

where C is a chromoionophore, L is a ligand (ionophore) for targeted ion I^+, and R^- is a lipophilic ion. This process illustrates that the optode response is not governed by the changes in the boundary potential at the sample/membrane interface but involves a total exchange of the ionic content between the membrane bulk and a surrounding solution. However, a complete ion exchange results in a transmembrane mass transfer that may lengthen the response time of a sensor. Nevertheless, the application of optical sensors within different scientific fields has been continuously rising since the optodes do not require any wiring to the additional electronic devices and therefore they can be easily miniaturized [154]. Until recently, the optode membranes have worked mainly under the conditions of thermodynamic equilibrium (passive mode).

5.4.4 Dynamic Optodes

As determined by the ongoing research, some optical sensors can be designed to work in an active mode [168]. This means that the process of ion exchange between a sample and the sensing membrane can be controlled during the optode operation and storage. Shvarev reported that a photochemical acid generator (PGA), when incorporated into the sensing membrane matrix, could influence the optode response toward the targeted ions [169]. Photogeneration of a hydrochloric acid molecule (HCl) that was initiated during the membrane exposure to the UV radiation was proved to cause protonation of the chromoionophore. However, the degree of protonation decreased with the diffusion of HCl from the matrix interface into the aqueous solution. Though, the optical sensors containing PGA could be successfully switched "on" and "off," their application remained largely limited due to the irreversible photolysis of the PGAs. In 2011, a research group led by Bakker noted that the photoinduced acidity change of the indicator dye (spiropyran) might strongly influence the equilibrium conditions of an ion-sensing membrane [168]. They observed that illumination of the spiropyran-based membranes with both UV and visible light could induce large shifts in the pK_a values inside the membrane (over six orders of magnitude). As the optical sensor was exposed to the UV radiation (λ - 365 nm), a chromoionophore was converted into a more basic form causing a diffusion of metal ions to the aqueous solution and a simultaneous uptake of the hydrogen ions into the membrane matrix. A reverse process was observed upon the illumination with a visible light (the excitation wavelength of 410 nm). However, the sensor was rendered inactive at very high pH (>10). This can be attributed to the very low concentration of H^+ at this high pH. Metal ions that are more abundantly present within the aqueous solution would enter the optode membrane and form complexes with the ionophore. As a consequence, an ejection of protons from the membrane matrix would occur (even in the presence of basic merocyanine). Although a substantial drift from the expected response may occur at extremely low or high pH values, the spiropyran-based sensors are still suitable for performing measurements on the majority of biological and environmental samples [160]. A slightly different approach was taken by Xie and coworkers during the design of chloride selective sensors [161]. They discovered that the hydrogen ion uptake that is governed by the presence of a chromoionophore (merocyanine) was accompanied by the simultaneous diffusion of chloride ions into the membrane bulk. Therefore, the irradiation with UV light initiates the sensing process rather than "switching it off." These findings illustrate a potential of optical sensors containing spiropyran to detect both cations and anions during the same measurement.

Liang et al. [170] described a potentiometric sensor based on a molecularly imprinted polymeric membrane ISE for determination of melamine in milk. The membrane electrode shows near-Nernstian response (54 mV/decade) to the protonated melamine over the concentration range of 5.0×10^{-6} to 1.0×10^{-2} M. A potentiometric sensor was also fabricated for the determination of 2-aminopyridine (2-AP). The membrane potential increased with 2-AP concentration over a wide range of concentration (5 μM to 100 mM) with a near-Nernstian response of 54.1 mV/decade and detection limit of 2.0 μM. The utility of the sensor was successfully tested by examining of 2-AP in piroxicam as a potential impurity.

5.5 CHALLENGES AND OPPORTUNITIES

The breadth of analytes that can now be quantified using stimuli-responsive polymer hydrogels has increased exponentially in the last 20 years. This is primarily due to the development of new synthetic methods to produce complex, well-defined, polymeric matrices and to a better fundamental understanding of the processes governing the dynamic equilibria controlling reversible stimuli responses. As the production of well-controlled structures and surfaces with features at the nanoscale becomes more cost-effective and widely spread, new, more complex, sensing layer architectures will become possible. These new materials and surfaces will allow the simultaneous study of many analytes within the same sample using a variety of probing methods. As the miniaturization of these new sensors continues, it will also become possible to simultaneously study the time-dependent evolution of all the species involved in complex reactions, instead of the concentration variation of only one or two species. This will increase the spatial resolution of our chemical analyses, thus enabling the study of many phenomena as function of location, which is vital for the better understanding of many processes in biochemistry and materials science.

Many challenges still face the development of the next generation of multiresponsive smart sensors, including mass and energy transport limitations, the need for better understanding of surface phenomena at the nanoscale, the need for better-defined functional nanostructures and surfaces, and the identification of synthetic macromolecules capable of molecular recognition at the exquisite levels found in nature. The materials science and analytical chemistry communities have made encouraging progress in all these areas in the last few decades. A compilation of their results, as seen through the prism of what is needed to generate even smarter sensing layers, should help inspire the research approaches required to overcome the challenges ahead.

REFERENCES

1. Hulanicki, A., Glab, S., Ingman, F. (1991) Chemical sensors: definitions and classification. In *Pure Appl. Chem.*, 63: 1247–1250.
2. Byrne, R., Coleman, S., Fraser, K. J., Raduta, A., MacFarlane, D. R., Diamond, D. (2009) Photochromism of nitrobenzospiropyran in phosphonium based ionic liquids. *Phys. Chem. Chem. Phys.* 11 (33): 7286–7291.
3. Gil, E. S., Hudson, S. M. (2004) Stimuli-responsive polymers and their bioconjugates. *Prog. Polym. Sci.* 29 (12): 1173–1222.
4. Murthy, N., Campbell, J., Fausto, N., Hoffman, A. S., Stayton, P. S. (2003) Bioinspired pH-responsive polymers for the intracellular delivery of biomolecular drugs. *Bioconjug. Chem.* 14 (2): 412–419.
5. Magnusson, J. P., Khan, A., Pasparakis, G., Saeed, A. O., Wang, W., Alexander, C. (2008) Ion-sensitive "isothermal" responsive polymers prepared in water. *J. Am. Chem. Soc.* 130 (33): 10852–10853.
6. Lvov, Y., Antipov, A. A., Mamedov, A., Moehwald, H., Sukhorukov, G. B. (2001) Urease encapsulation in nanoorganized microshells. *Nano Lett.* 1 (3): 125–128.
7. Akhoury, A., Bromberg, L., Hatton, T. A. (2011) Redox-responsive gels with tunable hydrophobicity for controlled solubilization and release of organics. *ACS Appl. Mater. Interfaces* 3 (4): 1167–1174.
8. Yan, Q., Yuan, J., Cai, Z., Xin, Y., Kang, Y., Yin, Y. (2010) Voltage-responsive vesicles based on orthogonal assembly of two homopolymers. *J. Am. Chem. Soc.* 132 (27): 9268–9270.
9. Schild, H. G. (1992) Poly(N-isopropylacrylamide): experiment, theory and application. *Prog. Polym. Sci.* 17 (2): 163–249.
10. Zhao, Y., He, J. (2009) Azobenzene-containing block copolymers: the interplay of light and morphology enables new functions. *Soft Matter* 5 (14): 2686–2693.
11. Weder, C. (2009) Mechanochemistry: polymers react to stress. *Nature (London)* 459 (7243): 45–46.
12. Hoogenboom, R. (2014) 2 - Temperature-responsive polymers: properties, synthesis and applications. In *Smart Polymers and Their Applications*, Aguilar, M. R., Román, J. S., Eds. Woodhead Publishing, Cambridge: 15–44.
13. Lendlein, A., Kelch, S. (2002) Shape-memory polymers. *Angew. Chem. Int. Ed.* 41 (12): 2034–2057.
14. Li, M.-H., Keller, P. (2006) Artificial muscles based on liquid crystal elastomers. *Philos. Trans. R. Soc. A.* 364 (1847): 2763–2777.

15. Sage, I. (2011) Thermochromic liquid crystals. *Liq. Cryst.* 38 (11–12): 1551–1561.
16. Maeda, Y., Nakamura, T., Ikeda, I. (2001) Changes in the hydration states of poly (N-alkylacrylamide)s during their phase transitions in water observed by FTIR Spectroscopy. *Macromolecules* 34 (5): 1391–1399.
17. Idziak, I., Avoce, D., Lessard, D., Gravel, D., Zhu, X. X. (1999) Thermosensitivity of aqueous solutions of poly(N,N-diethylacrylamide). Macromolecules 32 (4): 1260–1263.
18. Lutz, J.-F. (2008) Polymerization of oligo(ethylene glycol) (meth)acrylates: toward new generations of smart biocompatible materials. J. Polym. Sci., Part A: Polym. Chem. 46 (11): 3459–3470.
19. Hoogenboom, R. (2009) Poly(2-oxazoline)s: a polymer class with numerous potential applications. Angew. Chem. Int. Ed. 48 (43): 7978–7994.
20. Roy, D., Brooks, W. L. A., Sumerlin, B. S. (2013) New directions in thermoresponsive polymers. *Chem. Soc. Rev.* 42 (17): 7214–7243.
21. Jochum, F. D., Theato, P. (2013) Temperature- and light-responsive smart polymer materials. *Chem. Soc. Rev.* 42 (17): 7468–7483.
22. Roy, D., Cambre, J. N., Sumerlin, B. S. (2010) Future perspectives and recent advances in stimuli-responsive materials. *Prog. Polym. Sci.* 35 (1–2): 278–301.
23. Zrinyi, M., Szabo, D., Kilian, H.-G. (1999) Kinetics of the shape change of magnetic field sensitive polymer gels. *Polym. Gels Networks* 6 (6): 441–454.
24. Filipcsei, G., Feher, J., Zrinyi, M. (2000) Electric field sensitive neutral polymer gels. *J. Mol. Struct.* 554 (1): 109–117.
25. Szabo, D., Szeghy, G., Zrinyi, M. (1998) Shape transition of magnetic field sensitive polymer gels. *Macromolecules* 31 (19): 6541–6548.
26. Korth, B. D., Keng, P., Shim, I., Bowles, S. E., Tang, C., Kowalewski, T., Nebesny, K. W., Pyun, J. (2006) Polymer-coated ferromagnetic colloids from well-defined macromolecular surfactants and assembly into nanoparticle chains. *J. Am. Chem. Soc.* 128 (20): 6562–6563.
27. Czaun, M., Hevesi, L., Takafuji, M., Ihara, H. (2008) A novel approach to magneto-responsive polymeric gels assisted by iron nanoparticles as nano cross-linkers. *Chem. Commun. (Camb., UK)* (18): 2124–2126.
28. Kumar, G. S., Neckers, D. C. (1989) Photochemistry of azobenzene-containing polymers. *Chem. Rev.* 89 (8): 1915–1925.
29. Irie, M. (1990) Properties and applications of photoresponsive polymers. *Pure Appl. Chem.* 62 (8): 1495–1502.
30. Ciardelli, F., Pieroni, O., Fissi, A., Carlini, C., Altomare, A. (1989) Photoresponsive optically active polymers - a review. *Br. Polym. J.* 21 (2): 97–106.
31. Matejka, L., Ilavsky, M., Dusek, K., Wichterle, O. (1981) Photomechanical effects in crosslinked photochromic polymers. *Polymer* 22 (11): 1511–1515.
32. Wang, G., Tong, X., Zhao, Y. (2004) Preparation of azobenzene-containing amphiphilic diblock copolymers for light-responsive micellar aggregates. *Macromolecules* 37 (24): 8911–8917.
33. Sugiyama, K., Sono, K. (2001) Characterization of photo- and thermoresponsible amphiphilic copolymers having azobenzene moieties as side groups. *J. Appl. Polym. Sci.* 81 (12): 3056–3063.
34. Li, Y., Deng, Y., Tong, X., Wang, X. (2006) Formation of photoresponsive uniform colloidal spheres from an amphiphilic azobenzene-containing random copolymer. *Macromolecules* 39 (3): 1108–1115.

35. Ravi, P., Sin, S. L., Gan, L. H., Gan, Y. Y., Tam, K. C., Xia, X. L., Hu, X. (2005) New water soluble azobenzene-containing diblock copolymers: synthesis and aggregation behavior. *Polymer* 46 (1): 137–146.

36. Desponds, A., Freitag, R. (2003) Synthesis and characterization of photoresponsive N-isopropylacrylamide cotelomers. *Langmuir* 19 (15): 6261–6270.

37. Junge, D. M., McGrath, D. V. (1997) Photoresponsive dendrimers. *Chem. Commun. (Camb.)* 9: 857–858.

38. Konak, C., Rathi, R. C., Kopeckova, P., Kopecek, J. (1997) Photoregulated association of water-soluble copolymers with spirobenzopyran-containing side chains. *Macromolecules* 30 (18): 5553–5556.

39. Moniruzzaman, M., Sabey, C. J., Fernando, G. F. (2007) Photoresponsive polymers: an investigation of their photoinduced temperature changes during photoviscosity measurements. *Polymer* 48 (1): 255–263.

40. Ivanov, A. E., Eremeev, N. L., Wahlund, P. O., Galaev, I. Y., Mattiasson, B. (2002) Photosensitive copolymer of N-isopropylacrylamide and methacryloyl derivative of spirobenzopyran. *Polymer* 43 (13): 3819–3823.

41. Lee, H.-I., Wu, W., Oh, J. K., Mueller, L., Sherwood, G., Peteanu, L., Kowalewski, T., Matyjaszewski, K. (2007) Light-induced reversible formation of polymeric micelles. *Angew. Chem. Int. Ed.* 46 (14): 2453–2457.

42. Laschewsky, A., Rekai, E. D. (2000) Photochemical modification of the lower critical solution temperature of cinnamoylated poly(N-2-hydroxypropylmethacrylamide) in water. *Macromol. Rapid Commun.* 21 (13): 937–940.

43. Goodwin, A. P., Mynar, J. L., Ma, Y., Fleming, G. R., Frechet, J. M. J. (2005) Synthetic micelle sensitive to IR light via a two-photon process. *J. Am. Chem. Soc.* 127 (28): 9952–9953.

44. Jiang, J., Tong, X., Morris, D., Zhao, Y. (2006) Toward photocontrolled release using light-dissociable block copolymer micelles. *Macromolecules* 39 (13): 4633–4640.

45. Theato, P. (2008) Synthesis of well-defined polymeric activated esters. *J. Polym. Sci., Part A: Polym. Chem.* 46 (20): 6677–6687.

46. Ito, Y., Casolaro, M., Kono, K., Imanishi, Y. (1989) An insulin-releasing system that is responsive to glucose. *J. Control. Release* 10 (2): 195–203.

47. Chu, L.-Y., Li, Y., Zhu, J.-H., Wang, H.-D., Liang, Y.-J. (2004) Control of pore size and permeability of a glucose-responsive gating membrane for insulin delivery. *J. Control. Release* 97 (1): 43–53.

48. Huang, H. Y., Shaw, J., Yip, C., Wu, X. Y.(2008) Microdomain pH gradient and kinetics inside composite polymeric membranes of pH and glucose sensitivity. *Pharm. Res.* 25 (5): 1150–1157.

49. Kang, S. I., Bae, Y. H. (2003) A sulfonamide based glucose-responsive hydrogel with covalently immobilized glucose oxidase and catalase. *J. Control. Release* 86 (1): 115–121.

50. Podual, K., Doyle, F. J., 3rd, Peppas, N. A. (2000) Dynamic behavior of glucose oxidase-containing microparticles of poly(ethylene glycol)-grafted cationic hydrogels in an environment of changing pH. *Biomaterials* 21 (14): 1439–1450.

51. Uchiyama, T., Kiritoshi, Y., Watanabe, J., Ishihara, K. (2003) Degradation of phospholipid polymer hydrogel by hydrogen peroxide aiming at insulin release device. *Biomaterials* 24 (28): 5183–5190.

52. Cheng, S. Y., Gross, J., Sambanis, A. (2004) Hybrid pancreatic tissue substitute consisting of recombinant insulin-secreting cells and glucose-responsive material. *Biotechnol. Bioeng.* 87 (7): 863–873.

53. Miyata, T., Jikihara, A., Nakamae, K., Hoffman, A. S. (2004) Preparation of reversibly glucose-responsive hydrogels by covalent immobilization of lectin in polymer networks having pendant glucose. *J. Biomater. Sci. Polym. Ed.* 15 (9): 1085–1098.

54. Springsteen, G., Wang, B. (2002) A detailed examination of boronic acid-diol complexation. *Tetrahedron* 58 (26): 5291–5300.

55. Miyata, T., Uragami, T., Nakamae, K. (2002) Biomolecule-sensitive hydrogels. *Adv. Drug Deliv. Rev.* 54 (1): 79–98.

56. Hisamitsu, I., Kataoka, K., Okano, T., Sakurai, Y. (1997) Glucose-responsive gel from phenylborate polymer and poly(vinyl alcohol): prompt response at physiological pH through the interaction of borate with amino group in the gel. *Pharm. Res.* 14 (3): 289–293.

57. Kataoka, K., Miyazaki, H., Okano, T., Sakurai, Y. (1994) Sensitive Glucose-induced change of the lower critical solution temperature of poly[N,N-(dimethylacrylamide)-co-3-(acrylamido)-phenylboronic acid] in physiological saline. *Macromolecules* 27 (4): 1061–1062.

58. Kikuchi, A., Suzuki, K., Okabayashi, O., Hoshino, H., Kataoka, K., Sakurai, Y., Okano, T. (1996) Glucose-sensing electrode coated with polymer complex gel containing phenylboronic acid. *Anal. Chem.* 68 (5): 823–828.

59. Matsumoto, A., Ikeda, S., Harada, A., Kataoka, K. (2003) Glucose-responsive polymer bearing a novel phenylborate derivative as a glucose-sensing moiety operating at physiological pH conditions. *Biomacromolecules* 4 (5): 1410–1416.

60. Pellon, J., Schwind, L. H., Guinard, M. J., Thomas, W. M. (1961) Polymerization of vinyl monomers containing boron. II. p-vinylbenzeneboronic acid. *J. Polym. Sci.* 55: 161–167.

61. Lennarz, W. J., Snyder, H. R. (1960) Arylboronic acids. III. Preparation and polymerization of p-vinylbenzeneboronic acid1. *J. Am. Chem. Soc.* 82 (9): 2169–2171.

62. Kato, M., Kamigaito, M., Sawamoto, M., Higashimura, T. (1995) Polymerization of methyl methacrylate with the carbon tetrachloride/dichlorotris- (triphenylphosphine) ruthenium(II)/methylaluminum bis(2,6-di-tert-butylphenoxide) initiating system: possibility of living radical polymerization. *Macromolecules* 28 (5): 1721–1723.

63. Chiefari, J., Chong, Y. K., Ercole, F., Krstina, J., Jeffery, J., Le, T. P. T., Mayadunne, R. T. A., Meijs, G. F., Moad, C. L., Moad, G., Rizzardo, E., Thang, S. H. (1998) Living free-radical polymerization by reversible addition-fragmentation chain transfer: the RAFT process. *Macromolecules* 31 (16): 5559–5562.

64. Ulijn, R. V. (2006) Enzyme-responsive materials: a new class of smart biomaterials. *J. Mater. Chem.* 16 (23): 2217–2225.

65. Lutolf, M. P., Raeber, G. P., Zisch, A. H., Tirelli, N., Hubbell, J. A. (2003) Cell-responsive synthetic hydrogels. *Adv. Mater. (Weinheim, Ger.)* 15 (11): 888–892.

66. Plunkett, K. N., Berkowski, K. L., Moore, J. S. (2005) Chymotrypsin responsive hydrogel: application of a disulfide exchange protocol for the preparation of methacrylamide containing peptides. *Biomacromolecules* 6 (2): 632–637.

67. Lu, Z.-R., Kopeckova, P., Kopecek, J. (2003) Antigen responsive hydrogels based on polymerizable antibody Fab' fragment. *Macromol. Biosci.* 3 (6): 296–300.

68. Jocelyn, P. C. (1987) Chemical reduction of disulfides. *Methods Enzymol.* 143 (Sulfur Sulfur Amino Acids): 246–256.

69. Jones, D. P., Carlson, J. L., Samiec, P. S., Sternberg, P., Jr., Mody, V. C., Jr., Reed, R. L., Brown, L. A. (1998) Glutathione measurement in human plasma. Evaluation of sample collection, storage and derivatization conditions for analysis of dansyl derivatives by HPLC. *Clin. Chim. Acta* 275 (2): 175–184.

70. Li, Y., Lokitz, B. S., Armes, S. P., McCormick, C. L. (2006) Synthesis of reversible shell cross-linked micelles for controlled release of bioactive agents. *Macromolecules* 39 (8): 2726–2728.

71. Ghosh, S., Yesilyurt, V., Savariar, E. N., Irvin, K., Thayumanavan, S. (2009) Redox, ionic strength, and pH sensitive supramolecular polymer assemblies. *J. Polym. Sci., Part A: Polym. Chem.*. 47 (4): 1052–1060.

72. Kungwatchakun, D., Irie, M. (1988) Photoresponsive polymers: photocontrol of the phase separation temperature of aqueous solutions of poly-[N-isopropylacrylamide-co-N-(4-phenylazophenyl)acrylamide]. *Makromol. Chem., Rapid Commun.* 9 (4): 243–246.

73. Roth, P. J., Wiss, K. T., Zentel, R., Theato, P. (2008) Synthesis of reactive telechelic polymers based on pentafluorophenyl esters. *Macromolecules (Washington, DC)* 41 (22): 8513–8519.

74. Lee, H.-I., Pietrasik, J., Matyjaszewski, K. (2006) Phototunable temperature-responsive molecular brushes prepared by ATRP. *Macromolecules* 39 (11): 3914–3920.

75. Plate, N. A., Lebedeva, T. L., Valuev, L. I. (1999) Lower critical solution temperature in aqueous solutions of N-alkyl-substituted polyacrylamides. *Polym. J. (Tokyo)* 31 (1): 21–27.

76. El-Ejmi, A. A. S., Huglin, M. B. (1996) Characterization of N,N-dimethylacrylamide/2-methoxyethyl acrylate copolymers and phase behavior of their thermotropic aqueous solutions. *Polym. Int.* 39 (2): 113–119.

77. Su, W., Han, K., Luo, Y., Wang, Z., Li, Y., Zhang, Q. (2007) Formation and photoresponsive properties of giant microvesicles assembled from azobenzene-containing amphiphilic diblock copolymers. *Macromol. Chem. Phys.* 208 (9): 955–963.

78. Jochum, F. D., Thermo- and light responsive micellation of azobenzene containing block copolymers. *Chem. Commun.*, 2010, 46, 6717.

79. Jochum, F. D., Theato, P. (2009) Temperature- and light-responsive polyacrylamides prepared by a double polymer analogous reaction of activated ester polymers. *Macromolecules (Washington, DC)* 42 (16): 5941–5945.

80. Crano, J. C., Guglielmetti, R. J. (2002) *Organic photochromic and thermochromic compounds*. Kluwer, New York: Vol. 1.

81. Mitra, S., Tamai, N. (2003) Dynamics of photochromism in salicylideneaniline: a femtosecond spectroscopic study. *Phys. Chem. Chem. Phys.* 5 (20): 4647–4652.

82. Suzuki, T., Kato, T., Shinozaki, H. (2004) Photo-reversible Pb2+-complexation of thermosensitive poly(N-isopropyl acrylamide-co-spiropyran acrylate) in water. *Chem. Commun. (Camb., UK)* (18): 2036–2037.

83. Sumaru, K., Ohi, K., Takagi, T., Kanamori, T., Shinbo, T. (2006) Photoresponsive properties of poly(N-isopropylacrylamide) hydrogel partly modified with spirobenzopyran. *Langmuir* 22 (9): 4353–4356.

84. Shiraishi, Y., Miyamoto, R., Hirai, T. (2009) Spiropyran-conjugated thermoresponsive copolymer as a colorimetric thermometer with linear and reversible color change. *Org. Lett.* 11 (7): 1571–1574.

85. Shiraishi, Y., Sumiya, S., Manabe, K., Hirai, T. (2011) Thermoresponsive copolymer containing a coumarin-spiropyran conjugate: reusable fluorescent sensor for cyanide anion detection in water. *ACS Appl. Mater. Interfaces* 3 (12): 4649–4656.

86. Joseph, G., Pichardo, J., Chen, G. (2010) Reversible photo-/thermoresponsive structured polymer surfaces modified with a spirobenzopyran-containing copolymer for tunable wettability. *Analyst (Camb., UK)* 135 (9): 2303–2308.
87. Achilleos, D. S., Vamvakaki, M. (2010) Multiresponsive spiropyran-based copolymers synthesized by atom transfer radical polymerization. *Macromolecules (Washington, DC)* 43 (17): 7073–7081.
88. Jin, Q., Liu, G., Ji, J. (2010) Micelles and reverse micelles with a photo and thermo double-responsive block copolymer. *J. Polym. Sci., Part A: Polym. Chem.* 48 (13): 2855–2861.
89. Li, L., Xing, X., Liu, Z. (2012) Triply-responsive (thermo/light/pH) copolymeric hydrogel of N-isopropylacrylamide with an azobenzene-containing monomer. *J. Appl. Polym. Sci.* 124 (2): 1128–1136.
90. He, J., Tong, X., Zhao, Y. (2009) Photoresponsive nanogels based on photocontrollable cross-links. *Macromolecules (Washington, DC)* 42 (13): 4845–4852.
91. Kuckling, D., Doring, A., Krahl, F., Arndt, K. F. (2012) In *Polymer Science: A Comprehensive Reference*, Matyjaszewski, K., Moller, M., Eds. Elsevier, Amsterdam: Vol. 8: 377.
92. Tanaka, T., Fillmore, D. J. (1979) Kinetics of swelling of gels. *J. Chem. Phys.* 70 (3): 1214–1218.
93. Buchmeiser, M. R. (2003) *Polymeric Materials in Organic Synthesis and Catalysis*. VCH-Wiley, Weinheim.
94. Bergbreiter, D. E., Tian, J., Hongfa, C. (2009) Using soluble polymer supports to facilitate homogeneous catalysis. *Chem. Rev. (Washington, DC)* 109 (2): 530–582.
95. Bergbreiter, D. E., Case, B. L., Liu, Y.-S., Caraway, J. W. (1998) Poly(N-isopropylacrylamide) soluble polymer supports in catalysis and synthesis. *Macromolecules* 31 (18): 6053–6062.
96. Haraguchi, K. (2007) Nanocomposite gels: new advanced functional soft materials. *Macromol. Symp.*. 256 (Functional and Biological Gels and Networks: Theory and Experiment): 120–130.
97. Mohan, Y. M., Lee, K., Premkumar, T., Geckeler, K. E. (2007) Hydrogel networks as nanoreactors: a novel approach to silver nanoparticles for antibacterial applications. *Polymer* 48 (1): 158–164.
98. Sivudu, K. S., Reddy, N. M., Prasad, M. N., Raju, K. M., Mohan, Y. M., Yadav, J. S., Sabitha, G., Shailaja, D. (2008) Highly efficient and reusable hydrogel-supported nano-palladium catalyst: evaluation for Suzuki–Miyaura reaction in water. *J. Mol. Catal. A Chem.* 295 (1–2): 10–17.
99. Hamamoto, H., Suzuki, Y., Takahashi, H., Ikegami, S. (2007) A new solid-phase reaction system utilizing a temperature-responsive catalyst: oxidative cyclization with hydrogen peroxide. *Adv. Synth. Catal.* 349 (17+18): 2685–2689.
100. Ikegami, S., Hamamoto, H. (2009) Novel recycling system for organic synthesis via designer polymer-gel catalysts. *Chem. Rev. (Washington, DC)* 109 (2): 583–593.
101. Sorrell, C. D., Serpe M. J., Reflection order selectivity of color-tunable poly (N-isopropylacrylamide) microgel based etalons. *Adv. Mater.* 2011, 23, 4088.
102. Carter, M. C. D., Sorrell, C. D., Serpe, M. J. (2011) Deswelling kinetics of color tunable Poly(N-Isopropylacrylamide) microgel-based etalons. *J. Phys. Chem. B* 115 (49): 14359–14368.

103. Johnson, K. C. C., Mendez, F., Serpe, M. J. (2012) Detecting solution pH changes using poly (N-isopropylacrylamide)-co-acrylic acid microgel-based etalon modified quartz crystal microbalances. *Anal. Chim. Acta* 739: 83–88.

104. Doring, A., Birnbaum, W., Kuckling, D. (2013) Responsive hydrogels - structurally and dimensionally optimized smart frameworks for applications in catalysis, micro-system technology and material science. *Chem. Soc. Rev.* 42 (17): 7391–7420.

105. Baeissa, A., Dave, N., Smith, B. D., Liu, J. (2010) DNA-Functionalized monolithic hydrogels and gold nanoparticles for colorimetric DNA detection. *ACS Appl. Mater. Interfaces* 2 (12): 3594–3600.

106. Tian, E., Wang, J., Zheng, Y., Song, Y., Jiang, L., Zhu, D. (2008) Colorful humidity sensitive photonic crystal hydrogel. *J. Mater. Chem.* 18 (10): 1116–1122.

107. Wang, Z., Zhang, J., Li, J., Xie, J., Li, Y., Liang, S., Tian, Z., Li, C., Wang, Z., Wang, T., Zhang, H., Yang, B. (2011) Colorful detection of organic solvents based on responsive organic/inorganic hybrid one-dimensional photonic crystals. *J. Mater. Chem.* 21 (4): 1264–1270.

108. Schulz, V., Zschoche, S., Zhang, H. P., Voit, B., Gerlach, G. (2011) Macroporous smart hydrogels for fast-responsive piezoresistive chemical microsensors. *Procedia Eng.* 25: 1141–1144.

109. Jiang, H., Markowski, J., Sabarinathan, J. (2009) Near-infrared optical response of thin film pH-sensitive hydrogel coated on a gold nanocrescent array. *Opt. Express* 17 (24): 21802–21807.

110. Vasylevska, A. S., Karasyov, A. A., Borisov, S. M., Krause, C. (2007) Novel coumarin-based fluorescent pH indicators, probes and membranes covering a broad pH range. *Anal. Bioanal. Chem.* 387 (6): 2131–2141.

111. Griffete, N., Frederich, H., Maitre, A., Chehimi, M. M., Ravaine, S., Mangeney, C. (2011) Photonic crystal pH sensor containing a planar defect for fast and enhanced response. *J. Mater. Chem.* 21 (34): 13052–13055.

112. Kharlampieva, E., Kozlovskaya, V., Zavgorodnya, O., Lilly, G. D., Kotov, N. A., Tsukruk, V. V. (2010) pH-responsive photoluminescent LbL hydrogels with confined quantum dots. *Soft Matter* 6 (4): 800–807.

113. Luo, Q., Guan, Y., Zhang, Y., Siddiq, M. (2010) Lead-sensitive PNIPAM microgels modified with crown ether groups. *J. Polym. Sci., Part A: Polym. Chem..* 48 (18): 4120–4127.

114. Joseph, K. A., Dave, N., Liu, J. (2011) Electrostatically directed visual fluorescence response of DNA-functionalized monolithic hydrogels for highly sensitive Hg2+ detection. *ACS Appl. Mater. Interfaces* 3 (3): 733–739.

115. Arunbabu, D., Sannigrahi, A., Jana, T. (2011) Photonic crystal hydrogel material for the sensing of toxic mercury ions (Hg2+) in water. *Soft Matter* 7 (6): 2592–2599.

116. Hong, W., Li, W., Hu, X., Zhao, B., Zhang, F., Zhang, D. (2011) Highly sensitive colorimetric sensing for heavy metal ions by strong polyelectrolyte photonic hydrogels. *J. Mater. Chem.* 21 (43): 17193–17201.

117. Jacobi, Z. E., Li, L., Liu, J. (2012) Visual detection of lead(II) using a label-free DNA-based sensor and its immobilization within a monolithic hydrogel. *Analyst (Camb., UK)* 137 (3): 704–709.

118. Jiang, H., Zhu, Y., Chen, C., Shen, J., Bao, H., Peng, L., Yang, X., Li, C. (2012) Photonic crystal pH and metal cation sensors based on poly(vinyl alcohol) hydrogel. *New J. Chem.* 36 (4): 1051–1056.

119. Dave, N., Chan, M. Y., Huang, P.-J. J., Smith, B. D., Liu, J. (2010) Regenerable DNA-functionalized hydrogels for ultrasensitive, instrument-free mercury(II) detection and removal in water. *J. Am. Chem. Soc.* 132 (36): 12668–12673.
120. Lin, G., Chang, S., Hao, H., Tathireddy, P., Orthner, M., Magda, J., Solzbacher, F. (2010) Osmotic swelling pressure response of smart hydrogels suitable for chronically implantable glucose sensors. *Sens. Actuators, B* 144 (1): 332–336.
121. Tierney, S., Volden, S., Stokke, B. T. (2009) Glucose sensors based on a responsive gel incorporated as a Fabry-Perot cavity on a fiber-optic readout platform. *Biosens. Bioelectron.* 24 (7): 2034–2039.
122. Wu, W., Zhou, T., Aiello, M., Zhou, S. (2010) Construction of optical glucose nanobiosensor with high sensitivity and selectivity at physiological pH on the basis of organic-inorganic hybrid microgels. *Biosens. Bioelectron.* 25 (12): 2603–2610.
123. Ye, G., Wang, X. (2010) Glucose sensing through diffraction grating of hydrogel bearing phenylboronic acid groups. *Biosens. Bioelectron.* 26 (2): 772–777.
124. Zenkl, G., Mayr, T., Klimant, I. (2008) Sugar-responsive fluorescent nanospheres. *Macromol. Biosci.* 8 (2): 146–152.
125. Ehrick, J. D., Luckett, M. R., Khatwani, S., Wei, Y., Deo, S. K., Bachas, L. G., Daunert, S. (2009) Glucose responsive hydrogel networks based on protein recognition. *Macromol. Biosci.* 9 (9): 864–868.
126. Farooqi, Z. H., Khan, A., Siddiq, M. (2011) Temperature-induced volume change and glucose sensitivity of poly[(N-isopropylacrylamide)-co-acrylamide-co-(phenylboronic acid)] microgels. *Polym. Int.* 60 (10): 1481–1486.
127. Heo, Y. J., Shibata, H., Okitsu, T., Kawanishi, T., Takeuchi, S. (2011) Long-term in vivo glucose monitoring using fluorescent hydrogel fibers. *Proc. Natl. Acad. Sci. U. S. A.* 108 (33): 13399–13403, S13399/1-S13399/6.
128. Shibata, H., Heo, Y. J., Okitsu, T., Matsunaga, Y., Kawanishi, T., Takeuchi, S. (2010) Injectable hydrogel microbeads for fluorescence-based in vivo continuous glucose monitoring. *Proc. Natl. Acad. Sci. U. S. A.* 107 (42): 17894–17898, S17894/1-S17894/6.
129. Tierney, S., Hasle Falch, B. M., Hjelme, D. R., Stokke, B. T. (2009) Determination of glucose levels using a functionalized hydrogel-optical fiber biosensor: toward continuous monitoring of blood glucose in vivo. *Anal. Chem. (Washington, DC)* 81 (9): 3630–3636.
130. Sorrell, C. D., Serpe, M. J. (2012) Glucose sensitive poly(N-isopropylacrylamide) microgel based etalons. *Anal. Bioanal. Chem.* 402 (7): 2385–2393.
131. Vilozny, B., Schiller, A., Wessling, R. A., Singaram, B. (2011) Multiwell plates loaded with fluorescent hydrogel sensors for measuring pH and glucose concentration. *J. Mater. Chem.* 21 (21): 7589–7595.
132. Ancla, C., Lapeyre, V., Gosse, I., Catargi, B., Ravaine, V. (2011) Designed glucose-responsive microgels with selective shrinking behavior. *Langmuir* 27 (20): 12693–12701.
133. Honda, M., Kataoka, K., Seki, T., Takeoka, Y. (2009) Confined stimuli-responsive polymer gel in inverse opal polymer membrane for colorimetric glucose sensor. *Langmuir* 25 (14): 8349–8356.
134. Wu, W., Mitra, N., Yan, E. C. Y., Zhou, S. (2010) Multifunctional hybrid nanogel for integration of optical glucose sensing and self-regulated insulin release at physiological pH. *ACS Nano* 4 (8): 4831–4839.
135. Zhang, X., Guan, Y., Zhang, Y. (2012) Ultrathin hydrogel films for rapid optical biosensing. *Biomacromolecules* 13 (1): 92–97.

136. Guenther, M., Kuckling, D., Corten, C., Gerlach, G., Sorber, J., Suchaneck, G., Arndt, K. F. (2007) Chemical sensors based on multiresponsive block copolymer hydrogels. *Sens. Actuators, B.* 126 (1): 97–106.

137. Schulz, V., Gerlach, G., Gunther, M., Magda, J. J., Solzbacher, F. (2010) Piezoresistive pH microsensors based on stimuli-sensitive polyelectrolyte hydrogels. *Tech. Mess.* 77 (3): 179–186.

138. Guenther, M., Gerlach, G., Corten, C., Kuckling, D., Sorber, J., Arndt, K. F. (2008) Hydrogel-based sensor for a rheochemical characterization of solutions. *Sens. Actuators, B.* 132 (2): 471–476.

139. Sridhar, V., Takahata, K. (2009) A hydrogel-based passive wireless sensor using a flex-circuit inductive transducer. *Sens. Actuators, A* 155 (1): 58–65.

140. Matsumoto, A., Sato, N., Sakata, T., Yoshida, R., Kataoka, K., Miyahara, Y. (2009) Chemical-to-Electrical-Signal Transduction Synchronized with Smart Gel Volume Phase Transition. *Adv. Mater. (Weinheim)* 21 (43): 4372–4378.

141. Tierney, S., Hjelme, D. R., Stokke, B. T. (2008) Determination of Swelling of Responsive Gels with Nanometer Resolution. Fiber-Optic Based Platform for Hydrogels as Signal Transducers. *Anal. Chem. (Washington, DC)* 80 (13): 5086–5093.

142. Zourob, M., Ong, K. G., Zeng, K., Mouffouk, F., Grimes, C. A. (2007) A wireless magnetoelastic biosensor for the direct detection of organophosphorus pesticides. *Analyst (Camb., UK)* 132 (4): 338–343.

143. Griffete, N., Frederich, H., Maitre, A., Schwob, C., Ravaine, S., Carbonnier, B., Chehimi, M. M., Mangeney, C. (2011) Introduction of a planar defect in a molecularly imprinted photonic crystal sensor for the detection of bisphenol A. *J. Colloid Interface Sci.* 364 (1): 18–23.

144. Yuan, Y., Li, Z., Liu, Y., Gao, J., Pan, Z., Liu, Y. (2012) Hydrogel photonic sensor for the detection of 3-pyridinecarboxamide. *Chem.–Eur. J.* 18 (1): 303–309, S303/1-S303/2.

145. Wu, Z., Hu, X., Tao, C.-a., Li, Y., Liu, J., Yang, C., Shen, D., Li, G. (2008) Direct and label-free detection of cholic acid based on molecularly imprinted photonic hydrogels. *J. Mater. Chem.* 18 (45): 5452–5458.

146. Hu, X., Li, G., Li, M., Huang, J., Li, Y., Gao, Y., Zhang, Y. (2008) Ultrasensitive specific stimulant assay based on molecularly imprinted photonic hydrogels. *Adv. Funct. Mater.* 18 (4): 575–583.

147. Ran, D., Wang, Y., Jia, X., Nie, C. (2012) Bovine serum albumin recognition via thermosensitive molecular imprinted macroporous hydrogels prepared at two different temperatures. *Anal. Chim. Acta.* 723: 45–53.

148. Islam, M. R., Lu, Z., Li, X., Sarker, A. K., Hu, L., Choi, P., Li, X., Hakobyan, N., Serpe, M. J. (2013) Responsive polymers for analytical applications: a review. *Anal. Chim. Acta* 789: 17–32.

149. Xuan, R., Wu, Q., Yin, Y., Ge, J. (2011) Magnetically assembled photonic crystal film for humidity sensing. *J. Mater. Chem.* 21 (11): 3672–3676.

150. Pan, Z., Ma, J., Yan, J., Zhou, M., Gao, J. (2012) Response of inverse-opal hydrogels to alcohols. *J. Mater. Chem.* 22 (5): 2018–2025.

151. Zhao, Y., Zhao, X., Tang, B., Xu, W., Li, J., Hu, J., Gu, Z. (2010) Quantum-dot-tagged bioresponsive hydrogel suspension array for multiplex label-free DNA detection. *Adv. Funct. Mater.* 20 (6): 976–982.

152. Dole, M. (1980) The early history of the development of the glass electrode for pH measurements. *J. Chem. Educ.* 57 (2): 134.
153. Bakker, E., Bühlmann, P., Pretsch, E. (1997) Carrier-based ion-selective electrodes and bulk optodes. 1. General characteristics. *Chem. Rev.*. 97 (8): 3083–3132.
154. Buhlmann, P., Chen, L. (2012) Ion-selective electrodes with ionophore-doped sensing membranes. In *Supramolecular Chemistry: From Molecules to Nanomaterials*, Gale, P. A., Steed, J. W., Eds. John Wiley & Sons, New York: 2539–2579.
155. Lundland, R. L. (2007) *Biochemistry and Molecular Biology Compendium*. 1st Edition ed., CRC Press, Boca Raton.
156. Stefanac, Z., Simon, W. (1968) Electromotive properties of a gallium-silicon-potassium glass for measurement of potassium ion activities. *Helv. Chim. Acta* 51 (1): 74–78.
157. Cattrall, R. W., Tribuzio, S., Freiser, H. (1974) Potassium ion responsive coated wire electrode based on valinomycin. *Anal. Chem.* 46 (14): 2223–2224.
158. Van Gent, J., Lambeck, P. V., Kreuwel, H. J. M., Gerritsma, G. J., Sudhölter, E. J. R., Reinhoudt, D. N., Popma, T. J. A. (1989) Chromoionophores in optical ion sensors. *Sens. Actuators* 17 (1–2): 297–305.
159. Klajn, R. (2014) Spiropyran-based dynamic materials. *Chem. Soc. Rev.* 43 (1): 148–184.
160. Mistlberger, G., Xie, X., Pawlak, M., Crespo, G. A., Bakker, E. (2013) Photoresponsive ion extraction/release systems: dynamic ion optodes for calcium and sodium based on photochromic spiropyran. *Anal. Chem.* 85 (5): 2983–2990.
161. Xie, X., Mistlberger, G., Bakker, E. (2012) Reversible photodynamic chloride-selective sensor based on photochromic spiropyran. *J. Am. Chem. Soc.* 134 (41): 16929–16932.
162. Wagner, K., Byrne, R., Zanoni, M., Gambhir, S., Dennany, L., Breukers, R., Higgins, M., Wagner, P., Diamond, D., Wallace, G. G., Officer, D. L. (2011) A multiswitchable poly(terthiophene) bearing a spiropyran functionality: understanding photo- and electrochemical control. *J. Am. Chem. Soc.* 133 (14): 5453–5462.
163. Baillet, G., Giusti, G., Guglielmetti, R. (1993) Comparative photodegradation study between spiro[indoline—oxazine] and spiro[indoline—pyran] derivatives in solution. *J. Photochem. Photobiol. A Chem.* 70 (2): 157–161.
164. Matsushima, R., Nishiyama, M., Doi, M. (2001) Improvements in the fatigue resistances of photochromic compounds. *J. Photochem. Photobiol. A Chem.* 139 (1): 63–69.
165. Florea, L., Diamond, D., Benito-Lopez, F. (2012) Photo-responsive polymeric structures based on spiropyran. *Macromol. Mater. Eng.* 297 (12): 1148–1159.
166. Paramonov, S. V., Lokshin, V., Fedorova, O. A. (2011) Spiropyran, chromene or spirooxazine ligands: insights into mutual relations between complexing and photochromic properties. *J. Photochem. Photobiol., C: Photochem. Rev.* 12 (3): 209–236.
167. Natali, M., Soldi, L., Giordani, S. (2010) A photoswitchable Zn (II) selective spiropyran-based sensor. *Tetrahedron* 66 (38): 7612–7617.
168. Bakker, E., Crespo, G., Grygolowicz-Pawlak, E., Mistlberger, G., Pawlak, M., Xie, X. (2011) Advancing membrane electrodes and optical ion sensors. *CHIMIA Int. J. Chem.* 65 (3): 141–149.
169. Shvarev, A. (2006) Photoresponsive ion-selective optical sensor. *J. Am. Chem. Soc.* 128 (22): 7138–7139.
170. Liang, R., Zhang, R., Qin, W., Potentiometric sensor based on molecularly imprinted polymer for determination of melamine in milk. *Sens. Actuators, B* 2009, 141 (2), 544–550.

CHAPTER 6

Self-Stratifying Polymers and Coatings

Jamil Baghdachi, H. Perez, and Punthip Talapatcharoenkit
Coatings Research Institute, Eastern Michigan University, Ypsilanti, MI, USA

6.1 INTRODUCTION

Nature constructs composite materials efficiently and with the purpose of functionality of its constituent parts in tandem with the whole. The surface chemistry, physics, and functionality of some organic and inorganic materials are different from the bulk. For example, the surface of leaves and/or stems is more hydrophobic and contains nanostructures making it distinctly different from the bulk. In certain coatings, the pigment particles are not homogeneously distributed throughout the coating film. Depending on the resin and pigment type and chemistry and their levels as well as processing conditions, the immediate top surface may contain more resin than the bulk. In some special cases, where several curing agents and resins are present in a homogenous or semi-homogeneous solution, the reactivity and solidification sequence may also result in stratification.

In general, the solution of incompatible polymers using mutually compatible solvents can separate into distinct layers upon the evaporation of solvent(s) or water. Generally, the stratification and therefore surface and bulk heterogeneity happen during the material blending and film formation. Such separation, that is, stratification, can simply be due to insolubility of polymers in each other, differential densities, surface tensions, and other contributing chemical and physical factors. In principle, the solution of two polymers with large specific gravity differences (Table 6.1) should separate into two layers. As can be seen in Table 6.1, the specific gravities of most polymers used in coatings are within a narrow range and considering the effect of many other coating ingredients such as pigments and additives, phase separation of polymers due to specific gravity is not very likely.

Functional Polymer Coatings: Principles, Methods, and Applications, First Edition.
Edited by Limin Wu and Jamil Baghdachi.

TABLE 6.1 The Average Specific Gravities of Common Coating Polymers

Polymer Type (No Solvent)	Specific Gravity[a]
Polystyrene	1.05
Bisphenol A liquid epoxy	1.11
Polyurethanes	1.11–1.25
Polyvinyl acetate	1.19
Polymethylmethacrylate	1.19
Cellulose butyrate acetate	1.21
Polyvinyl chloride	1.25
Polytetrafluoroethylene	2.27

[a] Approximate specific gravities.

On the other hand, a blend of water/oil, certain types of incompletely emulsified water/polymer, etc., is known to separate into its constituent layers due to differences in surface tensions. Organic coatings are applied on various surfaces for a number of reasons including decoration; mechanical and physical protection; insulation or conductivity; and functions such as corrosion protection, antimicrobial properties, radar absorbency, and many other stimuli-responsive properties.

Ideally, a coating would combine all protective, decorative, and functional features of adhesion, corrosion protection, low water permeability, color, and desired surface properties in a single layer, and dried or cured in one step. Some coatings are applied in multiple layers, where each layer provides certain functionality such as adhesion, corrosion resistance, barrier, gloss, and in some cases stimuli-responsive properties. Multilayer coatings are commonly applied in the automotive, aerospace, general industrial, architectural, and transportation industries. In most cases, an intermediate coat or a primer coat is applied, dried or cured to provide foundation for good adhesion or to function as a barrier/corrosion control layer. In some cases such as automotive and aircraft applications, the primer layer is also sanded to provide a smooth surface for subsequent coating layers. Coatings with certain chemistries are compatible or provide the required intercoat adhesion properties if and only if they are applied within a specified time, which mandates careful attention to processing parameters such as optimal application condition, scheduling, and time management. These multilayered systems require complex application, curing/drying procedures, and extensive time for processing.

Except for temporary and strippable coatings, all other coatings function by surface attachment [1, 2]. Surface attachment is a property essential to corrosion protection by intact-type coatings such as polymer paint coatings versus nonintact coatings such as zinc-rich coatings. In some cases, adhesion to the substrate is maintained by the application of a separate coating layer, that is, primer or intermediate coat. In most cases, the chemical composition of primer is different from that of the top coat. The primer layer also requires separate application time and cure schedule than the top coat. The cure schedule depends on the chemistry of the coatings and can range from a mere 30 min prior to top-coating to 24 h at ambient temperature. Epoxy-based primers used in aerospace industry usually must be top-coated within a specific time after their application. The primary reason for this time interval is that the overcured primer is thought to be impermeable to top coat thereby resulting in inferior adhesion.

In systems in which the paint contains corrosion-inhibiting pigment, the preferred location of the corrosion-inhibiting agent within the film is at the interface of the coating and the metallic substrate, rather than at the coating–air interface or dispersed uniformly throughout the coating. Excess of such pigments negatively impacts formulation economics and is detrimental to other coating properties such as critical pigment volume concentration (CPVC), permeability, and elongation. However, in a self-stratifying coating, the corrosion-inhibiting agent can be preferentially directed to surfaces where corrosion inhibition is required.

Some surface-functional coatings such as antimicrobial, oil, and grease barrier coatings; superhydrophobic, and those with sensory functions such as stealth and explosive detection either contain expensive components that are currently dispersed throughout the coating or are applied as a separate layer. The most common method of preparing antibacterial coatings is to include a certain amount of bactericidal agents in the formulation [3, 4]. The activity of such systems is based upon the slow release of antibacterial agents such as silver ions [5–8], quaternary ammonium compounds [9, 10], or other antibiotics into the coating-air interface over time. Although this approach has proven to be successful for contact-killing coating systems, it requires that an excessive quantity of costly agents be added to the bulk of the coating to ensure sufficient concentration at the coating–air interface for the desired bactericidal activity. Self-stratifying coatings in which the majority of antimicrobial agent is directed to top layers of the coating surface will most likely increase both the economy and the efficacy of the antimicrobial activity of the coating.

Certain hydrophobic, oleophobic coatings are prepared by incorporating fluorinated polymers such as polytetrafluoroethylene (PTFE), polyvinylidene fluoride (PVDF), perfluoropolyethers, or alkyl vinyl ether polymers in coating formulations [11–14]. While coatings using such polymers have shown promising properties, their synthesis, formulation, and high costs are among challenges to be addressed. As with the aforementioned materials, smart coatings with sensory properties such as explosive detection or radar signal absorption are prepared by enriching the bulk of the coating with responsive agents, even though they only provide surface functionality.

Self-stratifying coatings have potential technological and economic advantages. It is conceivable that instead of two distinct and layers of coatings applied one after the other, one could apply a self-stratifying coating that can phase-separate into two distinct layers with different concentrations of resins and pigments and other functional additives. The concentration of materials in each phase (layer) would provide all the advantages of a conventional application but would potentially eliminate the intercoat boundary, which can be the point of failure. Additionally, through manipulation of coating ingredients, one may be able to force a higher concentration of materials either in coatings and air interface or coatings and substrate interface. In all cases, self-stratifying coatings would have economic advantages over the conventional systems.

Most product-finishing, automotive, aerospace, even architectural coatings require multi-coat applications. In general, initially a primer coat (intermediate coat) is applied and cured before application of the top coat. Both application and curing requires time, manpower, and energy. Therefore, coatings applied as a single layer that separates into layers upon curing to form an adhesion promoter/primer/intermediate coat and a top coat are an interesting concept that may address some of the challenges of sustainability, economy, and durability of coatings. Conceptually, the self-stratifying coating is applicable to all coating classes: conventional solvent-based, waterborne, powder coatings, and UV-curable systems.

6.2 BASIC CONCEPTS OF SELF-STRATIFICATION

Self-stratifying coatings concept allows the formulation of liquid and powder coatings based on partially incompatible polymer blends to produce multilayer coating structures by one-coating application of a single coating composition, advantageously useful in a number of industrial applications.

Several types of liquid binders can be blended with each other and dissolved in common solvents or a mixture of solvents thereby producing a homogeneous, thermodynamically stable liquid composition. In order to be able to stratify, phase separation of the binder assembly needs to occur during or immediately after the application of such composition to the substrate as a result of evaporation of the common solvents or reactions between the polymers where a latent curing catalyst and high temperature have been used.

In general and in case of thermoplastic systems, blending immiscible polymers into one liquid composition can result in the following:

- A thermodynamically stable microemulsion where a cosolvent or an emulsifier exists
- A kinetically stable polymer/polymer macroemulsion
- An unstable polymer/polymer emulsion

Among the aforementioned three cases, the unstable emulsion most readily separates into two layers as applied unto substrate provided that the cosolvents or emulsifier agents also leave the film or change their role as a function of applied heat or other parameters. Conceptually, in coatings containing all the necessary ingredients, the following four types of forces either singularly or in combination can drive stratification:

1) Evaporation/adsorption/absorption
2) Surface tension gradient
3) Substrate-wetting forces
4) Kinetically controlled reaction

6.2.1 Evaporation Effect

In principle, two incompatible polymers both dissolved in a common solvent should phase-separate upon the evaporation of the common solvent [15]. Factors such as solvent type (polar or nonpolar), evaporation rate, polymer molecular weights and types, phase separation behavior, and surface energy relationships are important factors that either independently or in combination with other factors influence the self-stratification and its quality. For example, the phase separation of epoxy/acrylic systems is favorable when the molecular weight of epoxy and acrylic resins increases where 50/50 solvent mixture of methyl isobutyl ketone and xylene was used; however, the same relationship does not hold between acrylic and alkyd [16]. Self-stratifying system where one of the polymers is an amine-cured epoxy resin while the other is a physically drying acrylic resin, an increase in incompatibility may be caused both by solvent evaporation and by a change of chemical structure of epoxy resin [17].

Mezzenga et al. [18] and Page [19] have examined amine-cured epoxy systems blended with various reactive epoxidized dendritic hyperbranched polymer modifiers. They have developed a procedure for calculating solubility parameters of epoxy resin during polymerization and a model for predicting free energy of mixing being a function of mixture composition, temperature, and a degree of polymerization.

Despite of the importance of evaporation rate and any consequent instability [20] on film quality, it is not easy to study and predict the effect of evaporation rate on film formation [15, 21–23]. In some applications such as spin coating, the effect of solution viscosity, intramolecular interactions, and glass transition must be taken into account.

Self-stratification based on evaporation is much more complex than those of two well-known epoxy and acrylic polymers. Evaporation of polar solvents from, for example, polyvinyl alcohol is more difficult than evaporation from a saturated polyester resin. In general, the higher the molecular weight and therefore the extent of chain entanglements, the more the retention of solvents by almost all polymers. Thus, predicting self-stratification through solvent evaporation for all polymers does not appear to be an easy task and may not always be reliable.

6.2.2 The Surface Tension Gradient

The surface tension gradient results from the difference in surface tensions of the polymers originally present in between the polymers but overcome due to the presence of solvents or emulsifiers in the homogeneous solution. According to the well-known lubrication approximation [24], the velocity of a liquid flow (v) driven by a surface tension gradient $d\sigma/dx$ and averaged over the local thickness (h) can be expressed as follows:

$$v = \frac{d\sigma / dx \cdot h}{\eta}$$

where η is the phase viscosity. Known as the Marangoni flow effect (spreading of a lower-surface tension component over a high-surface tension liquid), the surface tension gradient as a driving force is applicable mainly to the low-viscosity, solvent-based or thin-layer thermally curable coating assemblies.

Funke [25, 26] proposed that the primary driving force for stratification of incompatible polymer blends is the difference between the surface energies of simple resin systems. A 100% stratification can be represented by Fig. 6.1.

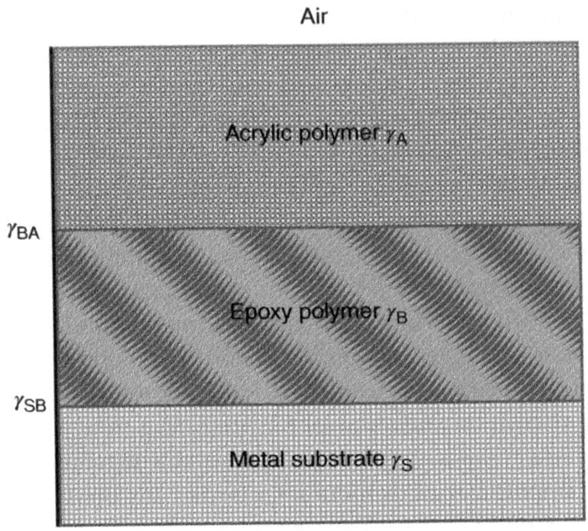

FIG. 6.1 Schematic representation of a stratified coating.

TABLE 6.2 Examples of High and Low Surface Free Energy Resins

High Surface Free Energy >35 mN/m		Low Surface Free Energy <30 mN/m	
Melamine formaldehyde resin	57	Polytetrafluoroethylene	18–22
Polyvinyl butyral	54	Silicone resins	20–25
Bisphenol A epoxy resin	42–47	Fluorine-modified resins	22–25
Acrylic resin	40–44	Silicone-modified resins	22–28
Polyvinyl acetate	38–42	Alkyd resins	25–28
Polyester	35–40	Saturated polyester resins	28–30

Based on theories expressed [27], self-stratification will occur if the three conditions presented here are met:

$$\gamma_{SA} - \gamma_{SB} - \gamma_{BA} \geq 0 \quad (1)$$
$$\gamma_{SA} - \gamma_{A} - \gamma_{SB} + \gamma_{B} > 0 \quad (2)$$
$$\gamma_{S} - \gamma_{A} - \gamma_{SB} - \gamma_{BA} > 0 \quad (3)$$

where γ_A and γ_B are the surface tensions of acrylic and epoxy resins and γ_{SA} and γ_{SB} are the interfacial tensions of surface/acrylic and surface/epoxy resins, respectively. As can be seen from Table 6.2, the difference in approximate surface energies of acrylic and epoxy resins at the high end is about 7 mN/m; however; the difference may increase if the epoxy resin is cured due to contribution of polarity component. Again, the contribution of molecular weight increase will add another factor that most likely will favor incompatibility of two polymers resulting in system stratification.

6.2.3 The Substrate-Wetting Force

The substrate-wetting force is the basis of the selective substrate-wetting mechanism [28]. According to this mechanism, the ability to stratify can be deduced from the balance of surface energy. In order for a liquid to spread over a solid surface, the work of adhesion (W_a) must be greater than the work of cohesion (W_c) and the spreading coefficient S must be positive. Therefore,

$$S = \gamma_{sv} - (\gamma_{lv} + \gamma_{ls}) = W_a - W_c > 0$$

where γ_{sv}, γ_{lv}, and γ_{ls} are the interfacial tensions associated with the substrate/vapor, liquid/vapor, and liquid/substrate interfaces, respectively. Accordingly, the spreading coefficient for the base polymeric phase must be greater than that for the stratifying phase. For example, self-stratification in acrylic latex/alkyd emulsion blends can be achieved due to a more rapid penetration of alkyd oligomer into a porous substrate such as wood [29]. Schematic representation of the two surface tension gradients and preferential surface wetting is shown in Fig. 6.2.

Stages and the contribution of the three factors of evaporation, surface tension gradient, and surface wetting is shown in Fig. 6.3. In this case, a blend of incompatible polymers is originally homogeneous due to dissolution of coating ingredients. The primary phase separation may begin immediately after the application as a result of evaporation of solvents. If properly controlled, such a primary phase separation may produce fluid phases capable of migration to upper or lower layers and solidification on further evaporation of solvent. If one of the resins preferentially wets the substrate surface more than the other resin, such resin can adhere to the substrate while losing more solvent during the film formation process. Once the first layer

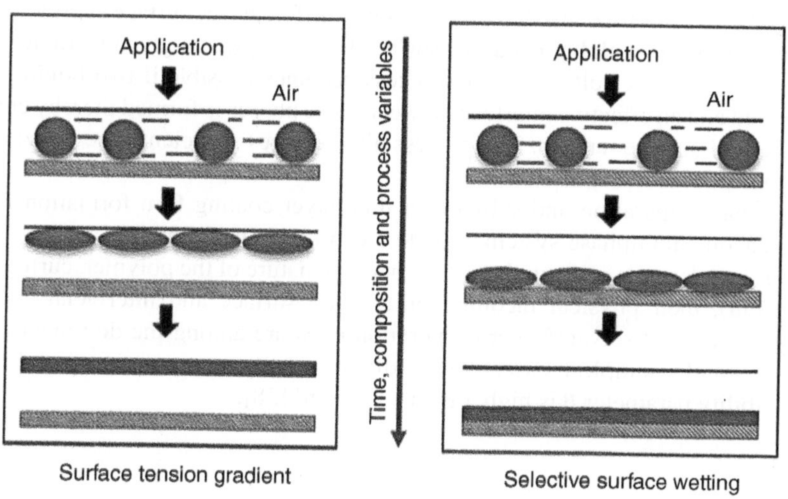

FIG. 6.2 Schematic representation of surface tension gradient and substrate-wetting force mechanisms.

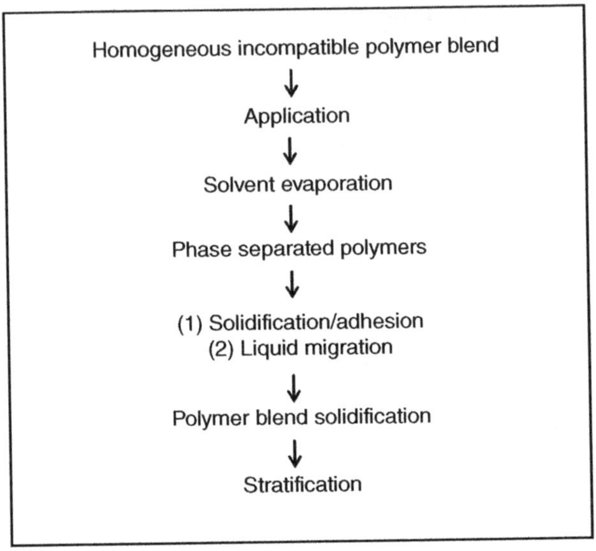

FIG. 6.3 Schematic representation of self-stratifying coating structure from initially homogeneous solutions of incompatible polymer blends.

(bottom layer) is formed, the remainder of the resin can in principle migrate to top surface and solidify as a stratified layer.

Spreading of a lower-surface tension liquid over a higher-surface tension counterpart and selective wetting of substrate by a higher-surface tension resin are the most common mechanisms to produce self-stratifying coatings from liquid thermoplastic-type coating composition. However, both mechanisms require that the incompatible resins used also have a sufficiently different free surface energy. Typical examples of resins with high and low surface free energies are given in Table 6.2.

In most cases, the self-stratifying process becomes possible if two liquid phases are formed and an effective interface between these phases is formed thereby creating a gradient of surface tension. The process also requires sufficient time for viscosity buildup and mass transfer.

The phase separation and subsequent multilayer coating film formation from a homogeneous multiphase system is quite a challenge in a complex system such as coatings and depends on a number of factors. The nature of the polymer, curing agent (if present), their physical incompatibility, total surface and interfacial energies, reaction rates, and solvent/water evaporation rates are among the dominant factors [30–37]. In the simplest case, the condition for phase separation is when the compatibility parameter β is higher than 0.07 J/cm^3 [38].

$$\beta = (\delta_1 - \delta_2)^2 > 0.07$$

where δ_1 and δ_2 are the Hildebrand solubility parameters of the two resins.

As a general principle, all liquid ingredients of coatings—polymers, additives, and solvents—must be mutually compatible so that a homogeneous wet film is formed upon application and a cohesive and smooth film is formed upon drying, curing, or coalescence. In practice, certain components necessary to particular formulations such as microgels, defoamers [16, 39], and polymeric particles must be incompatible with some of the coating's ingredients to provide their properties. Depending upon the nature of the incompatible material, it may disperse uniformly throughout the coating or it may preferentially migrate to the base or surface of the film. The latter two cases can be considered a special situation of stratification due to material incompatibility. Funke [25, 26], perhaps, was the first, and later Murase [40], to investigate the use of incompatible polymeric systems and demonstrated the utility of such polymer blends in organic coating formulations to produce self-stratifying coatings. The concept was further expanded by a series of formulations that included alkyd, silicone, epoxy, and vinyl resins [28, 41–45].

In principle, simply by using two incompatible liquid polymers, either in a homogeneous form in the presence of solvent or heterogeneous form, upon drying or curing one can obtain a multilayered film. Such films, however, may suffer from lack of adhesion between layers, and the uppermost layer may not possess the necessary properties as top coat.

6.2.4 Kinetically Controlled Reactions

It is possible that the two incompatible polymers cross-link with reactive agents at two distinct rates, times, and temperatures. Examples of such systems are the reaction of epoxy resin with a thiol cross-linking agent and a typical reaction of a polyester/acrylic resin with either isocyanate compounds or a melamine formaldehyde-curing agent. A special case can be the reaction of a mixture of amine-containing polymers and various polyols. It is well known the reaction of epoxy resin with thiol or aliphatic amine compounds is quite fast at room temperature. However, the reaction of a polyester polyol with melamine formaldehyde requires general or specific acid catalyst and temperatures above room temperature. Since the aforementioned compounds as far as solubility and polarity are concerned are somewhat incompatible materials, have different surface tensions, and require specific cross-linking agents, a formulation may be prepared and processed at gradient temperatures to afford self-stratifying composition. In such systems, while the major contributor is the preferential cross-linking of one polymer versus the other, previously discussed factors of evaporation, surface tension gradient, and substrate wetting still play a role in stratification rate, percent, and possibly its quality. Thus, similar to the thermoplastic system, it is conceivable that one of the polymers is selected from the low-surface energy class such as fluorine-modified acrylic/polyester/vinyl ether polyols while the other polymer is from the high-surface energy materials such as epoxy or acrylic resin categories. Thermosetting self-stratifying systems offer advantages of broad industrial applications where intercoat adhesion as well as timing between the coats is critical. In addition, reduced application and reduced processing time and temperature as well as formulation time can be considered as a sustainable coating

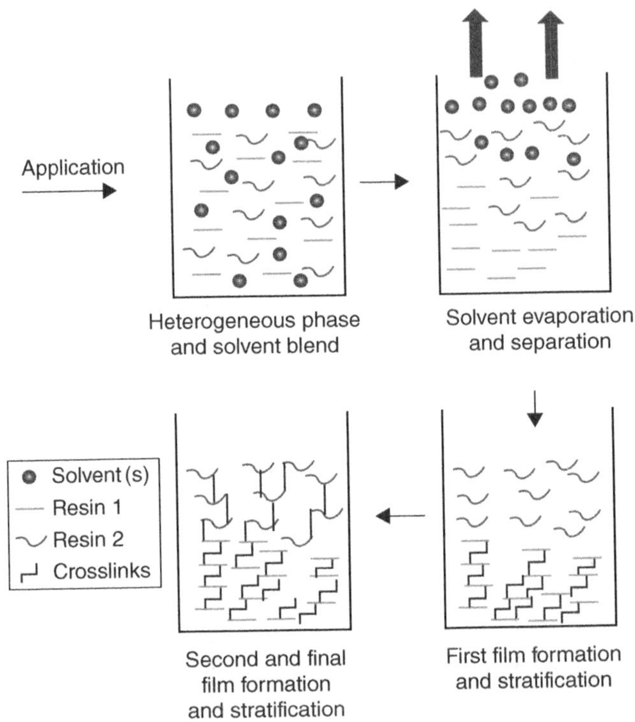

FIG. 6.4 Schematic representation of kinetically controlled self-stratifying coating structure from initially homogeneous solutions of incompatible polymer blends.

technology. A schematic representation of kinetically controlled self-stratifying coating structure from initially homogeneous solutions of incompatible polymer blends is shown in Fig. 6.4.

In the past four decades, a number of researchers and authors have explored, investigated, and discussed the various possible mechanisms of coating film stratification phenomena [14, 15, 18–30]. Most of these investigations provide models for the self-stratification of simple binary polymer systems with or without the presence of solvents. However, a typical industrial coating is a complex system with many interactive and reactive materials, and their properties are affected by application, storage, and environmental conditions. A number of authors have demonstrated that the use of solubility parameters to predict stratification is not always reliable [32, 33]. In their opinion, this lack of correlation between theoretical prediction and experimental results is due to the Flory–Huggins theory that does not take into account the free volume variation and possible specific interactions between the components of a coating. Funke's [25, 26] model posited that the primary driving force for stratification of incompatible polymer blends is the difference between the surface energies of the resins used. Based upon the results of this investigation, the Funke model has shortcomings in that it does not consider the interaction of coating ingredients, solvent evaporation, and chemical reactions, as well as the presence of pigments.

In their investigations, Kuczynska et al. [21] indicate that the differences in surface energies of epoxy resins examined in their study are small and do not exceed 8%, which may be the limit of the surface tension gradient for stratification of polymers. In our opinion, in a complex system such as a multicomponent coating, there are other important criteria that must be met before stratification is realized.

In recent years, certain thermoplastic or mixed thermoplastic and thermosetting prototype coatings have been reported in the literature. Self-stratifying coatings for metallic substrates [17] described the preparation and self-stratification of coatings that contain amine-cured epoxy as base coat and acrylic or fluoroethylene/vinyl ether copolymer as top coat resins. The most favorable conditions were found to be a resin ratio of 1:1. Extension to systems containing a protective pigment in the layer at the substrate interface has shown some success although the addition of the second pigment such as titanium dioxide interfered with stratification and restricting it to bottom layer proved to be difficult. It was also shown that the information gained from solubility parameter data alone cannot adequately predict resin compatibility as an aid to the prediction of stratification and that the compatibility of resin pairs in common solvents must be experimentally determined.

Langer [46] reported self-stratifying coatings containing barrier and active anti-corrosive pigments using a 7:3 ratio of diamine-cured bisphenol A epoxy and copolymer of methyl methacrylate and ethyl acrylate, respectively. The researchers conclude that the most important properties of pigments that contributed to pigment distribution were average particle size of pigment dispersed in epoxy resin and solubility parameter of pigment dispersed in acrylic resin. Additionally, the presence of chrome oxide green and zinc phosphate impeded resin stratification. The distribution of pigments in stratified coating is shown in Fig. 6.5.

Self-stratification process can be employed to produce coating films that are rich in one of the surface functionalities of coatings such as hydrophobicity, antimicrobial, reflectivity, and corrosion resistance. For example, matting agents such as special silica particles and fluorinated polymers are dispersed in coatings to provide matting [47, 48].

Since such materials are dispersed in the bulk of the coating, only a low concentration of the material is present on the surface where its functionality is

FIG. 6.5 Cross-section of coating containing titanium dioxide and natural lamellar iron oxide (MIO_X) [46]. From Ref. 46. © Elsevier.

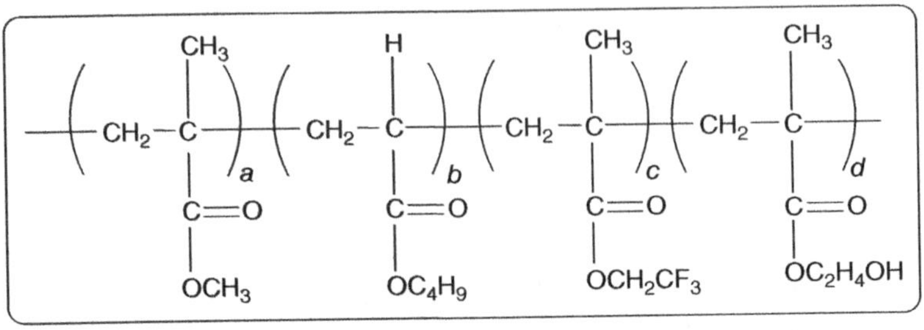

FIG. 6.6 Representative molecular structure of fluorinated acrylic copolymer [52]. From Ref. 52. © Springer.

needed. Ming et al. [49] has reported the development of low-surface energy cross-linked films on the basis of partially fluorinated hydroxyl-end-capped solvent-less coatings either thermally cured [50] or photocured [51] after the introduction of acrylic unsaturation. The investigators conclude that the low surface energy of about 12 mN/m was achieved with less than 1% weight of fluorine content.

A totally thermosetting system containing modified epoxy resin and acrylic polymer has been investigated [52]. In this system, the investigators explored the effect of polar (tetraethoxysilane, TEOS, and phosphoric acid) and nonpolar (conjugated linoleic acid) modification of liquid epoxy bisphenol A resin. The acrylic resin was also modified by 2, 2, 2-trifluoroethylmethacrylate at 5 and 10% volume levels [53], Fig. 6.6.

Incorporation of fluorine in the acrylic copolymer was designed to drive the acrylate to the film–air interface during film formation. Likewise, the siloxane and phosphate ester–modified epoxide resin were designed to be driven to the interface of film–substrate by the chemical bond formed from the co-condensation of Si–OH, P–OH, and metal OH producing Si–O–metal and P–O–metal linkage [54, 55]. The formulation of upper layer and lower layer and proposed reaction with melamine formaldehyde resin is shown in Fig. 6.7.

The investigators reported that the combination of 5% fluorinated medium-molecular-weight acrylic copolymer and fatty acid ester–modified epoxy cross-linked with melamine formaldehyde resin failed to produce any stratification. A composition containing 5% low molecular weight fluorinated acrylic resin and TEOS/phosphate modified epoxy resin once cured, showed partial film stratification. While a composition consisting of 10% fluorinated low-molecular-weight acrylic copolymer and TEOS modified epoxy cured under the same condition resulted in self-organized two-phase system with diffuse phase boundary rather than a distinct self-stratifying coating system.

Recent investigation by Baghdachi et al. [22, 56–58] proposed kinetically controlled cross-linking reactions as a mechanism of self-stratification. The kinetically controlled reactions between various polymers and cross-linking agents result in phase separation presumably where the faster reactions result in rapid rise in glass transition temperature, and hence reduced mobility, to produce the layer adjacent to the substrate. In the next step, the slower-reacting species migrate to the surface and cure to produce the self-stratifying coating. The components and surface energies of chemicals used in prototype formulations are shown in Table 6.3.

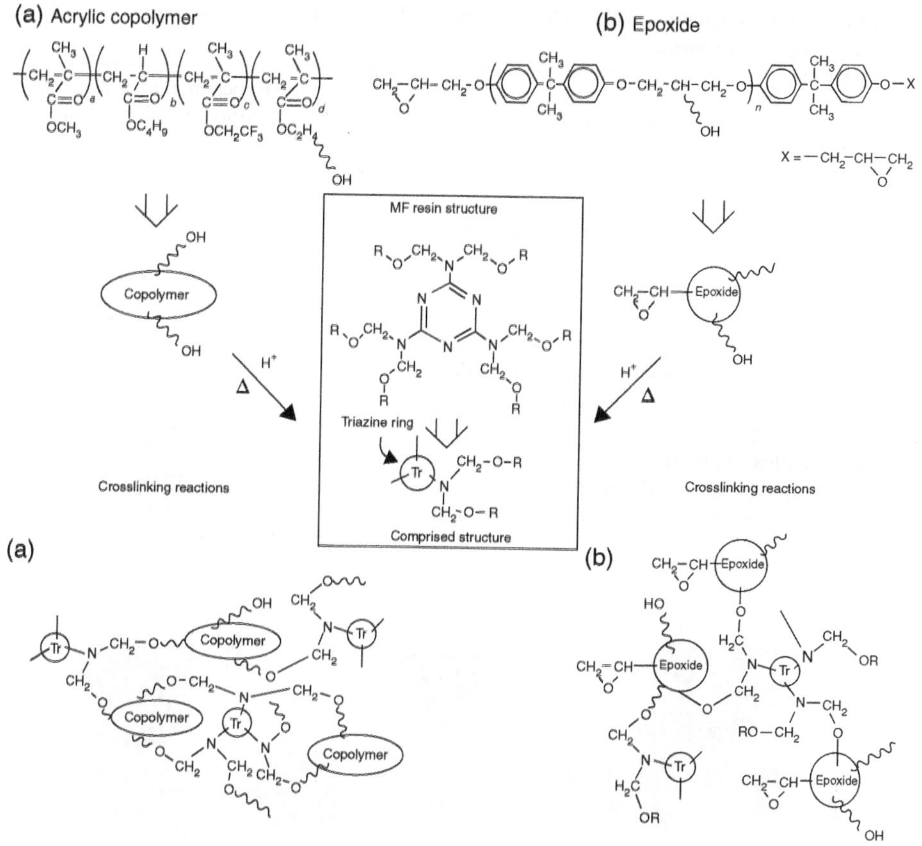

FIG. 6.7 Cross-linking reactions between Melamine formaldehyde (MF) resin and –OH functionality of organic phases: (a) acrylic and (b) epoxide [52]. From Ref. 52. © Wiley.

According to the authors, upon the addition of the cross-linker, it was observed that the homogeneous hazy solution gradually began to exhibit phase separation. After the mixture was allowed to stand overnight at 25°C, three distinct phases—including the solvent mixture—were noted. Prolonged storage (24 and 48 h) accompanied by gentle agitation appeared to assist the reaction, resulting in two seemingly connected semisolid layers and a solvent layer that evaporated on exposure to 100°C for 30 min (Fig. 6.8).

It is believed that the terminal primary hydroxyl groups of the isophorone triisocyanurate urethane dendrimer, shown in Fig. 6.9, react at faster rates with hexamethylene diisocyanate than crowded, high-molecular-weight pendent hydroxyl groups of the Lumiflon®, a fluoroethylene alkyl vinyl ether polymer. Presumably in the absence of other cross-linking reactions, the reaction of diisocyanate cross-linker and dendrimer produces polyurethane polymers with increased glass transition temperature and reduced mobility that tend to phase-separate from the remainder of the solution. Conceivably, reduced availability of dendrimer for further reaction permits the diisocyanate to react with the fluoropolymer, albeit slowly, to form the second layer.

TABLE 6.3 Experimental Surface Energy Values of Materials Used in Prototype Formulation

	Chemicals	Experimental Surface Energy at 23°C, mJ/m^2
1	Hydroxyl functional dendimer; 60% in DMA	32.33
2	Tetrahydrofuran	26.40
3	Lumiflon®; 60% in xylene	25.11
4	*N,N*-Dimethyl acetamide	36.70
5	Xylene(s)	28.6
6	Hexamethylene diisocyanate	37.89
7	Epoxy functional silsequioxane; 75% in THF	31.92
8	Epoxypropoxypropyl terminated PDMS	28.00
9	Trisilanol isooctyl silsesquioxane	29.85
10	Octa epoxy POSS	30.15
11	K-Flex® polyester	32.02
12	Joncryl® 920 acrylic polyol	32.95
13	Boltron® H 2004 dendrimer	33.11

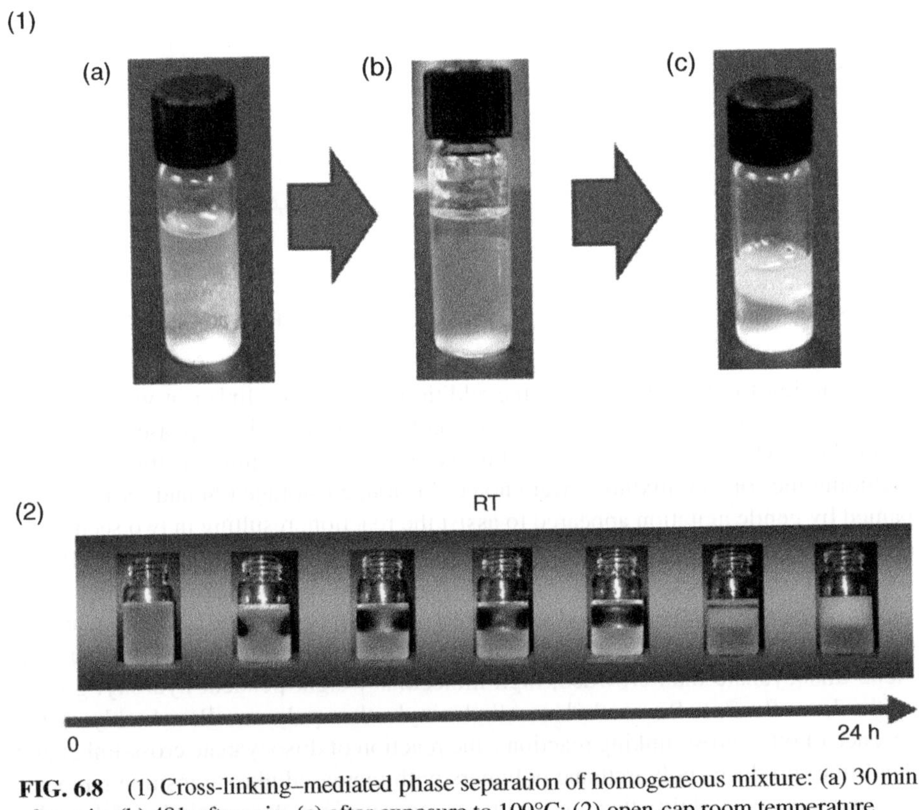

FIG. 6.8 (1) Cross-linking–mediated phase separation of homogeneous mixture: (a) 30 min after mix, (b) 48 h after mix, (c) after exposure to 100°C; (2) open-cap room temperature cross-linking.

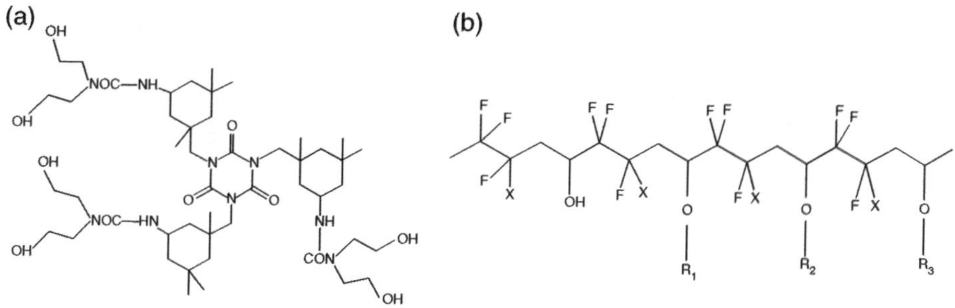

FIG. 6.9 Proposed chemical structures of (a) first-generation polyurethane and (b) hydroxyl functional fluoroethylene alkyl vinyl ether (Lumiflon®).

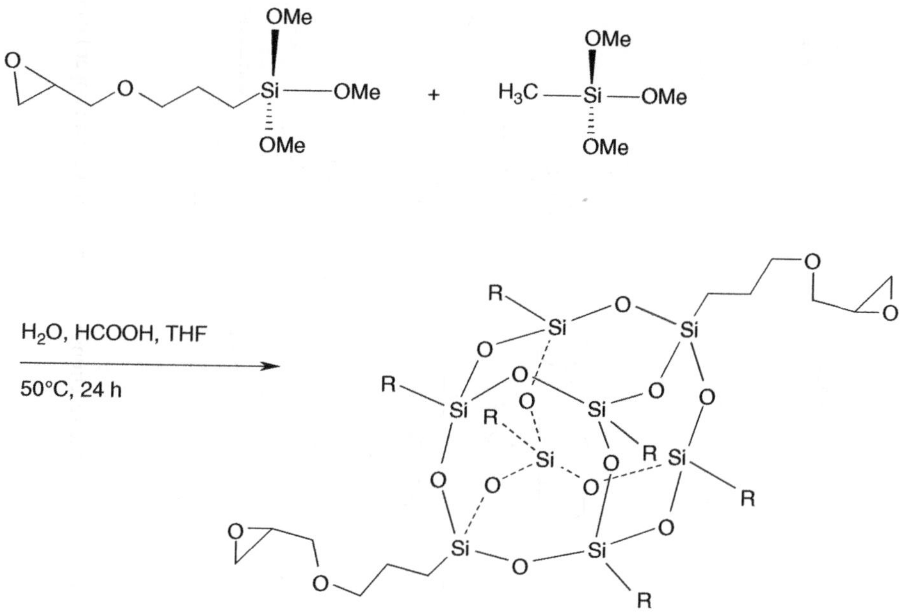

FIG. 6.10 Schematic representation of glycidoxypropyl functional silsesquioxane.

Upon continued reactions between all the reactants, first layer and second layer are bonded together through hexamethylene bridge of the diisocyanate cross-linker.

The full formulation also contained a 30% solution of diglycidoxypropyl functionalized polyhedral oligomeric silsesquioxane (di-epoxy POSS) in tetrahydrofuran, a reaction product of glycidoxypropyltrimethoxysilane and methyltrimethoxysilane following the procedure described in US pat. [22] (Fig. 6.10).

Schematic reactions of various components of the self-stratifying formulation is shown in Fig. 6.11.

In general, the most consistent results were obtained by baking the systems at two different temperatures of 100°C for 30 min followed by another 15 min at 145°C,

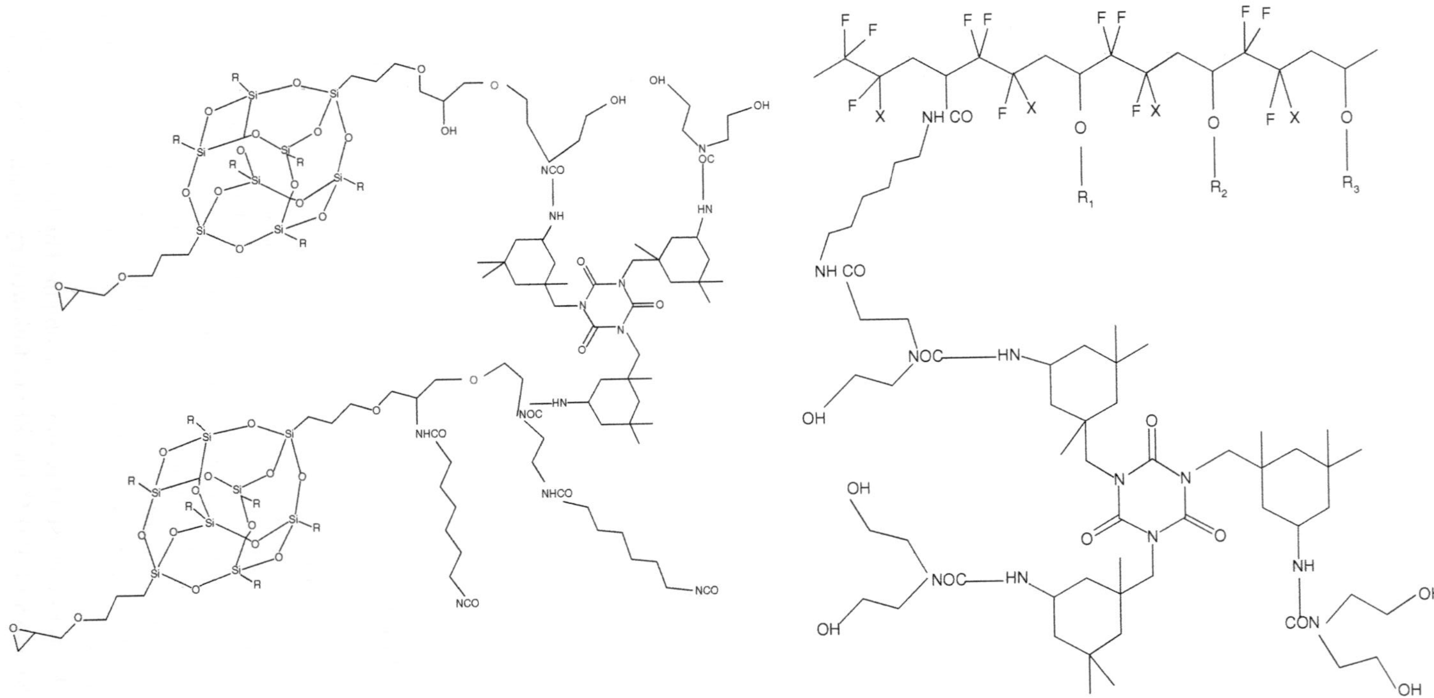

FIG. 6.11 Schematic representation of the reactions representing lower and upper layers of the stratified coating.

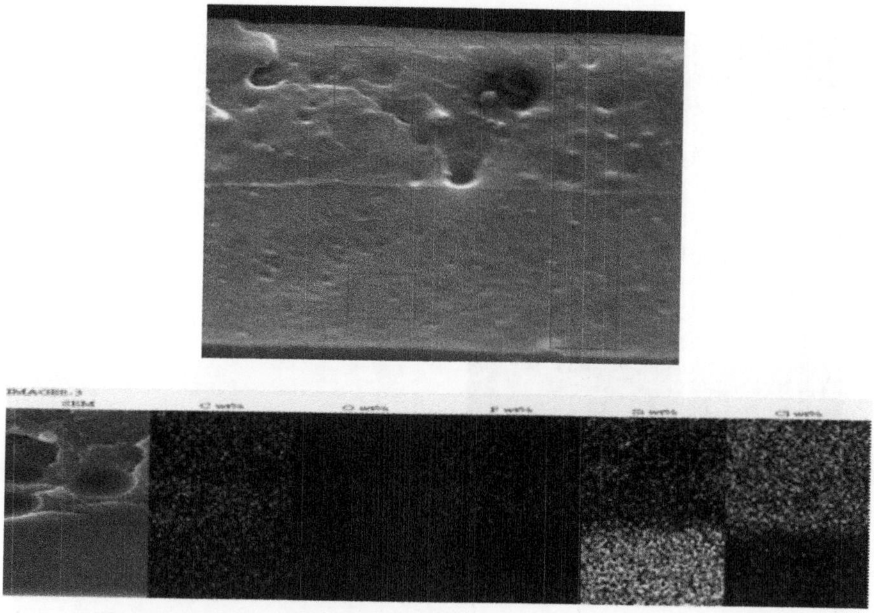

FIG. 6.12 SEM image and EDX elemental analysis of stratified coating.

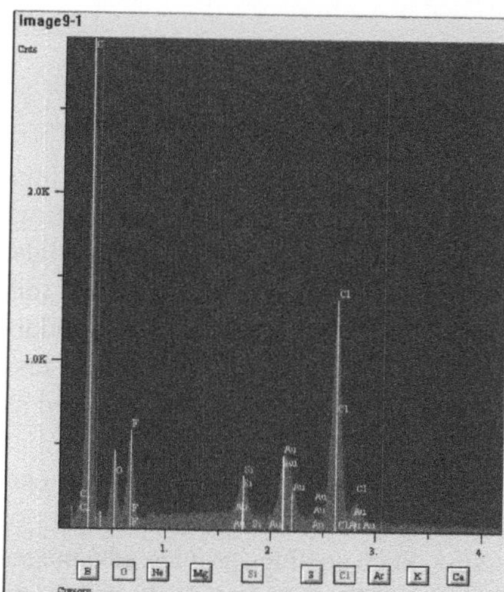

Elt.	Line	Intensity (c/s)	Error 2-sig	Atomic %	Atomic Ratio	Conc		K-Ratio
C	Ka	279.34	5.747	76.196	5.7867	64.813	wt.%	0.6694
N	Ka	0.00	0.000	0.000	0.0000	0.000	wt.%	0.0000
O	Ka	46.89	2.355	13.167	1.0000	14.920	wt.%	0.0654
F	Ka	63.81	2.747	8.065	0.6125	10.851	wt.%	0.0584
Si	Ka	35.70	2.055	0.373	0.0283	0.741	wt.%	0.0171
Cl	Ka	210.40	4.988	1.923	0.1460	4.828	wt.%	0.1230
Au	La	6.27	0.861	0.276	0.0209	3.847	wt.%	0.0667
				100.000		100.000	wt.%	Total

kV 20.0
Takeoff Angle 35.0°
Elapsed Livetime 60.0

FIG. 6.13 Analysis of area 1 in Fig. 6.12, top layer of the stratified system.

respectively. Scanning electron photomicrographs of the aforementioned formulation are shown in Fig. 6.12.

Furthermore, the energy dispersive X spectroscopy (EDX) results of areas 1 and 2 in Fig. 6.12 are shown in Figs. 6.13 and 6.14 respectively. The EDX data in Fig. 6.12

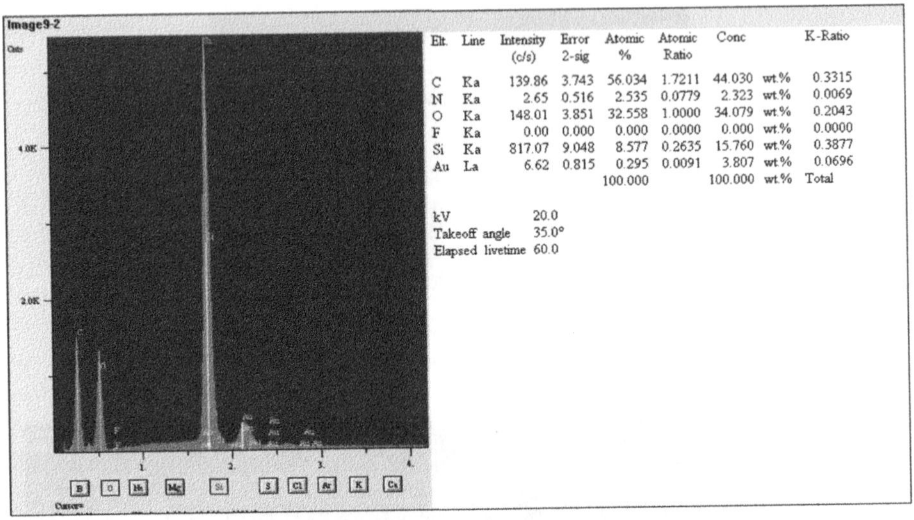

FIG. 6.14 Analysis of area 2 in Fig. 6.12, bottom layer of the stratified system.

indicates that the top layer contains 10.85 wt% fluorine and 0.741 wt% silicon, while the reverse is true in the bottom layer, with 0.45 wt% fluorine and 15.76 wt% silicon, shown in Figs. 6.13 and 6.14.

6.3 CONCLUSIONS

Self-stratifying coating as described earlier can potentially incorporate two coatings in one formulation, that is, an intermediate coat or primer and a top coat coating. Obviously, such formulations require shorter formulation times, less raw materials, shorter processing times, and perhaps more durable coatings. In addition, this approach can be used to prepare surface-rich oleophobic, antimicrobial, radar-absorbing, highly reflective, and similar surface-active coatings with much less needed active ingredients. Hence, a stratifying coating can potentially be regarded as a sustainable coating technology.

Self-stratifying coating technology offers potential industrial and architectural coating formulators a tool to design coatings that can separate into two distinct layers of primer and top coat after application and processing. Such coatings provide advantages and cost-savings to both formulators and coating users. Thermoplastic self-stratifying can be prepared by combining incompatible polymers and most of other coating ingredients using common solvents that upon evaporation yield a composite coating system. Thermosetting systems can be prepared by combining incompatible polymers and common cross-linking agents that react with respective polymers at varied rates. The faster-cured component reaches high glass transition temperatures at early stages and forms the bottom layer. Following the solidification of the first layer, the slower-reaction species migrate to top surfaces and cure subsequently, forming the stratified layer.

REFERENCES

1. Baghdachi, J., Fundamentals of adhesion, *JCT*, 69 870 85–91 (1997).
2. Baghdachi, J. *Adhesion Aspects of Polymeric Coatings*, FSCT, Pennsylvania, (1997).
3. Piozzi, A. Francolini, I., Occhiaperti, L., Venditti, M., Marconi, W., Antimicrobial activity of polyurethanes coated with antibiotics: a new approach to the realization of medical devices exempt from microbial colonization, *J. Pharm.*, 280 173 (2004).
4. Vartianen, J., Tuominen, M., Nattinen, K., Bio-hybrid nanocomposite coatings from sonicated chitosan and nanoclay, *J. Appl. Polym. Sci.*, 116 3638 (2010).
5. Baghdachi, J., Clemans, D., Xu, Q., Rhoades, J., Kendzorski, J., Evaluating antimicrobial activities of surface coatings, *Am. Chem. Soc. Symp. Ser.*, 957 23–41 (2007).
6. Konwara, U., Karaka, N., Mandal, M., Vegetable oil based highly branched polyester/clay silver nanocomposites as antimicrobial surface coating materials, *Prog. Org. Coat.*, 68 265 (2010).
7. Ho, C.H., Tobis, J., Sprich, C., Thomann, R., Tiller, J.C., Nanoseparated polymeric networks with multiple antimicrobial properties, *Adv. Mater.*, 16 957 (2004).
8. Sambhy, V., MacBride, M.M., Peterson, R.P., Sen, A., Silver bromide nanoparticle/polymer composites: dual action tunable antimicrobial materials, *J. Am. Chem. Soc.*, 128 9798 (2006).
9. Massi, M., Guittard, F., Geribaldi, S., Levy, R., Duccini, Y., Antimicrobial properties of highly fluorinated bis-ammonium salts, *Int. J. Antimicrob. Agents*, 21 20 (2003).
10. Harney, M.B., Pant, R.R., Fulmer, P.A., Wynne, J.H., Surface self-concentrating amphiphilic quaternary ammonium biocides as coating additives, *ACS Appl. Mater.*, 1 39 (2009).
11. Hougham, G., Cassidy, P.E., Johns, K., Davidson, T. (Eds.), *Fluoropolymers 1 and 2*, Kulwer Academic Publishers, Dordrecht Plenum Press, New York, (1999).
12. Scheirs, J. (Ed.), *Modern Fluoropolymers*, Wiley, New York, (1997).
13. Lousenberg, R.D., Shoichet, M.S., Radical copolymerization of novel trifluorovinyl ethers with ethyl vinyl ether and vinyl acetate: estimating reactivity ratios and understanding reactivity behavior of the propagating radical, *J. Polym. Sci. Part A: Polym. Chem.*, 38 1344 (2000).
14. Ming, W., Melis, F., van de Grampel, R.D., van Ravenstein, L., Tian, M., van der Linde, R., Low surface energy films based on partially fluorinated isocyanates: the effects of curing temperature, *Prog. Org. Coat.*, 48 316–321 (2003).
15. Vink, P., Bots, T.L., Formulation parameters influencing self-stratification of coatings, *Prog. Org. Coat.*, 28 173–181 (1996).
16. Sebba, F., *Foams and Biliquid Foams-Aphrons*, Wiley, New York, (1987).
17. Benjamin, S., Carr, C., Walbridge, D.J., Stratifying coatings for metallic substrates, *Prog. Org. Coat.*, 28 197–207 (1996).
18. Mezzenga, R., Boogh, L., Manson, J.A.E., Thermodynamic model for thermoset polymer blends with reactive modifiers, *J. Polym. Sci. Part B*, 38 1883 (2000).
19. Page, S.A., Surface energetics evolution during processing of epoxy resins, *J. Colloid Interface Sci.*, 222 55 (2000).
20. de Gennes, P.G., Instabilities during the evaporation of a film: non-glassy polymer + volatile solvent, *Eur. Phys. J. E*, 6 421 (2001).
21. Kuczynska, H., Langer, E., Kaminska-Tarnawski, E., Kulikov, D.A., Indeikin, E.A., Study of self-stratifying compositions, *J. Coat. Technol.*, 6(3) 345–352 (2009).

22. Baghdachi, J., Perez, H., Templeman, C., Self-stratifying automotive topcoat compositions and processes. US Patent, 7,863,375 (2010).
23. Mokarian-Tabari, P., Geoghegan, M. Howse, J., Heriot, S., Thompson, R., Jones, R., Quantitative evaluation of evaporation rate during spin-coatings of polymer blend films: control of film structure through defined-atmosphere solvent-casting, *Eur. Phys. J. E*, 33 283–289 (2010).
24. Levich, V.G., *Physiochemical Hydrodynamics*, Printice-Hall Scripta Technica, Inc., Englewood Cliffs (1962).
25. Funke, W., Preparation and properties of paint films with special morphological structure, *J. Oil Color Chem. Assoc.*, 59 398–403 (1976).
26. Funke, W., New aspects of paint films with inhomogeneous structure, *Prog. Org. Coat.*, 2 289 (1974).
27. Carr, C., Wallström, E., Theoretical aspects of self-stratification, *Prog. Org. Coat.*, 28 161–171 (1996).
28. Verkholantsev, V.V., Self-stratifying coatings, *Eur. Coat. J.*, 12 24, 26, 28, 30–33 (2000).
29. Hofland, A., Water-borne coatings for decorative and protective coatings: a comparative survey, *Surf. Coat. Int.*, 77(7) 270–276 (1994).
30. Verkholantsev, V.V., Nonhomogeneous-in-layer coatings, *Prog. Org. Coat.*, 13 71 (1985).
31. Misev, T.A., Thermodynamic analysis of phase separation in selfstratifying coatings—solubility parameters approach, *J. Coat. Technol.*, 63(795) 23–27 (1991).
32. Hopkinson, I., Myatt, M., Phase separation in ternary polymer solutions induced by solvent loss, *Macromolecules*, 35 5153–5160 (2002).
33. Mitov, Z., Kumacheva, E., Convection-induced patterns in phase-separating polymer fluids, *Phys. Rev. Lett.*, 81 3427–3430 (1998).
34. Kumacheva, E., Li, L., Winnik, M.A., Shinozaki, D.M., Cheng, P.C., Direct imaging of surface and bulk structures in solvent cast polymer blend films, *Langmuir*, 13 2483–2489 (1997).
35. Verkholantsev, V.V., Flavian, M., Polymer structure and properties of heterophase and self-stratifying coatings, *Prog. Org. Coat.*, 29 239–246 (1996).
36. Verkholantsev, V.V., Functional variety. Effects and properties in surface-functional coatings, *Eur. Coat. J.*, 9 18–25 (2003).
37. Verkholantsev, V.V., Self-stratufying coatings for industrial applications, *Pigm. Resin Technol.*, 32(5) 300–306 (2003).
38. Hildebrand, J., Scott, R.L., *Solubility of Non-Electrolytes*, 3rd edn., Reinhold, New York, 21–44, (1950).
39. Mueller, B., Poth, A., *Coating Formulation*, Vincentz Network, Hanover, (2006).
40. Murase, H., Funke, W., 15th FATIPEC Congress, No 2, 387–409 (1980).
41. Krylova, I.A., Sazonova, S.V., Morozova, N.I., Composite coatings obtained from aqueous polymeroligomeric dispersions by electrodeposition, *Prog. Org. Coat.*, 21 1–15 (1992).
42. Verkholantsev, V.V., Coatings based on polymer/polymer composites, *J. Coat. Technol.*, 64(809) 51–59 (1992).
43. Verkholantsev, V.V., Heterophase and self-stratifying polymer coatings, *Prog. Org. Coat.*, 26(1) 31–52 (1995).
44. Verkholantsev, V.V., Flavian, M., Epoxy thermoplastic heterophase and self-stratifying coatings, *Mod. Paint Coating*, 85(11) 100–106 (1995).

45. Carr, C., Benjamin, S., Walbridge, D.J., Fluorinated resin in self-stratifying coatings, *Eur. Coat. J.*, 4 262–266 (1995).

46. Langer, E., Kuczyńska, H., Kamińska-Tarnawska, E., Łukaszczyk, J., Self-stratifying coatings containing barrier and active anticorrosive pigments, *Prog. Org. Coat.*, 71 162–166 (2011).

47. Bauer, F., Flyunt, R., Czihal, F., Langguth, H., Mehnert, R., Schubert, R., Buchmeiser, M., UV curing and matting of acrylate coatings reinforced by nano-silica and micro-corundum particles, *Prog. Org. Coat.*, 60(2) 121–126 (2007).

48. Gunde, M., Kunaver, M., Čekada, M., Surface analysis of matt powder coatings, *Dyes Pigments*, 74(1) 202–207 (2007).

49 Ming, W., Tian, M., van de Grampel, R., Melis, F., Jia, X., Loos, J., van der Linde, R., Low surface energy polymeric films from solventless liquid oligoesters and partially fluorinated isocyanates, *Macromolecules* 35 6920–6929 (2002).

50. Ming, W., Laven, J., der Linde, R., Synthesis and surface properties of films based on solventless liquid fluorinated oligoester, *Macromolecules*, 33 6886 (2000).

51. Ming, W., van Ravenstein L., van de Grampel, R., Van Gennip, W., Krupers, M., Niemantsverdriet, H., van der Linde, R., Low surface energy polymeric films from partially fluorinated photocurable solventless liquid oligoesters, *Polym. Bull. (Berlin)*, 47 321 (2001).

52. Alyamac, E., Gu, H., Soucek, M.D., Interface-driven phase-separated coatings, *J. Coat. Technol. Res.*, 11(5) 665–683 (2014).

53. Alyamac, E., Soucek, M.D., Acrylate-based fluorinated copolymers for high solids coatings, *Prog. Org. Coat.*, 71(3) 213–224 (2011).

54. Alyamac, E., Gu, H., Soucek, M.D., Qiu, S., Buchheit, R.G., Alkoxysilane oligomer modified epoxide primers, *Prog. Org. Coat.*, 74 67–81 (2012).

55. Witucki, G.L., A silane primer: chemistry and applications of alkoxy silanes, *J. Coat. Technol. Res.*, 65(822) 57–60 (1993).

56. Baghdachi, J., Perez, H., Talapatcharoenkit, P., Wang, B., Design and development of self-stratifying systems as sustainable coatings, Prog. Org. Coat., 78 464 (2015).

57. Baghdachi, J., Templeman, C., Talapatcharoenkit, P., Self-stratifying coatings. US Patent, 8,299, 170 B2 (2012).

58. Baghdachi, J., Perez, H., Templeman, C., He, H., Methods and compositions for pigmented self-stratifying coatings. US Patent, 8,044,140 B2 (2011).

CHAPTER 7

Surface-Grafted Polymer Coatings: Preparation, Characterization, and Antifouling Behavior

Marc A. Rufin[1] and Melissa A. Grunlan[1,2]

[1]Department of Biomedical Engineering, Texas A&M University, College Station, TX, USA

[2]Department of Materials Science and Engineering, Texas A&M University, College Station, TX, USA

7.1 INTRODUCTION

Material selection is often a complex part of the design process. Several characteristics of a material must be typically considered before one is ultimately chosen. Often, material selection is based on bulk properties (e.g., mechanical, thermal, and electrical) as well as cost [1]. However, materials with excellent bulk properties may exhibit poor surface properties that would limit their ultimate performance. These situations often arise in the medical device industry. For example, most central venous catheters are made with silicones or polyurethanes, which provide favorable strength and flexibility [2]. Unfortunately, these polymers have a high affinity for protein adsorption onto their surfaces, which makes the corresponding catheters prone to infection and thrombosis [3]. In another example, superparamagnetic iron oxide nanoparticles used as magnetic resonance imaging contrast agents are of interest for new diagnostic imaging techniques [4, 5]. However, it has been shown that in aqueous conditions, such particle surfaces oxidize and produce cytotoxic reactive species [6].

Surface modification techniques may be employed to potentially mitigate undesirable surface properties of materials. With surface modification, the nature of the surface is altered whereas the bulk properties are largely retained. Surface-grafted polymer coatings represent a particularly powerful strategy to achieve surface modification. This approach involves the chemical attachment of polymer (or oligomer) chains to a material surface and is highly versatile in its range of potential applications

Functional Polymer Coatings: Principles, Methods, and Applications, First Edition.
Edited by Limin Wu and Jamil Baghdachi.

[7–10]. This chapter will discuss methods of polymer surface-grafting, characterization of such modified surfaces, and utility of surface-grafted polymer coatings for antifouling applications.

7.2 SURFACE-GRAFTING METHODS

Surface-grafting via covalent attachment typically involves two discrete steps: activation of the surface and grafting [11]. Surface activation serves to introduce reactive sites onto the surface. Grafting may be further classified according to the nature of the grafting strategy as either "grafting-from" (i.e., surface-initiated polymerization) or "grafting-to." These two strategies are depicted in Fig. 7.1.

7.2.1 "Grafting-From" Method

In the case of "grafting-from," an initiator is immobilized or generated on a surface and followed by *in situ* surface-initiated polymerization with monomer. There are several common examples of activation of a surface with initiators. First, initiators may be immobilized onto a surface, including azobisisobutyronitrile [12], phenylpyrenylmethyl groups, [13] and silane coupling agents [14]. Initiator-containing self-assembled monolayers (SAMs) may also be immobilized onto surfaces [15–19]. Finally, surface initiators may be generated by treatment with oxidizing agents [20, 21], ozone [22, 23], and high-energy sources [24–26]. A variety of *in situ* surface-initiated polymerization strategies have been utilized, including conventional radical,

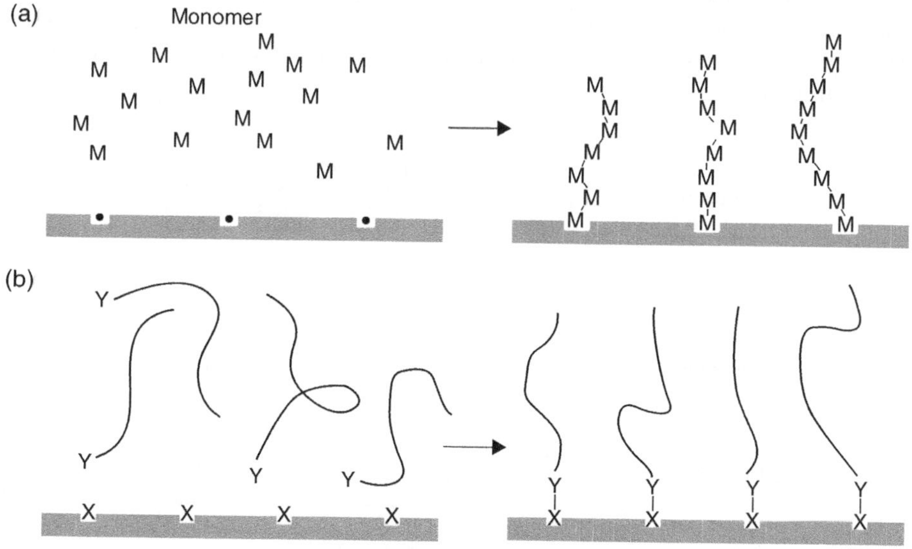

FIG. 7.1 Surface-grafting of polymers via (a) "grafting-from" (i.e., surface-initiated polymerization) and (b) "grafting-to." Adapted from Ref. 7. © Springer.

controlled radical, carbocationic, anionic, ring-opening metathesis, and atom transfer radical polymerization (ATRP) [11, 18].

Efforts have been made to improve control over the graft density and chain length. For example, during ATRP "graft-from" polymerization, graft density may be controlled by adding to the substrate "dummy initiators" that have the same anchoring group as the initiator but lack the ability to actually initiate polymerization [27]. Also, the length of the grafted chains can be controlled by adjusting the rate of termination through the use of "sacrificial" initiator molecules [28]. These initiators are added to the polymerization mixture and produce free-floating chains that can terminate propagation of grafted chains. This also helps to prevent grafted chains from terminating with their neighbors, which can produce loop structures that may be undesirable. Any "nongrafted" chains remaining upon completion of the reaction can easily be washed away. Additionally, it can be assumed that these nongrafted polymers are similar in size and composition to those grown on the surface and thus can be characterized to obtain information about the grafted chains.

7.2.2 "Grafting-To" Method

In the case of "grafting-to," an end-functionalized polymer is synthesized and subsequently adsorbed onto a substrate to form a chemical bond (i.e., chemisorption). Such end-functionalized polymers may be formed via a variety of polymerization strategies [29–31]. In contrast to "grafting-from" strategies, polymers may be conveniently characterized prior to grafting. However, it can be difficult to achieve high graft density due to steric effects, particularly as the chain length increases. The most notable "grafting-to" strategy in which high graft chain density is typically achieved is SAMs, which are classically defined as ordered molecular assemblies that form spontaneously onto substrates [8, 10]. As illustrated in Fig. 7.2, they consist of three main components: the head group, the functional group, and the assembling structure.

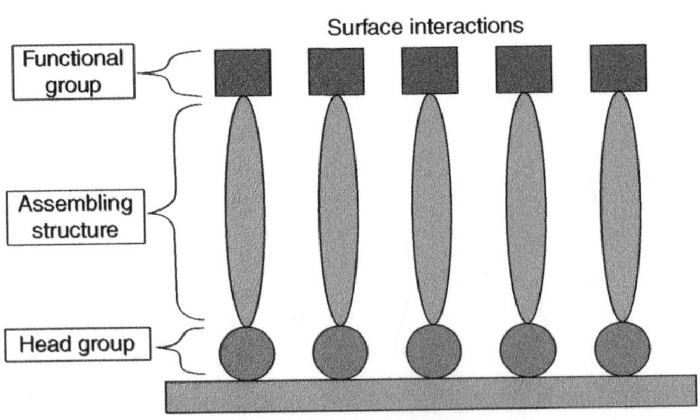

FIG. 7.2 Schematic illustration of a self-assembled monolayer (SAM).

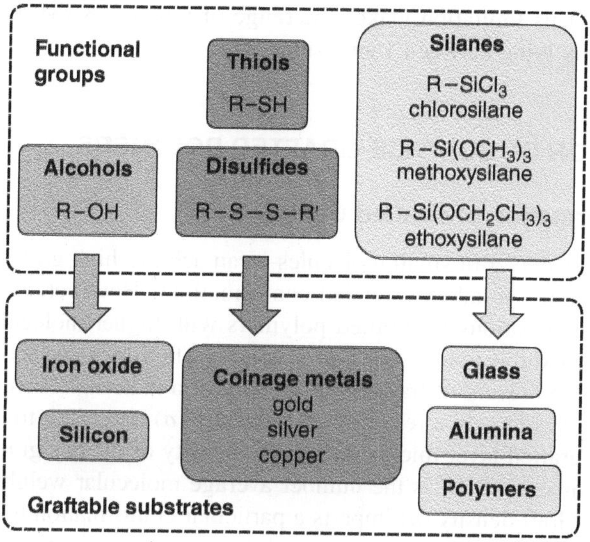

FIG. 7.3 Polymer head groups and corresponding substrates that can be used in graft-to approaches.

The head group of a SAM has a high affinity for and absorbs onto the intended substrate. In this way, SAMs are typically denoted by the head group and substrate "pair." Examples of these pairs are illustrated in Fig. 7.3. The choice of head group depends on the substrate material [32]. While the most commonly studied SAMs are thiols on gold substrates and silane-based systems on silicon wafer substrates [33, 34], a variety of SAM pairs have been studied. For instance, when grafting-to coinage metals such as gold, silver, and copper, alkane thiol (R–SH) and disulfide (R–S–S–R′) head groups may be used [35–38]. Alcohol (R–OH) head groups are useful for attaching chains to iron oxide and silicon surfaces [39–41]. If the surface is rich in hydroxyl groups (e.g., silica and glass alumina), silane-functionalized polymers may be effectively surface-grafted [42, 43]. For example, polymers containing an R–Si–X_3 end group (where X is –Cl, –OCH_3, or –OCH_2CH_3) may be grafted to a hydroxylated surface [32]. In addition, acid and plasma treatments may be used to increase the concentration of hydroxyl groups on many substrates, including polymer substrates [44–46].

To provide specific functionality to SAMs, various terminal functional groups may be utilized (e.g., oligomers [47–50], bioactive molecules [51, 52], and free radical initiators [15–19]). Finally, the assembling structure is responsible for the regular conformation of SAMs and is typically composed of an alkyl chain [32]. The high graft density of the SAMs facilitates van der Waals interactions that can lead to crystallization of neighboring alkyl chains. Furthermore, the length of the alkyl chain is imperative to its function as an assembling structure [32]. Specifically, it has been found that fewer than nine CH_2 groups are difficult to assemble because they do not provide sufficient intermolecular forces. Conversely, if there are too many CH_2 groups (> ~24), then defects in the crystal structure occur that also prevent efficient assembly [32]. This limitation makes the range of possible SAM thicknesses quite narrow, generally lying between 1 and 3 nm [10].

7.3 BEHAVIOR OF SURFACE-GRAFTED POLYMERS

7.3.1 Conformation of Grafted Chains

SAMs—as a monolayer of small molecules—can achieve high graft density (σ) as a result of their ability to adopt extended conformations. However, entropic penalties forbid such conformations for grafted polymers with higher molecular weights [9]. Polymer chains exhibit unique behaviors when grafted to a surface as opposed to when dissolved in solvent or trapped in the bulk of a material. The conformation of end-tether chains is largely a result of graft density (σ) according to $\sigma=(h\,\rho\,N_A)/M_n$, where h is the grafted layer thickness, ρ is the density of the dry grafted layer, N_A is Avogadro's number, and M_n is the number average molecular weight of the grafted chain [53–57]. Graft density (σ) imparts a particular conformation to a grafted chain [57, 58]. A random coil conformation (mushroom regime) occurs when the grafting distance (D) is greater than $2R_f$ (twice the Flory radius; $D>2R_f$) and a more extended conformation (brush regime) is observed when D is less than $2R_f$ [59]. The grafting distance (D) may be calculated as $D=(4/\pi\sigma)^{1/2}$ [55]. The degree to which grafted chains stretch from a surface (i.e., thickness, h) is dependent on the graft density (σ) as well as their affinity for the surrounding environment [9]. In thermodynamically good solvents, grafted polymers in the mushroom regime have a height (h) that scales as $h \propto N\sigma^0$ (i.e., h is independent of graft density, σ), where N is the average degree of polymerization of the grafted chain (Fig. 7.4) [60]. In the brush regime, graft density (σ) is greater such that lateral motion is restricted by neighboring chains and as a result, the graft layer as a whole extends further from the surface, leading to greater thickness (h).

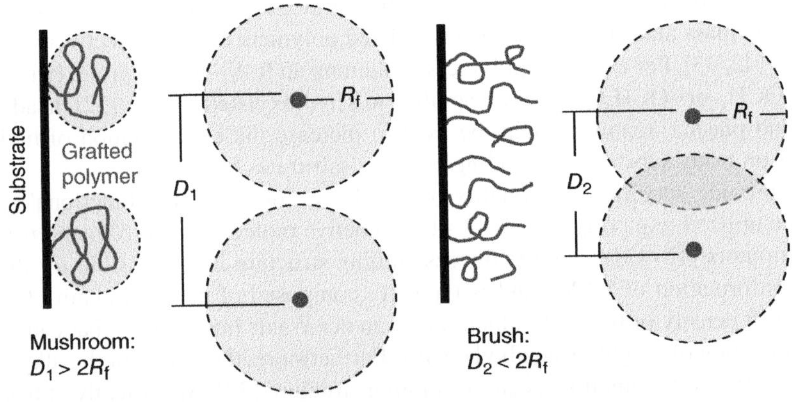

FIG. 7.4 Grafted polymer chains are in the mushroom regime if the distance between chains (D) is greater than twice the Flory radius (R_f), otherwise they are in a brush conformation.

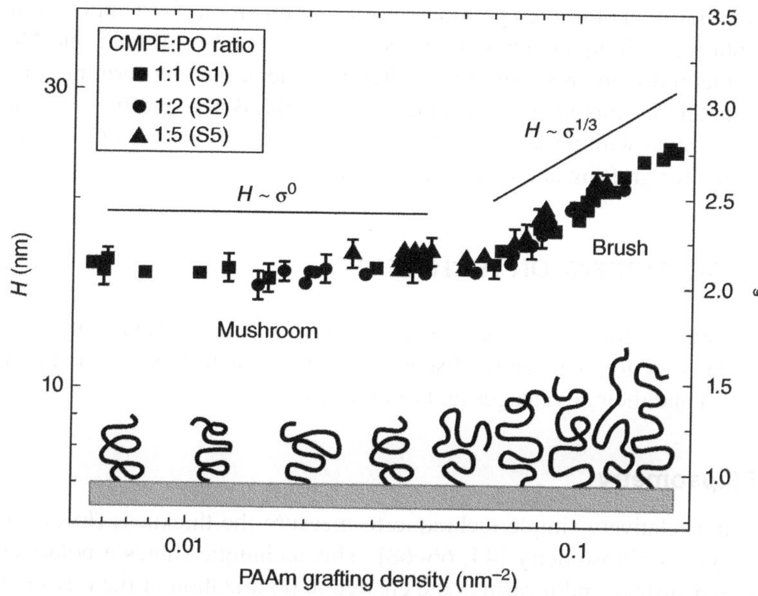

FIG. 7.5 Plot of thickness of grafted polyacrylamide (PAAm) versus graft density that demonstrates the transition from mushroom to brush regime in a good solvent. Samples were prepared on substrates containing different initiator gradients made of 1-trichlorosilyl-2-(*m*/*p*-chloromethylphenyl) ethane (CMPE) in paraffin oil (PO). The cartoon illustrates the change in polymer conformations as graft density increases. Adapted from Ref. 60. © American Chemical Society.

When brush regime grafted polymers are in a good solvent, the thickness (h) scales as $h \propto N\sigma^{1/3}$ (i.e., h is dependent on graft density, σ) (Fig. 7.5). "Graft-from" strategies can be used to readily obtain the high density needed for brush regimes, but this can be difficult to achieve with "graft-to" strategies for polymers [61].

7.3.2 Chain Migration

Surface-grafted polymers formed on polymer substrates are frequently observed to migrate from the surface into the bulk [46, 62–64]. This is a thermodynamically driven process to minimize interfacial energy [32]. Surface migration is particularly pronounced for polymer substrates that are above their glass transition temperature (T_g) and can occur rather soon after grafting, thereby reducing graft surface concentration [65]. The environment to which the grafted surface is exposed also plays an important role in surface rearrangement [62]. For a hydrophilic polymer grafted onto a hydrophobic substrate, exposure to an aqueous environment would be expected to retain the grafted chains at the surface. Conversely, if hydrophobic chains were to be grafted onto a hydrophilic polymer, in an aqueous environment the chains would likely be driven into the bulk. Given sufficient time, the hydrophobic polymer may be absent from the surface entirely.

If chain migration of grafted polymers is undesirable for a given application, there are preventative techniques that can be used as discussed by Ratner and Hoffman [32]. One method is to cross-link the grafted polymers, thereby forming a network that is physically incapable of penetrating into the bulk. Before grafting, the substrate can also be coated with an impermeable layer (e.g., metal coating) that serves as a new substrate for grafting and blocks migration.

7.4 CHARACTERIZATION TECHNIQUES

There are many useful surface analysis techniques for characterizing surface-grafted polymers. This section will briefly discuss some common techniques, including how they are used and their advantages and limitations.

7.4.1 Ellipsometry

A useful and relatively simple technique to measure the thickness (h) of a grafted polymer layer is ellipsometry [44, 66–68]. This technique shines a polarized light source onto a surface and measures the change in polarization of the reflected beam [69, 70]. The exact change in polarization is a result of certain properties of the surface including thickness and refractive index. Ellipsometry can be used to measure film thickness ranging from a few angstroms to several micrometers, which is useful for polymer grafts. Furthermore, it can be used on surfaces submerged in water thereby permitting observation of swelling behavior of grafted polymers [71].

Some limitations exist for utilizing ellipsometry to analyze surface-grafted polymers [70]. First, the optical properties of the substrate must be known. However, most ellipsometry analytical software packages include a library of commonly used and well-characterized substrates. Second, ellipsometry is limited to films that are optically homogenous and isotropic. As a result, the thickness of low-density grafted polymers in mushroom regimes may be difficult to accurately measure.

7.4.2 Contact Angle

The behavior of a drop of liquid on a surface is a simple but effective method to characterize surfaces [72, 73]. The behavior of the liquid drop is dependent on the cohesive forces of the liquid and the adhesive forces between the liquid and the surface. The affinity that a surface has for a liquid is referred to as its wettability, and it can be quantified by the contact angle (θ_C). It is geometrically defined as the angle formed by a liquid drop at the three-phase boundary where the liquid, gas, and solid meet. The relationship between the contact angle (θ_C) and the interfacial energies between the three phases is defined by Young's equation: $\cos \theta_C = (\gamma_{SV} - \gamma_{SL})/\gamma_{LV}$, where γ_{SV}, γ_{SL}, and γ_{LV} represent the interfacial tensions at the solid/vapor, solid/liquid, and liquid/vapor interfaces [74].

Contact angles may be measured statically or dynamically. A static contact angle reflects a system in equilibrium and is largely dependent on the recent history of

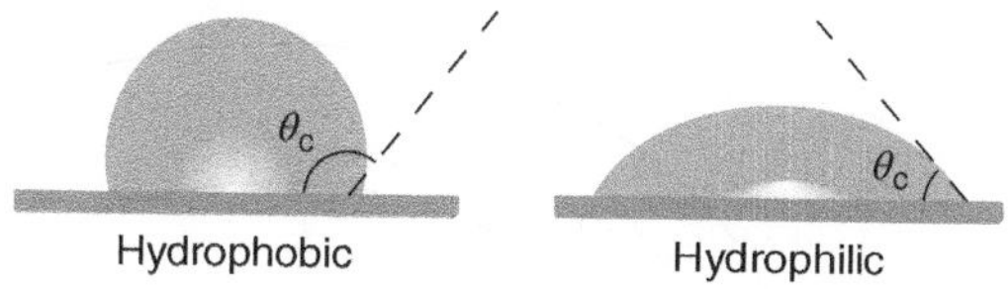

FIG. 7.6 Illustration of static contact angles on a hydrophobic and hydrophilic surface.

interaction. Goniometry is commonly used to measure the shape of a sessile (free) drop of liquid on a solid surface and determine the contact angle (θ_C) (Fig. 7.6). In the simplest of terms, a drop of water on a hydrophilic surface would be expected to "spread out" whereas on a hydrophobic surface it would "bead up." Thus, if the contact angle (θ_C) of water is greater than 90°, then the surface is generally considered to be hydrophobic [75].

Advancing (θ_{adv}) and receding (θ_{rec}) dynamic contact angles are produced when the three-phase (solid/liquid/vapor) boundaries are set in motion; θ_{adv} is measured as the meniscus of the liquid moves across a "nonwetted" surface and θ_{rec} is subsequently measured as the meniscus moves back over the already wetted surface; θ_{adv} provides information about the hydrophobic or low-surface energy nature of a solid whereas hydrophilic or high-surface energy properties of the film are derived from θ_{rec} [76]. Hysteresis (θ_Δ) is defined as the difference between the advancing and receding contact angles ($\theta_\Delta = \theta_{adv} - \theta_{rec}$).

Contact angle analysis is a convenient tool to characterize grafted polymer surfaces [44, 73]. Since contact angles are very sensitive to changes in surface chemistry, it is effective to confirm successful grafting [25, 44, 77]. Second, the temporal change of static contact angles as well as hysteresis (i.e., $\theta_{adv} - \theta_{rec}$) may be utilized to evaluate restructuring of surface-grafted polymers [63, 76–78]. While hysteresis may be indicative of surface restructuring, users must be aware that it may also be attributed to surface chemical heterogeneity, surface roughness, surface hydration, or contamination of the solid or liquid phase [78].

7.4.3 X-ray Photoelectron Spectroscopy

When an X-ray is focused on a sample, it causes emission of core-level electrons in the atoms near the surface [79, 80]. This is known as the photoelectric effect, and it is the phenomenon on which X-ray photoelectron spectroscopy (XPS) or electron spectroscopy for chemical analysis (ESCA) is based. For this technique, kinetic energies of the emitted electrons are measured by the XPS instrument and subtracted from the known energy of the X-rays to determine the energies of binding to their respective atoms (as illustrated in Fig. 7.7). Since each element has a unique range of binding energies, individual peaks in the spectrometer signal can be attributed to the presence of certain elements. The areas under each of the peaks also correlate to the respective ratios of those elements on the sample surface and can therefore be used for quantitative compositional analysis.

XPS is very effective in characterizing the elemental composition of surfaces. This makes it particularly useful for identifying and confirming the presence of grafted polymers on a substrate [8, 33, 44]. There are many features of XPS that

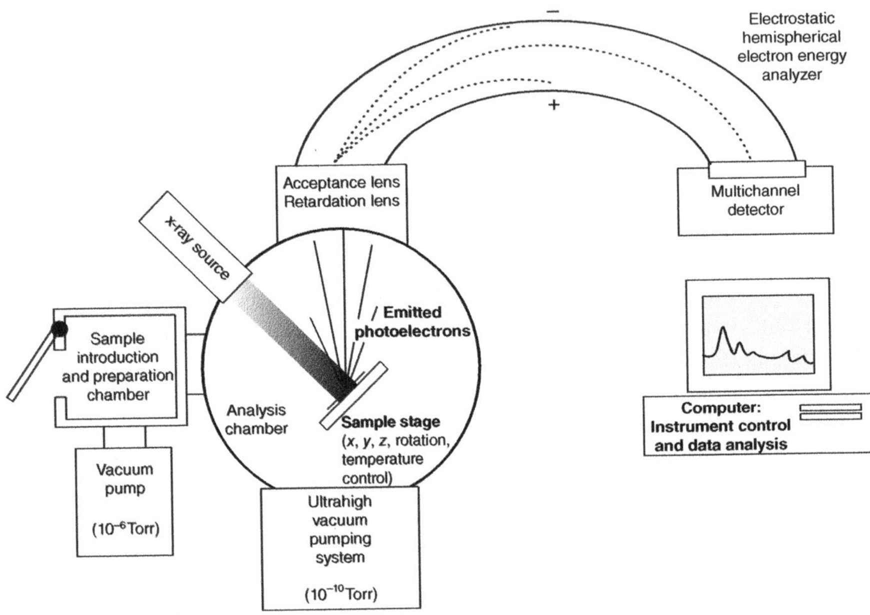

FIG. 7.7 Schematic diagram of XPS instrument. Adapted from Ref. 1. © Elsevier.

make it desirable for compositional surface analysis. It provides high information content, surface-localized measurements, relatively fast analysis and minimal sample damage and can be used on almost any dry surface with minimal sample preparation required [1]. Compared with contact angle analyses, XPS is an expensive technique that requires complex instrumentation and experienced operators. Samples must also be vacuum-compatible, meaning they cannot contain any liquids or volatile materials. Finally, there is risk of sample damage caused by the X-rays if analysis times are too long [1].

7.4.4 Scanning Probe Microscopies

To determine the topographical features of a given surface, scanning probe microscopy (SPM) techniques are typically employed [81]. With any SPM, a probe moves across a surface while measuring discrete changes in a specific surface property to generate a spatial image. The two main scanning probe techniques used to study surface-grafted polymers are scanning tunneling microscopy (STM) [33, 82, 83] and atomic force microscopy (AFM) [84–86].

The surface properties measured by STM and AFM are quite different. The STM probe has a metal tip terminating in a single atom. The atom is brought within 5–10Å of a surface, which is close enough that the electron cloud of the atom at the probe tip will overlap with the electron clouds of atoms on the surface [1]. A potential is generated between the probe and the surface to generate an electron tunneling current, which increases as the tip is brought closer to an atom. STM can therefore be

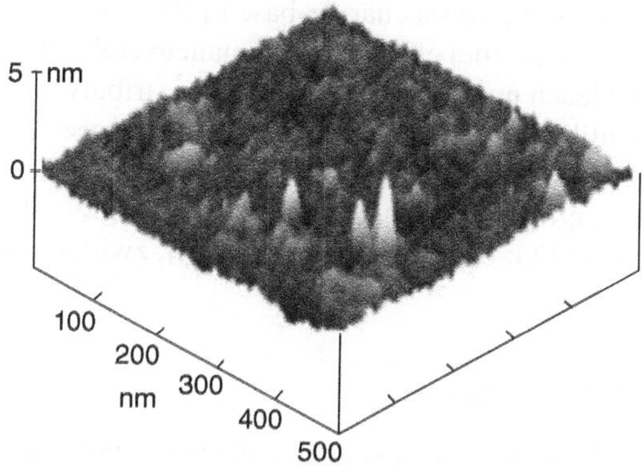

FIG. 7.8 AFM image of poly(dimethylsiloxane)-*block*-polystyrene-*block*-poly(dimethylmethylsilyl)propyl acrylate brushes on a glass surface. Adapted from Ref. 87. © Wiley.

used to map a surface based on electron density. A disadvantage of this technique is that it can only be used on electrically conductive and semiconductive surfaces. AFM probes have a similar tip as STM but are mounted on a flexible cantilever. As the tip is moved across the surface, van der Waals and electrostatic attraction and repulsion cause the tip to deflect. These small deflections cause the cantilever to flex, which can be observed simply by reflecting a laser beam off a mirror onto the cantilever arm. AFM is much more versatile in the types of surfaces it can image compared with STM and is therefore often preferred. An example of an AFM image of a surface-grafted polymer obtained by Pyun *et al.* is shown in Fig. 7.8 [87]. As can be seen, SPM techniques can be used to view nanoscale structures and organization of grafted polymers.

AFM also has the unique capability of measuring the affinity between a surface and certain molecules [88]. To do this, the tip is coated with a molecule that is expected to adhere to the grafted surface, such as a protein, and the physical force that binds it to the grafted polymers can be measured. Such studies are highly applicable for evaluating antifouling surfaces discussed later.

7.5 ANTIFOULING COATINGS

Surface-grafted polymers have been widely studied to control material interactions with the biological environment [89]. A notable example is the design of antifouling coatings for medical, marine, and industrial applications [90–93]. When a medical device is implanted, blood plasma proteins rapidly deposit onto the device surface and set off a cascade of negative events including thrombosis [94–97]. Clinically, attempts to control thrombosis on devices such as dialysis catheters have largely relied on pharmaceutical intervention (e.g., heparinization), which is associated with bleeding complications [98, 99]. On submerged structures, biofouling by marine organisms is facilitated by secreted biomolecules (i.e., bioadhesives), which may be protein-, glycoprotein- and polysaccharide-based [100]. Marine biofouling on ship hulls creates drag that lowers fuel efficiency and maneuverability [101]. Traditionally, ablative coatings that leach nonspecific toxicants (e.g., tributyl tin, copper, or organic biocides) have been utilized but are of environmental concern [102]. Surface-grafted polymers represent a promising strategy to create improved antifouling materials. Later, we discuss emerging trends in the use of surface-grafted polymers based on poly(ethylene glycol) [PEG; or poly(ethylene oxide)], zwitterions, or a combination of both.

7.5.1 Surface-Grafted PEG

PEG remains one of the most widely studied antifouling polymers [103]. The exceptional protein resistance of PEG is attributed to its high water content as well as its high chain mobility, which leads to an "exclusion effect" by which proteins are repelled and also an entropic penalty if protein adsorption were to occur [53, 104–107]. PEG has been grafted onto a variety of surfaces and its resistance to proteins evaluated [50, 108–117]. Unfortunately, the *in vivo* performance of PEG has been disappointing [118–120], leading to efforts to improve the antifouling efficiency of surface-grafted PEG. For instance, our group has prepared "PEG-silane amphiphiles" in which the PEG segment is distanced from the reactive triethoxy-silane head group by a siloxane tether and having the general formula $(EtO)_3Si(CH_2)_2$–oligodimethylsiloxane$_m$–*block*–$(OCH_2CH_2)_8$–OCH_3, where $m=0$, 4, and 13 (Fig. 7.9) [121]. These distinguish themselves from conventional PEG-silanes in which the PEG segment is separated from the alkoxysilane group by a short, less flexible alkane space [e.g., propyl for $(RO)_3Si$–$(CH_2)_3$–$(OCH_2CH_2)_n$–OCH_3] [65, 119, 122–125]. When surface-grafted onto silicon wafer, protein adsorption under static conditions was generally reduced for PEG-silane amphiphiles with longer siloxane tethers [44]. In addition, these PEG-silane amphiphiles were introduced into silicones prior to curing such that they are grafted throughout the entire resulting network [121, 126, 127]. Based on contact angle analyses, we demonstrated that the PEG segments effectively migrated to the water–film interface to enhance surface hydrophilicity. Moreover, as the length of the siloxane tether increased, surface hydrophilicity increased. As a result, resistance to protein adsorption [121] and to marine bacteria and diatoms [127] increased with siloxane tether length.

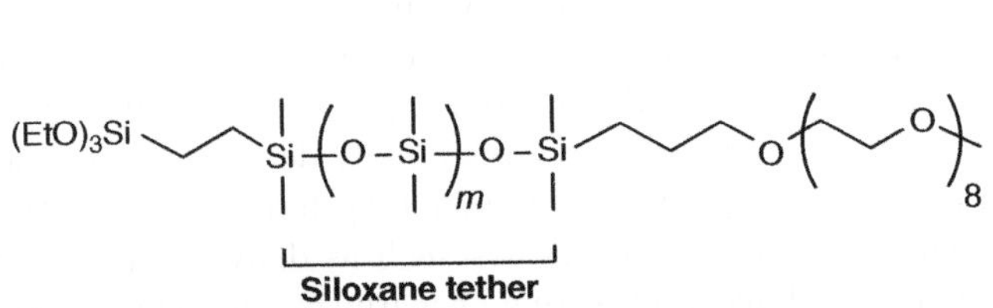

FIG. 7.9 PEG-silane amphiphiles have been studied to reduce biofouling ($m=0$, 4, and 13) [44, 121].

7.5.2 Surface-Grafted Zwitterionic Polymers

Zwitterions contain positively and negatively charged groups but have an overall neutral charge and comprise the majority of typical mammalian cell surfaces [128]. Because of the known protein resistance of zwitterions, materials such as those based on phosphobetaine, sulfobetaine, carboxybetaine, and amino acids have been increasingly studied to enhance antifouling behavior (Fig. 7.10) [129].

Many of these studies have utilized strategies based on formation of zwitterionic SAMs on gold and other substrates. For instance, several groups have evaluated the protein resistance of SAMs formed from phosphorylcholine-alkane thiols [47, 130, 131]. Tanaka *et al.* evaluated the efficacy of SAMs formed from phosphorylcholine-OEG-alkane thiols to suppress protein resistance [132]. Zhang *et al.* compared zwitterionic SAMs and high-graft density zwitterionic polymer brushes prepared via surface-initiated ATRP [47]. While all surfaces exhibited low single protein adsorption from buffer, it was shown that zwitterionic brushes adsorbed less protein from human blood plasma versus SAMs. Several other examples of zwitterionic polymer brushes formed on gold and other substrates and exhibiting antifouling properties have been reported. For instance, Liu *et al.* prepared surface-grafted poly(serine methacrylate) onto gold substrates via surface-initiated photoiniferter-mediated polymerization [133]. In addition, poly(sulfobetaine methacrylate)- and poly(carboxybetaine methacrylate)- containing catechol residues were surface-grafted onto gold and glass substrates [134, 135]. Prepared from methacrylated monomers, Feng *et al.* reported the formation of grafted polymers bearing oligo(ethylene glycol) [p(OEGMA)] and 2-methacryloyloxyethyl phosphorylcholine [p(MPC)] side chains onto silicon wafer via surface-initiated ATRP [136]. Still others have grafted zwitterionic polymer brushes onto polymer substrates. For example, Liu *et al.* grafted

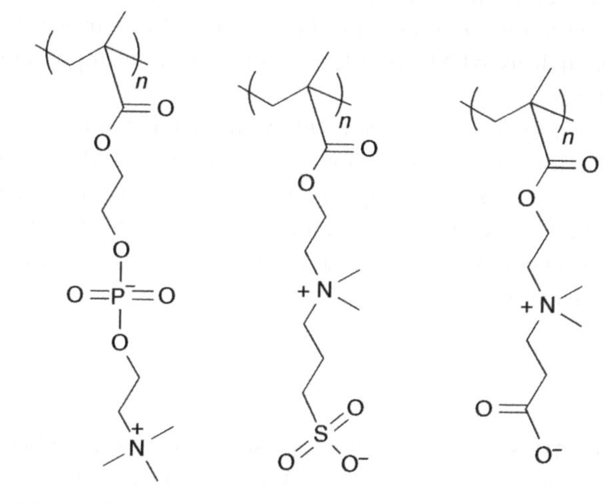

FIG. 7.10 Examples of polymers made from methacrylated zwitterions.

poly(phosphobetaine) [137] and poly(carboxybetaine) [138] onto silicone via ozone-induced random radical polymerization. This same group more recently reported several zwitterionic polymers formed onto silicone via ATRP [139]. Interestingly, at low graft density (for a given graft length), p(OEGMA) exhibited enhanced resistance to fibrinogen and platelets versus p(MPC) whereas at high graft density, their resistance was similar. This was speculated to be due to the enhanced surface coverage of p(OEGMA) afforded by greater length of side chains.

7.6 SUMMARY

Surface-grafted polymers provide a method to alter surface properties without changing the bulk properties of a material. "Graft-from" and "graft-to" methods may be utilized to prepare surface-grafted polymers. The "grafting-from" strategy lends itself to achieving high graft density whereas the "graft-to" strategy more readily permits characterization of the grafted chains. For grafted polymers, a relatively low graft density results in a mushroom regime conformation rather than a brush regime conformation, with the latter providing better surface coverage. SAMs may be grafted to a substrate at very high graft densities, but their relatively low molecular weight limits the graft layer thickness. Compared with physically stable substrates (e.g., gold and silicon wafer), chains grafted onto polymers are often prone to restructuring, particularly when transferred from one environment to another (e.g., air to water).

Numerous techniques are useful to analyze surface-grafted layers with each providing different types of information. For instance, ellipsometry can be used to measure graft thickness and density whereas contact angle analysis provides information about wettability and surface restructuring. XPS is useful to evaluate the changes in surface atomic composition before and after surface-grafting. Finally, SPM techniques such as AFM provide information about topographic features of surface-grafted layers.

Surface-grafted polymers have been widely studied to control biofouling on medical, marine, and industrial applications. Surface-grafted PEG has been widely studied for its resistance to biofouling. To enhance this behavior, PEG-silane amphiphiles in which the PEG segment is distanced from the grafting group by a flexible siloxane tether have been prepared. In addition, zwitterionic grafted chains have also been studied to create efficient antifouling surfaces.

REFERENCES

1. Ratner, B.D. (2004) Surface Properties and Surface Characterization of Materials, in *Biomaterials Science: An Introduction to Materials in Medicine*, B.D. Ratner, *et al.*, Editors. Elsevier Academic Press, New York, p. 40–59.
2. Tal, M.G., Ni, N. (2008) Selecting optimal hemodialysis catheters: material, design, advanced features, and preferences. *Tech. Vasc. Interv. Radiol.* 11(3): 186–191.

3. McGee, D.C., Gould, M.K. (2003) Preventing complications of central venous catheterization. *N. Engl. J. Med.* 348(12): 1123–1133.
4. Zhao, M., *et al.* (2003) Magnetic sensors for protease assays. *Angew. Chem. Int. Ed. Engl.* 115(12): 1413–1416.
5. Thorek, D., *et al.* (2006) Superparamagnetic iron oxide nanoparticle probes for molecular imaging. *Ann. Biomed. Eng.* 34(1): 23–38.
6. Buyukhatipoglu, K., Clyne, A.M. (2011) Superparamagnetic iron oxide nanoparticles change endothelial cell morphology and mechanics via reactive oxygen species formation. *J. Biomed. Mater. Res. A.* 96A(1): 186–195.
7. Uyama, Y., Kato, K., Ikada, Y. (1998) *Surface Modification of Polymers by Grafting*, H. Galina, *et al.*, Editors. Springer, Berlin/Heidelberg, p. 1–39.
8. Vericat, C., Vela, M.E., Salvarezza, R.C. (2005) Self-assembled monolayers of alkanethiols on Au(111): surface structures, defects and dynamics. *Phys. Chem. Chem. Phys.* 7(18): 3258–3268.
9. Zhou, F., Huck, W.T.S. (2006) Surface grafted polymer brushes as ideal building blocks for "smart" surfaces. *Phys. Chem. Chem. Phys.* 8(33): 3815–3823.
10. Love, J.C., *et al.* (2005) Self-assembled monolayers of thiolates on metals as a form of nanotechnology. *Chem. Rev.* 105(4): 1103–1170.
11. Brittain, W.J., Minko, S. (2007) A structural definition of polymer brushes. *J. Polym. Sci. Pol. Chem.* 45(16): 3505–3512.
12. Bergbreiter, D.E., Caraway, J.W. (1996) Thermoresponsive polymer-bound substrates. *J. Am. Chem. Soc.* 118(25): 6092–6093.
13. Bergbreiter, D.E., Bandella, A. (1995) pH and solvent responsive reactivity of surface-grafted polyethylene films. *J. Am. Chem. Soc.* 117(42): 10589–10590.
14. Boven, G., *et al.* (1991) Radical grafting of poly(methyl methacrylate) onto silicon wafers, glass slides and glass beads. *Polym. Commun.* 32(2): 50–53.
15. Hyun, J., Chilkoti, A. (2001) Surface-initiated free radical polymerization of polystyrene micropatterns on a self-assembled monolayer on gold. *Macromolecules*, 34(16): 5644–5652.
16. Prucker, O., Rühe, J. (1998) Synthesis of poly(styrene) monolayers attached to high surface area silica gels through self-assembled monolayers of azo initiators. *Macromolecules*, 31(3): 592–601.
17. Prucker, O., Rühe, J. (1998) Mechanism of radical chain polymerizations initiated by azo compounds covalently bound to the surface of spherical particles. *Macromolecules*, 31(3): 602–613.
18. Edmondson, S., Osborne, V.L., Huck, W.T.S. (2004) Polymer brushes via surface-initiated polymerizations. *Chem. Soc. Rev.* 33(1): 14–22.
19. Ista, L.K., *et al.* (2001) Synthesis of poly(N-isopropylacrylamide) on initiator-modified self-assembled monolayers. *Langmuir* 17(9): 2552–2555.
20. Bergbreiter, D.E., Zhou, J. (1992) New grafting chemistry for functionalized polyethylene films. *J. Polym. Sci. Pol. Chem.* 30(9): 2049–2053.
21. Bamford, C.H., Al-Lamee, K.G. (1994) Studies in polymer surface functionalization and grafting for biomedical and other applications. *Polymer* 35(13): 2844–2852.
22. Loh, F.C., *et al.* (1995) Near-u.v. radiation induced surface graft copolymerization of some O3-pretreated conventional polymer films. *Eur. Polym. J.* 31(5): 481–488.

23. Bucheñska, J. (1995) Modification of polyamide fibers (PA6) by grafting polyacrylamide (PAM). *J. Appl. Polym. Sci.* 58(10): 1901–1911.
24. Svarfvar, B.L., *et al.* (1996) Electron-beam graft-modified membranes with externally controlled flux. *Polym. Adv. Technol.* 7(11): 839–846.
25. Sheu, M.S., *et al.* (1993) Immobilization of polyethylene oxide surfactants for non-fouling biomaterial surfaces using an argon glow discharge treatment. *J. Adhes. Sci. Technol.* 7(10): 1065–1076.
26. Tian, J., et al. (1995) Plasma induced grafting of styrene onto nascent polyethylene powder. *Eur. Polym. J.* 31(8): 755–760.
27. Matyjaszewski, K., Tsarevsky, N.V. (2009) Nanostructured functional materials prepared by atom transfer radical polymerization. *Nat. Chem.* 1(4): 276–288.
28. Matyjaszewski, K., et al. (1999) Polymers at interfaces: using atom transfer radical polymerization in the controlled growth of homopolymers and block copolymers from silicon surfaces in the absence of untethered sacrificial initiator. *Macromolecules* 32(26): 8716–8724.
29. Zhao, B., Brittain, W.J. (2000) Polymer brushes: surface-immobilized macromolecules. *Prog. Polym. Sci.* 25(5): 677–710.
30. Hilf, S., Kilbinger, A.F.M. (2009) Functional end groups for polymers prepared using ring-opening metathesis polymerization. *Nat. Chem.* 1(7): 537–546.
31. Tasdelen, M.A., Kahveci, M.U., Yagci, Y. (2011) Telechelic polymers by living and controlled/living polymerization methods. *Prog. Polym. Sci.* 36(4): 455–567.
32. Ratner, B.D., Hoffman, A.S. (2004) Physicochemical Surface Modification of Materials Used in Medicine, in *Biomaterials Science: An Introduction to Materials in Medicine*, B.D. Ratner, et al., Editors. Elsevier Academic Press, New York, p. 201–218.
33. Schreiber, F. (2000) Structure and growth of self-assembling monolayers. *Prog. Surf. Sci.* 65(5–8): 151–257.
34. Schreiber, F. (2004) Self-assembled monolayers: from 'simple' model systems to biofunctionalized interfaces. *J. Phys. Condens. Matter* 16(28): R881–R900.
35. Laibinis, P.E., et al. (1991) Comparison of the structures and wetting properties of self-assembled monolayers of n-alkanethiols on the coinage metal surfaces, copper, silver, and gold. *J. Am. Chem. Soc.* 113(19): 7152–7167.
36. Motte, L., Pileni, M.P. (1998) Influence of length of alkyl chains used to passivate silver sulfide nanoparticles on two- and three-dimensional self-organization. *J. Phys. Chem. B*, 102(21): 4104–4109.
37. Templeton, A.C., Wuelfing, W.P., Murray, R.W. Monolayer-protected cluster molecules. *Acc. Chem. Res.* 33(1): 27–36.
38. Shon, Y.-S., et al. (2002) Monolayer-protected bimetal cluster synthesis by core metal galvanic exchange reaction. *Langmuir* 18(10): 3880–3885.
39. Boal, A.K., et al. (2002) Monolayer exchange chemistry of γ-Fe2O3 nanoparticles. *Chem. Mater.* 14(6): 2628–2636.
40. Zharnikov, M., et al. (2003) Aromatic self-assembled monolayers on hydrogenated silicon. *Langmuir* 19(11): 4682–4687.
41. Niederhauser, T.L., et al. (2002) Arrays of chemomechanically patterned patches of homogeneous and mixed monolayers of 1-alkenes and alcohols on single silicon surfaces. *Angew. Chem. Int. Ed.* 41(13): 2353–2356.

42. Mao, G., Castner, D.G., Grainger, D.W. (1997) Polymer immobilization to alkylchlorosilane organic monolayer films using sequential derivatization reactions. *Chem. Mater.* 9(8): 1741–1750.

43. Xiao, S.J., et al. (1997) Immobilization of the cell-adhesive peptide Arg–Gly–Asp–Cys (RGDC) on titanium surfaces by covalent chemical attachment. *J. Mater. Sci. Mater. Med.*, 8(12): 867–872.

44. Murthy, R., Shell, C.E., Grunlan, M.A. (2009) The influence of poly(ethylene oxide) grafting via siloxane tethers on protein adsorption. *Biomaterials*, 30(13): 2433–2439.

45. Krishnan, S., et al. (2006) Comparison of the fouling release properties of hydrophobic fluorinated and hydrophilic PEGylated block copolymer surfaces: attachment strength of the diatom navicula and the green alga ulva. *Biomacromolecules* 7(5): 1449–1462.

46. Owen, M.J., Smith, P.J. (1994) Plasma treatment of polydimethylsiloxane. *J. Adhes. Sci. Technol.* 8(10): 1063–1075.

47. Zhang, Z., et al. (2008) Blood compatibility of surfaces with superlow protein adsorption. *Biomaterials* 29(32): 4285–4291.

48. Chapman, R.G., et al. (2000) Preparation of mixed self-assembled monolayers (SAMs) that resist adsorption of proteins using the reaction of amines with a SAM that presents interchain carboxylic anhydride groups. *Langmuir* 16(17): 6927–6936.

49. Deng, L., Mrksich, M., Whitesides, G.M. (1996) Self-assembled monolayers of alkanethiolates presenting tri(propylene sulfoxide) groups resist the adsorption of protein. *J. Am. Chem. Soc.* 118(21): 5136–5137.

50. Prime, K.L., Whitesides, G.M. (1993) Adsorption of proteins onto surfaces containing end-attached oligo(ethylene oxide): a model system using self-assembled monolayers. *J. Am. Chem. Soc.* 115(23): 10714–10721.

51. Adden, N., et al. (2006) Phosphonic acid monolayers for binding of bioactive molecules to titanium surfaces. *Langmuir* 22(19): 8197–8204.

52. Choi, S., Murphy, W.L. (2008) Multifunctional mixed SAMs that promote both cell adhesion and noncovalent DNA immobilization. *Langmuir* 24(13): 6873–6880.

53. Lee, J.H., Lee, H.B., Andrade, J.D. (1995) Blood compatibility of polyethylene oxide surfaces. *Prog. Polym. Sci.* 20(6): 1043–1079.

54. Sharma, S., Johnson, R.W., Desai, T.A. (2003) Ultrathin poly(ethylene glycol) films for silicon-based microdevices. *Appl. Surf. Sci.* 206(1–4): 218–229.

55. Zdyrko, B., Klep, V., Luzinov, I. (2003) Synthesis and surface morphology of high-density poly(ethylene glycol) grafted layers. *Langmuir* 19(24): 10179–10187.

56. Feng, W., Brash, J.L., Zhu, S. (2006) Non-biofouling materials prepared by atom transfer radical polymerization grafting of 2-methacryloloxyethyl phosphorylcholine: separate effects of graft density and chain length on protein repulsion. *Biomaterials* 27(6): 847–855.

57. Allen, C., et al. (2002) Controlling the physical behavior and biological performance of liposome formulations through use of surface grafted poly(ethylene glycol). *Biosci. Rep.* 22(2): 225–250.

58. Ligoure, C., Leibler, L. (1990) Thermodynamics and kinetics of grafting end-functionalized polymers to an interface. *J. Phys. Paris* 51(12): 1313–1328.

59. Unsworth, L.D., et al. (2005) Chemisorption of thiolated poly(ethylene oxide) to gold: surface chain densities measured by ellipsometry and neutron reflectometry. *J. Colloid Interface Sci.* 281(1): 112–121.

60. Wu, T., et al. (2003) Formation and properties of anchored polymers with a gradual variation of grafting densities on flat substrates. *Macromolecules* 36(7): 2448–2453.
61. Jones, D.M., Huck, W.T.S. (2001) Controlled surface-initiated polymerizations in aqueous media. *Adv. Mater.* 13(16): 1256–1259.
62. Yoon, S.C., Ratner, B.D. (1988) Surface and bulk structure of segmented poly(ether urethanes) with perfluoro chain extenders. 3. Effects of annealing, casting solvent, and casting conditions. *Macromolecules* 21(8): 2401–2404.
63. Garbassi, F., et al. (1989) Dynamics of macromolecules: a challenge for surface analysis. *Surf. Interface Anal.* 14(10): 585–589.
64. Kang, E.T., et al. (1996) Surface structures and adhesion characteristics of poly(tetrafluoroethylene) films after modification by graft copolymerization. *J. Adhes. Sci. Technol.* 10(8): 725–743.
65. Delamarche, E., et al. (2003) Microcontact printing using poly(dimethylsiloxane) stamps hydrophilized by poly(ethylene oxide) silanes. *Langmuir* 19(21): 8749–8758.
66. Wasserman, S.R., et al. (1989) The structure of self-assembled monolayers of alkylsiloxanes on silicon: a comparison of results from ellipsometry and low-angle x-ray reflectivity. *J. Am. Chem. Soc.* 111(15): 5852–5861.
67. Fan, X., Lin, L., Messersmith, P.B. (2006) Cell fouling resistance of polymer brushes grafted from Ti substrates by surface-initiated polymerization: effect of ethylene glycol side chain length. *Biomacromolecules*, 7(8): 2443–2448.
68. Ingall, M.D.K., et al. (1999) Surface functionalization and imaging using monolayers and surface-grafted polymer layers. *J. Am. Chem. Soc.* 121(15): 3607–3613.
69. McCrackin, F.L., et al. (1963) Measurement of the thickness and refractive index of very thin films and the optical properties of surfaces by ellipsometry. *J. Res. Natl. Bur. Stand.* 67A(4): 363–377.
70. Azzam, R.M.A., Bashara, N.M. (1977) *Ellipsometry and Polarized Light*. Elsevier, North-Holland.
71. Motornov, M., et al. (2006) Nonwettable thin films from hybrid polymer brushes can be hydrophilic. *Langmuir* 23(1): 13–19.
72. Kwok, D.Y., Neumann, A.W. (1999) Contact angle measurement and contact angle interpretation. *Adv. Colloid Interface Sci.* 81(3): 167–249.
73. Morra, M., Occhiello, E., Garbassi, F. (1990) Knowledge about polymer surfaces from contact angle measurements. *Adv. Colloid Interface Sci.* 32(1): 79–116.
74. Young, T. (1805) An essay on the cohesion of fluids. *Trans. Roy. Soc.* 95: 65–87.
75. Sangermano, M., et al. (2003) Fluorinated epoxides as surface modifying agents of UV-curable systems. *J. Appl. Polym. Sci.* 89(6): 1524–1529.
76. Davies, J., et al. (1996) Use of dynamic contact angle profile analysis in studying the kinetics of protein removal from steel, glass, polytetrafluoroethylene, polypropylene, ethylenepropylene rubber, and silicone surfaces. *J. Colloid Interface Sci.* 182(2): 437–443.
77. Freij-Larsson, C., Jannasch, P., Wesslén, B. (2000) Polyurethane surfaces modified by amphiphilic polymers: effects on protein adsorption. *Biomaterials*, 21(3): 307–315.
78. Johnson, R.E., Dettre, R.H. (1964) Contact angle hysteresis. III. Study of an idealized heterogeneous surface. *J. Phys. Chem.* 68(7): 1744–1750.
79. Dilks, A. (1981) *X-Ray Photoelectron Spectroscopy for the Investigation of Polymer Surfaces, in Photon, Electron, and Ion Probes of Polymer Structure and Properties.* American Chemical Society, Washington, DC, p. 293–317.

80. Ratner, B.D., Castner, D.G. (2009) *Electron Spectroscopy for Chemical Analysis, in Surface Analysis—The Principal Techniques*. John Wiley & Sons, Ltd., New York, p. 47–112.

81. Bottomley, L.A., Coury, J.E., First, P.N. (1996) Scanning probe microscopy. *Anal. Chem.* 68(12): 185–230.

82. Poirier, G.E. (1997) Characterization of organosulfur molecular monolayers on Au(111) using scanning tunneling microscopy. *Chem. Rev.* 97(4): 1117–1128.

83. Delamarche, E., et al. (1996) Golden interfaces: the surface of self-assembled monolayers. *Adv. Mater.* 8(9): 719–729.

84. Xu, S., et al. (1998) In situ studies of thiol self-assembly on gold from solution using atomic force microscopy. *J. Chem. Phys.* 108(12): 5002–5012.

85. Torrelles, X., et al. (2004) New Insights in the c(4×2) reconstruction of hexadecanethiol on Au(111) revealed by grazing incidence X-ray diffraction. *Langmuir* 20(21): 9396–9402.

86. Barrena, E., et al. (2003) The role of intermolecular and molecule—substrate interactions in the stability of alkanethiol nonsaturated phases on Au(111). *J. Am. Chem. Soc.* 126(1): 385–395.

87. Pyun, J., Kowalewski, T., Matyjaszewski, K. (2003) Synthesis of polymer brushes using atom transfer radical polymerization. *Macromol. Rapid Commun.* 24(18): 1043–1059.

88. Weinman, C.J., et al. (2010) Protein adsorption resistance of anti-biofouling block copolymers containing amphiphilic side chains. *Soft Matter* 6(14): 3237–3243.

89. Raynor, J.E., et al. (2009) Polymer brushes and self-assembled monolayers: versatile platforms to control cell adhesion to biomaterials (review). *Biointerphases* 4(2): FA3–FA16.

90. Rana, D., Matsuura, T. (2010) Surface modifications for antifouling membranes. *Chem. Rev.* 110(4): 2448–2471.

91. Charnley, M., Textor, M., Acikgoz, C. (2011) Designed polymer structures with antifouling–antimicrobial properties. *React. Funct. Polym.* 71(3): 329–334.

92. Krishnan, S., Weinman, C.J., Ober, C.K. (2008) Advances in polymers for anti-biofouling surfaces. *J. Mater. Chem.* 18(29): 3405–3413.

93. Kang, G.-D., Cao, Y.-M. (2012) Development of antifouling reverse osmosis membranes for water treatment: a review. *Water Res.* 46(3): 584–600.

94. Horbett, T.A. (1982) Protein Adsorption on Biomaterials, in *Biomaterials: Interfacial Phenomena and Applications*, S.L. Cooper, N.A. Peppas, Editors. American Chemical Society, Washington, DC, p. 233–244.

95. Wu, Y., et al. (2005) The role of adsorbed fibrinogen in platelet adhesion to polyurethane surfaces: a comparison of surface hydrophobicity, protein adsorption, monoclonal antibody binding, and platelet adhesion. *J. Biomed. Mater. Res. A* 74A(4): 722–738.

96. Raad, I.I., et al. (1994) The relationship between the thrombotic and infectious complications of central venous catheters. *J. Am. Med. Assoc.* 271(13): 1014–1016.

97. Lloyd, D.A., et al. (1993) Does the fibrin coat around a central venous catheter influence catheter-related sepsis? *J. Pediatr. Surg.* 28(3): 345–349.

98. Levine, M.N., et al. (2004) Hemorrhagic complications of anticoagulant treatment. *Chest* 126(Suppl. 3): 287S–310S.

99. Rogacka, R., et al. (2008) Dual antiplatelet therapy after percutaneous coronary intervention with stent implantation in patients taking chronic oral anticoagulation. *JACC. Cardiovasc. Interv.* 1(1): 56–61.

100. Railkin, A.I. (2004) *Marine Biofouling Colonization Processes and Defenses*. CRC Press LLC, Boca Raton.
101. Townsin, R.L. (2003) The ship hull fouling penalty. *Biofouling* 19(Suppl. 1): 9–15.
102. Chambers, L.D., et al. (2006) Modern approaches to marine antifouling coatings. *Surf. Coat. Technol.* 201(6): 3642–3652.
103. Banerjee, I., Pangule, R.C., Kane, R.S. (2011) Antifouling coatings: recent developments in the design of surfaces that prevent fouling by proteins, bacteria, and marine organisms. *Adv. Mater.* 23(6): 690–718.
104. Jeon, S.I., et al. (1991) Protein—surface interactions in the presence of polyethylene oxide: I. Simplified theory. *J. Colloid Interface Sci.* 142(1): 149–158.
105. Jeon, S.I., Andrade, J.D. (1991) Protein—surface interactions in the presence of polyethylene oxide: II. Effect of protein size. *J. Colloid Interface Sci.* 142(1): 159–166.
106. Knoll, D., Hermans, J. (1983) Polymer-protein interactions. Comparison of experiment and excluded volume theory. *J. Biol. Chem.* 258(9): 5710–5715.
107. Österberg, E., et al. (1993) Comparison of polysaccharide and poly(ethylene glycol) coatings for reduction of protein adsorption on polystyrene surfaces. *Colloids Surf. A* 77(2): 159–169.
108. Lee, S.-W., Laibinis, P.E. (1998) Protein-resistant coatings for glass and metal oxide surfaces derived from oligo(ethylene glycol)-terminated alkyltrichlorosilanes. *Biomaterials*, 19(18): 1669–1675.
109. Zhang, M., Ferrari, M. (1998) Hemocompatible polyethylene glycol films on silicon. *Biomed. Microdevices* 1(1): 81–89.
110. Du, H., Chandaroy, P., Hui, S.W. (1997) Grafted poly-(ethylene glycol) on lipid surfaces inhibits protein adsorption and cell adhesion. *Biochim. Biophys. Acta* 1326(2): 236–248.
111. Feldman, K., et al. (1999) Probing resistance to protein adsorption of oligo(ethylene glycol)-terminated self-assembled monolayers by scanning force microscopy. *J. Am. Chem. Soc.* 121(43): 10134–10141.
112. Malmsten, M., Emoto, K., Van Alstine, J.M. (1998) Effect of chain density on inhibition of protein adsorption by poly(ethylene glycol) based coatings. *J. Colloid Interface Sci.* 202(2): 507–517.
113. Gombotz, W.R., Guanghui, W., Hoffman, A.S. (1989) Immobilization of poly(ethylene oxide) on poly(ethylene terephthalate) using a plasma polymerization process. *J. Appl. Polym. Sci.* 37(1): 91–107.
114. Sofia, S.J., Premnath, V., Merrill, E.W. (1998) Poly(ethylene oxide) grafted to silicon surfaces: grafting density and protein adsorption. *Macromolecules* 31(15): 5059–5070.
115. Pale-Grosdemange, C., et al. (1991) Formation of self-assembled monolayers by chemisorption of derivatives of oligo(ethylene glycol) of structure HS(CH2)11(OCH2CH2)mOH on gold. *J. Am. Chem. Soc.* 113(1): 12–20.
116. Jo, S., Park, K. (2000) Surface modification using silanated poly(ethylene glycol)s. *Biomaterials*, 21(6): 605–616.
117. Chen, H., et al. (2008) Effect of chain density and conformation on protein adsorption at PEG-grafted polyurethane surfaces. *Colloids Surf. B* 61(2): 237–243.
118. Park, K., et al. (2000) In vitro and in vivo studies of PEO-grafted blood-contacting cardiovascular prostheses. *J. Biomater. Sci. Polym. Ed.* 11(11): 1121–1134.

119. Sui, G., et al. (2006) Solution-phase surface modification in intact poly(dimethylsiloxane) microfluidic channels. *Anal. Chem.* 78(15): 5543–5551.

120. Du, Y.J., et al. (2005) In vivo rabbit acute model tests of polyurethane catheters coated with a novel antithrombin-heparin covalent complex. *Thromb. Haemost.* 94(8): 366–372.

121. Murthy, R., et al. (2007) Protein-resistant silicones: incorporation of poly(ethylene oxide) via siloxane tethers. *Biomacromolecules* 8(10): 3244–3252.

122. Papra, A., et al. (2001) Microfluidic networks made of poly(dimethylsiloxane), Si, and Au coated with polyethylene glycol for patterning proteins onto surfaces. *Langmuir* 17(13): 4090–4095.

123. Chen, H., Brook, M.A., Sheardown, H. (2004) Silicone elastomers for reduced protein adsorption. *Biomaterials* 25(12): 2273–2282.

124. Chen, H., et al. (2005) Surface properties of PEO–silicone composites: reducing protein adsorption. *J. Biomater. Sci. Polym. Ed.* 16(4): 531–548.

125. Chen, H., et al. (2005) Protein repellant silicone surfaces by covalent immobilization of poly(ethylene oxide). *Biomaterials* 26(15): 2391–2399.

126. Hawkins, M.L., Grunlan, M.A. (2012) The protein resistance of silicones prepared with a PEG-silane amphiphile. *J. Mater. Chem.* 22(37): 19540–19546.

127. Hawkins, M.L., et al. (2014) Bacteria and diatom resistance of silicones modified with PEO-silane amphiphiles. *Biofouling*, 30(2): 247–258.

128. West, S.L., et al. (2004) The biocompatibility of crosslinkable copolymer coatings containing sulfobetaines and phosphobetaines. *Biomaterials* 25(7–8): 1195–1204.

129. Jiang, S., Cao, Z. (2010) Ultralow-fouling, functionalizable, and hydrolyzable zwitterionic materials and their derivatives for biological applications. *Adv. Mater.* 22(9): 920–932.

130. Chen, S., et al. (2005) Strong resistance of phosphorylcholine self-assembled monolayers to protein adsorption: insights into nonfouling properties of zwitterionic materials. *J. Am. Chem. Soc.* 127(41): 14473–14478.

131. Chen, S., Liu, L., Jiang, S. (2006) Strong resistance of oligo(phosphorylcholine) self-assembled monolayers to protein adsorption. *Langmuir* 22(6): 2418–2421.

132. Tanaka, M., et al. (2009) Synthesis of phosphorylcholine–oligoethylene glycol–alkane thiols and their suppressive effect on non-specific adsorption of proteins. *Tetrahedron Lett.* 50(28): 4092–4095.

133. Liu, Q., Singh, A., Liu, L. (2012) Amino acid-based zwitterionic poly(serine methacrylate) as an antifouling material. *Biomacromolecules* 14(1): 226–231.

134. Li, G., et al. (2008) Ultra low fouling zwitterionic polymers with a biomimetic adhesive group. *Biomaterials* 29(35): 4592–4597.

135. Gao, C., et al. (2010) Functionalizable and ultra-low fouling zwitterionic surfaces via adhesive mussel mimetic linkages. *Biomaterials* 31(7): 1486–1492.

136. Feng, W., et al. (2011) Methacrylate polymer layers bearing poly(ethylene oxide) and phosphorylcholine side chains as non-fouling surfaces: in vitro interactions with plasma proteins and platelets. *Acta Biomater.* 7(10): 3692–3699.

137. Xu, J., et al. (2003) Ozone-induced grafting phosphorylcholine polymer onto silicone film grafting 2-methacryloyloxyethyl phosphorylcholine onto silicone film to improve hemocompatibility. *Colloids Surf. B* 30(3): 215–223.

138. Zhou, J., et al. (2005) Platelet adhesion and protein adsorption on silicone rubber surface by ozone-induced grafted polymerization with carboxybetaine monomer. *Colloids Surf. B* 41(1): 55–62.

139. Liu, P., et al. (2013) Facile surface modification of silicone rubber with zwitterionic polymers for improving blood compatibility. *Mater. Sci. Eng. C Mater. Biol. Appl.* 33(7): 3865–3874.

CHAPTER 8

Partially Fluorinated Coatings by Surface-Initiated Ring-Opening Metathesis Polymerization

G. Kane Jennings and Carlos A. Escobar

Department of Chemical and Biomolecular Engineering, Vanderbilt University, Nashville, TN, USA

8.1 BASIC CONCEPTS

Partially fluorinated polymers are those that contain fluorocarbon substituents, often as side chains, on a polymer that is otherwise based on a different chemistry [1, 2]. When these polymers are deposited onto substrates as films and coatings, the fluorocarbon components often partition to the air interface (Fig. 8.1) to reduce the interfacial free energy well below that of the diluent components [2]. If the fluorocarbon components are perfluoroalkyl groups that terminate in CF_3, the enrichment of CF_3 groups at the surface can lead to interfacial free energies that are even well below those of fully fluorinated polymer coatings such as poly(tetrafluoroethylene) (PTFE) [2, 3]. Thus, partially fluorinated polymer coatings offer a key advantage over all-fluorocarbon systems in terms of the ability to use less fluorocarbon to reduce cost and environmental impact and still achieve superior properties. Efficient methods to deposit or grow partially fluorinated polymers can lead to the realization of these high-performance materials in applications that require a combination of hydrophobicity, oleophobicity, low adhesion, and ultralow surface energies.

In contrast to classical deposition approaches where extra material must often be deposited to ensure complete surface coverage, surface-initiated polymerizations (SIPs) provide the ability to obtain smooth polymer films with thicknesses ranging from several nanometers up to a few micrometers [4–6]. SIPs are defined based on the mechanism of propagation such that a specific method and its associated chemistry can be selected to produce polymer films with the optimal compositions

Functional Polymer Coatings: Principles, Methods, and Applications, First Edition.
Edited by Limin Wu and Jamil Baghdachi.

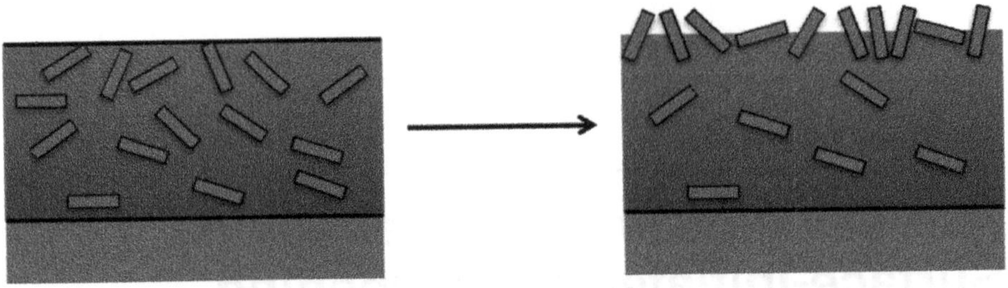

FIG. 8.1 Surface partitioning of fluorocarbon components in a partially fluorinated polymer film. A partially fluorinated polymer film with fluorocarbon (cylinders) and nonfluorocarbon components, before and after removal of the film from the polymerization solution and exposure to air.

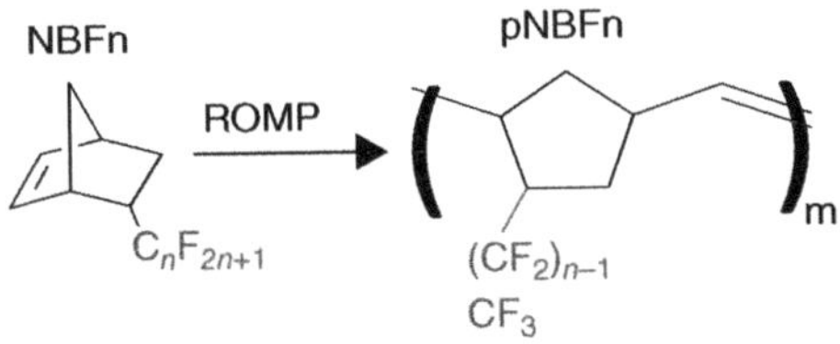

FIG. 8.2 Schematic showing the structures of the NBF*n* monomers and, after ROMP, the pNBF*n* polymer.

and architectures for targeted applications [7]. Of the various SIP approaches, surface-initiated ring-opening metathesis polymerization (SI-ROMP) is a particularly convenient strategy due to its rapid kinetics of film growth at room temperature, the use of commercially available, high-performance initiators, and the ability to access a wide range of surface compositions through functionalized norbornene monomers, many of which may be synthesized by straightforward Diels–Alder reactions [3, 4, 7]. SI-ROMP has been employed by some groups to fabricate micron-thick dielectric layers [4] and functionalized chromatographic supports [8].

A series of SI-ROMP of 5-(perfluoroalkyl)norbornenes (NBF*n*, where *n* denotes the number of carbons in the perfluoroalkyl side chain and is equal to 4, 6, 8, and 10; Fig. 8.2) has been studied by our group over the past years. The least wettable film of the series (pNBF8) exhibits a critical surface tension of 9 mN/m for $pNBF_8$, well below the value of PTFE (18 mN/m), and high contact angles for both water (123°) and hexadecane (78°). The film can be initiated by attaching Grubbs second-generation catalyst (1,3-bis-(2,4,6-trimethylphenyl)-2-(imidazolidinylidene) (dichlorophenylmethylene) (tricyclohexylphosphine) ruthenium) onto a self-assembled monolayer (SAM) [3] or within a functionalized polymer to form a surface-anchored macroinitiator [9] to produce coatings with thicknesses up to several micrometers after 15 min of SI-ROMP. Clearly, the ability to achieve robustly anchored films with thicknesses exceeding a few micrometers by this approach is attractive for coating applications. This approach importantly retains the olefin structure of the backbone, which can be modified to enable a broader range of functionality, including sulfonic acid groups for applications in proton transfer [10].

This chapter reviews the SI-ROMP of pNBF*n* films over the past 5 years. We first describe the surface chemistry for growing pNBF*n* from different substrates, including smooth substrates, nanoporous membranes, and fibrous mats, as well as the kinetics for film growth as dependent on monomer concentration. Then, we examine the dispersive surface free energy of these partially fluorinated materials and further describe the ability to produce microtextured coatings through a combination of micromolding a master and SIP of NBF*n*.

8.2 SURFACE CHEMISTRY

The SI-ROMP of NBF*n* procedure has been to produce partially fluorinated polymer films from a variety of substrates, including gold [3], carbon paper [11], nanoporous gold [9], platinum-modified gold [10], nanoporous alumina [9], and poly (2-hydroxyethylmethacrylate) (pHEMA) [12]. The ability to grow these films from a wide variety of porous and fibrous substrates is consistent toward applications in the production of membranes. In practice, virtually any substrate could be employed as long as the surface can be modified to contain vinyl- or noborneny1-terminated groups. On gold and nanoporous gold, as shown in Fig. 8.3, a hydroxyl-terminated SAM is first attached by adsorption of 4-mercapto-1-butanol, followed by exposure to *trans*-3,6-endomethylene-1,2,3,6-tetrahydrophthaloyl chloride , known as NBDAC because it is a norbornene with two pendant acid chloride groups that are used for an acylation reaction with the hydroxyl groups of the SAM. Subsequent exposure to a solution of Grubbs second-generation catalyst followed by the monomer solution results in the growth of a pNBF*n* film.

Escobar *et al.* successfully grew pNBF6 within a composite membrane consisting of an outer layer of nanoporous gold attached to an anodized alumina membrane [9]. By using the approach in Fig. 8.3, pNBF6 could be grown on both the gold and alumina regions because NBDAC reacts with both the hydroxyl-terminated SAM on gold

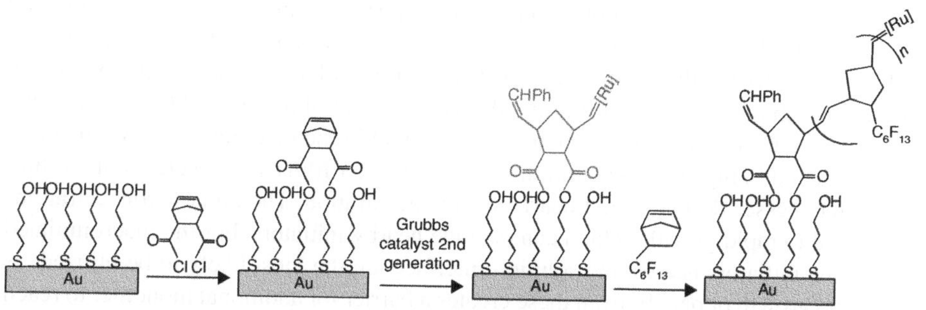

FIG. 8.3 Steps employed for the SI-ROMP of NBF6 to form a pNBF6 film onto gold. [Ru] is used to represent the remaining components of the initiator after surface attachment.

and the nanoporous alumina substrate to produce nobornenyl-terminated groups throughout the composite membrane for effective attachment of the initiator. In contrast, when a vinyl-terminated thiol was used to functionalize the gold followed by exposure to the initiator, pNBF6 was only grown in the nanoporous gold region. This approach avoids NBDAC and results in a faster process that is reduced by one step.

To prepare carbon paper for SI-ROMP, the fibrous electrode was first exposed to an oxygen plasma to generate reactive groups for further functionalization [11]. Subsequent exposure to NBDAC followed by initiator attachment enabled the formation of conformal pNBF*n* films along the carbon fibers of the paper. The hydrophobicity of the carbon paper was amplified by the coatings as the advancing water contact angle increased from 134 to 154°. Further, the film provided a barrier to the surface of the carbon paper, exhibiting a resistance that was higher than commercial PTFE-coated carbon paper.

NBDAC was similarly used to acylate the many hydroxyl groups within a 280-nm-thick pHEMA film, which upon further functionalization with the initiator served as a macroinitiator for SI-ROMP [12]. Through analysis with X-ray photoelectron spectroscopy, around 38% of the hydroxyl groups in the outer approximately 10 nm of the pHEMA film were functionalized by NBDAC and 7% of the norbornenyl groups were able to bind the initiator. Rutherford backscattering spectrometry revealed that 3.7×10^{14} Ru atoms were attached per square centimeter of the film [12], which is a coverage that is approximately 30 times greater than that achieved by attaching the initiator via the approach of Fig. 8.3.

8.3 KINETICS OF FILM GROWTH

The kinetics of SI-ROMP of NBF8 was compared from both an initiator-terminated SAM and from an initiator-modified, 280-nm-thick pHEMA film (macroinitiator) [12]. Both initiation processes lead to thicker films when grown at higher temperatures since the activation energy for the propagation reaction is higher than that for the termination reaction. After 15 min of polymerization at room temperature, films with thicknesses ranging from 1 to 12 μm can be produced, depending on the concentration of the monomer and the method of initiation, as indicated by Fig. 8.4. With a normalized growth rate over the first 5 min of approximately 15 μm/min M for the macroinitiated case at 50 mM, these kinetic rates are among the fastest reported for surface-initiated films [12]. $pNBF_8$ films initiated from the modified pHEMA macroinitiator exhibit faster growth rates and higher limiting thicknesses than films initiated from SAMs due to the greater concentration of bound initiators within the macroinitiator. As shown in Fig. 8.4a and b, higher concentrations produce thicker films, as expected, but the limiting thicknesses do not scale linearly with concentration. The highest concentrations may lead to rapid growth, which can occlude nearby initiators. For the macroinitiated films, the initiators nearest the solution interface are the most likely to be utilized, as the rapid growth of $pNBF_8$ from these creates a barrier for additional monomer to reach buried initiation sites. In this sense, polymer growth from the macroinitiator begins to resemble a monolayer-based initiation as monomer concentrations increase.

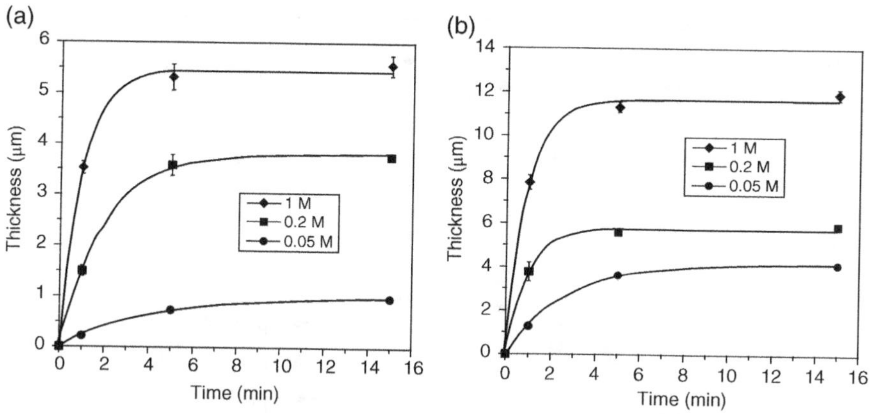

FIG. 8.4 Monomer concentration effects on film thickness for (a) a monolayer initiation and (b) macroinitiation. The data points represent thickness measurements for pNBF8 films grown in 0.05, 0.2, and 1 M monomer solutions. For the macroinitiator case, an average thickness for pHEMA of 280 nm was subtracted from the measured thickness. Solid curves represent fits of the data using a kinetic model consisting of first-order propagation in terms of both monomer and active chains and first-order termination [12].

8.4 SURFACE ENERGY OF pNBF*n* FILMS

A key advantage in the use of partially fluorinated polymer films is that the perfluoroalkyl components partition to the surface to reduce the surface energy [2]. The pNBF*n* films exhibit critical surface tensions of 19, 13, 9, and 11 mN/m for $n=4, 6, 8$, and 10 carbons, respectively, in the perfluoralkyl side chain by measuring the contact angles of dispersive liquids (alkanes) on the surfaces of these films and extrapolating to complete wetting ($\cos\theta = 1$). Here, we have replotted that same data as a function of the inverse square root of the surface tension of the alkane probe liquid. By combining Young's equation,

$$\cos\theta = \frac{\gamma_{sv} - \gamma_{sl}}{\gamma_{lv}} \tag{8.1}$$

where γ_{sv} is the solid surface energy (equivalent to $\gamma_s^{\,d}$ for a purely dispersive surface), γ_{sl} is the solid–liquid interfacial free energy, and γ_{lv} is the surface tension of the liquid, with Fowkes relation for dispersive solid–liquid interfaces,

$$\gamma_{sl} = \gamma_s^d + \gamma_{lv} - 2\left(\gamma_s^d \gamma_{lv}\right)^{1/2} \tag{8.2}$$

we are able to derive the relationship between the measured contact angle (θ) and the dispersive component of the surface energy of the polymer film ($\gamma_s^{\,d}$) and the surface tension of the dispersive alkane:

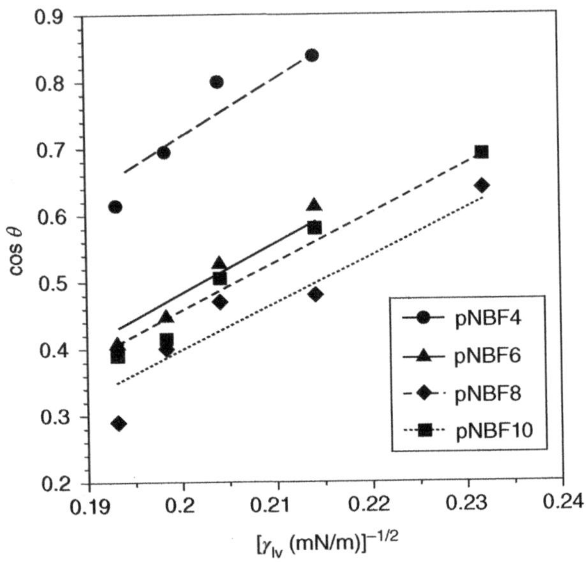

FIG. 8.5 Effect of surface tension on the cosine of the contact angle for pNBF*n* films. Probe liquids include tetradecane, dodecane, decane, octane, and hexane. The best fit lines are constrained to intersect cos $\theta = -1$.

TABLE 8.1 Dispersive Surface Energies of Each Polymer Film as Estimated from Fig. 8.2

Film	pNBF4	pNBF6	pNBF8	pNBF10
$\gamma_s^d (\mathrm{mJ/m^2})$	18	14	12	13

$$\cos\theta = 2\left[\frac{\gamma_s^d}{\gamma_{lv}}\right]^{\frac{1}{2}} - 1 \tag{8.3}$$

Thus, by plotting cosθ versus $(\gamma_{lv})^{-1/2}$ in Fig. 8.5, we can obtain a line with slope of $2(\gamma_s^d)^{1/2}$ that is constrained to intersect cos$\theta = -1$. From the linear fits, we have estimated the values of the dispersive component of surface energy for the four pNBF*n* films (Table 8.1). The values here ranging from 12 mJ/m^2 for $n=8$ to 18 mJ/m^2 for $n=4$ are well below ($n=6$, 8, 10) or comparable with ($n=4$) the reported dispersive surface energy for PTFE (18.4 mJ/m^2) [13]. For $n=6-10$, the results are consistent with a polymer surface that is rich in perfluoromethyl ($-CF_3$) groups with essentially no contributions from hydrocarbon groups. The shorter fluorocarbon chain for $n=4$ may lead to more $-CF_2-$ exposure, thus providing a surface energy similar to PTFE. The remarkably low surface energies for these films are congruent with applications where nonwettability and low adhesion are critical.

8.5 MICROMOLDING SIP

The ultralow surface energies for pNBF*n* coatings make them useful for many applications provided that the surface topography can be tuned to achieve distinct wetting states. We can further develop a process referred to as micromolding surface-initiated polymerization (μMSIP) to achieve various microscale topographies with pNBF8 films [14]. Figure 8.6 shows the process for μMSIP. First, a mold is prepared of a master surface by using hard polydimethylsiloxane (h-PDMS). After curing the mold, it is exposed to a dilute solution of the initiator in dichloromethane (DCM). The DCM evaporates leaving a thin layer of initiator on the surface of the mold. This step for initiator deposition can be skipped if the reproduced features of the coating are designed to be fine-tuned in height to a level smaller than the master's features [14]. The initiator-coated mold is then overfilled with monomer and rapidly placed against a substrate that has been preactivated for SI-ROMP, whether an SAM or a polymer film laden with initiator (macroinitiator). The overfilling allows for the fabrication of a coating underneath the microtopography with thickness that is proportional to the amount of monomer added. After 1 h, the mold is removed to reveal a pNBF8 coating with a surface topography that reproduces the surface of the original master.

Figures 8.7a and b show SEM images of the master, which is an array of diamond field emission pyramidal tips and the coating prepared through μMSIP, respectively. The μMSIP process is able to reproduce the microscopically rough surface well

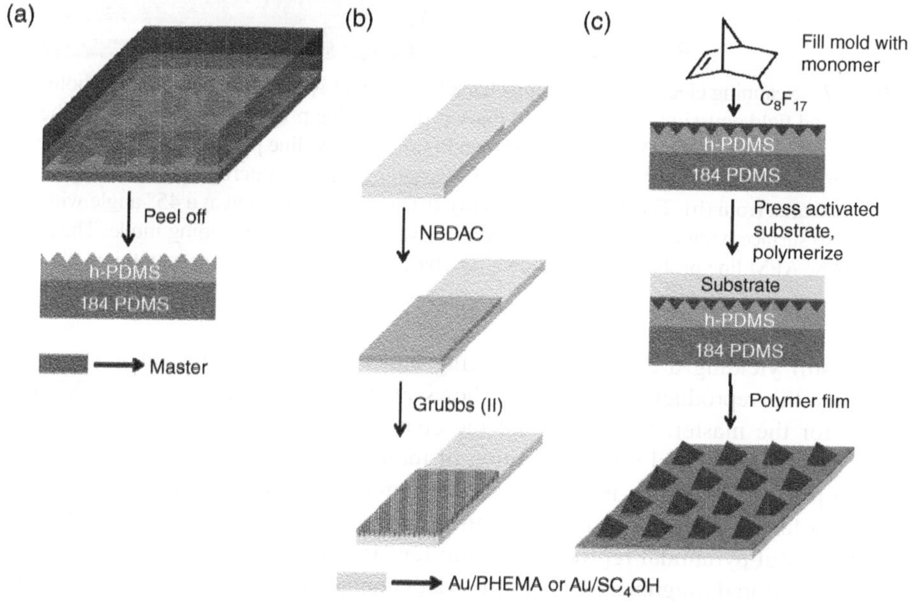

FIG. 8.6 The μMSIP process used to produce microtextured pNBF8 films, which consists of (a) molding a master surface with hard PDMS, (b) preparing an activated substrate for SI-ROMP, and (c) placing a monomer-filled mold against the activated substrate. Initiator is added to the mold prior to adding monomer in (c) if the surface features are desired to be replicas of the master features.

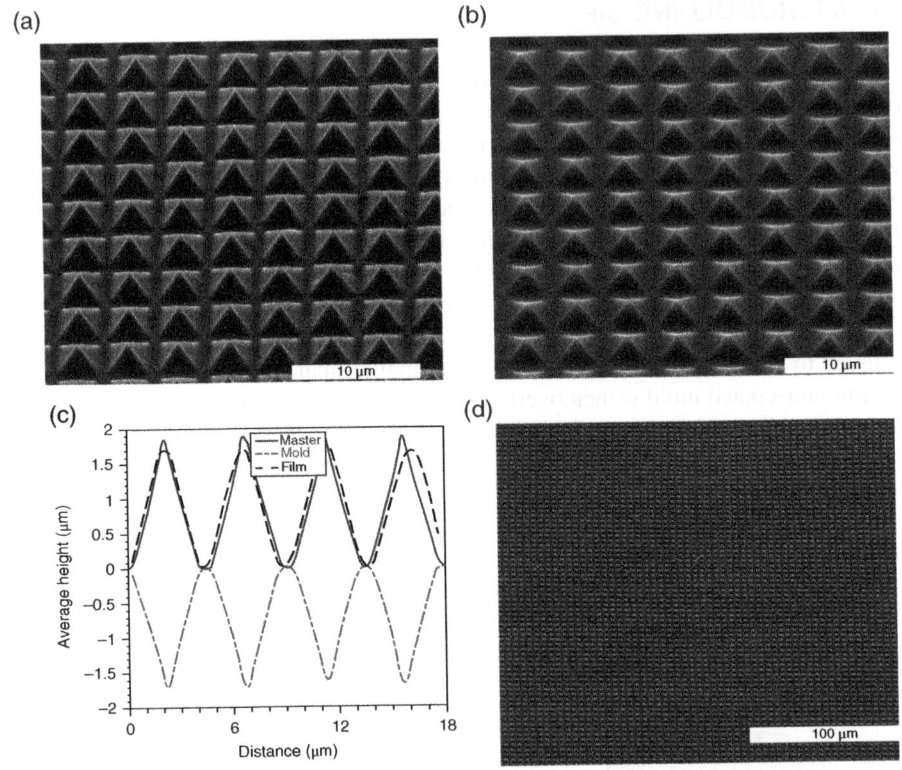

FIG. 8.7 Scanning electron microscopy characterization [14] of (a) the surface morphology of diamond field emission arrays (DFEA) master; (b) resulting pNBF8 film reproducing the surface topography of the DFEA master shown in (a); (c) AFM line profiles of DFEA master in (a), composite mold, and pNBF8 film in (b); (d) scanning electron micrographs showing an expanded area from (b). The SEM images in (a), (b), and (d) were taken at a 45° angle with respect to surface parallel. All AFM measurements were performed in tapping mode. The error bars of the AFM line profiles are not shown for clarity of the plot. The error is within 3.5%.

while still yielding a 20-μm-thick coating underneath the pyramidal features. To quantify the reproduction, Fig. 8.7c shows atomic force microscopy (AFM) line scans for the master, the mold, and the coating. The coating achieves complete filling of the mold and 93% replication of the master; the mold shrinks during the curing process to explain its smaller size as compared with the master. Figure 8.7d shows a larger-scale SEM image of the coating that demonstrates >60,000 μm^2 of successful pyramidal reproduction with few, if any, defects. We have shown that this well-tailored roughness elevates the water contact angle in exact agreement with predictions by the Wenzel equation [14].

The μMSIP process also enables the reproduction of inverted structures that yield recesses or cavities along the surface. Figure 8.8a shows the surface of a μMSIP coating that reproduces a Klarite® master, which consists of pyramidal cavities that are

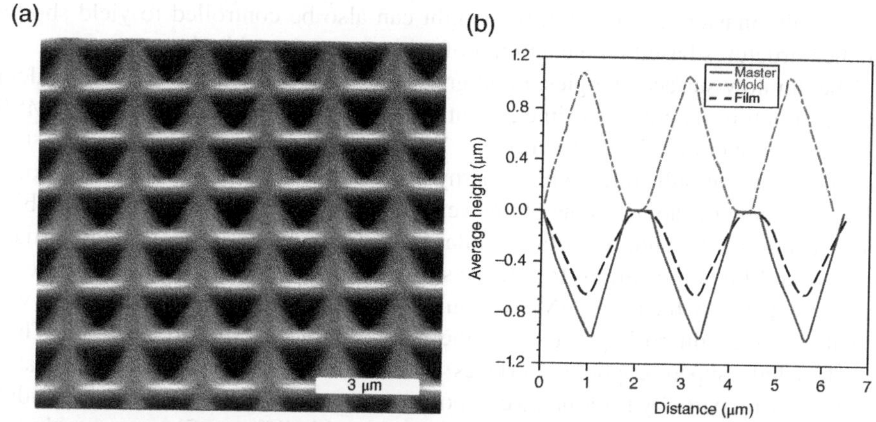

FIG. 8.8 (a) Scanning electron microscopy image of a polymer film prepared using a Klarite® master; (b) AFM line profiles of Klarite master, composite mold, and pNBF8 film. All AFM images were taken in tapping mode. The SEM image was taken at a 45° angle with respect to surface parallel. Error bars are not shown for clarity of the plot. The error is within 3.5% [14].

1 μm in depth. In this case, initiator is not added into the mold, so terminations of the purely surface-initiated growth limits the height of the walls of the pyramidal cavities in μMSIP, producing smaller cavities than those of the original master. This disparity in size is shown clearly by the AFM line scans of Fig. 8.8b as the μMSIP cavities are approximately 65% as large as the master cavities. The ability to achieve smaller feature heights than in the original master is one key advantage of μMSIP. Polymerization temperature with both monolayer-initiated and macro-initiated μMSIP can be used to fine tune the size of the topographical features, ranging from 20 to 100% of the heights of those on the original master [14]. Multigenerational molding, where the film produced in the first generation serves as the master for a subsequent μMSIP process, can also be employed to produce still smaller feature heights.

8.6 CONCLUSIONS AND OUTLOOK

Partially fluorinated coatings of a few to several micrometers in thickness can be grown within 15 min by SI-ROMP of 5-(perfluoro-*n*-alkyl)norbornenes. These films have ultralow critical surface tensions and dispersive surface energies that are well below that of PTFE for fluorocarbon chain lengths of 6–10 due to the partitioning of low-energy perfluoromethyl groups to the surface of the film. The films can be grown from many different substrates, including smooth, nanoporous, and fibrous mats, by anchoring either vinyl or norbornenyl groups to the surface prior to exposure to the initiator. The microtexture of these coatings can be controlled via μSIP in which the polymer chains growing from a surface are confined by a mold prepared from a master containing microfabricated features. The resulting surface features can be as large as the master features, but the height can also be controlled to yield shorter, smaller structures based on the amount of initiator employed in the process.

The ultralow surface energies, rapid growth rates, and topographical control demonstrated in this chapter are prime advantages for the use of pNBF*n* films and MSIP for the reproduction of various masters. We expect that more complicated masters, including natural substrates with superhydrophobicity, can be reproduced through μSIP to expand the applications of this versatile technique. Such approaches enable a reproduction of the natural surface while using the particularly convenient partially fluorinated film compositions to achieve surface energies far below those achieved in nature. In addition, the use of μMSIP would be advantageous to prepare coatings with specific surface microshapes to tune the coatings for specific applications. These could include superoleophobic surfaces with reentrant architectures or lubricating interfaces with partially fluorinated compositions or many other applications including biofouling resistance and antimicrobial properties with different film compositions.

ACKNOWLEDGMENTS

The authors greatly appreciate the financial support for this work provided by the National Science Foundation through the grants CBET-1134509 and CMMI-1300406. We thank Dr. Jonathan Jarvis and Dr. Yang Jiao for providing the diamond field emission arrays and Klarite substrates, respectively.

REFERENCES

1. Genzer, J., Sivaniah, E., Kramer, E. J., Wang, J., Xiang, M., Char, K., Ober, C. K., Bubeck, R. A., Shmakova, O. E., Lee, T. R. (2000) Molecular Orientation of Single and Two-Armed Monodendron Semifluorinated Chains on Soft and Hard Surfaces Studied Using NEXAFS, *Macromolecules* 33: 6068–6077.
2. Brantley, E. L., Jennings, G. K. (2004) Fluorinated Polymer Films from Acylation of ATRP Surface-Initiated Poly(hydroxyethyl methacrylate), *Macromolecules* 37: 1476–1483.
3. Faulkner, C. J., Fischer, R. E., Jennings, G. K. (2010) Surface-Initiated Polymerization of 5-(Perfluoro-n-alkyl)norbornenes from Gold Substrates, *Macromolecules* 43: 1203–1209.
4. Rutenberg, I. M., Scherman, O. A., Grubbs, R. H., Jiang, W., Garfunkel, E., Bao, Z. (2004) Synthesis of Polymer Dielectric Layers for Organic Thin Film Transistors via Surface-Initiated Ring-Opening Metathesis Polymerization, *J. Am. Chem. Soc.* 126: 4062–4063.
5. Jones, D. M., Huck W. T. S. (2001) Controlled Surface-Initiated Polymerizations in Aqueous Media, *Adv. Mater.* 13: 1256–1259.
6. Huang, W., Kim, J.-B., Bruening, M. L., Baker, G. L. (2002) Functionalization of Surfaces by Water-Accelerated Atom-Transfer Radical Polymerization of Hydroxyethyl Methacrylate and Subsequent Derivatization, *Macromolecules* 35: 1175–1179.
7. Jennings, G. K., Brantley, E. L. (2004) Physicochemical Properties of Surface-Initiated Polymer Films in the Modification and Processing of Materials, *Adv. Mater.* 16: 1983–1994.

8. Buchmeiser, M. R., Sinner, F., Mupa, M., Wurst, K. (2000) Ring-Opening Metathesis Polymerization for the Preparation of Surface-Grafted Polymer Supports, *Macromolecules* 33: 32–39.
9. Escobar, C. A., Zulkifli, A. R., Faulkner, C. J., Trzeciak, A., Jennings, G. K. (2012) Composite Fluorocarbon Membranes by Surface-Initiated Polymerization from Nanoporous Gold-Coated Alumina, *ACS Appl. Mater. Interfaces* 4: 906–915.
10. Berron, B. J., Faulkner, C. J., Fischer, R. E., Payne, P. A., Jennings, G. K. (2009) Surface-Initiated Growth of Ionomer Films from Pt-Modified Gold Electrodes, *Langmuir* 25: 12721–12728.
11. Faulkner, C. J., Payne, P. A., Jennings, G. K. (2010) Surface-Initiated Ring-Opening Metathesis Polymerization of 5-(perfluorohexyl)norbornene on Carbon Paper Electrodes, *J. Colloid Interface Sci.* 351: 248–253.
12. Escobar, C. A., Harl, R. R., Maxwell, K. E., Mahfuz, N. N., Rogers, B. R., Jennings, G. K. (2013) Amplification of Surface-Initiated Ring-Opening Metathesis Polymerization of 5-(Perfluoro-n-alkyl)norbornenes by Macroinitiation, *Langmuir* 29: 12560–12571.
13. Solid Surface Energy Data for Common Polymers. (2007) from www.surface-tension.de/solid-surface-energy.htm (accessed December 13, 2014).
14. Escobar, C. A., Cooksey, T. J., Spellings, M. P., Jennings, G. K. (2014) Micromolding Surface-Initiated Polymerization: A Versatile Route for Fabrication of Coatings with Microscale Surface Features of Tunable Height. *Adv. Mater. Interfaces*, 1: 1400055.

CHAPTER 9

Fabrication and Application of Structural Color Coatings

Zhehong Shen[1,2], Hao Chen[1,2], and Limin Wu[1]

[1]Department of Materials Science and Advanced Coatings Research Center of Ministry of Education of China, Fudan University, Shanghai, P.R. China

[2]Engineering Department of Zhejiang Agriculture and Forestry University, Hangzhou Linan, P.R. China

9.1 INTRODUCTION

Traditionally, the colors of coatings come from pigments and dyes. Unfortunately, the manufacture of most dyes and pigments usually involves substantial pollution and energy loss. Dyes and pigments can also exert toxic effects and degrade or fade easily. With the increasing attention paid to environmental pollution, the fabrication of environment-friendly colors has drawn worldwide interest.

Peacock feathers and natural opals are brilliantly colored without dyes or pigments. In these cases, light of a specific wavelength is selectively scattered by nanostructures with variations in refractive index with a length on the order of the wavelength of visible light [1]. This phenomenon is called structural color. There are two kinds of structural color in nature. The first is caused by periodic anisotropic nanostructures such as natural opals and peacock feathers. The other is caused by isotropic nanostructures such as parrot feathers. This chapter mainly deals with the first type of structural color.

Structural color has merits that traditional dyes and pigments do not share. First, the colors tend to be more brilliant. Second, they are more resistant to sun and will not fade over time because the color is caused by Bragg diffraction in nanostructures. Third, they are more environment-friendly, involving no toxic materials or pollutions. Structural color has considerable potential as a replacement for toxic dyes and pigments used in coatings, decorations, displays, and textiles, especially because

Functional Polymer Coatings: Principles, Methods, and Applications, First Edition.
Edited by Limin Wu and Jamil Baghdachi.

more and more pollution restrictions on dyes and pigments are likely to be put into effect in the future.

At present, fabricating photonic crystals (PCs) is the main method of producing structural colors. PCs are periodic arrays whose refractive index is modulated on a length scale comparable with the light wavelength of interest [2, 3]. Interference from light waves scattered from the dielectric lattice (i.e., Bragg scattering) leads to omnidirectional stop bands or photonic band gaps (PBGs), which are analogous to the electronic energy band gaps in a semiconductor [1]. If the band gaps of PCs are located in visible wavelength, they produce structural colors. For this reason, the fabrication of PCs has often been used to produce structural colors. The technique works by controlling and manipulating the propagation of photons in the dielectric lattice of PCs. For three-dimensional (3D) PCs, the diffraction wavelength and forbidden gap are determined by Bragg's law Equation (9.1):

$$m\lambda = 2\,nd_{hkl}\sin\theta \tag{9.1}$$

where m is the order of diffraction, λ is the wavelength of diffracted light (or the so-called forbidden band), n is the effective refractive index of the system, d_{hkl} is the interplanar spacing along the $[hkl]$ direction, and θ is the glancing angle between the incident light and diffraction crystal planes. This equation indicates that the position of the stop band is directly proportional to the interplanar spacing and the effective refractive index. Any variation in the spacing or effective refractive index can cause an observable shift in the diffraction peak. If the position of the stop band is located in the visible light range, corresponding colors can be seen. Different hues of structural colors, such as blues, greens, and reds, can be produced easily by controlling and manipulating interplanar spacing and the effective refractive index.

PCs have considerable applications in communications [4, 5], sensors and monitors [5, 6], templates [7–11], color display, and solar energy [12–14]. Methods developed to produce PCs for structural colors include both bottom–up and top–down processes [15]. Though top–down methods such as mechanical processing and holography can be used to fabricate PCs with few defects, they are time-consuming and involve significant tedium and cost [16–18].

Colloidal assembly is a widely used bottom–up method to fabricate PCs. By organizing spherical colloids such as polymer latexes and silica spheres into a colloidal crystal, it is possible to easily tail the functionality not only from the constituent material of the colloidal particles but also from the periodic structure associated with a crystalline lattice. It is a relatively simple and cost-efficient way to fabricate novel materials with advanced functionality [19–23]. In general, the wavelength of light diffracted from a 3D colloidal crystal is determined by the Bragg law Equation (9.2):

$$\lambda = 2n_{\mathrm{avg}}d\left(1-\sin^2\theta / n_{\mathrm{avg}}{}^2\right)^{1/2} \tag{9.2}$$

where d is the distance between crystalline planes in the (111) direction and related to the bead diameter D by $d=(2/3)^{1/2}D$; n_{avg} is the effective refractive index, which has a relationship with the filling volume ratio v_{i} of different refractive indexes n_{i} by

$n_{avg} = (\Sigma n_i^2 v_i)^{1/2}$; and θ is the angle between the incident light and the normal to the (*hkl*) planes. When light is normal to the crystalline planes, θ is equal to 0°, and the Equation (9.2) changes as follows:

$$\lambda = \left(\frac{8}{3}\right)^{1/2} D\left(\sum n_i^2 v_i\right)^{1/2} \tag{9.3}$$

In this way, the wavelength (and thus the color) of light diffracted from a colloidal PC is directly proportional to the lattice constant. Any variations, such as the refractive index contrast, filling ratio of components, the distance between crystalline planes in the (111) direction in the lattice constant might lead to an observable shift in the stop band position and thus the color displayed by the surface of this crystal.

9.2 GENERAL METHODS OF COLLOIDAL ASSEMBLY

Many physical methods based on optical instruments, such as holographic lithography [24], fem to second laser interference [25], and multibeam laser interference [26], have exhibited obvious advantages in the preparation of high-quality PCs for structural colors. However, high equipment expenses and time requirements limit their use in industrial settings. As mentioned earlier, colloidal self-assembly is based on the self-assembly of monodispersed colloidal particles into ordered crystalline structure. The periodicity of PCs can be tuned simply by varying the particle size from hundreds of nanometers to micrometers. Gravity sedimentation [27], centrifugation sedimentation [28], and vertical deposition [29] are the most popular methods of colloidal self-assembly in the fabrication of photonic films or structural color coating and have been demonstrated to be inexpensive, simple to implement, inherently parallel, and high-throughput. Some reviews had dealt with the aforementioned methods. In this section, the deposition methods are classified into two classes: flow-induced deposition and field-induced deposition, according to the dominant driving force during the ordering. There are also some classical and special preparation methods for various structural colors. These will be introduced in the following paragraphs.

Most high-quality colloidal crystals are assembled from monodisperse polystyrene (PS), poly(methylmethacrylate) and silica particles. Simulation shows that high-quality colloidal crystals can be formed if the particle size is within 2% of standard deviation. As the size distribution increases, the intensity reflectance by the stop band decreases rapidly [30]. The polymeric particles are typically prepared by emulsion polymerization and dispersion polymerization. The silica particles are typically prepared by sol–gel reaction of tetraorthosilicate precursors using the Stöber method [31].

9.2.1 Flow-Induced Deposition

Some typical colloidal assembly approaches to fabricate PCs for obtaining structural colors, such as vertical deposition, dip-coating, spin coating, doctor blade coating, spray coating, and so on, are classed as flow-induced deposition method. In this method, the flow is only used to achieve a high-concentration region that exceeds the concentration required for colloidal crystallization. The assembly-driving force comes from the flow dragging of the spheres. Colloidal spheres are often driven by the flow of suspension toward a growth front, as the solvent evaporates at an evaporating front (meniscus), and they stick together through a stochastic process at this front [1].

As a recently most widely used colloidal assembly method, vertical deposition is mainly based on synergistic reaction of colloidal self-gravity and capillary force induced by solvent evaporation [32–35]. In this approach, a small amount of colloid suspensions is needed, as-prepared samples with good assembly microstructure can be easily handled. Dip-coating can be considered as an important vertical deposition approach. This approach, as shown in Fig. 9.1, is processed with substrate being wetted by dropping the suspension on it and then the substrate is inclined vertically until the suspension is dried. The whole process is very simple. However, the film has strong thickness fluctuations because the suspension concentration increases as the solvent evaporates, and it needs tedious works to get multilayering colloidal PCs [33]. A modification fabrication procedure of dip-coating is shown in Fig. 9.2. A suspension of the colloids is diluted to a definite concentration using appropriate solvent. Then a suitable substrate is immersed vertically into the dispersion and lifted up at constant speed. This process is precisely controlled by a motor [29]. With solvent evaporation, large areas of 3D structural color film form on the substrate. This method is suitable for most colloids that contain organic and inorganic colloidal particles, and substrates of various shapes are available. Compared with previous simple

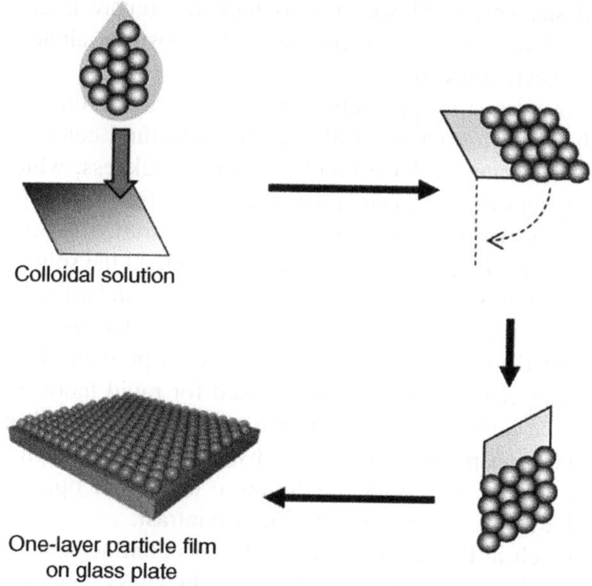

FIG. 9.1 Schematic of colloidal self-assembly by dip-coating. Reproduced from the permission of Ref. 33. © 2008 American Chemical Society.

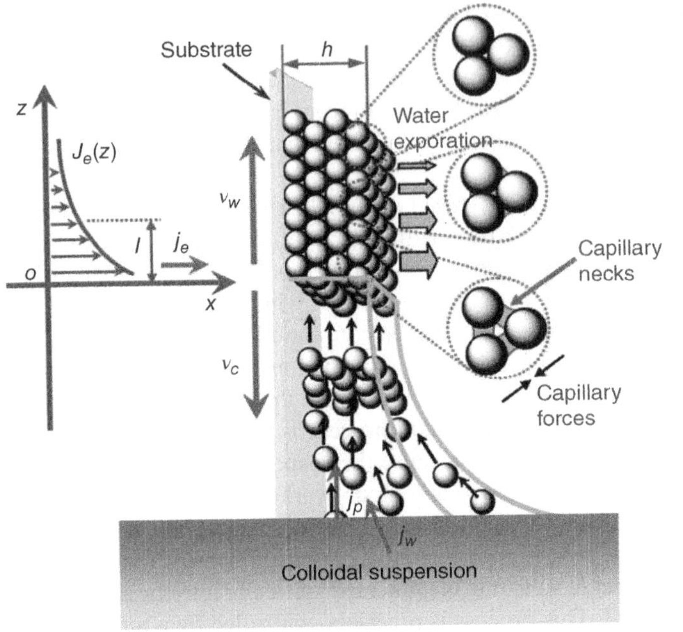

FIG. 9.2 Schematic of colloidal self-assembly by vertical deposition. Reproduced from the permission of Ref. 29. © Elsevier.

dip-coating, this modified approach can decrease the thickness gradient by lifting substrate out of suspension. However, it is hard to prepare large-area monolayer structural color films or coatings using vertical deposition alone. It is also time-consuming to fabricate thick films.

Spin coating is a versatile approach of film preparation in this group of methods. It can enable the rapid formation of 3D-ordered structural color films or coatings. The resultant planar samples have highly uniform thickness, which can be easily adjusted by changing the spin speed and time [36–38]. Fabrication procedures are as follows: colloidal suspension with appropriate concentration is first dispensed on a substrate. After tilting and rotating the substrate to spread the solution for full wafer coverage, the wafer is spin-coated at low rate on a standard spin-coater for one minute. After many seconds, the wafer is quickly accelerated to the desired spin speed and continued to spin for a specific period to produce the target thickness (Fig. 9.3). The spin coating method can be used for rapid fabrication of structural color coatings at the centimeter scale within minutes. However, this method required highly viscous latex suspension, and the redispersion of latex spheres was tedious [39]. The use of special dispersant was found to impair the optic properties of the resulting films due to the lower refractive index contrast.

Another approach in this group of methods uses confinement, which inherently leads to smooth surfaces and easily controllable thickness. Doctor blade coating is widely used in the textile, paper, photographic film, printing, and ceramic industries to create flat, highly uniform films over large areas [40, 41]. As a confining assembly

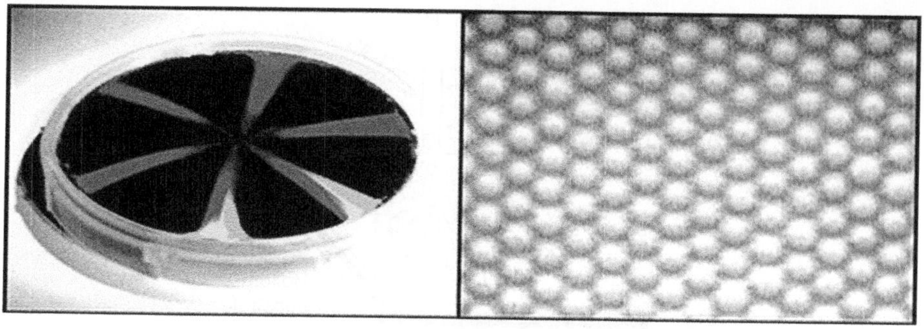

FIG. 9.3 Schematic of colloidal self-assembly by spin-coating and SEM image of obtained structural color coating. Reproduced from the permission of Ref. 36. © 2004 American Chemical Society.

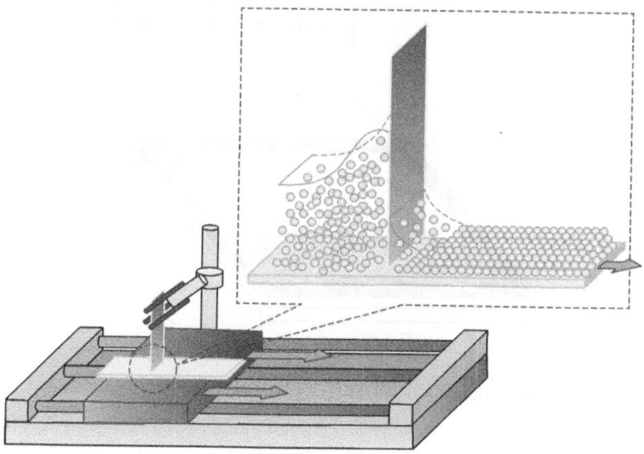

FIG. 9.4 Schematic of colloidal self-assembly by doctor blade coating. Reproduced from the permission of Ref. 41. © 2010 American Chemical Society.

approach, it is also used to assemble colloidal particles to manufacture structural colors [42, 43]. Fabrication procedures are as follows: An immobilized 90°-beveled razor blade is gently placed on a substrate. Colloidal suspension with appropriate concentration is dispensed along one side wall of the blade onto the substrate. The substrate is dragged with a syringe pump at a controlled speed. The blade could then spread the colloidal suspension uniformly on the substrate. Large-area structural color films or coatings form after solvent evaporation (Fig. 9.4). The thickness of the prepared film is correlated with the viscosity of the colloidal suspension, and the coating speed and the correlations can be qualitatively explained by adapting the mechanisms developed for conventional doctor blade coating [41].

Another interesting confining assembly approach can manufacture thick, controllable structural color coatings by confining colloidal deposition into the space between two parallel substrates, which are separated by thin microstructured spacer acting as a filter [8]. Figure 9.5 outlines the procedure of packing colloids into

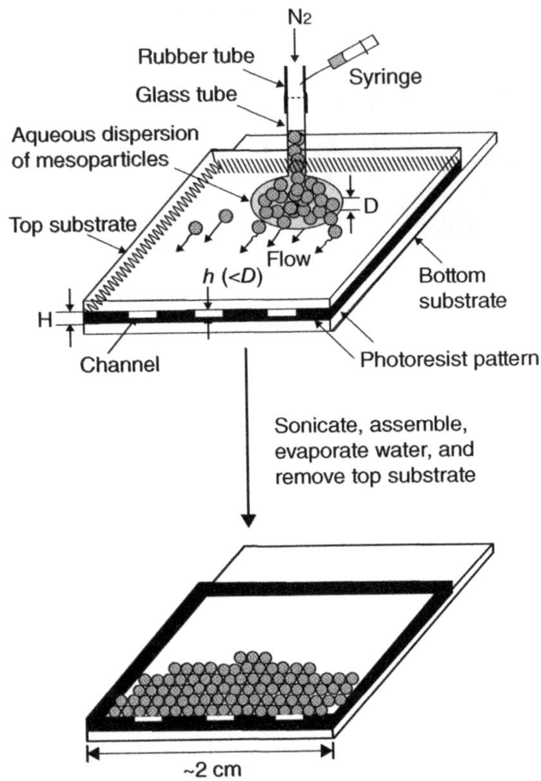

FIG. 9.5 Scheme of confining colloidal self-assembly. Reproduced from the permission of Ref. 8. © 1999 Wiley.

crystalline assemblies. A dispersion of monodispersed particles is injected into the cell through the rubber tube with a syringe, and then positive pressure N_2 is applied through the glass tube to force the solvent to flow through the channels. The particles accumulate at the bottom of the cell and form a close-packed structure under continuous sonication. When the particles are almost depleted, the cell is placed in an oven at 65°C to evaporate the remaining solvent. The top substrate is carefully removed, leaving behind a crystalline assembly of particles on the bottom substrate. The rate of packing of polymer beads increases as the pressure of nitrogen increases. The number of layers of structural color film or coating is determined by the ratio between the thickness of the cell and the diameter of the particles. Using this method, well-ordered films having a controlled thickness of several square centimeters in size have been successfully created. However, unavoidable fabrication of the confinement cell and the microstructured filter limits the application of this approach in rapid preparation of large-area structural color coatings.

Spray coating is a very simple and versatile approach to the fabrication of surface paints. It has also been used to produce structural color coatings as one of the flow-induced depositions [44–47]. Figure 9.6 presents the illustration of the fabrication

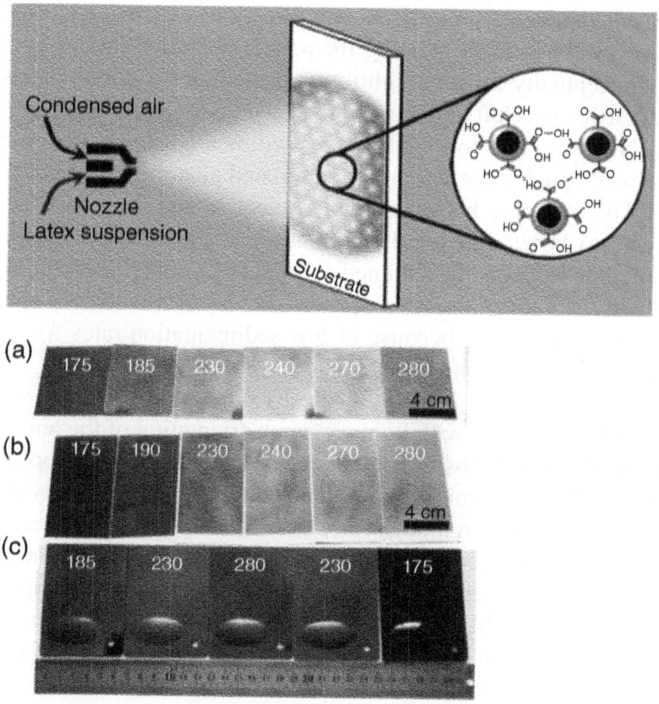

FIG. 9.6 Schematic of colloidal self-assembly by spray coating and the structural color coating on the substrates of paper (a), aluminum (b) and plastic (c) with color as blue, cyan, green, yellow-green, orange-red and red from left to right. Reproduced from the permission of Ref. 47. © 2009 American Chemical Society.

procedure for structural color coatings using spray coating method [47]. Simply, the latex suspension is first loaded into the spray gun, and then air pressure is used to eject it from the nozzle orifice toward the substrate. Upon evaporation of the water, the latex spheres subsequently are self-assembled on the substrates, primarily via the capillary and interparticle forces [46]. The mixing procedure among the latex suspension and air takes a fraction of a second, and the additional whiff procedure can accelerate the water evaporation rate. All of these greatly shorten the fabrication time of structural color coatings. However, this process also has unavoidable disadvantages: highly dispersed mist and volatile solvent not only pollute the environment and harm human health but also waste raw materials.

9.2.2 Field-Induced Deposition

Although flow-induced deposition method could produce the assembly arrangement of most colloidal particles, it is difficult to prepare structural color coating with homogeneous anisotropic structure of colloidal arrangement. Field-induced deposition methods have been proved to be effective approaches for the well-ordered assembly of the anisotropic colloidal particles [25, 48–51]. In these methods, the presence of gravity (or other external fields, such as electric force and magnetic force) can result in a density profile in one direction, which can strongly influence the colloidal crystal assembly. Among them, sedimentation driven by gravitational force is the conceptually most straightforward method for colloidal assembly. The colloidal suspension was left undisturbed in a cuvette for a prolonged period (usually about several months). The colloidal spheres sedimented slowly driven by gravity at a velocity defined by a Stokes law and accumulated near the bottom of the cuvette in a dense fan. The interplay between gravity, electrostatic repulsion, and Brownian motion appears to be crucial for the formation of the colloidal crystal sediment. Though the method is facile, the obtained colloidal PCs contained large concentration of lattice defects. Moreover, the whole process is very slow and only applicable for a limited range of sphere sizes because of low sedimentation rates for small spheres and bad ordering for large spheres [1, 50].

Other deposition methods induced by electric and magnetic fields were proposed for faster colloidal assembly with homogeneous orientation of the anisotropic structure. The electrophoretic deposition technique has a wide range of applications in the processing of advanced ceramic materials and coatings. It has recently drawn interest both from academia and from industry not only because of its high versatility, attributable to its ability to work with different materials and combinations of materials, but also because of its cost-effectiveness and the simplicity of the equipment required [51]. Using the electric field to drive the sedimentation velocity has been reported to produce structural color coatings very efficiently [49]. Figure 9.7 shows the procedure of self-assembly by electrophoretic deposition. The cell in which the electrophoresis was performed consisted of a cylindrical tube fixed to the basis, where the structural color coating would settle. This basis was obtained from a sheet substrate with good conductivity. Platinum sheets were used for the upper electrodes because they have

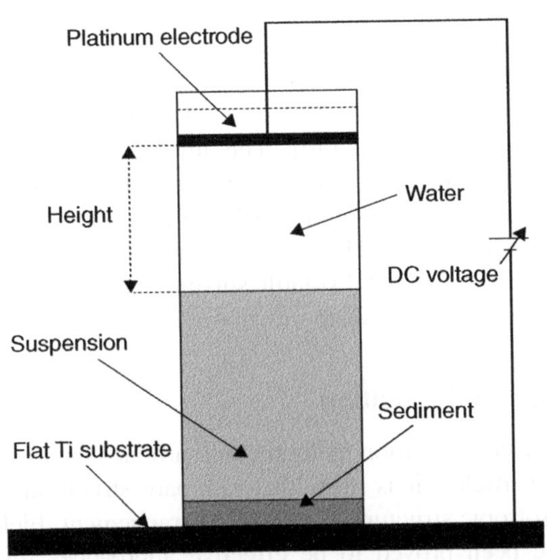

FIG. 9.7 Schematic of colloidal self-assembly by electrophoretic deposition. Reproduced from the permission of Ref. 49. © 1999 American Chemical Society.

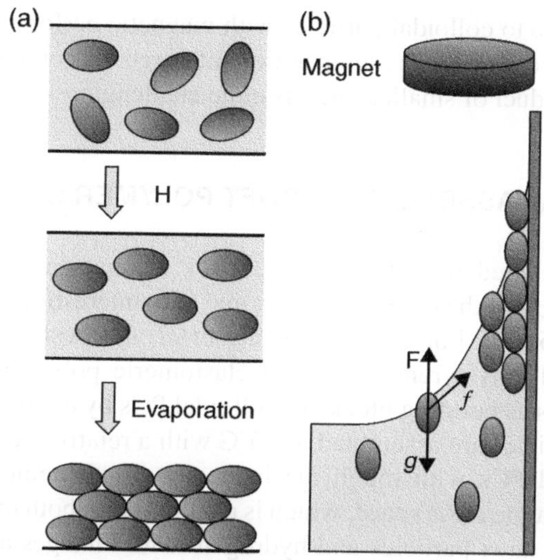

FIG. 9.8 Schematic of colloidal self-assembly in a magnetic field (a) the proposed assembly procedure and (b) the experimental setup for assembly. Reproduced from the permission of Ref. 54. © 2009 Wiley.

the highest redox potential. Then, both electrodes were connected to a direct-current source used to obtain an electrical field. Under a controllable electrical field, charged colloids in suspension were deposited on the substrate to form structural color coating. By this process, large-area structural color coating with thickness of a few monolayers to 1 mm can be produced. Electric field strength and colloid surface charge density are two of the important factors behind producing the desired coatings. Electrophoretic deposition is applicable to the self-assembly of various charged colloids with different sizes. However, structural color coatings from uncharged colloids or particles with relatively low charges cannot be produced using this method.

Three-dimensional PCs of ellipsoidal colloidal particles can be prepared by direct convective self-assembly from suspension with the aid of a magnet [52–54]. The magnetic field provides the orientational order, which is needed for colloidal particles that are not perfectly symmetrical in three dimensions. Figure 9.8 summarizes the procedure of crystallizing magnetically active ellipsoids in a magnetic field. The assembly is carried out by using the convective self-assembly method in a magnetic field [54]. The dispersion of ellipsoidal particles in solvent with an appropriate concentration is added into a clean container, followed by inserting a clean panel of glass substrate. A magnet with appropriate surface magnetic strength is placed above the container. Magnetic strength can be tuned by altering the distance between the magnet and the dispersion. The container is kept in an oven undisturbed at 65°C for about 24 h. Gravity, magnetic forces, and capillary forces induced by solvent evaporation cause magnetic ellipsoids to assemble with their main axes parallel to the magnetic field and form structural color coating on the substrate. The magnetic field–induced colloidal assembly exhibits obvious advantages of being quick and easy in fabricating structural color coatings with high-quality anisotropic structure. However, it is only applicable to colloidal particles with magnetic performance, and magnetic colloids are not common. The preparation of large uniform magnetic colloids, without the by-product of smaller ones, is quite challenging.

9.3 COLLOIDAL ASSEMBLY OF SOFT POLYMER SPHERES

Colloidal PCs prepared from hard spheres lack mechanical strength. In order to improve the strength of the colloidal PCs, some very interesting strategies have been developed [47, 55–57]. For example, Wang *et al.* synthesized polymer colloidal spheres with hard polystyrene cores and elastomeric poly(methyl methacrylate–acrylic acid) shells as building blocks of colloidal PCs by emulsion polymerization. The colloidal particles are assembled at 90°C with a relative humidity of 70%. The resultant colloidal PCs with tough mechanical strength demonstrate fully dense structure with separate close space, which is resulted from both the compact physical connection of the latex particles and hydrogen bond linkages among the carboxyl groups from the surface of the latex spheres [57]. McGrath *et al.* synthesized monodisperse colloidal particles composed of poly(styrene-*co*-*N*-isopropylacrylamide) (PS-*co*-NIPAm). As a suspension of the PS-*co*-NIPAm particles dries, the interactions of their soft, flexible peripheries assist in particle self-assembly to form a close-packed ordered structure. Interparticle adhesion occurs at these contacts as the final amount of solvent evaporates, enhancing the durability of the assembly. The dense core provides stability of the ordered structure subsequent to drying [56]. The aforementioned treatment processes are effective in enhancing the film strength, but the methods may be unsuitable for some materials that cannot resist temperatures higher than room temperature or bring about pollution resulted by organic solvent evaporation.

Wu *et al.* presented a novel method of fabricating structural color coatings from soft polymer and hard silica particle mixtures [58]. In this approach, the monodisperse soft polymer sphere colloids (Tg: 19°C) are synthesized using surfactant-free emulsion polymerization and then blended with colloidal silica to produce nanocomposite polymer latex. When this nanocomposite polymer latex is cast on a glass substrate and dried at room temperature, these soft polymer spheres can periodically be arranged themselves, directly forming 3D structural color films or coatings (Fig. 9.9). In comparison with other techniques, the primary differences and advantages of this new approach are as follows: (i) the process is very simple and the colloidal crystal films are easily obtained by casting, brush coating, spin coating or spray coating; (ii) the obtained crystal films have enough mechanical properties to be further processed discretionarily, for example, cutting, puckering, bending, and so on (Fig. 9.10); (iii) it can be easily used for mass production of crystal films with any sizes from square centimeters to square meters. These advantages provide a good basis for the practical application of structural color coatings.

Hard nanosilica colloids and the interaction between silica and polymer colloids play important roles in constructing 3D-ordered binary assembled films. As shown in Fig. 9.11, when the soft polymer sphere colloids are blended with colloidal silica beads, the latter can be adsorbed on the surfaces of the polymer spheres because of

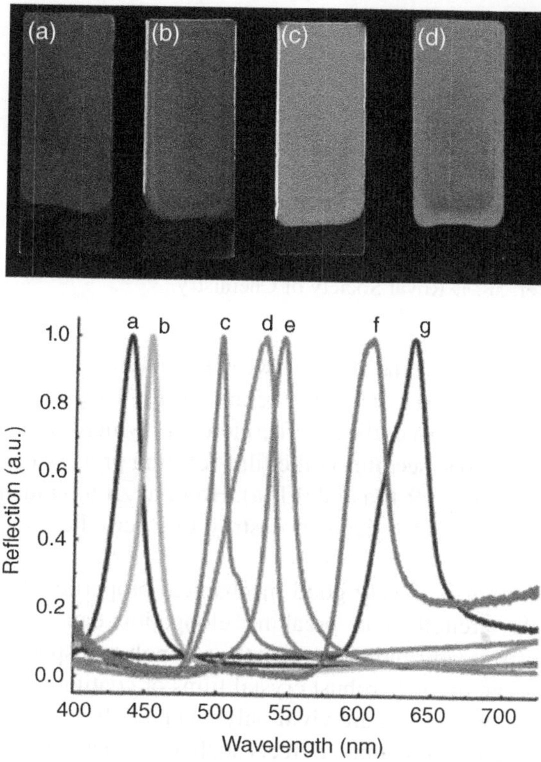

FIG. 9.9 Photographs and corresponding reflection spectra of structural color coatings prepared with polymer spheres of various sizes: (a) 180 nm, (b) 210 nm, (c) 240 nm, (d) 260 nm, (e) 270 nm, (f) 310 nm, and (g) 340 nm with colors as purple, cyan, blue, green, yellow-green, orange-red and red from left to right. Nanosilica content: 5 wt%. Reproduced from the permission of Ref. 58. © Royal Society of Chemistry.

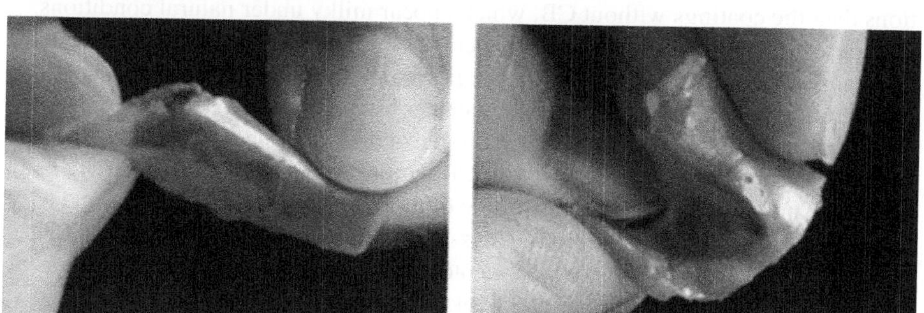

FIG. 9.10 Robust red structural color coatings prepared by binary colloidal assembly. Reproduced from the permission of Ref. 58. © Royal Society of Chemistry.

the hydrogen bonding interaction between the –SiOH groups of silica and –COOH groups of polymer spheres [59]. As the water evaporates, the deformation and coalescence of the soft polymer spheres during the film-forming process can be restrained by the surrounding nanosilica particles. This can guarantee these monodisperse and

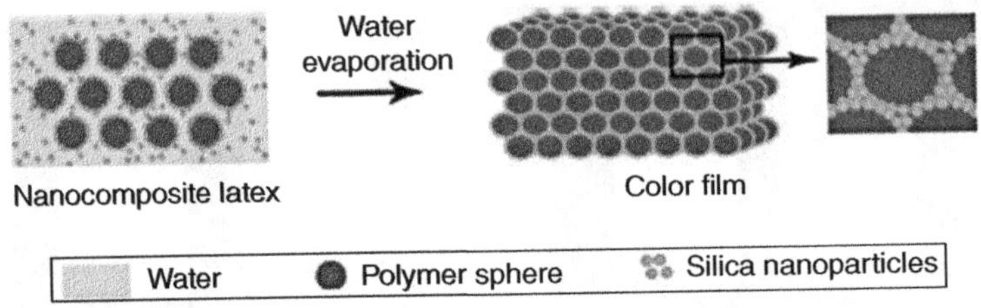

FIG. 9.11 Film-formation mechanism of 3D-ordered structure in binary colloidal coating. Reproduced from the permission of Ref. 58. © Royal Society of Chemistry.

spherical polymer spheres pile tightly and that the voids between the polymer spheres are filled with nanosilica particles, forming 3D-ordered structures (Figs. 9.12b–d and 9.13b–d). If no acrylic acid or nanosilica is used, these soft colloidal polymer spheres cannot be protected. They deform and coalesce during the film-forming process like a typical polymer latex would happen (Figs. 9.12a and 9.13a). However, if too much nanosilica is present, the aggregation of nanosilica can destroy this periodic array (Figs. 9.12e and 9.13e).

All the structural color films present relatively good mechanical properties; the maximum impact strength, tensile strength, and breaking elongation can reach 50 kg cm^{-1}, 3.19 mPa, and 372%, respectively, indicating that the soft polymer spheres combined with nanosilica particles can produce robust crystal films or coatings.

However, the colors of colloidal crystal films are vivid only when the light is at a certain angle. This is because, during self-assembly, defects such as dislocation and misplaced cavities always occur. This makes the color so weak that is barely observable under natural conditions. To overcome the shortcomings and improve the saturation of the structural color under normal conditions, carbon black (CB) was added to the dispersions of the polymer/silica as broadband wavelength absorber [60]. As shown in Fig. 9.14, the colors of the coatings with CB have higher chroma or saturations than the coatings without CB, which appear milky under natural conditions.

Figure 9.14a shows that the structural color coatings with CB have less reflectivity than those without CB. For example, the polymer/silica structural color coating prepared from polymer spheres 241 nm in diameter has about 79.0 and 74.0% reflectivity in the band gap and on the long-wavelength side of the stop band, respectively, while the structural color coating with 0.15 wt% of CB has only 22.0 and 7.0% reflectivity in the stop band and on the long-wavelength side of the stop band, decreases of 57 and 67%, respectively. The structural color coatings from polymer spheres 200 and 280 nm in diameter show the same phenomenon. Photographs show that the structural color coatings without CB have low chroma and weak colors that do not match the vivid colors of natural species. Doping CB into the structural color coatings can enhance the chroma, causing brilliant colors (see photographs). For ideal, defect-free CCs (Colloidal crystals), the light must have very strong reflectance in the stop band but very high transmittance in the other wavelengths from the point of view of physical foundation if it is to display brilliant structural colors. However, as stated earlier, certain unavoidable defects in the artificial CC not only decrease the reflectivity of the stop band but also scatter the

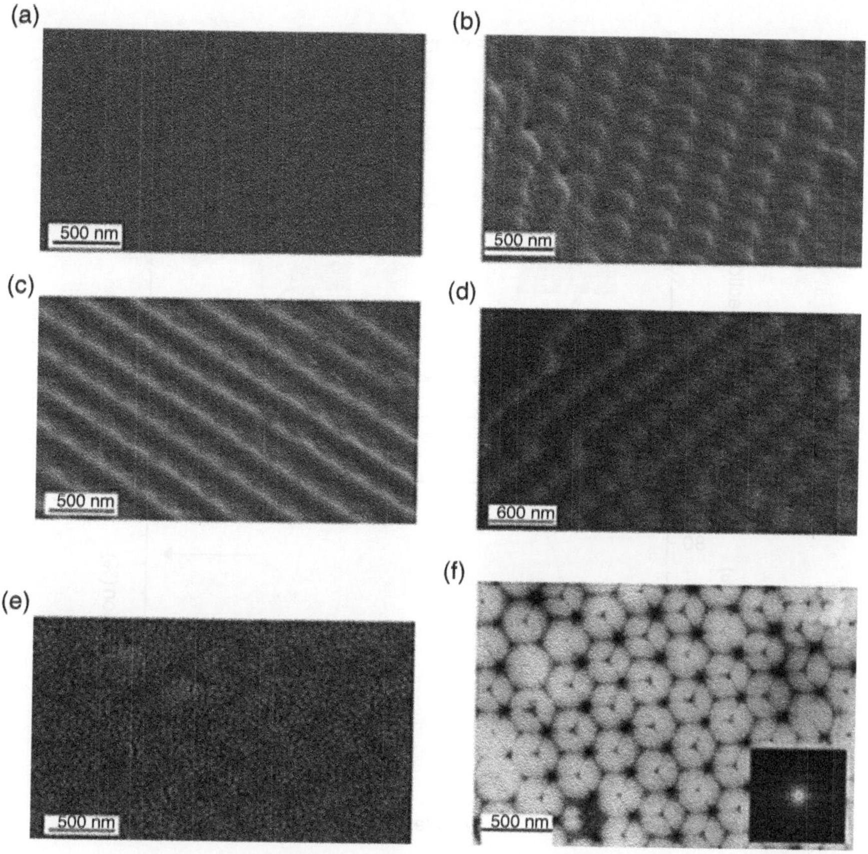

FIG. 9.12 Cross-section SEM images of structural color coatings prepared by binary colloidal assembly with soft polymer spheres and various nanosilica concentrations: (a) pure polymer, (b) 2 wt%, (c) 5 wt%, (d) 10 wt%, and (e) 20 wt%. (f) A typical TEM image of structural color coatings. Reproduced from the permission of Ref. 58. © Royal Society of Chemistry.

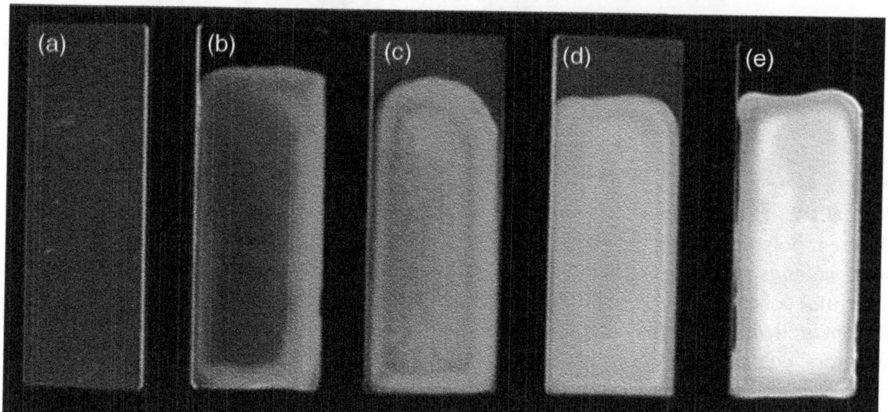

FIG. 9.13 Photographs of structural color coatings prepared by binary colloidal assembly with soft polymer spheres and various nanosilica concentrations: (a) pure polymer, (b) 2 wt%, (c) 5 wt%, (d) 10 wt%, and (e) 20 wt% with colors as transparent, weak blue, blue, white-blue and white from left to right. Reproduced from the permission of Ref. 58. © Royal Society of Chemistry.

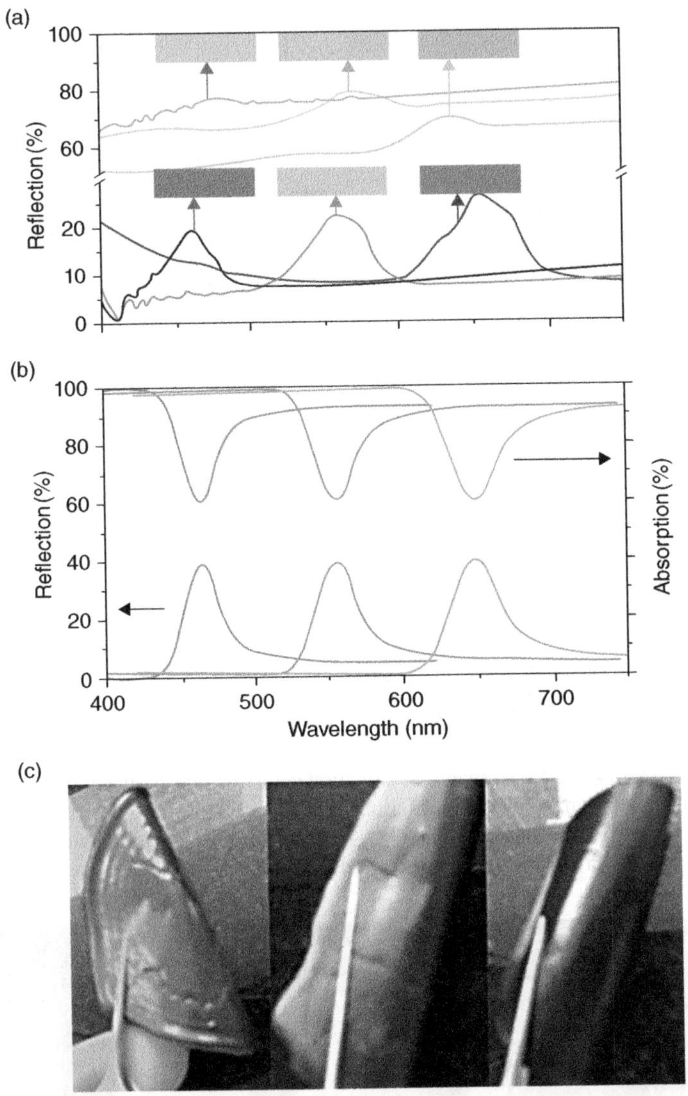

FIG. 9.14 Experimental and theoretical reflection spectra of structural color coatings with different polymer sphere diameters. (a) Top: polymer/silica structural color coatings with weak color as purple, green and red from left to right; bottom: polymer/silica/carbon black structural color coatings with brilliant color as purple, green and red from left to right. (b) Theoretical simulated absorption and reflection spectra of polymer/silica/carbon black structural color coatings. (c) Bent polymer/silica/carbon black structural color coatings with color as purple, green and red from left to right. Reproduced from the permission of Ref. 60. © Royal Society of Chemistry.

long- and short-wavelength light outside the stop band. In this way, the light that is usually observed not only contains the wavelength corresponding to the stop band but also unwanted wavelengths outside the stop band. Low chroma and milky white colors appear. When CB is doped, the overall reflectivity is reduced, but the unwanted wavelengths, which are usually scattered by defects, are absorbed by the CB to less than 10% reflectivity. Because the color saturation of the polymer films is dependent upon the contrast of reflectance intensity between the stop band and the wavelength outside the stop band rather than the absolute intensity of reflectance, the more the strength of the reflectance of the stop band exceeds that outside the stop band, the sharper the colors. The chroma of structural colors can also be evaluated quantitatively using the contrast between the reflection of the stop band and the wavelength side of the stop band, namely, the height of reflectance peak. The higher the peak, the greater the chroma of reflected light in the band gap and the more brilliant the visible colors. The peak heights of the polymer/silica/CB structural color coatings made of polymers of 200, 241, and 280 nm diameter are 12, 14, and 18%, respectively, which is much higher than the 1.0, 5.0, and 4.0% of their corresponding polymer/silica structural color coatings. In this way, introduction of CB dopant to the structural color coatings can markedly increase the chroma, producing very brilliant structural colors visible under natural lighting conditions, as shown by the large and robust structural color coatings in Fig. 9.14c.

Another interesting phenomenon is that the strong absorption characteristics of CB are very different in structural color coatings. The homogeneous CB system should cause black films due to its strong broad-spectrum absorption. However, when CB is doped into the structural color coatings, its absorption in the stop band becomes very low because the photonic density of state in the stop band is very low. To confirm this, three kinds of structural color coatings with different-sized polymers were produced using a modified Lay-KKR method [61]. In the simulation, the effective refractive index of matrix and polymer spheres is 1.15 and 1.53 according to experiment. For the simulated absorption of CB particles in the matrix, the imaginary part of refractive index of matrix is 0.15, thus the theoretical reflection and absorption spectra can be shown in Fig. 9.14b. Obviously, the theoretical reflection spectra can match the experimental spectra very well but also differ from the actual absorption spectra of CB. Thus, it can be inferred that CB in the CC films has very low photon-absorption ability in the stop band region but very strong absorption outside the stop band. It is this characteristic of CB in the CC films that the doped CB can restrain the light scattering of the unwanted wavelength effectively and keep the low absorption in the stop band. This can ensure that the CB–doped CC films have relative stronger reflectance in the stop band, enhancing the chroma of structural colors.

9.4 USES OF STRUCTURAL COLORS

9.4.1 Photonic Paper

As an important technical extension of responsive PCs, photonic paper and ink have recently become prevalent in both academic research and the marketplace [62–64]. Xia *et al.* demonstrated a photonic paper/ink system where color patterns could be conveniently generated on the surface of a thin film of colloidal crystal by writing with a Pilot pen, by screen printing, or by microcontact printing with an elastomer stamp. Such a colloidal crystal was generated by infiltrating the voids within an opaline lattice of polystyrene beads with a liquid prepolymer to poly(dimethylsiloxane), followed by thermal curing. Depending on how the rubber stamp was fabricated, colored patterns with an edge resolution as high as 50 μm could be routinely achieved. Depending on the vapor pressure of the ink liquid, the colored patterns could be programmed to last for intervals of time ranging from a few seconds to hundreds of hours. [63] Ge *et al.* produced a flexible photonic paper through the instant magnetic assembly of Fe_3O_4–SiO_2 core-shell colloids after a rapid photopolymerization to fix the photonic structures inside the poly(ethylene glycol) diacrylate (PEGDA) matrix (Fig. 9.15) [64]. The colloids formed chain-like structures with regular interparticle spacing of a few hundred nanometers along the external magnetic field so that the photonic paper strongly diffracted visible light. Solutions of LiCl, $MgCl_2$, and $CaCl_2$ were used as ink to print durable letters and patterns with color contrast based on the bumps in the polymer matrix (Based on Eq. 9.2, these solutions in polymer matrix can change d and n_{avg} values, thus the color changes as the result of variation in

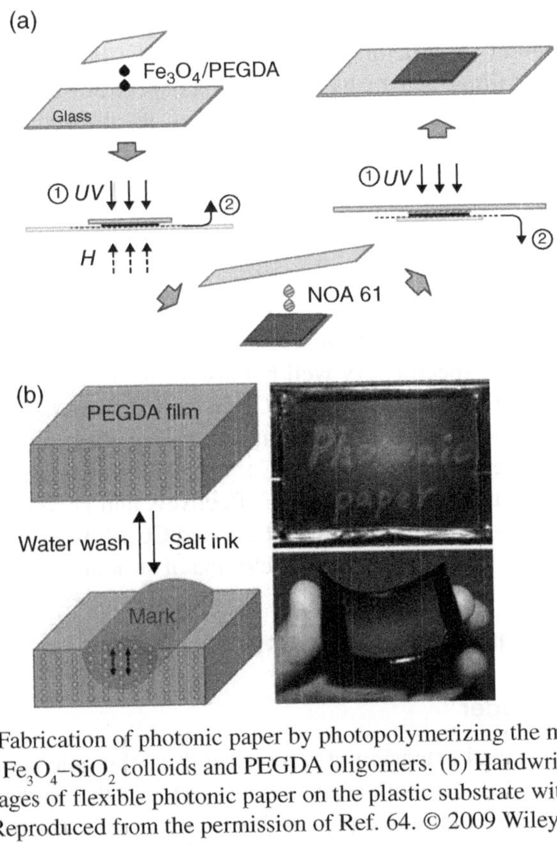

FIG. 9.15 (a) Fabrication of photonic paper by photopolymerizing the mixture of magnetically assembled Fe_3O_4–SiO_2 colloids and PEGDA oligomers. (b) Handwriting display and digital photo images of flexible photonic paper on the plastic substrate with and without letters printed. Reproduced from the permission of Ref. 64. © 2009 Wiley.

λ value). The photonic paper was rewritable and benign to the environment because the ink marks can be removed simply by washing with water.

9.4.2 Coloring and Protection of Substrates

As one of the most important properties of colloidal PCs, structural color has applications in coloring due to their brilliant and colorful appearance [65–69]. Shen *et al.* used structural color coatings on bamboo strand board (BSB) [69]. Figure 9.16a shows an unfinished BSB with a simple and homogenous surface. This limits its usefulness in interior decorations. Figure 9.16b and c show the BSBs coated with structural color coatings without CB films and with CB under natural conditions, respectively. CB doping gives the colors higher saturation under natural conditions, rendering the BSBs more useful in interior decoration and furniture. Structural color coatings with CB present not only vivid colors but also good protection of BSB against water wetting and weathering and endow BSB with good scratching resistance [70].

The structural colors of coatings on BSB can also be controlled by the sizes of the polymer spheres (Fig. 9.17). In addition, structural color coatings were also used successfully on the glass and polypropylene substrates (Fig. 9.18). However, there was less adhesion between coatings and glass or polypropylene substrates than between coatings and BSB. This was due to the strong molecular attractions such as hydrogen

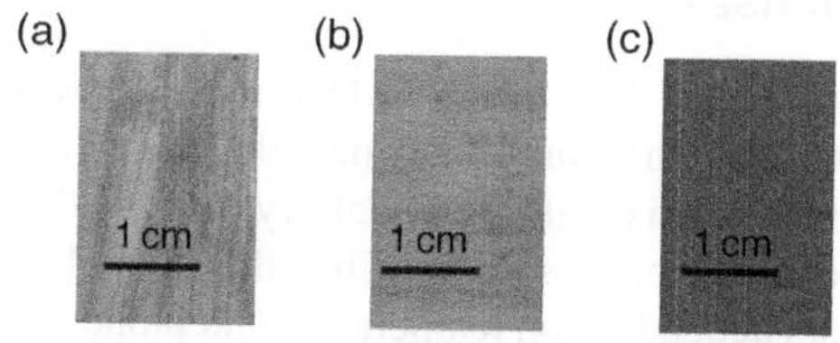

FIG. 9.16 Photographs of (a) unfinished BSB and finished BSB with red structural color coating (b) without carbon black and (c) with carbon black. Reproduced from the permission of Ref. 70. © Hindawi Publishing Corporation.

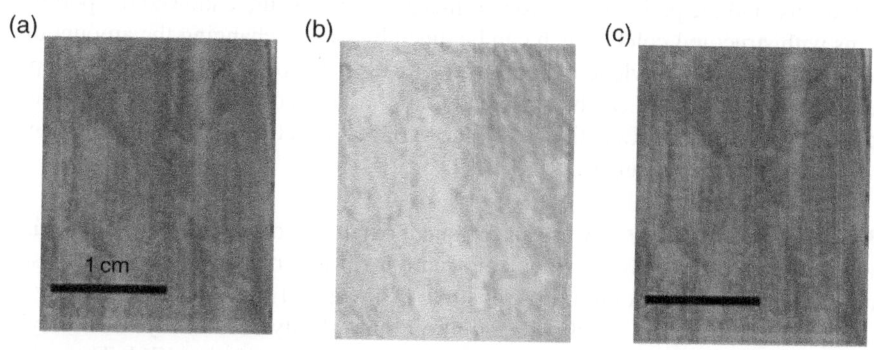

FIG. 9.17 Photographs of structural color coatings on BSB prepared from polymer spheres in diameters of (a) 205 nm for blue, (b) 241 nm for green, and (c) 265 nm for red. Reproduced from the permission of Ref. 69. © Royal Society of Chemistry.

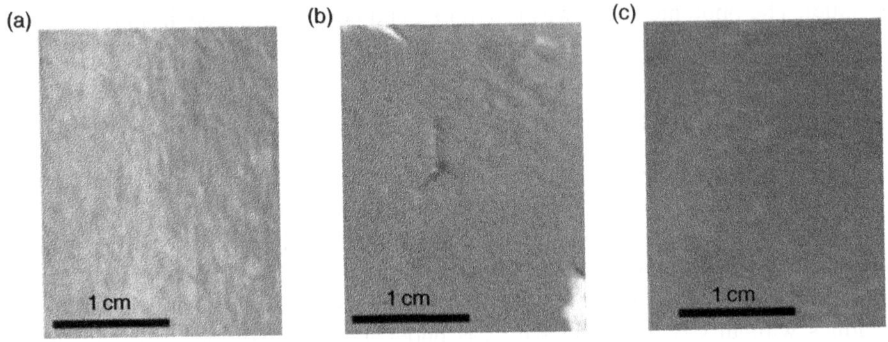

FIG. 9.18 Photographs of green structural color coatings on different substrates: (a) BSB, (b) glass, and (c) polypropylene. Reproduced from the permission of Ref. 69. © Royal Society of Chemistry.

bonds and van der Waals forces between coatings and BSB. There was so little adhesion between structural color coating and polypropylene that the film could be peeled off by hand. For practical applications, more attention should be paid to the modification of various functional groups on the surface of structural color coating.

9.4.3 Color Responses

9.4.3.1 To Temperatures Polymer-based thermosensitive PCs were probably the successful examples of using structural colors of colloidal PCs that combine a functional and responsive polymer matrix with colloidal crystals [71–73]. Takeoka *et al.* presented the synthesis of smart porous PC gel films that could be induced to exhibit different optical behaviors depending on temperature and monomer ratio (Fig. 9.19) [72]. A periodically ordered interconnecting porous structure could be created in the gels using a closest-packing silica CC as a template. The interconnecting porosity provided fast response to changes in temperature through the reversible swelling and shrinking of the gels, and the periodically ordered mesoscopic structure endowed the porous gel films with structural color, which can be tuned by simply changing the amount of the cross-linker (N,N-methylene-bis-acrylamide, BIS) in the pre-gel solutions. The optical response of temperature can be attributed to the variation of lattice spacing affected by temperature. Based on Equation (9.2), the variation of lattice spacing d can change λ value, thus the color changes.

9.4.3.2 To Humidity A useful technical extension of responsive systems is the humidity sensor, which provides information of water vapor content in gaseous atmospheres [74–77]. A humidity-sensitive PC hydrogel film was fabricated by Tian *et al.*, using infiltrating acrylamide solution into a P(Styrene–methyl methacrylate-acrylic acid) PC template and subsequently photopolymerization [76]. The color of the samples was sensitive to humidity: They could reversibly change from transparent to violet, blue, cyan, green, and red at various levels of humidity, covering the

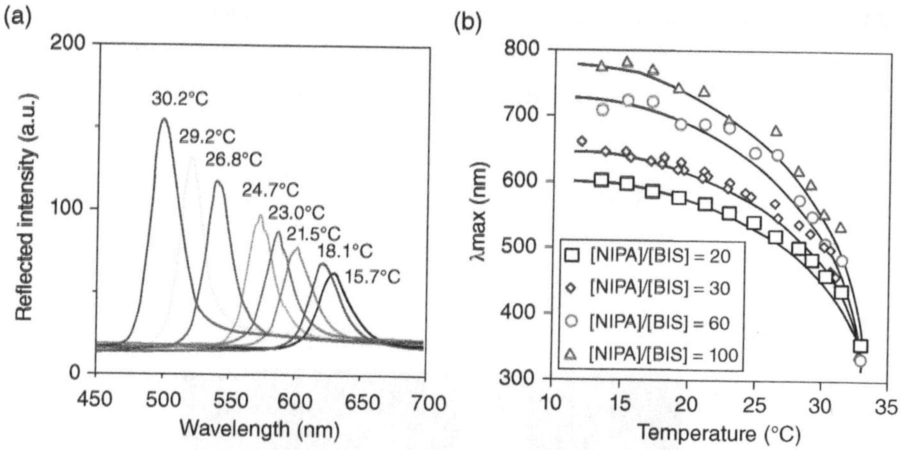

FIG. 9.19 (a) Temperature dependence of the reflection spectra of the porous *N*-isopropylacrylamide photonic crystal gel film. (b) Peak wavelength of the reflection spectra of the porous photonic crystal gel films as a function of temperature. Reproduced from the permission of Ref. 72. © 2003 American Chemical Society.

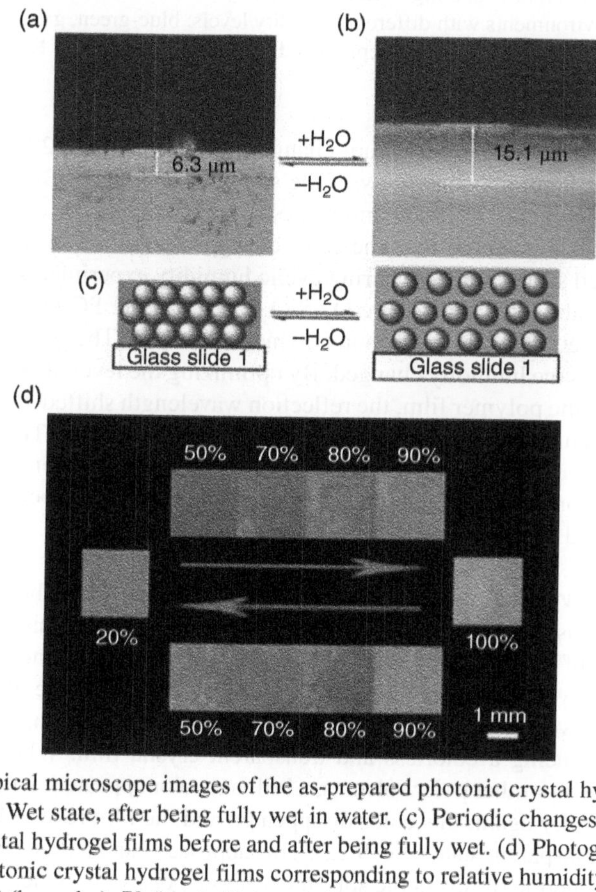

FIG. 9.20 Typical microscope images of the as-prepared photonic crystal hydrogel films (a) dry state. (b) Wet state, after being fully wet in water. (c) Periodic changes in the structure of photonic crystal hydrogel films before and after being fully wet. (d) Photographs of the as-prepared photonic crystal hydrogel films corresponding to relative humidity levels of 20 (transparent), 50 (lavender), 70 (blue), 80 (cyan), 90 (green), and 100% (red), respectively. Reproduced from the permission of Ref. 76. © 2008 Royal Society of Chemistry.

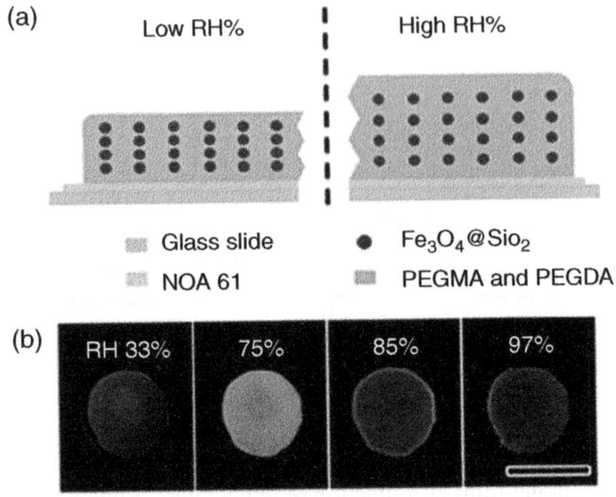

FIG. 9.21 (a) Structure of the humidity-sensing film and the mechanism by which the color changes between low- and high-relative humidity environments. (b) Digital photos of a typical film in environments with different humidity levels: blue-green, green, yellow, and red are visible. The scale bar is 1 cm. Reproduced from the permission of Ref. 77. © 2011 Royal Society of Chemistry.

whole visible range, with a maximum change of 240 nm (Fig. 9.20). This could be attributed to the humidity-sensitivity of the sample's stop band. Based on Equation (9.2), different contents of water molecules in polymer matrix can bring various changes in d and n_{avg} values, thus the color changes due to the shift of λ. The color response showed strong stability during cyclic humidity experiments.

Xuan *et al.* fabricated a novel Fe_3O_4–SiO_2/PEG acrylate PC humidity sensor by combining magnetic assembly and photopolymerization [77]. The sensor self-displayed brilliant colors as the humidity changed. By optimizing the level of cross-linkage and the thickness of the polymer film, the reflection wavelength shifted 160 nm in several minutes, and its intensity was visible to the naked eye (Fig. 9.21). The tested sensors demonstrated reversibility and durability during cycling experiments. Such PC sensors may have considerable potential as energy-efficient devices because no electric power is required for their operation.

9.4.3.3 *To Solvents* Wu *et al.* presented a facile and novel method for fabrication of solvent-responsive polymer colloidal crystal films [76, 78]. In this approach, when the suspension of monodisperse poly(styrene-butyl acrylate) polymer colloidal particles containing hydrophilic monomers was cast on a glass substrate, the polymer colloidal particles were found to self-assemble into a periodic arrangement as water evaporated, producing a colorless and transparent crystal film. When this film was wetted with water or other polar solvents, a brilliant color appeared; when this wet film was completely dry again, this color disappeared and the film reverted to the original colorless and transparent state (Fig. 9.22). This approach can be used to produce rapid, large-scale production of reversible solvatochrom-responsive polymer films.

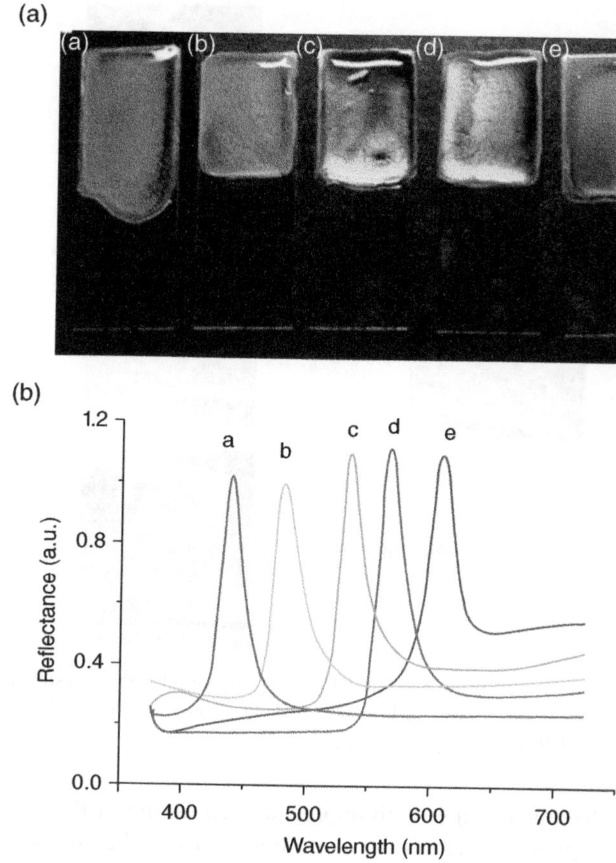

FIG. 9.22 Photograph and corresponding reflection spectra of the wet crystal films with polymer latexes of various particle sizes: (a) 159, (b) 195, (c) 215, (d) 253, and (e) 292 nm with corresponding color as purple, blue, grey green, green and red from left to right. Reproduced from the permission of Ref. 78. © Royal Society of Chemistry.

9.4.3.4 To Electricity The first demonstration of full-color tuning of a PC was based on an electroactive polymer-gel/silica opal composite whose reflected color could be electrically tuned through reversible expansion and contraction of its photonic lattice [79]. The problem with this system is the difficulty with which the electrolytes permeate through a contiguous space-filling opal lattice made of close-packed silica spheres embedded within a polymer-gel matrix. This construct impedes electron and ion charge transport, slows switching times, and increases the drive voltage needed to power the device, together negating the overall performance of the device [14].

Ozin *et al.* fabricated the first example of a high-performance electroactive inverse polymer-gel opal prepared by removing the silica colloids in the SiO_2/polyferrocenylsilane film and infusing electrolytes freely into the nanoporous lattice to shorten the response rate and increase the tuning range. The desired outcome is the reductions in electron and ion diffusion lengths, increase in switching speed, and

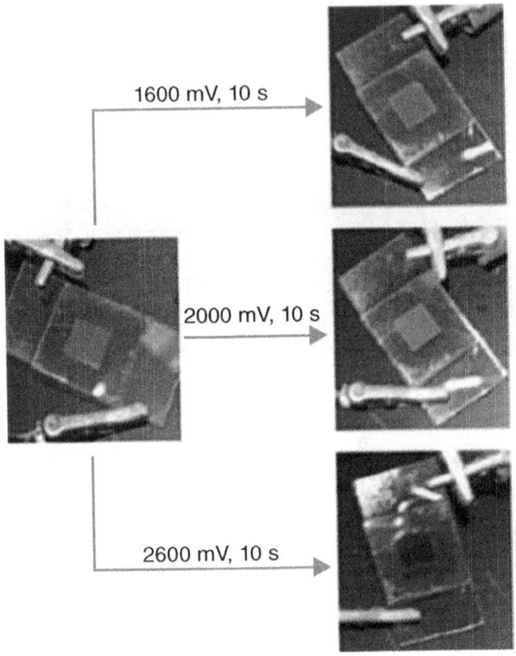

FIG. 9.23 Full-color tuning of an electric RPC (responsive photonic crystal) in an electrochemical cell with green, red and brown for 1600, 2000 and 2600 mV. Reproduced from the permission of Ref. 79. © 2011 Wiley.

decrease in the driving voltage, with unprecedented tuning of the wavelength and brightness of Bragg diffracted light continuously from the invisible ultraviolet through the visible to the invisible near infrared spectra. The voltage-dependent photonic properties make electrical PCs a natural choice for electrically responsive reflective display units. Figure 9.23 shows the color response of inverse opals within a voltage range from 1600 to 2800 mV, which decreases the voltage threshold for driving the device [79].

9.4.4 Structural Color Coatings with Lotus Effects and Superhydrophilicity

Colloidal PCs with controllable wettability could be achieved by either coating the hydrophilic or the hydrophobic materials upon the colloidal crystals or further modification of the roughness structure of the films [80–83]. Gu *et al.* succeeded in fabricating structural color coatings from inverse opal films with superhydrophobic properties using nanoparticles with the lifting method [84]. The contact angle (CA) of the structural color films is 155°. They also fabricated structural color films with superhydrophilic properties by preparing inverse opal films using TiO_2, which could be used as self-cleaning pigments (Fig. 9.24) [83]. Wang *et al.* demonstrated a facile strategy for finely controlling the wettability transition temperature of

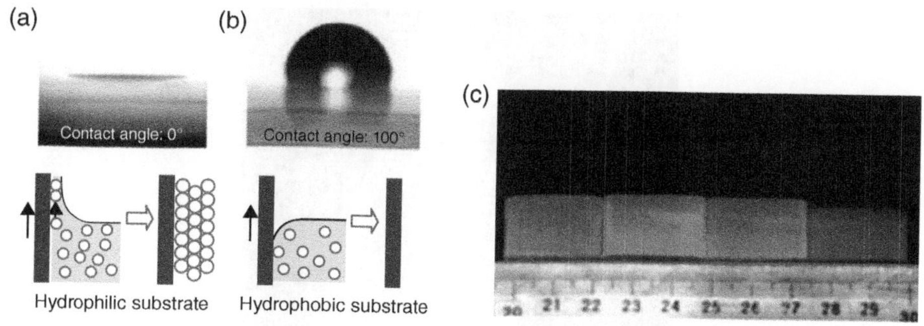

FIG. 9.24 (a) When its surface was hydrophilic, CC films formed on the substrate. (b) When its surface was hydrophobic, CC films did not form on the substrate. (c) Structural color coatings composed of polystyrene spheres of different sizes. The diameters of the spheres are 211, 233, 246, and 263 nm with corresponding color as blue-green, green, yellow and red from left to right, respectively. Reproduced from the permission of Ref. 83. © 2009 American Chemical Society.

colloidal-crystal films from superhydrophilic (water CA, 0°) to superhydrophobic (water CA, 150.5°). The colloidal-crystal films are assembled from poly(styrene-*n*-butyl acrylate–acrylic acid) amphiphilic latex spheres. Superhydrophobic films are achieved when assembled at 90, 80, 70, 60, 40, or even 20°C [82]. Yang *et al.* fabricated angle-independent colored films with superhydrophobicity from quasi-amorphous arrays of monodispersed fluorinated silica nanoparticles via one-step spray coating. The film exhibited a high CA (>150°C) and a low roll-off angle (~2°C) and the color could be tuned to blue, green, and moccasin by varying the size of the nanoparticles [44, 45, 81].

9.4.5 Structural Color as Effect Pigments

Usually colorants and color pigments rely on absorption of electromagnetic radiation in the visible spectrum, which evolve pollution and energy loss. An alternative possibility to the creation of colors is selective reflection by nanostructured materials, namely, structural colors [66, 85–87]. Allard *et al.* provided approach to building effect pigments directly on a porous substrate (especially paper) by spraying a colloidal dispersion (of polymer nanospheres) onto a paper substrate by airbrush technique. For the spraying experiments, polymer colloids with low-size polydispersity (1%) were prepared by emulsion polymerization. This is a novelty because effect pigments in the size range of 50–1000 mm cannot be applied in the form of small droplets [85]. Zentel *et al.* used structural colors from polymer beads as effect pigments in transparent industrial and automotive coatings [86]. Monodisperse colloids from fluorinated methacrylates were synthesized using surfactant-free emulsion polymerization. The fluorinated monomers, in combination with cross-linking, caused contrast in the refractive index, thermal stability, and solvent resistively, all of which are necessary for use as effective pigments. After crystallization of the

FIG. 9.25 Crystal flakes as effect pigments in a transparent coating. The diameter of the tube is 5 mm. Reproduced from the permission of Ref. 86. © 2004 Wiley.

monodisperse polymer beads, crystal flakes with iridescent colors were produced (Fig. 9.25). The crystal flakes could act as effective pigments in various transparent industrial and automotive coatings. Color flops up to 100 nm were observed due to the PC behavior of effective pigments.

9.5 CONCLUSIONS AND OUTLOOK

Compared with previous physical methods, colloidal self-assembly strategies exhibit some obvious advantages to fabricate PCs and structural color films or coatings. By tuning the refractive indexes of two periodic media, the lattice constants, and the filling factors, the colors and their saturation can be easily adjusted.

Although quite some ingenious methods have been developed to fabricate structural color coatings or pigments, some challenging works still need to be further explored to finally actualize the applications of structural color coatings or pigments. They include the improvement of chroma, the preparation of angle-independent structural colors, robust mechanical properties of coatings, etc. Another big challenge is how to massively manufacture and apply these structural color coatings and pigments. This means that large-area production and application can still assure the ordered structure of products.

REFERENCES

1. Marlow, F., Sharifi Muldarisnur, P., Brinkmann, R., Mendive, C. (2009) Opals: Status and Prospects. *Angew. Chem. Int. Ed.* 48: 6212.
2. Choi, K. K., Levine, B. F., Bethea, C. G., Walker, J., Malik, R. J. (1987) Photoexcited Coherent Tunneling in a Double-Barrier Superlattice. *Phys. Rev. Lett.* 59: 2459.

3. John, S. (1987) Strong localization of photons in certain disordered dielectric superlattices. *Phys. Rev. Lett.* 58: 2486.
4. Yamada, H., Nakamura, T., Yamada, Y., Yano, K. (2009) Colloidal-Crystal Laser Using Monodispersed Mesoporous Silica Spheres. *Adv. Mater.* 21: 1.
5. Zhao, X. W., Cao, Y., Ito, F., Chen, H. H., Nagai, K.., Zhao, Y. H., Gu, Z. Z. (2006) Colloidal Crystal Beads as Supports for Biomolecular Screening. *Angew. Chem. Int. Ed.* 45: 6835.
6. Lawrence, J. R., Shim, G. H., Jiang, P., Han, M. G., Ying, Y., Foulger, S. H. (2005) Dynamic Tuning of Photoluminescent Dyes in Crystalline Colloidal Arrays. *Adv. Mater.* 17: 2344.
7. Gu, Z. Z., Horie, R., Kubo, S., Yamada, Y., Fujishima, A., Sato, O. (2002) Fabrication of a Metal-Coated Three-Dimensionally Ordered Macroporous Film and its Application as a Refractive Index Sensor. *Angew. Chem. Int. Ed.* 41: 1153.
8. Park, S. H., Xia, Y. (1999) Assembly of Mesoscale Particles over Large Areas and Its Application in Fabricating Tunable Optical Filters. *Langmuir* 15: 266.
9. Meseguer, F., Blanco, A., Miguez, H., Garcia-Santamaria, F., Ibisate, M., Lopez, C. (2002) Synthesis of inverse opals. *Colloid Surface. Colloid Surf. A* 202: 281.
10. Wang, D. Y., Caruso, R. A., Caruso, F. (2001) Synthesis of Macroporous Titania and Inorganic Composite Materials from Coated Colloidal Spheres–A Novel Route to Tune Pore Morphology. *Chem. Mater.* 13: 364.
11. Velev, O. D., Lenhoff, A. M. (2000) Colloidal crystals as templates for porous materials. *Curr. Opin. Colloid Interface Sci.* 5: 56.
12. Fudouzi, H., Xia, Y. N. (2003) Photonic Papers and Inks: Color Writing with Colorless Materials. *Adv. Mater.* 15: 892.
13. Arsenault, A. C., Juez, H. M., Kitaev, V., Ozin, G. A., Manners, I. (2003) A Polychromic, Fast Response Metallopolymer Gel Photonic Crystal with Solvent and Redox Tunability: A Step Towards Photonic Ink (P-Ink). *Adv. Mater.* 15: 503.
14. Arsenault, A. C., Puzzo, D. P., Manners, I., Ozin, G. A. (2007) Photonic-crystal full-colour displays. *Nat. Photon.* 1: 468.
15. Moon, J. H., Yang, S. (2009) Chemical Aspects of Three-Dimensional Photonic Crystals. *Chem. Rev.* 110: 547.
16. Ho, K. M., Chan, C. T., Soukoulis, C. M. (1990) Existence of a photonic gap in periodic dielectric structures. *Phys. Rev. Lett.* 65: 3152.
17. Yablonovitch, E., Gmitter, T. J., Leung, K. M. (1991) Photonic band structure: The face-centered-cubic case employing nonspherical atoms. *Phys. Rev. Lett.* 67: 2295.
18. Míguez, H., Tétreault, N., Hatton, B., Yang, S. M., Perovicb, D., Ozin, G. A. (2002) Mechanical stability enhancement by pore size and connectivity control in colloidal crystals by layer-by-layer growth of oxide. *Chem. Commun.* 22: 2736.
19. Yue, L., Cai, W. P., Duan, G. T. (2008) Ordered Micro/Nanostructured Arrays Based on the Monolayer Colloidal Crystals. *Chem. Mater.* 20: 615.
20. Mitsuishi, M., Matsui, J., Miyashita, T. (2009) Photofunctional thin film devices composed of polymer nanosheet assemblies. *J. Mater. Chem.* 19: 325.
21. Debord, J. D., Lyon, L. A. (2000) Thermoresponsive Photonic Crystals. *J. Phys. Chem. B* 104: 6327.
22. Qu, X. S., Yang, H. K., Moon, B. K., Choi, B. C., Jeong, J. H., Kim, K. H. (2010) Preparation and Photoluminescence Properties of Gd_2O_3:Eu^{3+} Inverse Opal Photonic Crystals. *J. Phys. Chem. C* 114: 19891.

23. Li, Y., Sasaki, T., Shimizu, Y., Koshizaki, N. (2008) Hexagonal-Close-Packed, Hierarchical Amorphous TiO2 Nanocolumn Arrays: Transferability, Enhanced Photocatalytic Activity, and Superamphiphilicity without UV Irradiation. *J. Am. Chem. Soc.* 130: 14755.
24. Fan, H. Y., Lu, Y. F., Stump, A., Reed, S. T., Baer, T., Schunk, R., Perez-Luna, V., LoÂpez, G. P., Jeffrey Brinker, C. (2000) Rapid prototyping of patterned functional nanostructures. *Nature* 405: 56.
25. Divliansky, I. B., Shishido, A., Khoo, I.-C., Mayer, T. S., Pena, D., Nishimura, S., Keating, C. D., Mallouk, T. E. (2001) Fabrication of two-dimensional photonic crystals using interference lithography and electrodeposition of CdSe. *Appl. Phys. Lett.* 79: 3392.
26. Chien, F. S.-S., Wu, C.-L., Chou, Y.-C., Chen, T. T., Gwob, S. (1999) Nanomachining of (110)-oriented silicon by scanning probe lithography and anisotropic wet etching. *Appl. Phys. Lett.* 75: 2429.
27. Mayoral, R., Requena, J., Moya, J. S., López, C., Cintas, A., Miguez, H., Meseguer, F., Vázquez, L., Holgado, M., Blanco, Á. (1997) 3D Long-range ordering in ein SiO_2 submicrometer-sphere sintered superstructure. *Adv. Mater.* 9: 257.
28. Johnson, N. P., McComb, D. W., Richel, A., Treble, B. M., De La Rue, R. M. (2001) Synthesis and optical properties of opal and inverse opal photonic crystals. *Synthetic Met.* 116: 469.
29. Ko, Y. G., Shin, D. H. (2007) Effects of Liquid Bridge between Colloidal Spheres and Evaporation Temperature on Fabrication of Colloidal Multilayers. *J. Phys. Chem. B* 111: 1545.
30. Allard, M., Sargent, E. H. (2004) Impact of polydispersity on light propagation in colloidal photonic crystals. *Appl. Phys. Lett.* 85: 5887.
31. Stöber, W., Fink, A., Bohn, E. (1968) Controlled growth of monodisperse silica spheres in the micron size range. *J. Colloid Interface Sci* 26: 62.
32. Liu, G. Q., Wang, Z. S., Ji, Y. H. (2010) Influence of growth parameters on the fabrication of high-quality colloidal crystals via a controlled evaporation self-assembly method. *Thin Solid Films* 518: 5083.
33. Nagao, D., Kameyama, R., Matsumoto, H., Kobayashi, Y., Konno, M. (2008) Single- and multi-layered patterns of polystyrene and silica particles assembled with a simple dip-coating. *Colloid Surf. A* 317: 722.
34. Kitaev, V., Ozin, G. A. (2003) Self-Assembled Surface Patterns of Binary Colloidal Crystals. *Adv. Mater.* 15: 75.
35. Yin, Y., Lu, Y., Gates, B., Xia, Y. (2001) Template-assisted self-assembly: a practical route to complex aggregates of monodispersed colloids with well-defined sizes, shapes, and structures. *J. Am. Chem. Soc.* 123: 8718.
36. Jiang, P., McFarland, M. J. (2004) Large-Scale Fabrication of Wafer-Size Colloidal Crystals, Macroporous Polymers and Nanocomposites by Spin-Coating. *J. Am. Chem. Soc.* 126: 13778.
37. Ozin, G. A., Yang, S. M. (2001) The Race for the Photonic Chip: Colloidal Crystal Assembly in Silicon Wafers. *Adv. Funct. Mater.* 11: 95.
38. Xu, Y., Schneider, G. J., Wetzel, E. D., Prather, D. W. (2004) Centrifugation and spin-coating method for fabrication of three-dimensional opal and inverse-opal structures as photonic crystal devices. *J. Micro/Nanolithogr. MEMS MOEMS*, 3: 168.
39. Wang, D., Möhwald, H. (2004) Rapid Fabrication of Binary Colloidal Crystals by Stepwise Spin-Coating. *Adv. Mater.* 16: 244.

40. Kim, H. J., Krane, M. J. M., Trumble, K. P., Bowman, K. J. (2006) Analytical Fluid Flow Models for Tape Casting. *J. Am. Ceram. Soc.* 89: 2769.

41. Yang, H., Jiang, P. (2010) Large-Scale Colloidal Self-Assembly by Doctor Blade Coating. *Langmuir* 26: 13173.

42. Yang, H., Jiang, P. (2011) Macroporous photonic crystal-based vapor detectors created by doctor blade coating. *Appl. Phys. Lett.* 98: 011104.

43. Yang, H., Jiang, P. (2010) Self-Cleaning Diffractive Macroporous Films by Doctor Blade Coating. *Langmuir* 26: 12598.

44. Ge, D., Yang, L., Wu, G., Yang, S. (2014) Angle-independent colours from spray coated quasi-amorphous arrays of nanoparticles: combination of constructive interference and Rayleigh scattering. *J. Mater. Chem. C* 2: 4395.

45. Zhang, Y., Ge, D., Yang, S. (2014) Spray-coating of superhydrophobic aluminum alloys with enhanced mechanical robustness. *J. Colloid Interface Sci.* 423: 101.

46. Denkov, N. D., Velev, O. D., Kralchevsky, P. A., Ivanov, I. B., Yoshimura, H., Nagayama, K. (1993) Two-dimensional crystallization. *Nature* 361: 26.

47. Cui, L., Zhang, Y., Wang, J., Ren, Y., Song, Y., Jiang, L. (2009) Ultra-Fast Fabrication of Colloidal Photonic Crystals by Spray Coating. *Macromol. Rapid. Commun.* 30: 598.

48. Harris, M., Basaran, O. A. (1993) Capillary Electrohydrostatics of Conducting Drops Hangning from a Nozzle in an Electric Field. *J. Colloid Interface Sci.* 161: 389.

49. Holgado, M., García-Santamaría, F., Blanco, A., Ibisate, M., Cintas, A., Míguez, H., Serna, C. J., Molpeceres, C., Requena, J., Mifsud, A., Meseguer, F., López, C. (1999) Electrophoretic Deposition To Control Artificial Opal Growth. *Langmuir* 15: 4701.

50. Vlasov, Y. A., Astratov, V. N., Baryshev, A. V., Kaplyanskii, A. A., Karimov, O. Z., Limonov, M. F. (2000) Manifestation of intrinsic defects in optical properties of self-organized opal photonic crystals. *Phys. Rev. E.* 61: 5784.

51. Besra, L., Liu, M. (2007) A review on fundamentals and applications of electrophoretic deposition (EPD). *Prog. Mater. Sci.* 52: 1.

52. Ge, J., Hu, Y., Yin, Y. (2007) Highly Tunable Superparamagnetic Colloidal Photonic Crystals. *Angew. Chem. Int. Ed.* 46: 7428.

53. Ge, J., Hu, Y., Zhang, T., Huynh, T., Yin, Y. (2008) Self-assembly and field-responsive optical diffractions of superparamagnetic colloids. *Langmuir* 24: 3671.

54. Ding, T., Song, K., Clays, K., Tung, C.-H. (2009) Fabrication of 3D Photonic Crystals of Ellipsoids: Convective Self-Assembly in Magnetic Field. *Adv. Mater.* 21: 1936.

55. Tian, E. T., Cui, L. Y., Wang, J. X., Song, Y. L., Jiang, L. (2009) Tough Photonic Crystals Fabricated by Photo-Crosslinkage of Latex Spheres. *Macromol. Rapid Commun.* 30: 509.

56. McGrath, J. G., Bock, R. D., Cathcart, J. M., Lyon, L. A. (2007) Self-Assembly of "Paint-On" Colloidal Crystals Using Poly(styrene-co-N-isopropylacrylamide) Spheres. *Chem. Mater.* 19: 1584.

57. Wang, J., Wen, Y., Ge, H., Sun, Z., Zheng, Y., Song, Y., Jiang, L. (2006) Simple Fabrication of Full Color Colloidal Crystal Films with Tough Mechanical Strength. *Macromol. Chem. Phys.* 207: 596.

58. You, B., Wen, N., Shi, L., Wu, L. M., Zi, J. (2009) Facile fabrication of a three-dimensional colloidal crystal film with large-area and robust mechanical properties. *J. Mater. Chem.* 19: 3594.

59. You, B., Wen, N., Zhou, S. X., Wu, L. M., Zhao, D. (2008) Facile Method for Fabrication of Nanocomposite Films with an Ordered Porous Surface. *J. Phys. Chem. B* 112: 7706.
60. Shen, Z., Shi, L., You, B., Wu, L., Zhao, D. (2012) Large-scale fabrication of three-dimensional ordered polymer films with strong structure colors and robust mechanical properties. *J. Mater. Chem.* 22: 8069.
61. Stefanou, N., Yannopapas, V., Modinos, A. (1998) Heterostructures of photonic crystals: frequency bands and transmission coefficients. *Comput. Phys. Commun.* 113: 49.
62. Ozin, G. A., Arsenault, A. C. (2008) P-Ink and Elast-Ink from lab to market. *Mater. Today* 11: 44.
63. Fudouzi, H., Xia, Y. N. (2003) Colloidal Crystals with Tunable Colors and Their Use as Photonic Papers. *Langmuir* 19: 9653.
64. Ge, J., Goebl, J., He, L., Lu, Z., Yin, Y. (2009) Rewritable Photonic Paper with Hygroscopic Salt Solution as Ink. *Adv. Mater.* 21: 4259.
65. Pursiainen, O. L., Baumberg, J. J., Winkler, H., Viel, B., Spahn, P., Ruhl, T. (2007) Nanoparticle-tuned structural color from polymer opals. *Optic. Express* 15: 9553.
66. Lange, B., Fleischhaker, F., Zentel, R. (2007) Functional 3D photonic films from polymer beads. *Phys. Status Solidi A* 204: 3618.
67. Rastogi, V., Melle, S., Calderon, O. G., García, A. A., Marquez, M., Velev, O. D. (2008) Synthesis of Light-Diffracting Assemblies from Microspheres and Nanoparticles in Droplets on a Superhydrophobic Surface. *Adv. Mater.* 20, 4263.
68. Shen, Z. H., Zhu, Y., Wu, L. M., You, B., Zi, J. (2009) Fabrication of Robust Crystal Balls from the Electrospray of Soft Polymer Spheres/Silica Dispersion. *Langmuir* 26: 6604.
69. Shen, Z., Yang, Y., Lu, F., Bao, B., You, B. (2012) Self-assembly of binary particles and application as structural colors. *Polym. Chem.* 3: 2495.
70. Bao, B., Liu, D., Yang, Y., Shen, Z., You, B. (2013) Self-Assembly of Ternary Particles for Tough Colloidal Crystals with Vivid Structure Colors. *J. Nanomater.* 2013: 14.
71. Kanai, T., Lee, D., Shum, H. C., Shah, R. K., Weitz, D. A. (2010) Gel-Immobilized Colloidal Crystal Shell with Enhanced Thermal Sensitivity at Photonic Wavelengths. *Adv. Mater*. 22: 1521.
72. Takeoka, Y., Watanabe, M. (2003) *Langmuir* 19: 9104.
73. Weissman, J. M., Sunkara, H. B., Albert, S. T., Asher, S. A. (1996) Thermally switchable periodicities and diffraction from mesoscopically ordered materials. *Science* 274: 959.
74. Barry, R. A., Wiltzius, P. (2006) Humidity-sensing inverse opal hydrogels. *Langmuir* 22: 1369.
75. Karaman, M., Kooi, S. E., Gleason, K. K. (2008) Vapor deposition of hybrid organic–inorganic dielectric Bragg mirrors having rapid and reversibly tunable optical reflectance. *Chem. Mater.* 20: 2262.
76. Tian, E., Wang, J., Zheng, Y., Song, Y., Jiang, L., Zhu, D. (2008) Colorful humidity sensitive photonic crystal hydrogel. *J. Mater. Chem.* 18: 1116.
77. Xuan, R., Wu, Q., Yin, Y., Ge, J. (2011) Magnetically assembled photonic crystal film for humidity sensing. *J. Mater. Chem.* 21: 3672.
78. Duan, L., You, B., Zhou, S., Wu, L. (2011) Self-assembly of polymer colloids and their solvatochromic-responsive properties. *J. Mater. Chem.* 21: 687.
79. Puzzo, D. P., Arsenault, A. C., Manners, I., Ozin, G. A. (2009) Electroactive Inverse Opal: A Single Material for All Colors. *Angew. Chem. Int. Ed.* 48: 943.

80. Zhang, Y., Wang, J., Huang, Y., Song, Y., Jiang, L. (2011) Jiang, Fabrication of functional colloidal photonic crystals based on well-designed latex particles. *J. Mater. Chem.* 21: 14113.
81. Ge, D., Yang, L., Wu, G., Yang, S. (2014) Spray coating of superhydrophobic and angle-independent coloured films. *Chem. Commun.* 50: 2469.
82. Wang, J. X., Wen, Y. Q., Hu, J. P., Song, Y. L., Jiang, L. (2007) Fine control of the Wettability transition temperature of colloidal-crystal films: From superhydrophilic to superhydrophobic. *Adv. Funct. Mater.* 17: 219.
83. Sato, O., Kubo, S., Gu, Z.-Z. (2008) Structural Color Films with Lotus Effects, Superhydrophilicity, and Tunable Stop-Bands. *Acc. Chem. Res.* 42: 1.
84. Gu, Z.-Z., Uetsuka, H., Takahashi, K., Nakajima, R., Onishi, H., Fujishima, A., Sato, O. (2003) Structural Color and the Lotus Effect. *Angew. Chem. Int. Ed.* 42: 894.
85. Allard, D., Lange, B., Fleischhaker, F., Zentel, R., Wulf, M. (2005) Opaline effect pigments by spray induced self-assembly on porous substrates. *Soft Mater.* 3: 121.
86. Marc, E., Braun, L., Zentel, R., Tännert, K., Frese, P., Reis, O., Wulf, M. (2004) Artificial Opals as Effect Pigments in Clear-Coatings. *Macromol. Mater. Eng.* 289: 158.
87. Aguirre, C. I., Reguera, E., Stein, A. (2010) Colloidal Photonic Crystal Pigments with Low Angle Dependence. *ACS Appl. Mater. Interfaces* 2: 3257.

CHAPTER 10

Antibacterial Polymers and Coatings

Jamil Baghdachi and Qinhua Xu

Coatings Research Institute, Eastern Michigan University, Ypsilanti, MI, USA

10.1 INTRODUCTION

Improving hygienic standards in many parts of the world allows infectious diseases to be effectively controlled. About two million people acquire bacterial infections in the US hospitals, with 90,000 deaths reported each year [1]. In addition, almost half of the people in developing countries suffer from water-related diseases and more than three million people die annually from illnesses associated with unsafe drinking water due to the lack of disinfection [2].

Many infectious diseases can be acquired through casual contact with infected individuals and environmental sources. These environments include hospital and office surfaces, food- and drug-processing facilities, meeting facilities in schools; and private homes. Despite the best available sanitation and use of antimicrobial agents, many infectious diseases are acquired through contact with various surfaces (i.e., fomites). The contamination of fomites occurs through the aerosolization and deposition of various infectious agents (e.g., bacteria, viruses, and fungi) on inanimate objects such as door handles and counter tops. Casual contacts with these fomites spread the infectious agents from one person to another [3, 4].

Routine sanitation by the use of disinfecting agents is fairly successful in combating the spread of some infectious diseases. However, cleaning procedures can be costly and time-consuming and do not offer continuous protection or guard against fresh contamination. Therefore, numerous antimicrobial agents that can effectively inhibit the growth of microorganisms are being developed.

Cleaning supplies, disinfecting agents, and labor are costly, and the constant need to clean takes time away from other productive activities. It would be desirable to reduce these expenses by coating surfaces with antimicrobial coatings that fight fungi and bacteria upon contact. However, the coating is not a completely sterile

Functional Polymer Coatings: Principles, Methods, and Applications, First Edition.
Edited by Limin Wu and Jamil Baghdachi.

environment and will not be effective against some strains of bacteria and viruses, so the need for cleaning will not completely disappear.

There is a major gap in understanding of antimicrobial activities where continuous exposure to contaminated environment exists. One challenge is that the antimicrobial agents will need to be retained at the targeted areas for as long as possible in order to provide sustained antimicrobial activities and should not cause bacterial resistance. In addition, the antimicrobial agents must be resistant for mechanical removal and remain biologically active in the presence of complex chemicals such as variety of oils, protein, aqueous and nonaqueous cleaning products.

Another challenge facing activity, efficacy, and longevity of antimicrobial polymer coating is the multiplicity of the materials used in coatings. Materials such as surfactants that may contain counter ions can effectively bind with antimicrobial agent, neutralizing and rendering it ineffective.

10.2 BASIC CONCEPTS

Antimicrobial and antibacterial agents are materials capable of killing pathogenic microorganisms. Antibacterial agents may be divided into two groups according to their speed of action and residue production: The first group contains those that act rapidly to destroy bacteria but quickly disappear (by evaporation or breakdown) and leave no active residue behind, hence referred to as *non–residue-producing*. Examples of this type are the alcohols, halogen-releasing compounds, peroxides, and aldehydes. The second group consists mostly of newer compounds that leave long-acting residues on the surface to be disinfected and thus have a prolonged action and are referred to as *residue-producing*. Common examples of this group are quaternary ammonium (QA) compounds, bisphenols, triclosan, triclocarban, silver and mercury compounds, and benzalkonium chloride.

Surface modification that effectively kills microbes on contact without releasing a biocide represents a modern and efficient approach toward continuous and permanent sterile materials. Contact active surface modification can be achieved through either through polymerizing an active functional antimicrobial agent along with ordinary monomers or simply by chemical grafting of antimicrobial agents to common polymers. Obviously, in such manner, neither polymers nor antimicrobial agents are released and possibly effective against multiresistant bacteria.

Over the years, it has been known that many factors affect the antimicrobial properties of polymeric biocides. Factors such as molecular weight, spacer length between active sites and the backbone, hydrophobicity and hydrophilicity and their chain length as well as the length of bridge connecting the antimicrobial agent to polymer backbone and in some cases the counter ions play major roles.

Results have generally shown that molecular weights of about 5×10^4 to 5×10^5 are optimal for the required action. Also the longer the chain length extended from the backbone carrying the antimicrobial agent, the higher is the activity. Evidently, longer chains are acting similar to polymer brushes and primarily reside on the surface of the polymer aggregates.

Biocides are widely used in the coating industry to prevent or inhibit bacterial growth in storage containers as well as after application on substrate surfaces [5]. The ideal antimicrobial polymer coating should fulfill the following major requirements: (i) biocidal to a broad spectrum of pathogenic microorganisms; (ii) long-term activity; (iii) stable; (iv) environmentally friendly; (v) easily and inexpensively prepared; and (vi) nonleaching and insoluble in water.

Coatings can be designed to kill or inhibit the growth of bacteria through three mechanisms: (1) coatings that can resist the attachment of bacteria, (2) coatings that release biocides that will kill the bacteria, and (3) coatings that kill bacteria on contact. Coatings can also combine two or more of these mechanisms [6–8]. In all cases, while the ideal outcome would be to completely destroy bacteria or fungi, in practice, the reduction of colonies would still be acceptable.

10.2.1 Coatings that Resist Adhesion

Bacteria produce biofilms that are hospitable environments where they can grow without difficulty due to enhanced nutrient uptake and exchange of genetic information. The formation of biofilm requires an initial settling of bacteria on a surface that can form a colony and ultimately a biofilm. Coatings that prevent or resist the formation and growth of the biofilm are very promising agents to combat bacteria and fungi. To accomplish this, the surface must be hydrophobic. Early examples of such surfaces showed mixed results. This was due to the use of hydrocarbon or fluorocarbon additives in hydrophobic coatings that over time migrated from the coating. Inherently hydrophobic polymers such as fluoropolymers and organosilicone compounds have very low surface energies; therefore, high hydrophobic properties exist and can be formulated into coatings with or without antimicrobial agents. In recent years, polymer brushes that contain fluorocarbon and dimethyl siloxane pendent units have yielded superhydrophobic surfaces. These surfaces will not allow the collection of liquid water and hence are not favorable surface for bacterial settlement.

It is also possible to produce polymeric surfaces that result in antiadhesion through repelling mechanism. Repelling mechanisms are intricate and can be explained by steric repulsion and electrostatic repulsion. Electrostatic repulsion may be explained by an interaction between negatively charged surfaces such as specially prepared acrylic acid/styrene copolymer and bacteria with a negative transmembrane electrical potential relative to a moderate environment.

Hydrogels are the most important family of hydrophilic adhesion-resistant coatings. Hydrogels based on polyurethane chemistry are widely used in medicine to prevent the adhesion of cells to vascular implants and artificial joints. In practice, however, their use as surface coatings is limited by their comparatively poor resistance to abrasion and commercial cleaners.

10.2.2 Coatings that Release Toxins

The second family of antimicrobial coatings is those that release compounds that kill microorganisms. Many suitable bactericides and fungicides are available commercially, including antibiotics, QA compounds (“quats”), and other organic materials.

The bactericides commonly used, as in-can stabilizers for latex coatings, such as chloramines, triazines, ethylene oxide, or phenols, have no value here. The service life of these coatings is primarily determined by how long an effective release of toxin can be sustained. Simply mixing a toxin into a coating usually produces an inefficient, costly, short-lived, and relatively ineffective coating. The agent is released too quickly when the coating is new, wasting an expensive ingredient. Soon, the amount of agent released is too low to be effective.

Extensive open literatures and patents indicate that there are scores of chemical agents that can destroy certain microorganisms. The chemical makeups of such compounds are well known and range from organic compounds such as QA compounds, *N*-alkylated poly(4-vinyl-pyridine), active on both Gram-positive and Gram-negative bacteria [6–16].

When used in coatings, each class of compounds has some advances and some limitations. Organic compounds that are not bound to the backbone resin of the coating can leach out over time greatly reducing their useful life. It has been well documented that some microorganisms can develop resistance to or alters chemical properties of the antimicrobial agent(s). The lifetime of coatings containing bactericide as an unbound organic additive therefore may depend on the applied thickness, the frequency and intensity of cleaning, and bacterial resistance.

Zeolites are inorganic silicates that contain pores large enough to hold metal ions. Silver ions can be absorbed into zeolites and then released at a controlled rate. Coatings containing silver-filled zeolites may be suitable for controlling and preventing the growth of a wide variety of microorganisms. One specific agent is zirconium phosphate complex of silver ions, which has found commercial application as an antimicrobial agent in textile, filtration media, building materials, and adhesives.

10.3 POLYMERS AND ANTIMICROBIAL COATING BINDERS

Bioactive polymers are made of synthetic or natural polymers either with inherent antimicrobial properties or containing specific functional groups, which are designed to develop specific interactions with living organisms. They can be used to kill microorganisms or to heal damaged tissues. Antimicrobial coatings are composed either solely of antimicrobial polymer or of nonfunctional polymers in which certain types of bactericidal agents are dispersed in them. Biocidal polymers offer promise in helping curb the spread of infections by providing coatings for biomedical devices or molded articles. Such polymers may be classified as biocide release or nonrelease contact kill. Biocide release systems are the most advanced in that commercial products are available. A recent review covers recent advances in contact kill systems [17]. A brief description of types and functions of the four major categories of polymers and coatings are given later.

10.3.1 Polymeric Coatings with QA Groups

QA compounds are among the most widely used antibacterial agents for medical and public health applications. Since most bacterial cell walls are negatively charged, they can readily react with positively charged polymers. It is generally accepted that the mechanism of bactericidal action of the polycationic biocides involves destructive interaction with the cell wall and/or cytoplasmic membranes [18].

QA compounds have been shown to be effective against both Gram-negative and Gram-positive bacteria. The lethal mode of action is generally accepted to proceed by (i) adsorption of the positively charged QA compound on the negatively charged cell surface, (ii) disruption of the cell membrane by a lipophilic chain on the QA compound, and (iii) diffusion through the membrane leading to cell death. Cationic polymers with QA groups generally exhibit higher antimicrobial activities than their corresponding low-molecular-weight monomers. The higher activity is attributed to great attraction between the cell and polymer due to the greater charge density of the polymer.

The QA group–containing polymers were dissolved in either ethanol or methanol and were then sprayed at coverage of 5 mg polymer/in.^{2} onto stainless steel coupons previously painted with chemical agent–resistant coating. Coated coupons and control surfaces having no antibacterial polymeric coating were inoculated with liquid suspensions of vegetative bacteria including either Gram-positive *Bacillus globigii* or Gram-negative *Escherichia coli*. Test results showed that these coatings could kill 99.9% of both Gram-positive and Gram-negative bacteria on contact. Additionally, the material after 14 days of exposure to a simulated diurnal temperature cycle still could kill 87% of bacteria [19]. The same parent polymer, Brominated Tetramethyl bisphenol Polysulfone (Br-TMPS), could be treated in solution with tertiary amines that had either aldehydes or protected aldehydes attached to them.

10.3.2 Polymers with Quaternary Phosphonium Groups

These polymers were synthesized by immobilizing phosphonium salt onto the cross-linked copolymer: poly(Methyl methacrylate-vinylbenzyl chloride (MMA-VBC)), as shown in Figs. 10.1 and 10.2 [20].

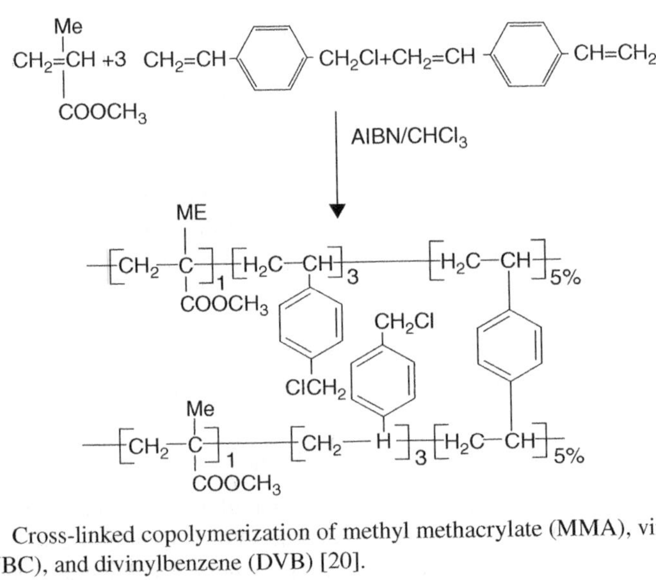

FIG. 10.1 Cross-linked copolymerization of methyl methacrylate (MMA), vinyl benzyl chloride (VBC), and divinylbenzene (DVB) [20].

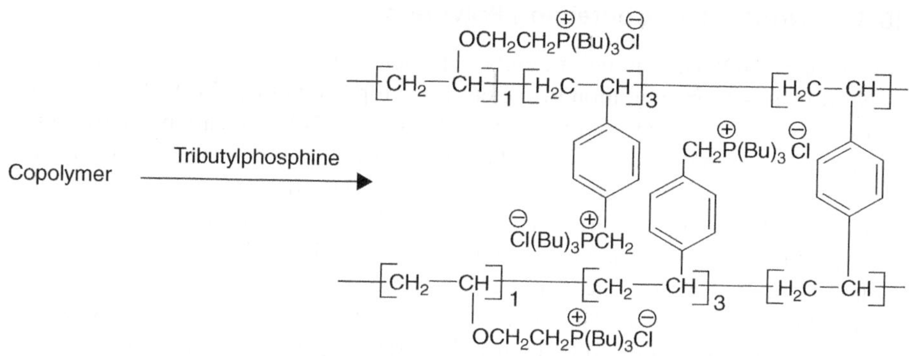

FIG. 10.2 Immobilization of phosphonium salt onto cross-linked copolymer [20].

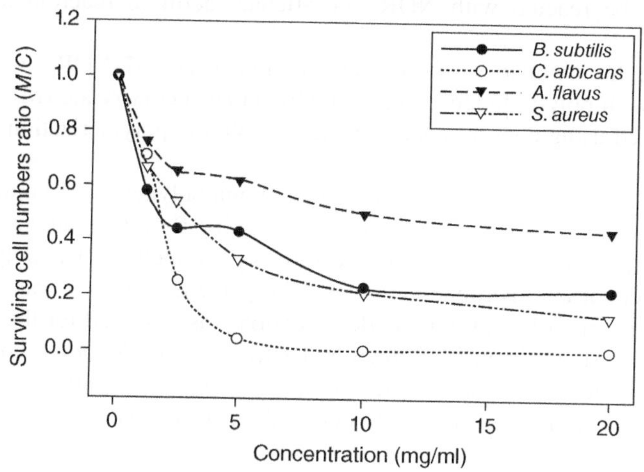

FIG. 10.3 Growth inhibition of different concentrations of polymer [20].

The antibacterial activities of modified poly(MMA-VBC) against *Candida albicans, Aspergillus flavus, Fusarium oxysporum, Bacillus subtilis, E. coli,* and *Staphylococcus aureus* were examined using cut plug method and visible count methods. The growth-inhibiting effect was quantitatively determined by the ratio (M/C) of the surviving cell number as shown in Fig. 10.3. The growth-inhibitory effect of polymer differed among the bacterial and fungal species. The inhibition becomes stronger in the order *A. flavus* < *B. subtilis* < *S. aureus* < *C. albicans.* The results show also that the inhibitory effect increased by increasing the concentration of the polymer. Generally, the potency of inhibition varied according to the polymer and the examined species. In general, the copolymers showed good antimicrobial activity against the tested microorganisms.

10.3.3 Norfloxacin-Containing Polymers

Norfloxacin (NOR) is a quinolone carboxylic acid derivative. The bactericidal action of NOR results from inhibition of the enzymes topoisomerase II (DNA *gyrase*) and topoisomerase IV, which are required for bacterial DNA replication, transcription, repair, and recombination. NOR is a broad-spectrum antibiotic that is active against both Gram-positive and Gram-negative bacteria. The fluorine atom at position 6 increases potency against Gram-negative organisms, and the piperazine moiety at position 7 is responsible for antipseudomonal activity [21, 22].

In general, two approaches can be utilized to obtain the antibacterial polymers with NOR moieties. The first approach involves the incorporation of the antibacterial agent to monomers, followed by their polymerization. The second approach, on the other hand, involves the linking of the antibacterial agents directly onto preformed functional polymers [23, 24].

In the approach, for example, 3-(Acryloyloxy)-2-hydroxypropyl methacrylate (AHM) can be reacted with NOR via Michael addition reaction as shown in Fig. 10.4.

Then the homopolymer of the new monomer (AHM–NOR) as well as its copolymer with poly-(ethylene glycol) methyl ether methacrylate (MPEGMA) can be synthesized using conventional free-radical solution polymerization as shown in Fig. 10.5.

Antimicrobial activities of the monomer, homopolymer, and copolymer were tested against *S. Aureus* and *E. coli* using the shaking flask method, and the number of viable cells was counted utilizing the spread plate method. The polymers tested showed excellent antimicrobial activities. The percent reduction was expressed as 100% since seven orders of magnitude reduction was observed for the compounds. The copolymer showed better antimicrobial activity against *S. aureus* compared with the homopolymer. Even though there is not a considerable difference, the better activity of the copolymer is probably due to the partial solubility of the copolymer in water.

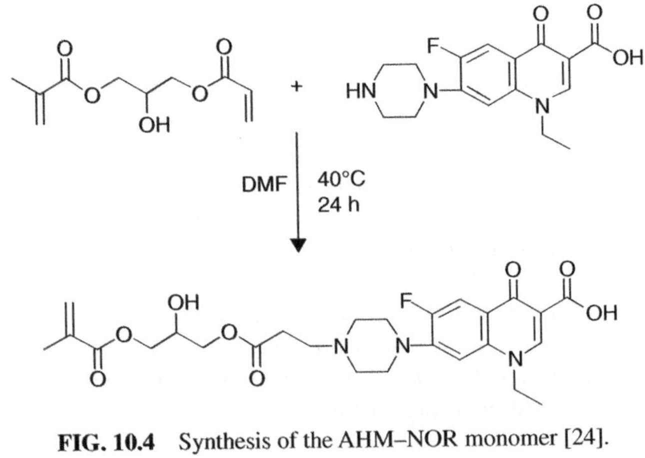

FIG. 10.4 Synthesis of the AHM–NOR monomer [24].

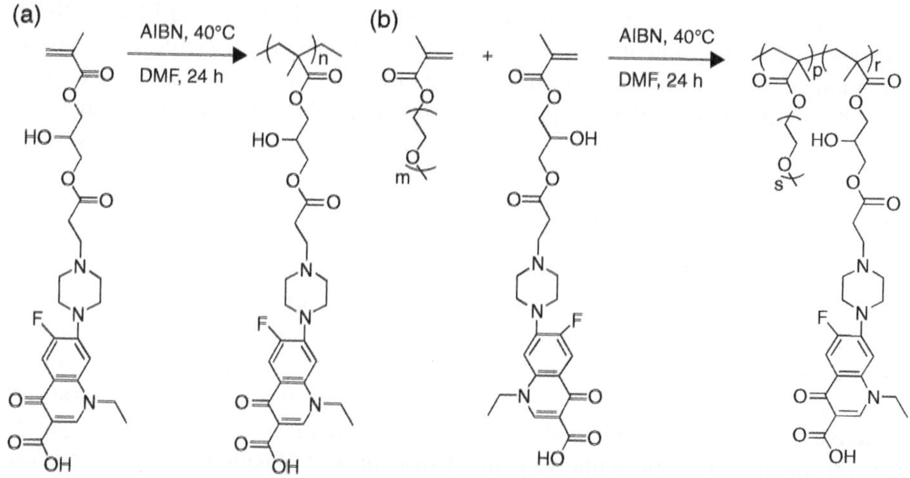

FIG. 10.5 Synthesis of the AHM–NOR homopolymer (a) and AHM–NOR/MPEGMA copolymer (b) [24].

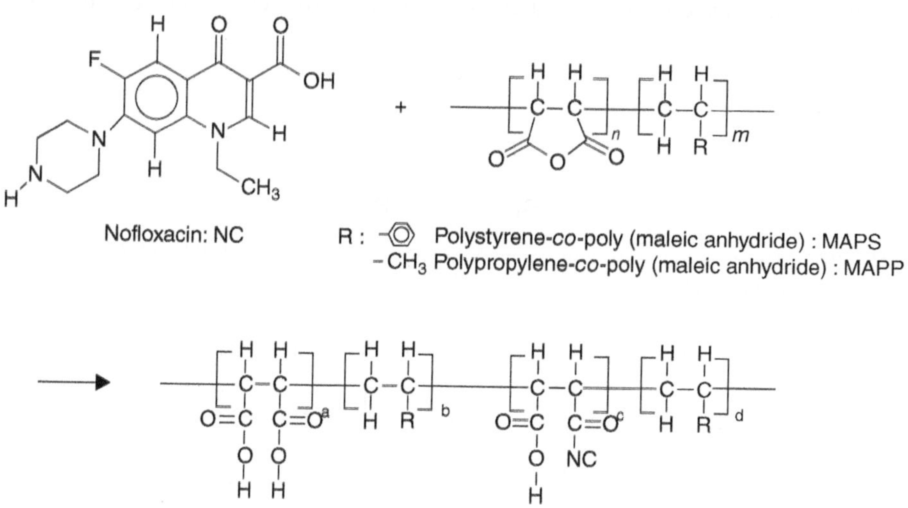

FIG. 10.6 Synthesis of the polymeric antibacterial agents [21].

NOR can also be grafted onto polypropylene-*graft*-maleic anhydride (MAPP) and poly-(styrene-*co*-maleic anhydride) (MAPS) to synthesize polymeric antimicrobial agents through a chemical reaction between the maleic group of the polymers and the amine group of NOR (Fig. 10.6).

The shake flask tests were carried out to examine the antibacterial activity of MAPP–NOR and MAPS–NOR. The addition of 1 wt% MAPP–NOR and MAPS–NOR reduced 99% of the viable cells of the native strains. In contrast, the viable cell number of *Bacillus cereus*-R, *Vibrio fluvialis*-R, and *Vibrio parahaemolyticus*-R did not decrease at all, even after contact with 1 wt% MAPP–NOR and MAPS–NOR. However, when the content of the polymeric antibacterial agents increased to 5 wt%, the number of the viable cells in the three NOR-tolerant strains was decreased by 73–98%, demonstrating that the polymeric antibacterial agents were somewhat bioactive against the bacteria even when the bacteria attained tolerance against NOR.

10.3.4 Polymeric *N*-Halamines

N-halamines are defined as compounds containing one or more nitrogen–halogen covalent bonds that are formed by the chlorination of imide, amide, or amine groups. The halogen is usually chlorine; however, bromine or iodine can function similarly. The biological properties are thought to be due to oxidation state of +1 halide atoms in chloramine or bromamine. These oxidizing halogens can act through direct transfer of active element to the biological receptor or through dissociation to free halogen in aqueous media. The representative general structures of *N*-halamine compounds are shown in Figs. 10.7, 10.8, and 10.9.

The corresponding polymers of such antimicrobial agents are one of the most promising antimicrobial candidates because they can provide fast and total kill against a wide range of microorganisms, without causing environmental concerns,

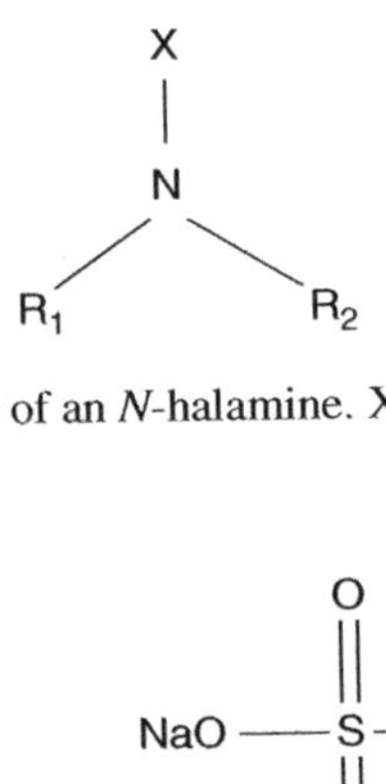

FIG. 10.7 Representative structure of an *N*-halamine. X=Cl, Br, or I. R1 and R2=organic, inorganic groups, or H.

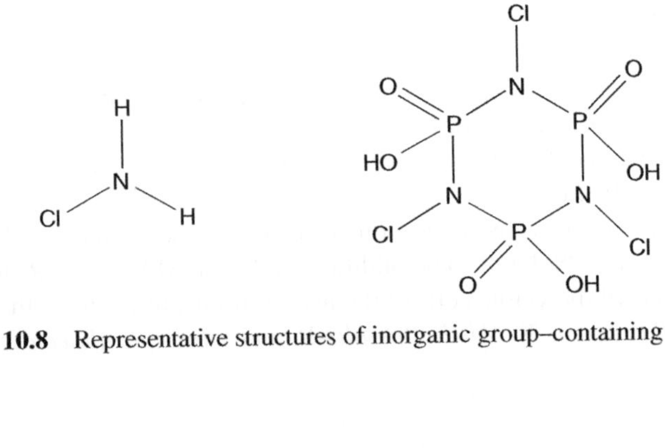

FIG. 10.8 Representative structures of inorganic group–containing halamines.

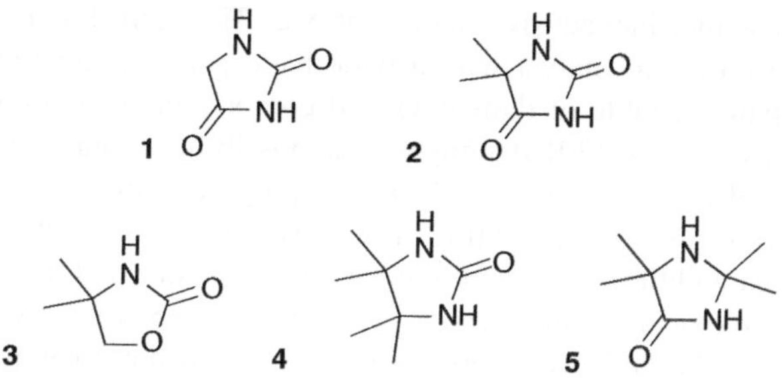

FIG. 10.9 Structures of cyclic organic molecules leading to *N*-halamines in presence of halogen-releasing agent: **1**, hydantoin (imidazolidone-2,4-dione); **2**, 5,5-dimethylhydantoin; **3**, 4,4-dimethyl-2-oxalinone; **4**, tetramethyl-2-imidazolidone; **5**, 2,2,5,5-tetramethylimidazolidin-4-one.

and it is highly unlikely for the microorganisms to establish resistance. The idea of *N*-halamine compounds was first proposed by Kovacic and coworkers, in 1969 [25]. In *N*-halamines, as a consequence of the halogen atoms covalently bonded to the nitrogen atoms, the free active halogen species (oxidative halogen Cl^+ or Br^+) are slowly released into the environment.

Sun *et al.* copolymerized 1-acryloyl-2,2,5,5-tetramethylimidazolidin-4-one, a cyclic-amine monomer with vinyl acetate (VA), acrylonitrile (AN), and methyl methacrylate (MMA) [26]. After regular chlorine bleach treatment, some of the N-3 amide groups could be transformed into stable *N*-halamine groups, which provided powerful, durable, and refreshable antibacterial efficacies. The tests against *E. coli* showed a 3-log (99.9%) reduction of 10^6–10^7 colony-forming units (CFU)/ml after 30 min of contact for copolymers with VA, whereas a 2-log (99%) reduction for MMA copolymers and a 1-log (90%) reduction for AN.

Later, Chen and Sun synthesized 2-vinyl-4,6-diamino-1,3,5-triazine monomer and copolymerized with styrene [27]. After chlorine bleach, these polymers were examined against *E. coli and S. aureus.* They found in all the copolymers a 7-log (99.99999%) reduction of 10^7–10^8 CFU/ml in both microorganisms, which means that with small monomer content, the copolymer achieves a powerful antimicrobial activity. Recently, new *N*-halamine acrylamide monomers were copolymerized and used as antimicrobial coating providing 8-log inactivation of both Gram-positive and Gram-negative bacteria in 5 min contact [28].

10.4 ADDITION OF INORGANIC PARTICLES

10.4.1 Titanium Dioxide

Photocatalytic disinfection in recent years has emerged as a successful technology and promising results have been obtained with significant number of microorganisms including bacteria, fungi, viruses and yeasts [29]. Photocatalysts are solid semiconductors that are able to absorb visible and/or UV light, chemically and biologically inert and photostable, inexpensive, and nontoxic. Titanium dioxide-anatase is the most broadly used photocatalyst with a wide band gap that generates energy-rich electron–hole pairs capable of damaging cell components of microorganisms rendering innocuous products [30]. Among several possibilities, titania can be supported on polymers. In the incorporation of TiO_2 into polymeric matrix, it is important to control the polymorphism, maintaining the anatase structure, which presents the biocidal function. In addition, control of the TiO_2 particle size and the homogeneity of the composite are crucial to limit scattering events, among others [31].

Recently, Kubacka and coworkers intensively tested the potential antimicrobial properties of polymer-based composite systems containing 2 wt% of TiO_2 as a biocidal agent [32, 33]. They described the preparation of nanocomposite films by the incorporation of TiO_2 nanoparticles into ethylene-vinyl alcohol copolymer via a melt process. These nanocomposites present extraordinary antibacterial properties against *Enterococcus faecalis* and *Pseudomonas aeruginosa*, up to 5- and 8-log reduction, respectively, of 10^9 CFU/ml.

The antimicrobial properties of TiO_2-polymer nanocomposite films have also been reported to be enhanced by extending the light absorption into visible regions through doping process using silver or by the addition of copper and zinc nitrate [34, 35]. Other valuable work concerns the incorporation of titania into a polyurethane resin modified with anionic surfactant. The presence of surfactant is said to increase the hydrophilicity of the final nanocomposite material. The system was tested against *E. coli* and *C. albicans* bacteria and *Aspergillus niger* fungus upon both UV and visible excitations. *E. coli* and *C. albicans* bacteria were efficiently eliminated; however, *A. niger* showed a higher resistance [36].

10.4.2 Zinc Oxide

Zinc oxide is also a semiconductor with a band gap very close to the TiO_2 anatase and with a similar batericidal mechanism of action. Exposure to light creates electron and hole pairs, which can possibly oxidize organic matter or oxidize water to yield hydrogen peroxide and radicals such as $OH^{\cdot}$ or $O_2^{\cdot-}$, with antibacterial properties. However, the action is not fully understood where participation of ionic form of zinc, Zn^{2+} is among possibilities.

Although the antibacterial activities of ZnO is well known, only a few examples of polymeric nanocomposites have been reported in the last decade. ZnO nanoparticles have been incorporated into polyamide 6 and low-density polyethylene [37]. In both cases the nanocomposites show great antimicrobial activity with low content of ZnO, 1% (w/w), which increases with increasing the concentration of ZnO. However, ZnO appears to be inactive toward *A. niger* fungus.

10.4.3 Silver Compounds

Silver is one of the oldest bactericidal and therapeutic agents that has been used for thousands of years in different parts of the world [38]. Because of its multitarget mechanism of action, a wide spectrum of bacteria, about 650 species of bacteria, can be inactivated using silver, whereas the antibacterial spectrum of antibiotics can be quite limited—in some cases, only 5–10 species of bacteria [39, 40].

Over the last 100 years, many studies have been published where the antimicrobial activity of silver ions has been quantified. Recent papers suggest that the mode of action of silver ions varies depending upon the target organism and can interact with functional groups on proteins containing sulfur, phosphorus, nitrogen, and oxygen. Further, silver ions have been shown to disrupt the TCA cycle, electron transport system, phosphate uptake, transport of various metabolites, and the polarization of bacterial membranes [41]. Recent studies have elucidated a broad spectrum of heavy metal–resistance mechanisms in both clinical and environmental isolates [42–49].

Bacterial resistance to heavy metals such as silver ions appears to be through the acquisition of various efflux pumps that pump the silver ions out of the cell [42–44]. These genes, however, are different from those that mediate resistance to various antibiotics. Silver-based products have been used extensively in Japan and in Europe. No reports of the failure of these products have been published.

The major challenge in using additives to coatings to impart antimicrobial activities is the possible reaction of such compounds with a wide variety of inorganic and organic compounds used in a coating formulation. The utility of encapsulated silver in a typical latex system was recently explored [50, 51].

The antimicrobial activities of these coatings on the test substrates were tested toward a selected panel of Gram-positive (*S. aureus* ATCC 6538) and Gram-negative (*E. coli* ATCC 11229) bacteria. These strains of bacteria have been used as standards for testing of disinfectant efficacy using the Association of Analytical Communities (AOAC) Use-Dilution Methods and represent common bacteria encountered in health-care and food-preparation facilities [52–57].

Antimicrobial coatings were formulated by incorporating various levels of encapsulated silver-containing antimicrobial agents in the appropriate resin systems. The antimicrobial coating formulations included both solvent-borne and waterborne coatings. The resins used in the formulations of the antimicrobial coatings included solvent-borne and water-based acrylic thermoplastic resins. Additionally, the formulations contained various additives, pigments, and solvents.

The coatings were applied onto glass, steel, and aluminum panels using drawdown rod to produce dry films in the range of 12–50 μm. Air-dry and waterborne coatings were allowed to dry at room temperature for a minimum of 7 days before physical and biological testing. Thermosetting coatings were cured depending on the cross-linking chemistry at 80°C × 30 min or 140°C × 30 min. Evaluations indicated that the air-dry waterborne coatings containing silver-based antimicrobials effectively killed the test strains of *E. coli* and *S. aureus* after 12 h of incubation at room temperature.

Most recently, Kayaman-Apohan *et al.* reported the preparation of UV-curable nanosilver-containing polyurethane-based organic–inorganic hybrid coatings [58]. The UV-cured hybrid coating samples were then inoculated with *E. coli and S. aureus* bacteria. It has been reported that the overnight incubation of the inoculated coating samples resulted in approximately 99.99% (10 CFU/ml) reduction in bacterial colonies compared with the control samples after 24 h.

10.5 CONCLUSIONS AND OUTLOOK

The use of polymers in medical field has become routine and undoubtedly has reduced the suffering of humans and improved the quality of life of so many patients. In recent years, antimicrobial polymers and coatings have gained interest from both academia research and as industrial products due to their potential to provide higher-quality life and safety. Polymeric coatings with added antimicrobial functionality present a wide range of possibilities of applications in many areas such as hospitals, food-processing facilities, grocery stores, schools, homes, home appliances, automobiles, airplanes, computer keyboards, and phones.

Self-disinfecting coatings while functioning as typical surface coating offer continuous protection against old and fresh contamination. In particular, antimicrobial coatings are nontoxic, nonirritant with improved and wide-range antimicrobial activities, compared with ordinary low-molecular-weight antibacterial agents, which may be volatile and usually have short-term activities. In addition, the increased use of disinfectants and antimicrobial agents in food and environment has increased the potential of microbial resistance.

As shown in this chapter, there are many strategies to design polymeric antimicrobial systems with diverse functions over microorganisms. Two major categories of antimicrobial polymer coatings exist: polymers that are inherently antimicrobial through structure or modification and those that become antimicrobial through addition of organic, inorganic antimicrobial agent or combination of both. While most of the research has been focused on polycationic systems, polymer coatings based on the combination of organic and inorganic antimicrobial agents are in the lead commercially. In the future, most likely the goals of addressing cytotoxicity, improving efficacy, long-term durability, and costs will be actively pursued.

The use of antimicrobial polymer coatings while not aimed at producing a totally sterile environment may serve a useful purpose in areas where a permanent and more sanitary environment is necessary. While there is no clear evidence that the routine use of antibacterial agents confer a health benefit, they are useful where the level of sanitation is critical and additional precautions need to be taken to prevent spread of disease. Thus, they are important in hospitals, day-care centers, and health-care facilities and other environments with high concentrations of infectious bacteria. In the home environment, they may be needed for the nursing care of sick individuals with specific infections or for those whose immune systems have been weakened by chronic disease, chemotherapy, or transplants. Under these circumstances, antibacterial agents should be used according to protocol, preferably under the guidance of a health-care professional.

REFERENCES

1. Gabriel, G.J., Som, A., Madkour, A.E., Eren, T., Tew, G.N. (2007) Infectious disease: connecting innate immunity to biocidal polymers, *Mater. Sci. Eng. Rep.* 57: 28–64.
2. Gleick, P.H., (2002) Dirty Water: Estimated Deaths from Water Related Diseases 2000–2020. Pacific Institute Research Report. Pacific Institute, Oakland, CA.

3. Beumer, R.R., Kusumaningrum, H. (2003) Kitchen hygiene in daily life, *Int. Biodeter. Biodegrad.* 51: 299–303.
4. Rayner, J., Veeh, R., Flood, J. (2004) Prevalence of microbial biofilms on selected fresh produce and household surfaces, *Int. J. Food Microbiol.* 95: 29.
5. Eknoian, M.W., Worley, S.D., Bickert, J., Williams, J.F. (1999) Novel antimicrobial N-halamine polymer coatings generated by emulsion polymerization, *Polymer* 40: 1367–1371.
6. Antelman, M.S. (2001) High Performance Cobalt (II,III) Oxide Antimicrobial Textile Articles. US Patent 6228491.
7. Herrera, M., Carrion, P., Baca, P., Liebana, J., Castillo, A. (2001) *In vitro* antibacterial activity of glass-ionomer cements, *Microbios* 104: 141.
8. Cho, D.L. Na, K., Shin, E.K., Kim, H.J., Lee, K.Y., Go, J.H., Choi, C.S., (2001) A study on the preparation of antimicrobial biopolymer film, *J. Microbiol. Biotechnol.* 11(2): 193–198.
9. Tiller, J.C., Liao, C.-J., Lewis, K., Klibanov, A.M. (2001) Designing surfaces that kill bacteria on contact, *Proc. Natl. Acad. Sci.* 98: 5981.
10. Lin, J., Tiller, J.C., Lee, S.B., Lewis, K., Klibanov, A.M. (2002) Insights into bactericidal action of surface-attached poly(vinyl-*N*-hexylpyridinium) chains, *Biotechnol. Lett.* 24: 801.
11. Tiller, J.C., Lee, S.B., Lewis, K., Klibanov, A.M., (2002) Polymer surfaces derivatized with poly(vinyl-*N*-hexylpyridinium) kill airborne and waterborne bacteria, *Biotechnol. Bioeng.* 79: 466.
12. Tew, G.N., Liu, D., Chen, B., Doerksen, R.J., Kaplan, J., Carroll, P.J., Klein, M.L., DeGardo, W.F. (2002) De novo design of antimicrobial polymers, *Proc. Natl. Acad. Sci.* 99: 5110.
13. Mackeen, C., Person, S., Warner, S.C., Snipes, W., Stevens, S.E. (1987) Silver-coated nylon fiber as an antibacterial agent, *Antimicrob. Agents Chemother.* 31: 93–99.
14. Schierholz, J.M., Wachol-Drewek, Z., Lucas, L.J., Pulverer, G. (1998) Activity of silver ions in different media, *Zentralbl Bakteriol.* 287: 411–420.
15. Spadaro, A., Berger, T.J., Barranco, S.D., Chapin, S.E., Becker, R.O. (1974) Antibacterial effects of silver electrodes with weak direct current, *Antimicrob. Agents Chemother.* 6: 637–642.
16. Thibodeau, E.A., Handelman, S.L., Marquis, R.E. (1978) Inhibition and killing of oral bacteria by silver ions generated with low density electric current, *J. Dent. Res.* 57: 922–926,
17. Bieser, A.M., Tiller, J.C. (2011) Mechanistic considerations on contact-active antimicrobial surfaces with controlled functional group densities, *Macromol. Biosci.* 11: 526.
18. Kenawy, E.R., Abdel-Hay, F.L., El-Shanshoury, A.E.R.P., El-newehy, M.H. (2002) Biologically active polymers. V. Synthesis and antimicrobial activity of modified of Poly(glycidyl methacrylate-co-2-hydroxyethyl methacrylate), *J. Polym. Sci Part A Polym. Chem.* 40: 2384–2393.
19. Hibbs, M., Barnhart, M., Allen, A., Wilson, M., Tucker, M. (2009) Antibacterial Polymer Coatings. Sandia Report. SAND2009-6171. Sandia National Laboratories, Albuquerque, NM.
20. Kenawy, E., Abdel-Hay, F., El-Magd, A., Mahmoud, Y. VII (2006) Biologically active polymers: VII. Synthesis and antimicrobial activity of some crosslinked copolymers with quaternary ammonium and phosphonium groups, *React. Funct. Polym.* 66: 419–429.
21. Kim, M., Lim, A., Yoon, J. (2005) Antibacterial activity of polymers with norfloxacin moieties against native and norfloxacin-tolerance-induced bacteria, *J. Appl. Polym. Sci.* 96: 936–943.

22. DrugBank Norfloxacin (DB01059). http://www.drugbank.ca/drugs/DB01059 (accessed December 9, 2014).
23. Dizman, B., Elasri, M., Mathias, L. (2009) Novel antibacterial polymers, *Smart Coat. Symp. Ser. II. Chapt* 2: 27–51.
24. Dizman. B., Elasri, M., Mathias, L. (2005) Synthesis, characterization, and antibacterial activities of novel methacrylate polymers containing norfloxacin, *Biomacromolecules* 6: 514–520.
25. Kovacic, P., Lowery, M.K., Field, K.W. (1970) Chemistry of N-bromamines and N-chloramine, *Chem. Rev.* 70: 639–665.
26. Sun, Y., Chen, T.-Y., Worley, S.D., Sun, G. J. (2001) Acrylic N-halamine polymeric biocidal films, *Polym. Sci. Part A Polym. Chem.* 39: 3073–3084.
27. Chen, Z.B., Sun, Y.Y. (2005) Antimicrobial polymers containing melamine derivatives. II. Biocidal polymers derived from 2-vinyl-4,6-diamino-1,3,5-triazine, *J. Polym. Sci. Part A Polym. Chem.* 43: 4089–4098
28. Kocer, H.B., Worley, S.D., Broughton, R.M., Huang, T.S. (2011) A novel N-halamine acrylamide monomer and its copolymers for antimicrobial coatings, *React. Funct. Polym.* 71: 561–568.
29. Muñoz-Bonilla, A., Fernández-García, M. (2012) Polymeric materials with antimicrobial activity, *Prog. Polym. Sci.* 37: 281–339.
30. Linkous, A.C., Carter, G.J., Locuson, D.B., Quellette, A.J., Slattery, D.K., Simitha, L.A. (2000) Photocatalytic inhibition of algae growth using TiO_2-WO_3 and cocatalyst modifications, *Environ. Sci. Technol.* 34: 4754–4758.
31. Carp, O., Huisman, C.L., Reller, A. (2004) Photoinduced reactivity of titanium dioxide, *Prog. Solid. State. Chem.* 32: 33–177.
32. Kubacka, A., Serrano, C., Ferrer, M., Luensdorf, H., Bielecki, P., Cerrada, M.L., Fernandez-Garcia, M. (2007) High-performance dual-action polymer-TiO_2 nanocomposite films via melting processing, *Nano. Lett.* 7: 2529–2534.
33. Cerrada, M.L., Serrano, C., Sanches-Chavez, M., Fernandez-Garcia, M., Fernandez-Martin, F., de Andres, A., Rioboo, R.J.J., Kubacka, A., Ferrer, M. (2008) Polymeric materials with antimicrobial activity from synthesis to application, *Adv. Mater.* 18: 1949–1960.
34. Kubacka, A., Cerrada, M.L., Serrano, C., Fernandez-Garcia, M., Ferrer, M. (2009) Plasmonic nanoparticle/polymer nanocomposites with enhanced photocatalytic antimicrobial properties, *Phys. Chem. C* 113: 9182–9190.
35. Kubacka, A., Ferrer, M., Fernandez-Garcia, M., Serrano, C., Cerrada, M.L. (2011) Tailoring polymer-TiO_2 properties by presence of metal (Ag, Cu, Zn) species: optimization of antimicrobial properties, *Appl. Catal. B Environ.* 104: 346–352.
36. Zhang, X., Su, H., Zhao, Y., Tan, T. (2008) Antibacterial and anti-mildew behavior of chitosan/nano-TiO_2 composite emulsion, *J. Photochem. Photobiol. A* 199: 123–129
37. Droval, D., Aranberri, I., Bilbao, A., German, L., Verelst, M., Dexpert-ghys, J. (2008) Antimicrobial activity of nanocomposites: poly(amide) 6 and low density polyethylene filled with zinc oxide, *e-Polymers* 128: 1–13.
38. Atiyeh, B.S., Costagliola, M., Hayek, S.N., Dibo, S.A. (2007) Effect of silver on burn wound infection control and healing: review of the literature, *Burns* 33: 139–148.
39. Lansdown, A.B. (2002) Silver. I: Its antibacterial properties and mechanism of action, *J. Wound Care.* 11: 125–130.

40. Thomas, S., McCubbin, P. (2003) A comparison of the antimicrobial effects of four silver-containing dressings on three organisms, *J. Wound Care.* 12: 101–107.
41. Dibrov, P., Dzioba, J., Gosink, K.K., Hase, C.C. (2002) Chemiosmotic mechanism of antimicrobial activity of Ag(+) in Vibrio cholerae, *Antimicrob. Agents Chemother.* 46: 2668–2670.
42. Gupta, A., Phung, L.T., Taylor, D.E., Silver, S. (2001) Diversity of silver resistance genes in IncH incompatibility group plasmids, *Microbiology* 147: 3393–3402.
43. Franke, S., Grass, G., Hies, D.H. (2001) The product of the ybdE gene of the *Escherichia coli* chromosome is involved in detoxification of silver ions, *Microbiology* 147: 965–972.
44. Nies, D.H. (1999) Microbial heavy metal resistance, *Appl. Microbiol. Biotechnol.* 51:730–750.
45. Gupta, A., Maynes, M., Silver, S. (1998) Effects of halides on plasmid-mediated silver resistance in *Escherichia coli, Microbiology* 64: 5042–5045.
46. Steven, L., Percival, B., Woods, E., Nutekpor, M., Bowler, P., Radford, A., Cochrane, C. (2008) Prevalence of silver resistance in bacteria isolated from diabetic foot ulcers and efficacy of silver-containing wound dressings, *Ostomy Wound Manage.* 54(3): 30–40.
47. Slawson, R.M., Lohmeier-Vogel, E.M., Lee, H., Trevors, J.T. (1994) Silver resistance in *Pseudomonas stutzeri, BioMetals* 7: 30–40.
48. Deshpande, L.M., Chopade, B.A. (1994) Plasmid mediated silver resistance in *Acinetobacter baumannii, BioMetals* 7: 49–56.
49. Hendry, A.T., Stewart, I.O. (1979) Silver-resistant Enterobacteriaceae from hospital patients, *Can. J. Microbiol.* 25: 915–921.
50. Baghdachi, J., Clemans, D., Rhoades, S., Kendzorski, J., Xu, Q. (2009) Performance testing of waterborne antibacterial and high-solids coatings, *ACS Symp. Ser.* 1002: 111–126.
51. Baghdachi, J., Clemans, D., Rhoades, S., Kendzorski, J., Xu, Q. (2007) Formulation and evaluation of organic antimicrobial coatings, *ACS Symp. Ser. Smart Coat.* 957: 27–42.
52. Barkley, W.E., Richards, J.H. (1994) Laboratory safety, in *Methods for General and Molecular Bacteriology*, P. Gerhardt, R.G.E. Murray, W.A. Wood, N.R. Krieg (Eds), American Society for Microbiology, Washington, DC, 715–734.
53. Luppens, S.B.I., Reij, M.W., van der Heijden, R.W.L., Rombouts, F.M., Abee, T. (2002) Development of a standard test to assess the resistance of *Staphylococcus aureus* biofilm cells to disinfectants, *Appl. Environ. Microbiol.* 168: 4194–4200.
54. Rubino, J.R., Bauer, J.M., Clarke, P.H., Woodward, B.B., Porter, F.C., Hilton, H.G. (1992) Hard surface carrier test for efficacy testing of disinfectants-collaborative study, *J. AOAC Int.* 75: 635–645.
55. Mariscal, A., Carnero-Varo, M., Gomez-Aracena, J., Fernandez-Crehuet, J. (1999) Development and testing of a microbiological assay to detect residual effects of disinfectant on hard surfaces, *Appl. Environ. Microbiol.* 65: 3717–3720.
56. Abrishami, S.H., Tall, B.D., Bruursema, T.J., Epstein, P.S., Shah, D.B., Bacterial, J. (1994) Bacterial adherence and viability on cutting board surfaces, *Food Saf.* 14: 153–172.
57. Miner, N., Armstrong, M., Carr, C.D., Maida, B., Schlotfeld, L. (1997) Modified quantitative association of official analytical chemists sporicidal test for liquid chemical germicides, *Appl. Environ. Microbiol.* 63: 3304–3307.
58. Toker, R.D., Kayaman, N., Kahraman, M.V. (2013) UV-curable nano-silver containing polyurethane based organic–inorganic hybrid coatings, *Prog. Org. Coat.* 76: 1243–1250.

CHAPTER 11

Novel Marine Antifouling Coatings: Antifouling Principles and Fabrication Methods

Yunjiao Gu and Shuxue Zhou
Department of Materials Science and Advanced Coatings Research Center of Ministry of Education of China, Fudan University, Shanghai, P.R. China

11.1 INTRODUCTION

Marine biofouling refers to the unfavorable attachment of a wide range of marine organisms to surfaces. This can cause serious safety problems, profound economic losses, and considerable environmental issues, in terms of fuel efficiency, vessel maneuverability and speed, engine lifetime, and marine ecology. Antifouling (AF) coatings are an effective way of eradicating fouling organisms. They are indispensable for maritime industries. Tributyl tin (TBT) self-polishing coatings (SPCs) are considered to be the most successful AF coating. These coatings have highly effective AF properties resulting from both TBT biocide and their self-polishing/smoothing performance. However, because of the high toxicity of TBT, use of these coatings has been prohibited since 2008. Currently, the most widely used AF coatings are Cu_2O-based SPCs. Although Cu_2O-based coatings are less toxic than TBT-based coatings, they are still harmful to marine ecology and will also be prohibited in the near future. The ban on TBT- and Cu_2O-based coatings, along with the increased cost of developing other new biocides (which require strict risk studies before registration), calls for novel AF coatings that are both effective and biologically safe.

The past decade has seen the advent of a variety of biocide-free AF strategies. Different from non-target chemically active biocidal ones, trends for novel AF coatings have proceeded in two directions: the targeted chemically active (or biodegrading) AF strategies and chemically inert (or biopreventing) AF strategies, as shown in Fig. 11.1. The first strategy, generally referring to enzyme-based coatings, is an advancement of

Functional Polymer Coatings: Principles, Methods, and Applications, First Edition.
Edited by Limin Wu and Jamil Baghdachi.

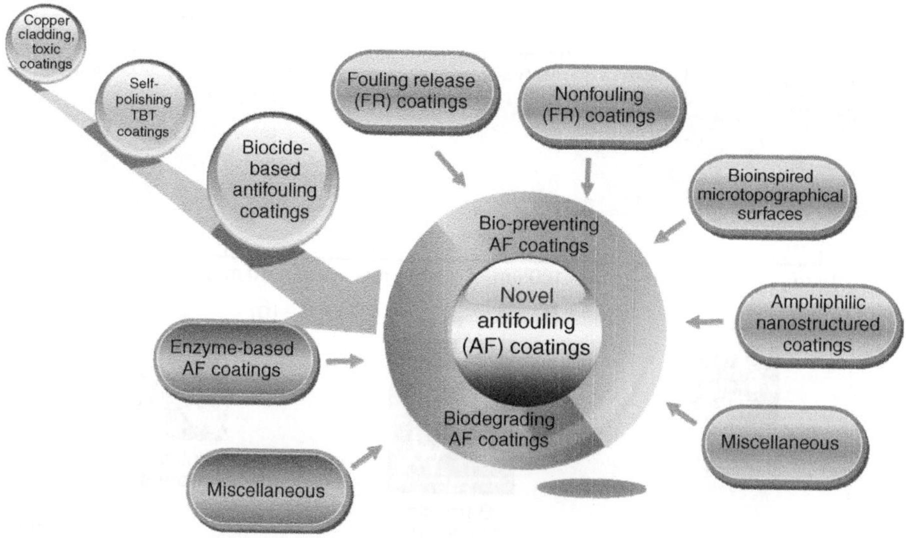

FIG. 11.1 Trend and types of current AF coatings.

biocide-based coatings. The later strategy is proposed based on the viewpoint that prevention is preferable to cure. They prevent foulants from approaching the surface rather than killing the ones that are already attached. Chemically inert AF strategies are produced by tuning the surface tensions and/or surface charges, mimicking natural surface topographies, and mixtures of these approaches.

This chapter seeks to present a thorough picture of the state-of-the-art of these novel nontoxic marine AF solutions developed in the past decade. These novel AF coatings include enzyme-based coatings, fouling release (FR) coatings, nonfouling (NF) coatings, bioinspired micro-topographical coatings, and amphiphilic nanostructure coatings. Their AF principles, fabrication methods, and AF performance are given in detail. The biological behaviors of fouling organisms are described to facilitate understanding of the fouling mechanisms of many different species. Summaries on the problems and future trends of the novel AF coatings are addressed at the end of this chapter.

11.2 MARINE BIOFOULING

Any surfaces submerged in aquatic environments are subjected to unwanted colonization of marine organisms such as bacteria, algae, and barnacles. This process is known as marine biofouling or fouling. Generally, biofoulers are sorted according to size (Fig. 11.2), among which bacteria, unicellular algae (e.g., diatoms), spores of macro algae (e.g., *Ulva*), and larvae of invertebrates (e.g., tubeworms, barnacles) are the most common microorganisms, and algae and invertebrates (e.g., soft corals, sponges, anemones, tunicates, hydroids, barnacles, mussels, and tubeworms) are

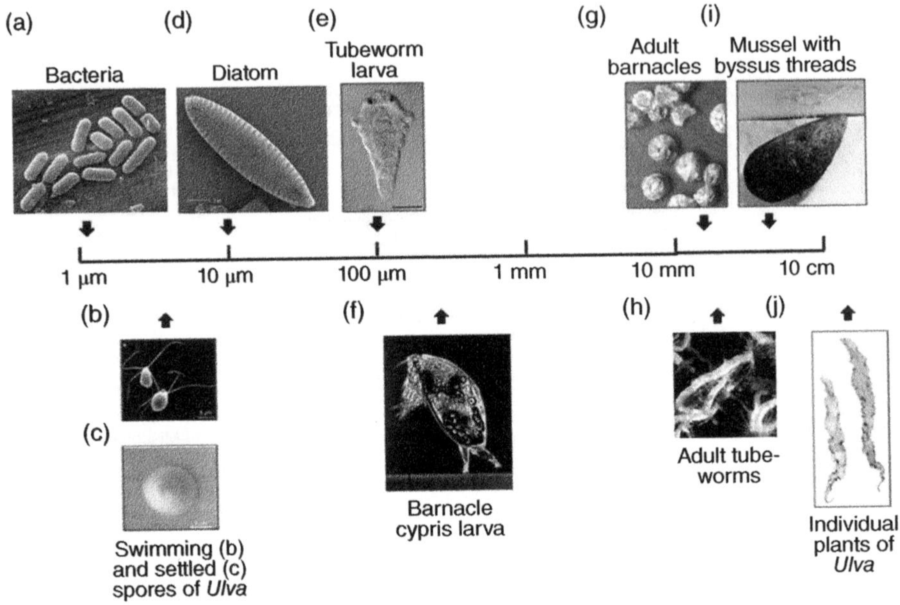

FIG. 11.2 Diversity and size scales of representative fouling organisms. (a) Bacteria; (b) false-color SEM of motile, quadriflagellate spores of the green alga (seaweed) *Ulva*; (c) false-color environmental SEM image of settled spore of *Ulva* showing secreted annulus of swollen adhesive; (d) SEM of diatom (*Navicula*); (e) larva of tube worm, *Hydroides elegans*; (f) barnacle cypris larva (*A. amphitrite*) exploring a surface by its paired antennules (image courtesy of N. Aldred); (g) adult barnacles; (h) adult tubeworms; (i) adult mussels showing byssus threads attached to a surface; (j) individual plants of the green alga (seaweed) *Ulva*. The diagram is intended to indicate relative scale rather than absolute size; individual species within a group can vary significantly in absolute size. Reprinted from a previous work with permission Ref. 1. © 2011 Nature Publishing Group.

categorized as macrofouling species [2]. The adhesion behaviors of the three most representative species—diatoms, *Ulva* spores, and barnacles—are discussed.

Diatoms are frequently selected microorganisms for biofouling assays because they predominate in biofilms and are very difficult to remove from nontoxic FR coatings. Diatoms are unicellular algae enclosed in an elaborate silica wall (called the frustule) composed of two overlapping valves [2]. They do not actively approach the surface; instead they encounter the surface through the action of currents or by settling under gravity [3]. They adhere by secreting sticky extracellular polymeric substances (EPS) whose major composition is polysaccharides. These attached diatoms then divide, rapidly growing into colonies that eventually coalesce to form a compact biofilm, which may reach 500 μm in thickness [2].

In macrofouling, the green algae and barnacles are the most typical flora and fauna. They respectively contribute to soft and hard fouling, and thus overgrow the previously formed microfouling film. The green alga *Ulva*, in the division Chlorophyta, is a thin, flat, green plant, with its margin somewhat torn and ruffled.

It covers rocks in the intertidal zone. This macrofouling alga are also used extensively in AF tests. It adheres to surfaces via the secretion of vast quantities of zoospores. These motile spores undergo transition from first-kiss adhesion (i.e., temporary attachment) to irreversible commitment to adhesion: Once a suitable surface is detected, spores secrete a hydrophilic glycoprotein bioadhesive by exocytosis [2]. Spore germinates within a few hours. Spore division and germination give rise to sporelings that grow to maturity.

Barnacles are arguably the most troublesome invertebrates because of their ability to adhere ship hulls. Nevertheless, adhesion behavior of these troublesome organisms is somewhat different from those of the other pests. Barnacles start their life as nauplius larvae. After a series of molts, they metamorphose into cypris larvae approximately 500 μm in length. Cyprids primarily swim freely in water, but in order to complete the transition to adult barnacles, they must attach themselves to hard substratum. In this process, cyprids, with a pair of antennules, explore a surface by "walking" and secreting temporary adhesives that enable them to attach to and later detach from the surface. During the exploratory period, cyprids attach and detach, leaving behind adhesive footprints that are neither water soluble nor biodegradable. These act as signaling molecules that attract additional cyprids. The cyprids secrete a proteinaceous cement that forms a discrete matrix within hours of selection of a suitable site [2]. The firmly attached juveniles then metamorphose into adults encased in hard calcareous shells. The growth of hard calcareous shell plates can damage coatings by cutting into them, which may lead to corrosion of the underlying metal [4]. For this reason, any successful AF coatings must first be able to suppress fouling of barnacles.

In the ocean, the process of biological fouling generally follows a sequential successional model. It is basically grouped into the following three periods: formation of a molecular conditioning film, microfouling, and macrofouling, as shown in Fig. 11.3. A clean surface physically absorbs organic molecules (e.g., proteins, polysaccharides, and glycoproteins) upon immersion into the aquatic environment. This initial stage takes place because of the extremely short distances between the molecules and surfaces so that passive short range interactions, such as hydrogen bonds and ionic and dipole interactions, bind molecules to the surface [6]. The formation of this conditioning film changes the physicochemical properties of

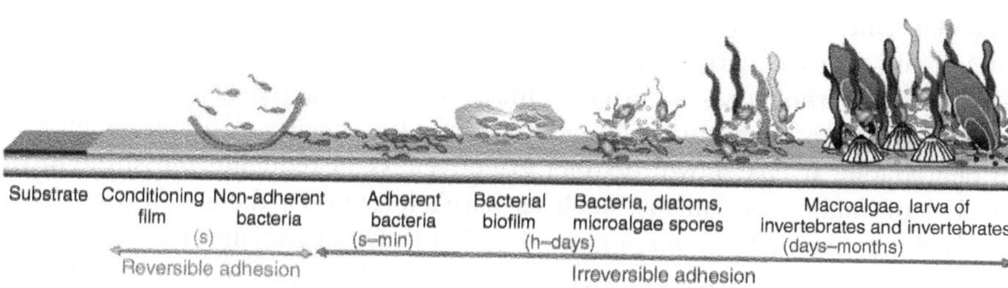

FIG. 11.3 Typical development processes of marine fouling. Reprinted from a previous work with permission Ref. 5. © 2012 American Chemical Society.

the pristine surface (e.g., hydrophobicity, charge density, surface chemistry). Bacteria colonize these films within seconds or hours, and some unicellular species follow (e.g., diatoms). Initially bacteria reversibly adhere to the surface via only weak and noncovalent bonds. However, the reversible adhesion becomes irreversible anchorage upon the secretion of a cellular appendix and exopolymers and by microfoulers at the interface [5]. All these early colonizers form a biofilm, also referred to as microfouling, or slime. The slime provides abundant food to make the colonization of multicellular species (known as macrofouling) possible. A macrofouling community develops and overgrows the microfouling. Generally, macrofouling consists of either soft fouling or hard fouling. Soft fouling comprises algae and invertebrates, such as soft corals, sponges, anemones, tunicates, and hydroids. Hard fouling involves invertebrates such as barnacles, mussels, and tubeworms [2]. At this point, marine fouling has developed to the fullest extent. It becomes a biologically active zone, where different organisms grow and fall dynamically and are no longer able to be completely removed easily.

The classical linear successional model outlined earlier does not apply to all marine organisms. For instance, presence of a biofilm is not necessary for later colonization by macrofoulers, such as zoospores of the alga *Ulva* linza and cyprids of the barnacle *Amphibalanus amphitrite* [7]. Further exploration in fouling mechanism is still in demand, because this would help the design of new AF coatings.

11.3 ENZYME-BASED COATINGS

Enzymes are chemically active proteins that are widespread in nature. Immobilization of enzymes on solid supports has found many applications ranging from medical/dental implants to food packaging materials to large-scale biocatalysis [8]. The idea of using enzymes for marine AF coatings emerged before the year 1985 [9]. In AF coating systems, they break down proteins and polysaccharides, directly target at fouling organisms, and produce other biocidal compounds. Generally, enzyme approaches can be divided into two main groups: direct enzymatic AF and indirect enzymatic AF [10]. The first works via degrading fouling organisms or adhesive components, and the second performs via catalytic production of repellent compounds [11]. Figure 11.4 shows the overall classification of four mechanisms of enzyme-based AF systems.

The enzymes most frequently used to decompose bioadhesives consisting of proteins and polysaccharides include proteases (also termed as peptidases) and glycosylases (carbohydrases). Among these, serine protease subtilisin A (trade name Alcalase®) is a typical proteolytic enzyme. It is readily available, nontoxic, and degradable. It has shown distinct effectiveness in reducing the strength of adhesion of fouling organisms, such as barnacle cyprids, the green alga *Ulva linza*, and the diatom *Navicula perminuta* [12–14]. Other enzymes that take effect include trypsin, dextranase, and α-chymotrypsin (α-CT). Combination of proteolytic and polysaccharide-degrading enzymes can broaden the types of biopolymers degraded [15, 16].

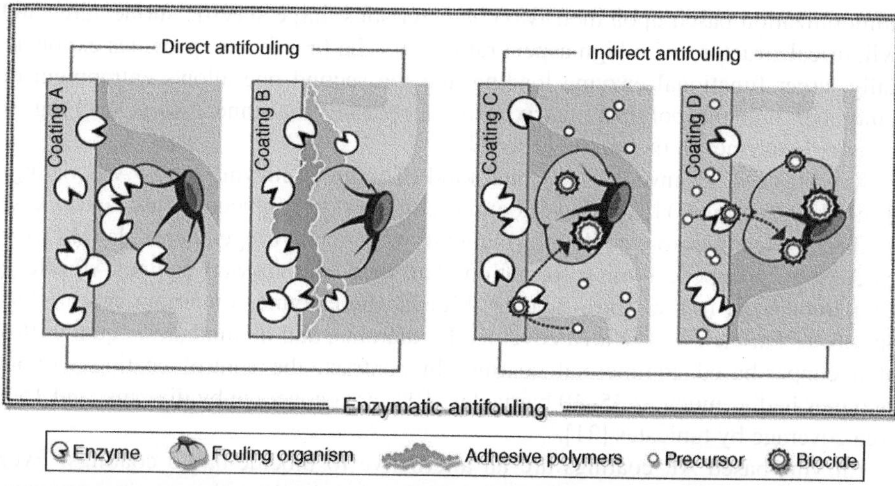

FIG. 11.4 Proposed mechanisms of enzyme-based AF coatings. Coatings A and B are based on direct enzymatic AF. Coatings C and D are based on indirect AF. (**A**) Enzymes attack the adhesive of settling organisms, thus preventing the settlement behavior. (**B**) Enzymes degrade the polymer matrix in the biofilm formed by attached fouling organisms. (**C**) Enzymes catalyze the secretion of AF compounds from aquatic environment. (**D**) Enzymes catalyze the release of AF matters from coatings *in situ*. Reprinted from a previous work with permission Ref. 6. © 2009 Woodhead publishing.

Covalent immobilization and sol–gel entrapment are two common ways of incorporating enzymes into polymeric AF coatings. These methods are more thermostable than enzymes dispersed in solution or chemically attached to the outer surface of poly(dimethylsiloxane) (PDMS) [17, 18]. Enzymes lose much of their activity when they encounter a solid surface due to unfavorable orientation, (partial) unfolding, and denaturation, so fixed binding should be minimized [19]. However, long-term stability is difficult to maintain if the attachment of the enzyme to the substrate is weak. To maintain the activity and stabilization of the enzymes to the greatest extent possible, the immobilization of enzymes has been investigated. Kim et al. reported that α-CT-containing PDMS films by sol–gel entrapment have excellent proteolytic activity and anti-protein performance [17]. Tasso et al. fixed subtilisin A to poly(ethylene-alt-maleic anhydride) (PEMA)-based coatings using covalent immobilization [14]. Due to the hydrophilic and swelling characteristics of PEMA, the swollen 3D structure of PEMA underwater allowed significantly more enzyme loading and better maintenance of activity. Liu et al. used a unique cysteinyl residue to immobilize 6-phospho-β-galactosidase (β-Gal) on a self-assembled maleimide monolayer (SAM) [19]. This did not reduce the activity of the immobilized enzyme.

Incorporation of enzymes at sufficiently high load while keeping them active is another issue for researchers. Asuri et al. largely increased the loading of enzymes (e.g., subtilisin Carlsberg) by successfully conjugating them onto single-walled carbon nanotubes (SWNTs) before being dispersed into poly(methyl methacrylate) (PMMA) AF paint [15]. SWNTs nanotubes are efficient as a support for enzyme immobilization based upon their two characteristics: large specific surface area and cylindrical structure with high aspect ratio [20]. The first feature provides exponentially larger functional enzyme loading and the second one allows entrapment of nanotubes into the polymer matrix, thus decreasing the unnecessary leaching of anchored enzymes in the matrix [15, 16].

Kristensen et al. invented a typical kind of indirect enzyme-based AF coatings (Fig. 11.4; Coating C) [21]. These coatings contain starch, glucoamylase, and hexose oxidase, which therefore could enzymatically generate hydrogen peroxide (H_2O_2) as a deterrent to impede fouling formation. The field trial showed distinct AF effect: after immersion in the North Sea for 97 days, the enzyme-containing coating had only 6–12 barnacles, 10% area coverage by diatoms, and no tunicates, equal to that of a copper-based commercial coating. In contrast, the control coating without enzymes had as much as 35–40 barnacles, 10% area coverage by diatoms, and 15% area coverage by tunicates [21].

Enzyme-based AF coatings are an alternative to biocide-based coatings. Even though the coatings are properly placed in the chemically active coating category, they are more benign to living creatures and the environment. However, long-term stability and efficacy of enzymes in a coating remains a challenge. In addition, according to the definition of biocides in the Biocidal Products Regulations (BPR, formerly known as BPD), an enzyme that biochemically cleaves the adhesive fouling organisms from the surfaces is determined to be within the scope of the regulation. Enzymes have to undergo registration procedures similar to those used for traditional biocides [5].

11.4 FOULING RELEASE COATINGS

Among the novel biocide-free AF coatings, FR coatings were the first to be commercialized. Early on, FR coatings were sidelined by the development of biocide-based SPC AF coatings, until nontoxicity and long service lifetime became required for marine coatings. Currently, FR coatings are considered the most practical type of nontoxic AF coating for ship hulls. FR coatings are a more mature technology than other chemically inert coatings. The two major hydrophobic polymeric materials, silicone and fluoropolymer, are generally used to fabricate FR coatings. Currently, FR coatings focus primarily on the hybridization of other AF-related polymers to bring about some new properties that neither of them possess. This section begins with a brief introduction on how FR coatings combat marine fouling and the related parameters. These can be controlled to increase AF performance, followed by a few recent novel reports concerning FR technology.

11.4.1 Principles of FR Coatings

The AF performance of FR coatings is based on a dual mode of action: nonstick properties and FR behavior [5]. The nonstick properties can be attributed to its hydrophobic nature. Low surface tension can reduce adhesion strength tremendously. The smoothness of the coating surface minimizes the fouling attachment. The reduced adhesion strength results in fewer permanent attachments and makes release easier. This also makes it easier to clean surfaces that are already fouled surface by hydrodynamic stress at different velocities during navigation or by mechanical means [22, 23].

These principles of FR coatings show that wettability, characterized by surface free energy, is the first factor considered. A well-known curve on the relationship of fouling retention degree with critical surface tension of substratum was reported by Baier, as shown in Fig. 11.5. The minimum in relative adhesion occurs at 22–24 mN/m rather than at the lowest critical surface energy. These results suggest that low surface tension is not the only factor contributing for low adhesion. Elastic modulus (E) also plays role in the degree of biofouling. According to a report by Brady et al., the minimum adhesion corresponds to the lowest value of elastic modulus but not to the lowest surface energy of the substrate [25]. The critical removal force (RFc) showed a positive linear relationship with the square root of $E\gamma_C$, correlating far better with either E or γ_C (γ_C is the critical surface energy and E the elastic modulus) [25]. This phenomenon can be explained according to the fracture mechanics. Peel, in-plane shear, and out-of-plane shear are the three kinds of crack-opening modes in fracture mechanics. Substrates with low elastic modulus allow failure in peel mode, which requires less energy than failure in other two shear modes. That is why fluorine-based polymers with low surface tension are inferior in FR performances to silcone-based polymers.

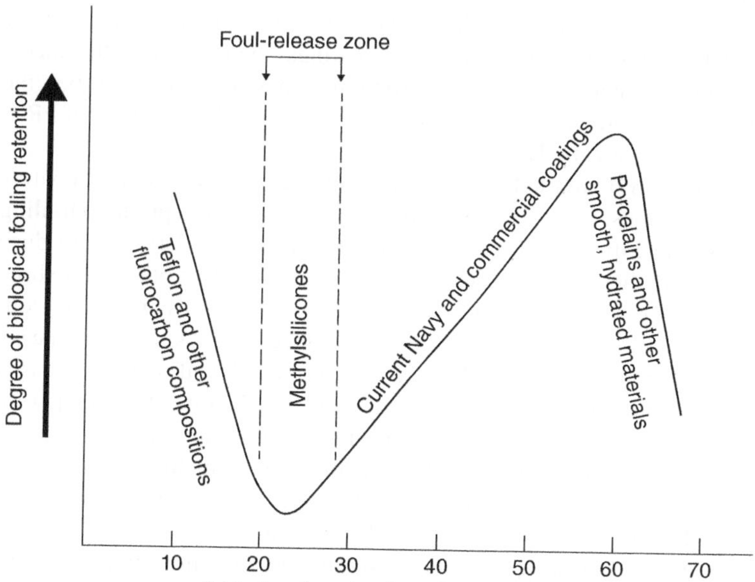

FIG. 11.5 Baier curve representing the empirical relationship between critical substratum surface tension and relative fouling retention. Reprinted from a previous work with permission Ref. 24. © 2006 Springer Science & Business Media B.V.

Thickness is another parameter of FR coatings that affects foulant removal. Singer et al. showed that the pull-off force, P_c, for removing pseudobarnacles from silicone decreased with increasing thickness [26]. A positive linear dependence, $P_c = f(\mathbf{t}^{-1/2})$, was observed between P_c and $\mathbf{t}^{-1/2}$, where **t** is an effective thickness that consists of both silicone top-coat and tie-coat [26]. However, P_c was independent of coating thickness for thick coatings.

Because the coating thickness influences elastic modulus and the chemical formulation determines both surface tension and elastic modulus, each parameter (surface tension, elastic modulus, coating thickness, etc.) interferes with the others, indicating a cooperation effect. Therefore, in order to enhance FR performance, all these factors have to be taken into consideration.

11.4.2 Hybrid Silicone-Based FR Coatings

Silicone-based coatings have many advantages, such as low surface energy, low elastic modulus, and low glass transition temperature (T_g). The low T_g endows the coatings with high local chain mobility. Low elastic modulus causes silicone-based coatings to perform better in AF than Teflon or other fluorocarbon composite coatings, even though their surface tension is not as low as fluoro-based ones. In the early 1970s, Milne patented a serendipitous discovery concerning commercially available silicone elastomers with enhanced FR properties [6]. This discovery later gave birth to the first commercially available silicone FR systems. Pure PDMS elastomer, despite its own FR properties, is no longer the best agent available. Nowadays, it is usually used as a control in experiments; it is always outperformed by other, more sophisticated agents. These silicone-based FR coatings include nanofiller-incorporated silicones, polyurethane- or epoxy-modified silicones, silicone-fluoropolymer coatings, and silicone hydrogels. A detailed summary of the development of FR coatings is available in a review by Lejars et al. [5].

Despite the current market dominance of hybrid silicone-based FR coatings, they still have limitations. First, PDMS-derived coatings are susceptible to fouling at low sailing velocities, the condition of which becomes particularly severe during idle periods. Second, they are not efficacious in the prevention of buildup of slime layers mainly composed of diatoms and bacteria [27]. Another disadvantage stems from their low elastic modulus, which results in poor mechanical strength of the soft coatings and lands in the quandary of being cut, scratched, and punctured easily. Plus, underwater cleaning has to be carried out gently with softer brushes [5]. Stability and maintenance of activity in water are another problem. Visible mass loss has been observed due to the release of small molecules, including cyclic siloxanes, uncrosslinked short linear molecules, additives, and fillers such as carbonate ($CaCO_3$) and silica (SiO_2) have been used as extending fillers to reduce the cost of the silicone elastomers [28]. This caused significant alterations in their morphology, such as increases in surface roughness. The depletion of molecules in the bulk increased the surface roughness and decreased water contact angle (CA) due to the drive of polar groups to the surface [28]. This could also cause FR performance to deteriorate.

11.4.3 Fluoropolymer-Based FR Coatings

Fluoropolymer is famous for its nonpolar nature and ultralow surface tension. At only approximately 10–20 mN/m, it is lower than that of silicone. As early as the 1970s, a polytetrafluoroethylene (PTFE)-filled fluorinated polyurethane paint was investigated [29]. Despite its strong UV resistance and hydrophobicity, it cannot deal with the damage caused by the sharp shell edges of barnacles. The roughened surface enables strong and permanent attachment of mature barnacles and prevents easy removal [5]. Emphasis later shifted to the developments of perfluoroalkyl polymers, perfluoropolyether polymers, and poly(ethylene glycol)-fluoropolymers (details in Section 7.2), among which a few were commercially available, for example, Intersleek 900 from International Paint. Nevertheless, weak repellence to mechanical stresses and damage, especially from macrofoulers like barnacles, is still a very real challenge.

Aizenberg et al. fabricated a multifunctional surface inspired by the carnivorous pitcher plants *Nepenthes*, which possess liquid surfaces rooted in porous architectures to capture prey [30]. This system, called SLIPS, works by infiltrating low-surface-energy porous solids (e.g., Teflon nanofibrous membranes) with lubricating liquids (e.g., 3 M Fluorinert FC-70). SLIPS exhibited a set of outstanding properties: repeatedly self-healable upon damage, extreme liquid repellency with very low CA hysteresis, good stability under hydraulic pressure, optical transparency, and practical applicability [31]. The lubricating film can restore the liquid-repellent function immediately after damage to the porous material by abrasion or impact. It is superior to solid AF fluoro-surfaces in this way. More recently, Yao et al. prepared a kind of omniphobic fluorogel elastomers using 2-perfluorooctylethyl acrylate (PFOEA), perfluoropolyether dimethacrylate (PFPEDMA), and lubricant [32]. The wet environment and the ends of the freely tethered polymer chain help reduce the interfacial friction and thus increase the slipperiness of the surface [32]. In experiments involving AF against proteins, the fluorogels performed similarly to an anti-biofouling surface benchmark (polyethylene glycol (PEG) hydrogels) [32].

11.5 NONFOULING COATINGS

The marine industry is not the only one that has been suffering from the problem of biofouling. In biomedical fields, biofouling (usually called biocontamination) is also a considerable problem. The combat of nonspecific protein adsorption, such as in surface-based diagnostic implantable devices and in protein/peptide therapeutics, fits exactly into the category of NF surface treatment [33, 34]. In PEGylated protein drugs, hydrophilic PEG moieties improve water solubility and increase native protein stability while simultaneously decreasing nonspecific protein adsorption [34]. This is a paradigmatic example of the AF practices in the biomedical industry. Analogies of AF principles exist between marine and biomedical sciences. Hydrophilic AF surface treatments, originally applied to biomedical fields, can also be used in marine AF. Herein, AF polymers that contain a large proportion of hydrophilic groups are referred to as NF (NF) coatings. In this section, four main categories of NF coatings are presented, namely, hybrid PEG-based NF coatings, poly(zwitterionic) NF coatings, NF hydrogels, and other hydrophilic NF coatings (e.g., peptoid-based coatings).

11.5.1 Principles of NF Coatings

Unlike FR coatings, NF coatings can prevent organisms from settling during the primary stage of marine biofouling [35]. As early as 1998, an interesting symposium on NF surface technologies was held at the University of Washington, where the general discussion was centered on the following two questions: "Is it polyethylenoxide or water that imparts and controls fouling resistance?" and "Is fouling resistance due to enthalpy or entropy?" [36]. In the past few years, various answers to these two questions have been reported to tackle the problems of bioadhesion and fouling resistance. Steric repulsion and the hydration layer are the two main theories involved in analyzing the mechanisms underlying the AF phenomenon.

The steric repulsion theory is graphically described from the physical school of thought [36]. Once the protein attaches to the substrate due to adsorption, polymers tethered at one end were subjected to compression. The compression forces compel these originally freely floating polymers to undergo unfavorable changes in configuration. This results in a nonspontaneous decrease in entropy. This entropically driven repulsion force prevents nonspecific protein adsorption. For this reason, this steric repulsion effect is also called an entropic effect or steric stabilization effects [33]. The steric repulsion theory is especially suitable for long-chained polymers tethered to the substrate because of the remarkable compression force. However, only the entropic effect is considered in the steric repulsion model [36]. Little or no room is left for interfacial chemistry and the water-soluble nature of molecular structure.

In aqueous environments, molecular chemistry of hydrophilicity is equally important. Hydrogen bonding is the most distinctive feature of water and water-soluble molecules. Grafted hydrophilic polymer chains like PEG would seem to attract water-soluble fouling biomolecules through hydrogen bonding, contrary to the previous assumption that this leads solely to repulsive forces between tethered layers and proteins [36]. However, because lower interfacial energies can form between hydrophilic surfaces and water molecules, the hydrated layer formed by the interactions between these two phases creates a strong energy barrier that resists the adsorption of other molecules. In other words, repulsion occurs when the orientational distribution is affected in adjoining water layers and the hydrogen bonding in the hydration layer is disrupted [36]. This is the universal interpretation of hydration layer theory. Unlike the entropic assumption of steric repulsion, this chemical school of thought is presented from the enthalpic point of view [36].

It is currently universally acknowledged that both the hydration layer and steric repulsion contribute to resistance to biofouling. In the surface modification with hydrophilic polymers, influencing factors like chemistry of hydrophilic functional groups, packing density, and chain length of polymer are all of considerable significance. FR coatings have disadvantages. NF hydrophilic coatings also have intrinsic limitations, specifically swelling, which reduces short-term stability.

11.5.2 PEG-Based NF Coatings

For decades in biomedical science, PEG-rich surfaces have long enjoyed the reputation for their outstanding performance of low thrombogenicity [37]. Due to their excellent nonspecific protein adsorption repellency, PEG-based coatings have also been noted for their effectiveness in marine AF, especially at the initial stage.

The most commonly used approaches to immobilization of PEG or EG moieties include physical adsorption, chemical adsorption, direct covalent attachment, and block or graft copolymerization. Yandi et al. established poly(HEMA-*co*-PEG_{10}MA) copolymer brushes on the SAM-treated surfaces via surface-initiated atom-transfer radical polymerization (SI-ATRP) [38]. Zhao et al. successfully built the non-swelling PEG molecular net-cloth for the first time on the polymeric substrates (as long as containing C–H groups, e.g., LDPE) in situ via a novel mild visible-light induced surface-controlled graft cross-linking polymerization (VSCGCP) technique (Fig. 11.6) [39]. Unlike traditional easily swelling PEG polymer brushes and hydrogels, this PEG net-cloth showed outstanding ultra-low-swelling behavior.

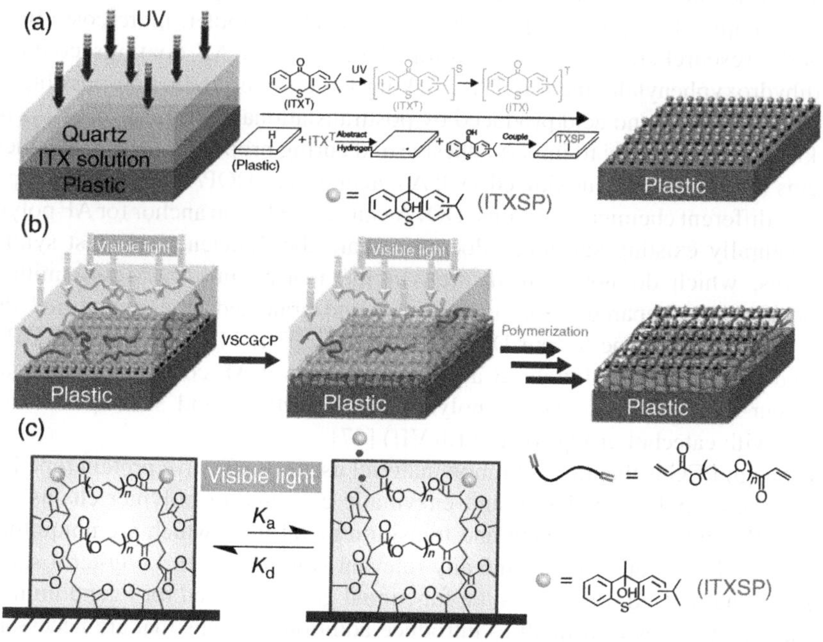

FIG. 11.6 Synthesis of PEG molecular net-cloth covalently attached on LDPE. (a) Scheme of photoreduction reaction of ITX on LDPE. (b, c) Scheme of VSCGCP process of PEGDA network. Adapted from a previous work with permission Ref. 39. © 2014 Nature Publishing Group.

Unfortunately, swelling-induced instability has always been a problem with many other traditional PEG polymer brushes or hydrogels. Swelling, in one sense, could be favorable as it indicates the fully hydrated interface and thereafter the AF properties. However, excessive swelling could greatly compromise mechanical properties and shorten the service life of coatings. For this reason, most PEG-based coatings were prepared with the addition of other polymer additives to relieve the swelling conditions.

With PEG and cardanol, Kim et al. synthesized a new star-shaped polymer by ATRP [40–42]. Its structure is shown in Fig. 11.7. Cardanol was used because of its favorable features, such as renewable use, easy separation, capacity for self-cross-linkage, capacity for chemically modification, and antibacterial properties [42, 43]. The polymer can be coated on ultra-filtration membranes after UV-curing and exhibited evident bio- and oil-fouling resistance. Gillich et al. concerned on the synthesis of second-generation, ethylene glycol dendrons covalently linked to a surface anchor containing catechol groups [44]. The dendron (Fig. 11.8 V) showed less hydrated absorbed mass, lower viscoelasticity, and similar or even better performance in fouling resistance, as compared to the control linear PEG sample (Fig. 11.8 VI) [44]. The molecular dendric structure provides a new way of increasing the density of PEG, which decreases hydration, reduces film thickness, and increases film stiffness while maintaining good AF properties [44].

The disadvantage of the two nonlinear PEG-based polymers described earlier is their complicated synthesis procedures. To develop simpler, more cost-effective methods, researchers have also developed PEG-based AF coatings containing 3,4-dihydroxyphenylalanine (DOPA), making large-scale production possible [45–47]. DOPA is an amino acid produced by posttranslational modification of Tyr [48]. It is known for its ability to impart adhesive and curing properties to mussel adhesive proteins (MAPs) [48]. Unoxidized DOPA catechol and DOPA quinone are capable of many different chemical reactions. DOPA can be used as an anchor for AF polymer. This naturally existing substance does not share the deficiency of most synthetic adhesives, which do not perform well in the marine milieu. Messersmith and coworkers have prepared a variety of linear and branched DOPA-modified PEGs (PEG-DOPAs) with one to four DOPA end groups (Fig. 11.8 I–IV) [45, 46]. Partly inspired by this work, Mizrahi et al. prepared a novel AF coating by aggregating short four-armed PEG-dopamine polymers into particles and subsequent surface binding with catechol groups (Fig. 11.8 VII) [47].

Although PEG is the most common material used for effective protein repellence, it undergoes rapid autoxidation and enzymatic cleavage of polymer chains, especially in the presence of oxygen and transition metal ions, which are ubiquitous in seawater and most other biochemically relevant solutions [49]. Long-term stability cannot be maintained. PEG lacks the functional groups needed for ligand immobilization, which makes it difficult to introduce additional groups that are necessary to adjusting its intrinsic AF properties [50]. This rendered PEG-based polymers less effective as AF materials in the complex marine biofouling. The aforementioned limitations pose great challenges for PEG to effectively cope with the marine fouling issue in terms of efficiency and durability.

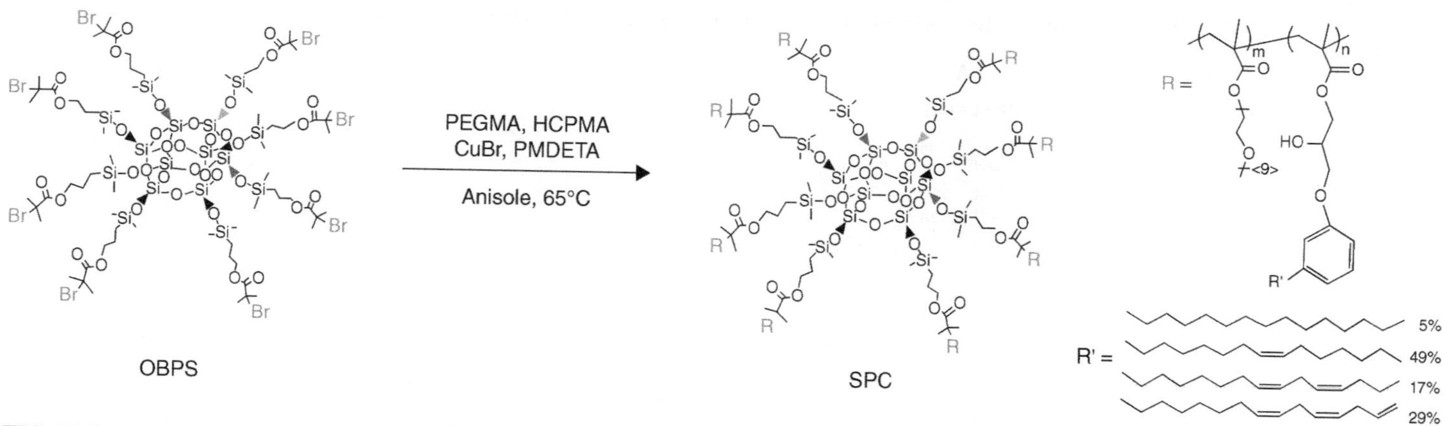

FIG. 11.7 Synthesis of star-shaped polymers with poly(ethylene glycol) and cardanol side groups via atom transfer radical polymerization using an octafunctional silsesquioxane initiator. Reprinted from a previous work with permission Ref. 42. © 2013 Royal Society of Chemistry.

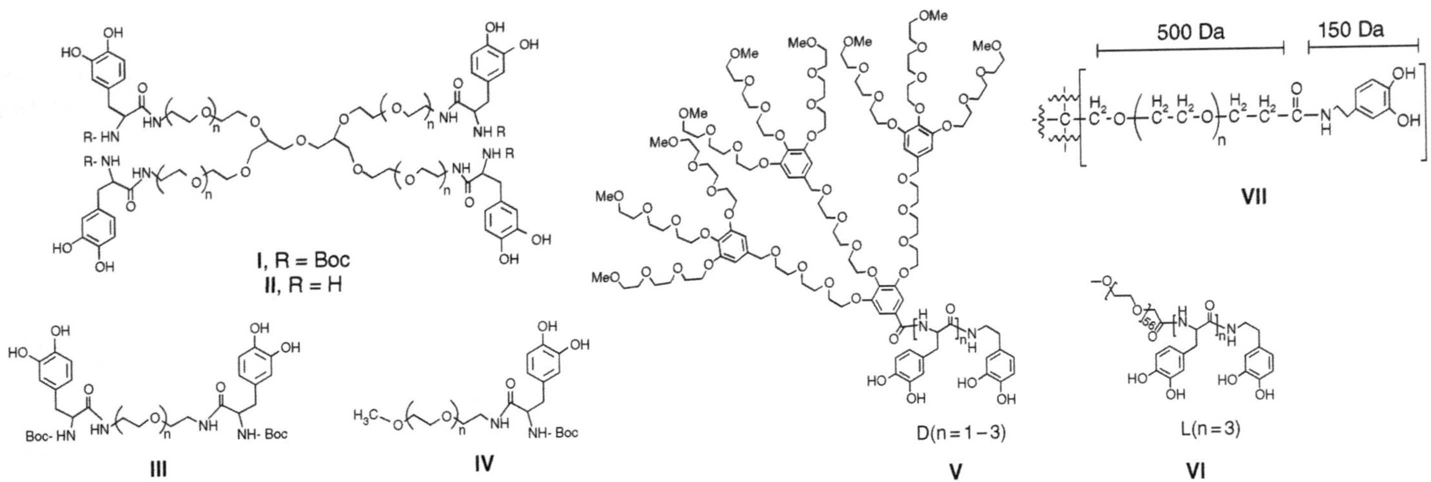

FIG. 11.8 Chemical structures of DOPA-modified PEGs. (I–IV) Adapted from a previous work with permission Ref. 46. © 2002 American Chemical Society. (V, VI) Adapted from a previous work with permission Ref. 44. © 2011 American Chemical Society. (VII) Adapted from a previous work with permission Ref. 47. © 2013 American Chemical Society.

11.5.3 Poly(Zwitterionic) NF Coatings

Recently, zwitterionic surface coatings have drawn attention as replacements for PEG-based coatings. They can be used as antibacterial agent, anticoagulant, biomedical diagnostic agent, as well as ultra-filtration membranes. Keefe et al. reported their use of poly(carboxybetaine) as an alternative to PEG, avoiding the need to compromise between stability and affinity in protein therapeutics [51]. It is now universally recognized that zwitterionic coatings could be applied for marine AF as well.

The AF principle of poly(zwitterionic) coatings is similar to that of PEG-based hydrophilic ones, specifically formation of a hydration layer by zwitterionic polymers or mixed charged groups, as depicted in Fig. 11.9. Nevertheless, the ionic solvation facilitates stronger hydration than PEG [53].

Generally, there are two types of zwitterionic polymer. One has both positively and negatively charged groups on the same monomer. These include phosphorylcholines and polybetaines. The other has differently charged components on separate monomers, creating a mixed charge complex. These are known as polyampholytes. Polybetaines can be further divided into three major groups based on the negatively charged groups: sulfonate-betaines (SB), carboxylate-betaines (CB), and phosphonate-betaines (PB) (Fig. 11.10).

Currently, various zwitterionic polymers have demonstrated resistance to various protein and cell adsorption, for example, fibrinogen, serum protein, platelet, lysozyme, bacteria (*Staphylococcus epidermidis*, *Pseudomonas aeruginosa*, *Escherichia coli*), and fibroblasts. However, only a few studies have placed emphasis on the development of zwitterionic NF coatings.

Of the first type of zwitterionic polymer, in which positive and negative charges coexist on the same monomers, poly(sulfobetaine methacrylate) (pSBMA) and poly(carboxybetaine methacrylate) (pCBMA) (Fig. 11.11b and c) are most intensively studied. For marine AF applications, Jiang et al. prepared a series of sulfobetaine-based polymers and carboxybetaine-based polymers (Fig. 11.11a) using the SI-ATRP method [49, 54]. Both types of materials show ultralow marine-fouling properties, as evaluated using green marine algae diatoms, sporelings of *Ulva*, and cypris barnacle larva as fouling organisms. Inspired by MAPs, Messersmith et al. synthesized a new bifunctional tripeptide bromide (BrYKY) as a mussel-mimetic peptide surface initiator, combining ATRP initiating alkyl bromide with DOPA and lysine [55]. This BrYKY-initiated graft-form polymerization of SBMA on substrate produced high-density AF pSBMA brushes, which could reduce bacterial adhesion for 24h and resist mammalian cell adhesion for at least 4 months, demonstrating long-term stability. Poly(zwitterionic) coatings are considered biocompatible. These new materials make it possible to develop environmentally benign, effective, durable, low-cost AF coatings. Unlike conventional ways of quantified evaluation, which involve settlement assays, AF properties is characterized by Aldred using EthoVision vedio software to record the 2D tracking of the cypris larvae of barnacles on pSBMA and pCBMA [54]. Results show that pSBMA cyprids explored the surface but did not settle, whereas on pCBMA, cyprids did not attempt exploration and left the surface quickly [1, 54]. This offers a new path for the characterization of AF performance.

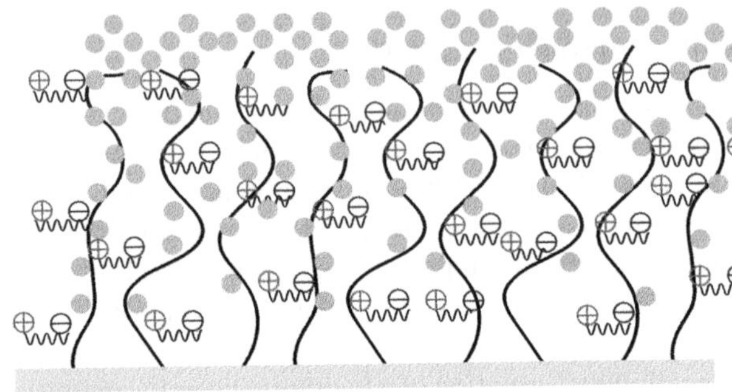

FIG. 11.9 Illustration of chain hydration and chain flexibility of zwitterionic polymers. Reprinted from a previous source with permission Ref. 52. © 2010 Elsevier Ltd.

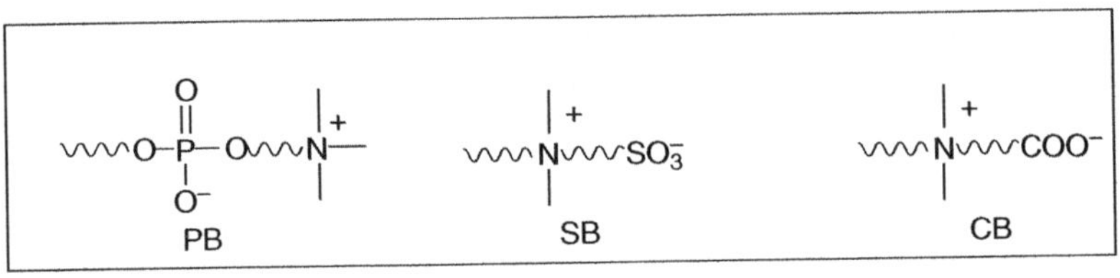

FIG. 11.10 Chemical structures of phosphonate-betaines (PB), sulfonate-betaines (SB), and carboxylate-betaines (CB). Reprinted from a previous source with permission Ref. 52. © 2010 Elsevier Ltd.

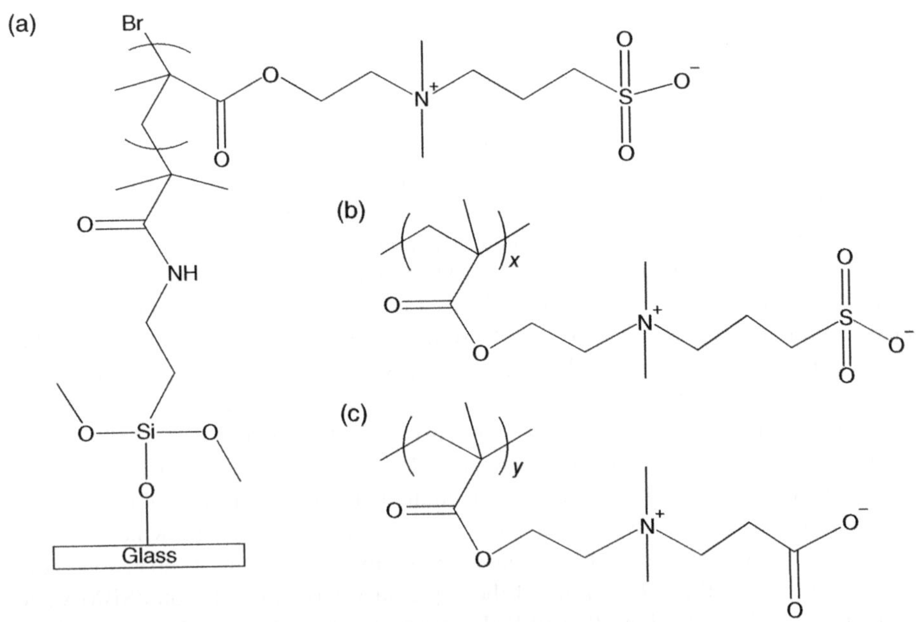

FIG. 11.11 Illustration of (a) surface grafting pSBMA on glass surfaces and chemical structures of (b) pSBMA and (c) pCBMA Ref. 49, 54.

By applying the mixed charge principle, Chen and Jiang prepared polyampholyte hydrogels from the mixture of positively charged compounds, such as aminoethyl methacrylate hydrochloride, and negatively charged compounds, such as 2-carboxyethyl acrylate [50]. The work demonstrated that a uniform charge distribution and homogeneous charge balance on the surface are key to producing AF properties, and this has become a goal in the design of new NF polyzwitterionic materials.

11.5.4 Other Hydrophilic NF Materials

11.5.4.1 Self-Generating Hydrogel Surfaces Hydrogels, a cross-linked hydrophilic polymer network, have been extensively studied because they are an environmentally benign marine AF material with strong hydration ability and low elastic modulus. Despite their considerable ability to inhibit germination of spores and settlement of barnacles, soft hydrogels in reality suffer from the fatal weaknesses of short-term stability and can be easily damaged due to their poor mechanical properties and poor adhesion to substrates. Even though it is not impossible, it is difficult to use these materials on actual ship hulls. Xue et al. designed an AF hydrogel coating with slippery hydrogel-released hydrous surface (SHRHS) by physical blending of sodium polyacrylate (PAAS) powder into the organic silicon resin [56]. The stain removal lab assays and marine field trials showed an effective attachment resistance to biofouling. It is explained as the synergetic effect of self-cleaning and self-regeneration abilities: PAAS can trap water and swell into a hydrogel layer on its surface as a water film to prevent attachment by foulants, and the superficial hydrolysis of PAAS resulted in hydrogel peeling simultaneously carrying away resident organisms (Fig. 11.12). Similarly, Wu et al. prepared self-generating hydrogel surfaces by mixing a polyfunctional axiridine cross-linking agent and a self-polishing resin containing the hydrolysable tributylsilyl methacrylate or triisopropylsilyl methacrylate [57, 58]. Unlike the self-polishing mechanism (Fig. 11.13a), the incorporation of the cross-linking agent makes it possible to hydrolyze at a relatively

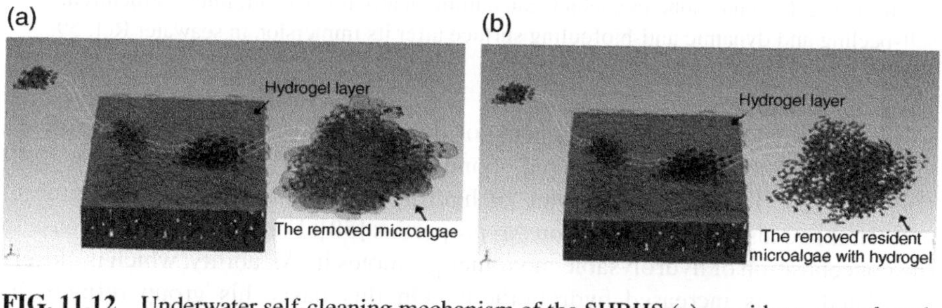

FIG. 11.12 Underwater self-cleaning mechanism of the SHRHS (a) provides a smooth and very slippery surface. The trapped water is retained by the hydrogel surface, rendering attachment difficult; (b) this produces a self-regeneration characteristic and the resident organisms can be carried off as the hydrogel peels. Adapted from a previous work with permission Ref. 56. © 2014 Elsevier Ltd.

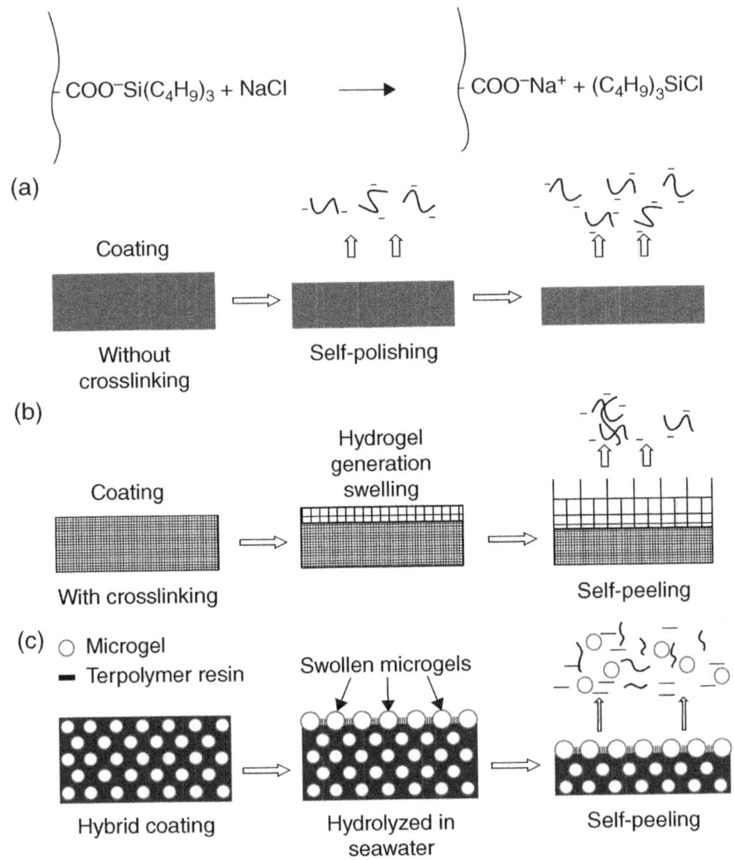

FIG. 11.13 Chemical reaction and schematic illustrations of (a) self-polishing and (b) hydrogel-generating and self-peeling of novel coatings immersed in seawater. Reprinted from a previous work with permission Ref. 57. © 2011 Elsevier Ltd. All rights reserved. Schematic of (c): a novel hybrid coating made of a hydrolysable polymer resin (gray) and small microgels (white spheres), which can automatically form a soft, micro-structured, self-peeling and dynamic anti-biofouling surface after its immersion in seawater Ref. 59. Ref. 59. © 2013 Royal Society of Chemistry.

slow, controlled rate, and the fragments of the gel network only wash away after a sufficient number of cross-linked bonds are hydrolyzed and broken [57] (Fig. 11.13b). The type and amount of hydrolyzable monomer and the degree of cross-linkage also have an effect on AF properties [58]. For instance, an increase in the concentration of hydrolysable monomer promotes its AF ability, which is closely correlated to its increased hydrophilicity. On this basis, his group refined the coatings by introducing a certain amount of narrow-size distributed cross-linkable PAAM-co-MAA microgels prepared via a two-step dispersion polymerization process (Fig. 11.13c) [59]. The cross-linked hard thin film formed during the coating process created a self-generated topographical microgel surface. In the system, both

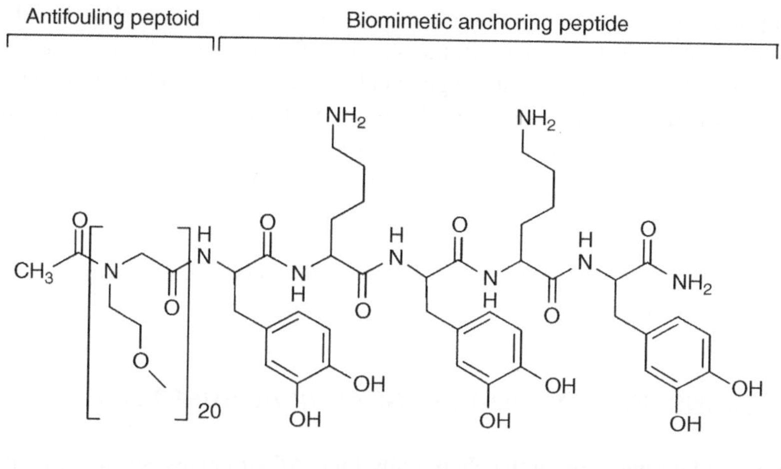

FIG. 11.14 AF peptidomimetic polymer. Reprinted from a previous work with permission Ref. 61. © 2005 American Chemical Society.

the self-peeling and topographically structured surface, characterized by small microgel bumps, can produce AF properties [59].

11.5.4.2 Peptoid-Based Protein Resistant Surfaces Peptoids were synthesized as potential PEG alternatives. They are a class of non-natural peptidomimetic molecules with protein-like backbone and side chains on the nitrogen atom instead of on the α-carbon [16, 48, 60]. To produce effective AF surfaces, the general principles in designing the side chains of peptoids are as follows: strong proton-accepting ability, lack of hydrogen bond donors, and water solubility [48, 61]. Messersmith et al. designed a peptidomimetic polymer consisting of two parts (Fig. 11.14) [61]. In the peptoid section, the hydrophilic methoxyethyl group was selected as the side chain because of its resemblance to the repeat unit of PEG and its lack of a hydrogen bond donor. This rendered the polymer water-soluble. An anchoring peptide consisting of a 5-mer peptide of alternating DOPA and Lys residues was used. This peptide directly mimics Mefp-5, one of the major mussel adhesive foot proteins. Lin et al. synthesized thiol-terminated poly(*N*-methyl-β-alanine) (PMeA) and poly(*N*-ethyl-β-alanine) (PEtA) via cobalt-catalyzed carbonylative polymerization and grafted onto Au surfaces [62]. Lau et al. modified the TiO_2 surfaces with polysarcosine (poly (*N*-methyl glycine)), a peptoid with a simple side chain [63]. Reverse-phase high-pressure liquid chromatography showed that polysarcosine is far more hydrophilic than its peptide analogue, polyalanine, indicating its stronger interaction with water [63]. The surface plasmon resonance (SPR), optical waveguide lightmode spectroscopy (OWLS), and ellipsometer measurements indicated that all these fabricated peptoid surfaces revealed high resistance to protein and cell adsorption similar to that of PEG under identical conditions, some of which may be comparable to the best AF materials [61–63].

Peptoids enjoy several additional important characteristics. The backbone is resistant to proteases due to the misattachment of the side chain from α-carbon to the amide nitrogen. Their molecular weight can be precisely controlled, they can be synthesized with high yield, and many different peptoid sequences can be produced through use of both natural and non-natural side chains [48, 61]. All these factors facilitate the creation of a new space for further molecular design of this kind of long-term AF polymers.

To conclude, despite the diversity of hydrophilic NF materials, most of these coatings are prepared based on the derivatives of PEG because they can decrease the limitations of PEG.

11.6 BIOINSPIRED MICRO-TOPOGRAPHICAL SURFACES

As explained earlier, among the chemically inert AF strategies, surface chemistry, in relation to surface energy, has been fully exploited to build surfaces with extreme wettability (either hydrophobic or hydrophilic). This section focuses on physically deterring fouling organisms, that is, the creation of microtopographical surfaces.

Microtopographical surfaces are often visible in nature. Through evolution, nature has arrived at what is optimal and has long been a school for scholars and researchers. In terms of surface topography, a variety of flora and fauna, such as lotus leaves, rice leaves, butterfly wings, mosquito eyes, moth eyes, cicada wings, red rose petals, gecko feet, desert beetles, spider silks, shark skins, and fish scales, have their own distinctive characteristics [65], for example, mosquito compound eyes for antifogging, van der Waals generated strong releasable adhesion by gecko feet, and desert beetles for water collection [66–68]. All these properties are associated with their unique topographies.

In marine environments, whales are often covered with barnacles, but sharks are not. Microstructure studies of the hydrophilic shark skin (the hydrophilicity is due to the secretion of mucus) have shown that there are ordered riblets on dermal denticles (Fig. 11.15). The riblets encourage anisotropic flow, lifting and pinning the generated vortices. Lifting reduces the shear stress and pinning reduces the perpendicular motion of fluid, both of which can reduce drag, increase the water flow rate, and hence facilitate FR and deter the colonization of foulants.

The generally accepted AF mechanisms of microtopographical surfaces and the approaches to microtopographical AF surfaces are presented and discussed in the following.

11.6.1 AF Principles of Bioinspired Microtopographical Surfaces

The microtopographies responsible for effective AF have been qualitatively explained by two theories: topography-induced wettability and attachment point theory.

Topography induces change in surface wettability, as indicated by Wenzel's equation (Chapter 2). Theoretically, if a flat substrate gives a CA higher than 90°, it becomes superhydrophobic when the roughness increases. However, a smooth

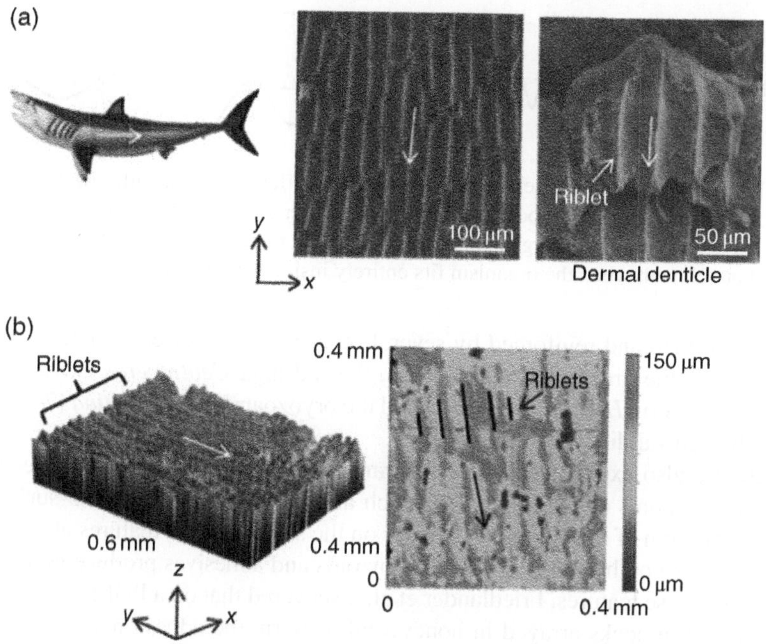

FIG. 11.15 (a) Digital photos and scanning electron microscope (SEM) images of shark skin. (b) Optical profiler height maps of each actual sample. Arrows indicate the direction of anisotropic water droplet movement. Reprinted from a previous work with permission Ref. 64. © 2012 Royal Society of Chemistry.

surface with a CA lower than 90° becomes superhydrophilic as roughness increases [69]. Topography-induced enhancement of AF behavior is the same as that of hydrophilic NF coatings and hydrophobic FR coatings.

The attachment point theory, first brought up by Scardino et al., has been well developed [70]. As has already been pointed out, quite a few microorganisms seek shelter against water currents by attaching in concaves of the surfaces like grooves and pits [71]. The attached areas they select for maximum protection are slightly larger than themselves to allow the fullest contact with the surface [72]. Based on this behavior, attachment point theory suggests attachment strength could be weakened by reduced attachment points between organisms and substrates. As described in Fig. 11.16, fouling organisms larger than the wavelength of the surface topography exhibit fewer attachment points depending on the feature wavelength. Only the feature wavelength, which is slightly smaller than the organisms themselves, lead to the most efficient deterrent AF surfaces (Fig. 11.16c). The decrease in feature wavelength generates more attachment points and causes AF performance to deteriorate (Fig. 11.16b). Surface contact (multiple contact points) took place when either the surface was smooth or the surface wavelength was larger than the size of organisms (Fig. 11.6a & d). In both cases, fouling was largely increased. This theory was first testified with four different diatom species against polyimide surfaces with a range of feature dimensions [70]. The theory was

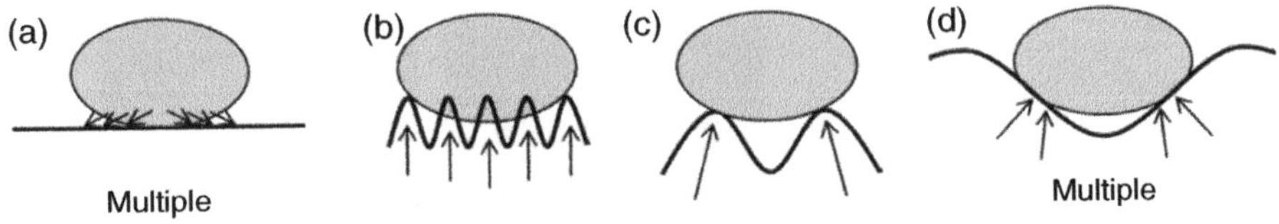

FIG. 11.16 Points of attachment on topographies of different wavelengths. (a) Multiple points of attachment on the smooth surface; (b)–(c) fewer points of attachment with increasing feature wavelength when the organism is bigger than the microstructure; and (d) multiple points of attachment when the organism fits entirely inside the microstructure Ref. 73.

further examined and reinforced by several other fouling species, namely the diatom *Amphora* sp., the green alga *Ulva rigida*, the red alga *Centroceras clavulatum*, the serpulid tube worm *Hydroides elegans*, and the bryozoan *Bugula neritina* [74]. The test results showed validity of the theory.

There are also exceptions to the attachment point theory model. Beigbeder et al. found that if spores of the alga *Ulva* attach and grow on topographical surfaces, the adhesion strength of sporelings are greater on the surfaces with features at micrometer scale than on smooth surfaces [75]. The rhizoids and adhesives produced may be able to penetrate these features. Friedlander et al. discovered that on a PDMS substrate with microscale hummocks arrayed in honeycomb pattern, the adhesion was significantly smaller than flat controls during the first 2 h, but significantly larger after longer periods of exposure [76]. This reversal is attributed to the bacterially induced wetting transitions and hence the promotion in adhesion to these wetted topographic surfaces [76].

A quantitative topographic model, the engineered roughness index model (ERI), was introduced by Schumacher et al. [77]. ERI is a dimensionless ratio that can be used to characterize the roughness of engineered surface topography. It can be calculated by the Wenzel's roughness factor (r), the degree of freedom to move around the surface (df), and the depressed surface fraction (f_d) according to Equation (11.1), among which f_d is equivalent to $1-\varphi_S$, where φ_S is the surface solid fraction [77]. Results showed that the attachment densities of spore settlement are roughly inverse with ERI_I. For refinement of ERI_I, ERI_{II} was brought up by substituting df with the number of distinct features in the surface design (n) [78], as presented in Equation (11.2). In this improved model, regression showed an excellent correlation with other organisms, and the parameter n was particularly useful for surfaces with more than one feature, such as Sharklet AF™ [73]. The level of spore settlement tested and predicted by ERI_{II} (Fig. 11.17) confirmed that the engineered nanoforce gradients may be an effective tool and predictive model for the design of unique nontoxic, nonfouling surfaces for marine applications [79].

$$\mathrm{ERI_I} = \frac{r \cdot df}{f_d} \quad (11.1)$$

$$\mathrm{ERI_{II}} = \frac{r \cdot n}{f_d} \quad (11.2)$$

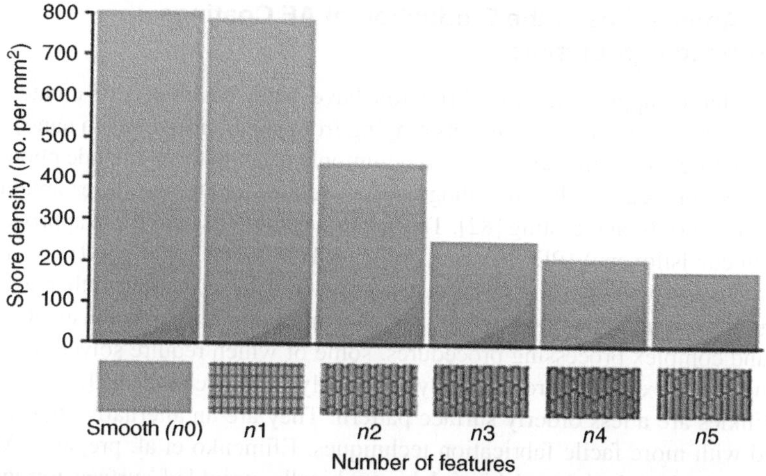

FIG. 11.17 Settlement of spores of *Ulva* on microengineered Sharklet AF patterns molded in PDMSe with increasing numbers of distinct topographic features. $n=0$ (smooth), 1–5, where n is the number of distinct topographic features. Reprinted from a previous work with permission Ref. 1. © 2011 Nature Publishing Group.

$$\mathrm{ERI_{II}} \cdot \mathrm{Re} = \frac{r \cdot n}{f_{\mathrm{d}}} \cdot \frac{\rho VL}{\mu} \tag{11.3}$$

By multiplying with the Reynolds number (Re) of organisms (Eq. 11.3), the model was proven to predict the attachment of both zoospores of *Ulva* and cells of the marine bacterium very well [80]. The Reynolds number was composed of the density (ρ) and viscosity (μ) of the assay solution, the velocity (V) of organisms in the solution, and the characteristic length (L) of organisms. The incorporation of this parameter for the first time enabled different organisms to be applied to this model in a uniform manner. It revealed a negative linear correlation between the attachment density and $\mathrm{ERI_{II}}$-R_{g}. The ERI series has evolved into a more comprehensive theory because it not only takes surface roughness into consideration but also addresses other influencing factors (e.g., the features of organisms), and hence being suitable for a wider variety of surface patterns.

By integrating attachment point theory and the ERI model, Decker produced a new surface energetic attachment model to predict the efficiency of AF topographies [81]. Generally, this model is based on both the properties of the interfaces and the dimensions and configuration of the substrate topographies against the organisms. A strong correlation was detected between experimental results and Monte Carlo simulation results for a wide range of marine foulants: *Ulva linza*, the diatom *Navicula incerta*, the marine bacterium *Cobetia marina*, and the barnacle cyprid *Balanus amphitrite* [81]. The advantages of this model are its ability to predict the enhancement of fouling attachment and its deterrence of fouling adhesion to a given topography [81].

11.6.2 Approaches to the Production of AF Coatings with Surface Topographies

Many different approaches to fabrication have been developed to create surface topographies with feature dimensions ranging from submicron (or even nano) scale to several hundred micrometers. The most commonly used methods include photolithography, electron beam (e-beam) lithography, reactive ion etching, laser ablation, hot embossing, molds, and casting [82]. The substrates often used in these approaches are poly(dimethylsiloxane) (PDMS), polyimide, polycarbonate, and poly(vinylchloride) (PVC) [5]. Even though the desired patterns could be generated rather accurately through these nanofabrication techniques, they normally require sophisticated equipment and complex processing procedures, some of which require solvent processing with the use of excess environmentally unfriendly etching chemicals [83].

Wrinkles are a less orderly surface pattern. They are an alternative that could be formed with more facile fabrication techniques. Efimenko et al. prepared AF coatings with both uniaxial and biaxial hierarchically wrinkled surface topographies (HWST) [84]. The hierarchical wrinkles are achieved using the following sequential steps: mechanical stretching (either uniaxially or biaxially), extended under ultraviolet/ozone exposure, modification with 1*H*,1*H*,2*H*,2*H*-perfluorodecyltrichlorosilane (tF8H2) vapor deposition, and releasing. Field tests showed that even after prolonged exposure to seawater for 18 months, the HWST coatings remained relatively free of biofouling [84]. For fine-tuning of wrinkle morphology, Evensen et al. used the interplay between wrinkle patterns and the thicknesses of argon plasma–treated PDMS slabs [85].

In order to make the surface functions (including adhesion control and wetting/dewetting behavior) adjustable, Dan et al. prepared a series of dynamic topologies (strips, island-in-sea, network) that can be switched between an on and an off state in response to thermal (Fig. 11.18) and UV stimuli (Fig. 11.19) [83, 88, 89].

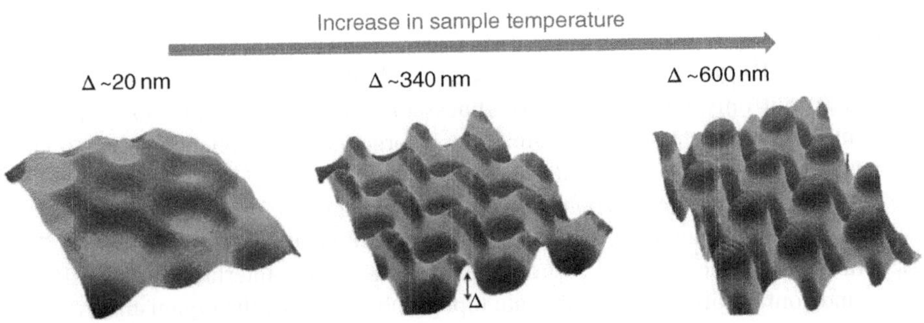

FIG. 11.18 Patterned liquid–crystal–polymer-based actuators can undergo radical shape deformations: regions of polymerized cholesteric liquid–crystal can rise among regions polymerized in the isotropic phase owing to differences in the thermal expansion coefficients (Δ) between the two phases. Reprinted from a previous work with permission Ref. 86. © 2007 Nature Publishing Group.

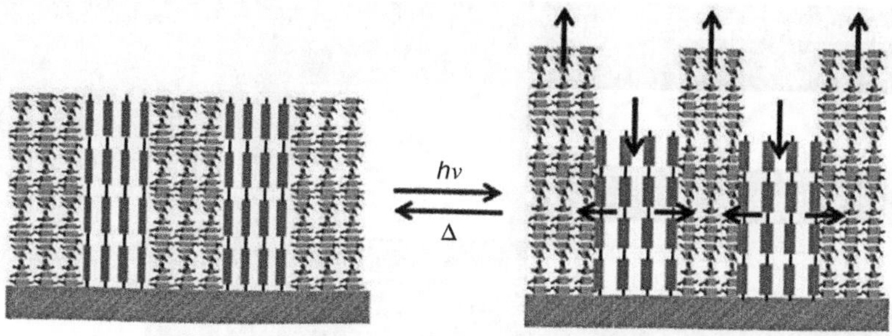

FIG. 11.19 Liquid–crystal networks containing striped patterns of alternating areas with chiral nematic order and homeotropic orientation. Upon exposure, the chiral nematic areas expand perpendicular to the plane of the film and the homeotropic areas contract in the perpendicular direction. In the plane of the film, the chiral nematic areas contract and the homeotropic areas expand. Reprinted from a previous work with permission Ref. 87. © 2012 WILEY-VCH.

In general, most patterns fabricated have only one feature dimension. Nevertheless, different fouling organisms vary considerably in shape and dimension during their initial colonization stage. The surface with one feature dimension cannot prevent the attachment of all marine organisms. Dealing with this problem, several ways have been brought up. The abovementioned hierarchically wrinkled substrate is conceptually one viable solution (Fig. 11.20) [84]. The wrinkles have lengths of different scales (generations) from tens of nanometers to a fraction of a millimeter, where every larger generation resides underneath and represents a scaled-up version of the smaller generation [84]. Schumacher et al. superimposed the smaller *Ulva*-specific topographies onto the larger barnacle-specific surfaces to form a hierarchical structure [82]. This provides new insight into the design of structures able to repel multiple settling species [82]. Epstein et al. developed a surface based on PDMS elastomer with microscale wrinkle topography [90]. Inspired by NF echinoderm skin, which has highly dynamic surface structures that can mechanically frustrate bio-attachment [90], they did so mainly by stretching and oxygen plasma treatment. In adhesion experiments performed on the bacteria *P. aeruginosa*, biofilm colonization decreased 54% with the introduction of submicron wrinkles [90]. For the surfaces with valleys at ≈1 μm, despite increased adsorption under static conditions, a sharp decrease in biomass was observed after dynamic strain was applied. This phenomenon shows that wider valleys provide more suitable shelters for bacterial colonization but can simultaneously frustrate the accumulated foulants because more vibration energy can be transferred with increased contact area [90]. The generation energy can be from the capture of passive sources such as wind or ocean waves to drive perpetual cyclic mechanical strain of the surface [90].

Creating bioinspired micro-topographical surfaces is a viable, effective, and nontoxic approach to AF. However, precise geometric control of surface textures requires costly and sophisticated techniques. This is not feasible for practical use on large vessels.

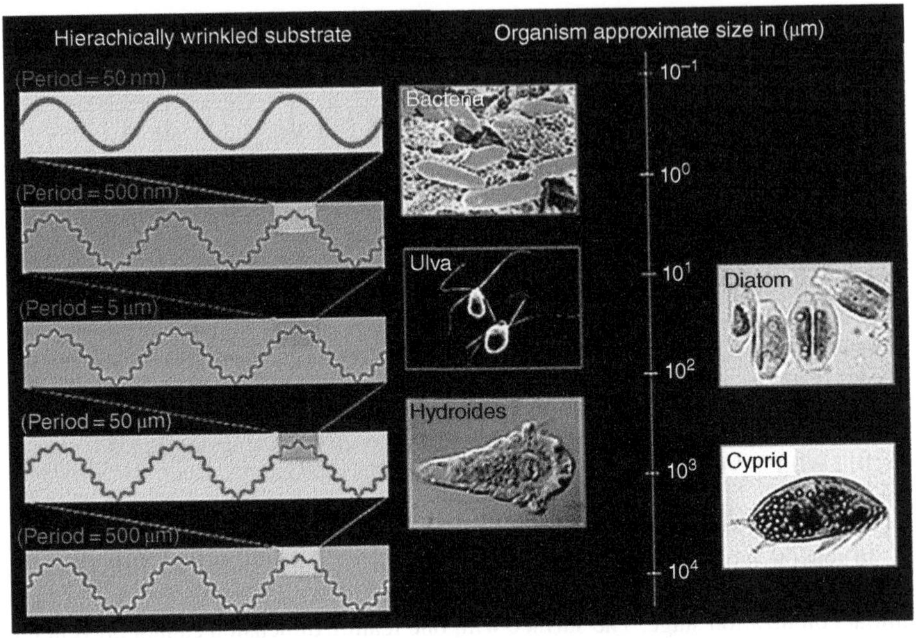

FIG. 11.20 Schematic depicting the structure of hierarchically wrinkled (HW) coatings for marine AF comprising nested wrinkled topographies from tens of nanometers to a fraction of a millimeter, corresponding to the right panel showing typical dimensions of selected marine organism. Reproduced from a previous work with permission Ref. 84. © 2009 American Chemical Society.

11.7 AMPHIPHILIC NANOSTRUCTURED COATINGS

Amphiphilicity, for polymeric surfaces, means the entities are composed of block copolymers, or occasionally polymer blends, with contrasting-chemistry domains (i.e., both hydrophilic and oleophilic) at the submicron or nanoscales.

As mentioned earlier, hydrophilic coatings are capable of preventing the settlement of foulants at the initial stage, and hydrophobic surfaces facilitate release due to their intrinsically low surface tension [35]. Nevertheless, neither of these two kinds of surface can deal with the adhesion of diverse marine foulants effectively, because different marine organisms use different adhesives and adhesion methods. For instance, macrofouling organisms generally adhere weakly to hydrophobic, low-modulus surfaces [91]. However, microfouling biofilms dominated by diatoms adhere more strongly to hydrophobic coatings (especially the silicone-based coatings) than hydrophilic ones [91]. If hydrophobicity and hydrophilicity are integrated into one coating, it might create a universal coating that could resist a wide range of foulants and enable their ready release through hydrodynamic forces. Evidence has shown that amphiphilic surfaces are effective inhibitors of platelet adhesion, which otherwise leads directly to lethal blood coagulation and thrombosis [92]. For this reason, amphiphilic surfaces have been used in biomedical applications. This outstanding biocompatibility is probably due to the fact that so many biological organisms are binary-wettability coexistent. For example, the structural characteristics of proteins, glycoproteins, and heteropolysaccharides in EPS are the balance between hydrophilic and hydrophobic segments [93]. Also the polar heads of the phopholipids and the hydrophobic moieties of the proteins make up the amphiphilic structure of cell membranes, giving them ideal nonthrombogenicity. In this sense, amphiphilic materials could probably be regarded as reminiscent of cytomembrane. A hypothesis was raised: amphiphilic coatings might also be effective in marine AF and FR applications, offering enhanced ability to deter various kinds of foulants.

Based upon this hypothesis, a variety of amphiphilic coatings have been designed and tested, and most of them exhibit efficacy in combating marine fouling. Amphiphilic nanostrutured coatings have emerged as one of the most promising candidates for AF and FR purposes.

11.7.1 Principles of Amphiphilic Nanostructured Coatings

Although many reports on the AF capabilities of amphiphilc coatings have been published, current knowledge on the chemistry of marine adhesives and their curing characteristics is still insufficient to ascribe a mechanism for the detachment of marine foulants from the amphiphilic coatings [93]. The mechanism must be explored further. Currently, the AF principle is generally described as the ability to confuse the approaching organisms into avoiding settlement due to their superficial chemical ambiguity [1, 35]. To be more specific, hybridization of polymeric composites leads to thermodynamically driven phase segregation of two (or more) kinds of immiscible polymers due to their mutual incompatibility [94]. However, phase segregation does not occur on the macroscale. Rather, it could be tuned to present optimal nanoscale heterogeneity in terms of chemical composition or even surface topography [95]. This is because kinetically the phase segregation is greatly trapped by the *in situ* cross-linking reaction of mixtures [94]. Such thermodynamically driven and kinetically trapped nanophase segregation leads to the nanoscale dimensions of the topographical features comparable to or somewhat smaller than that of marine adhesive biomacromolecules [94]. These become energetically unfavorable to the secreted glue, preventing the organism from adhering via either hydrophobic or hydrophilic interactions, thus weakening the interactions of the foulant with the surface [95]. All these are illustrated in Fig. 11.21. Krishnan et al. pointed out that to restrain the adsorption and unfolding of protein molecules on surface, the heterogeneities must be smaller than the protein molecules, which are within the range 1–10 nm, if the premise of chemical ambiguity is to be valid [35]. Finlay *et al.* found out that the *Ulva* spores were no longer able to distinguish between the fluorinated and PEGylated features on a patterned surface if the pattern dimension was below 20 μm, namely 2 or 5 μm in the experiment. The spores regarded these patterns as pure PEG and hence avoided settlement on them [97]. This example provides new conceptual guidance for the verification for the aforementioned explanation. The chemical ambiguity, referring to the nano-segregated amphiphilic surface, may lower the entropic and enthalpic driving forces for the adsorption of marine adhesive

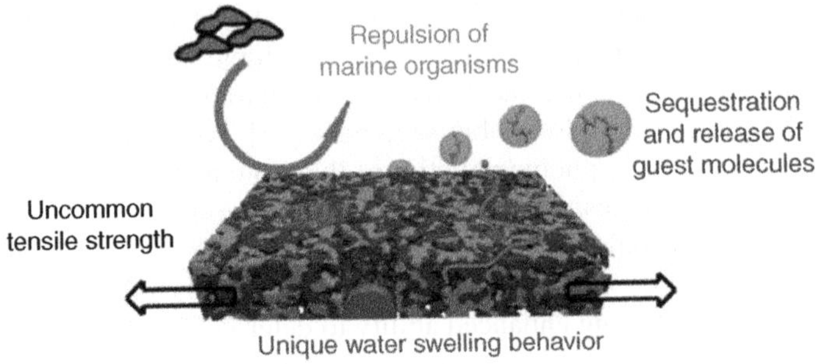

FIG. 11.21 A chemically heterogeneous, or mosaic-like, amphiphilic surface that repels proteins. Reprinted from a previous work with permission Ref. 96. © 2007 WILEY-VCH Verlag GmbH & Co. KGaA, Weinheim.

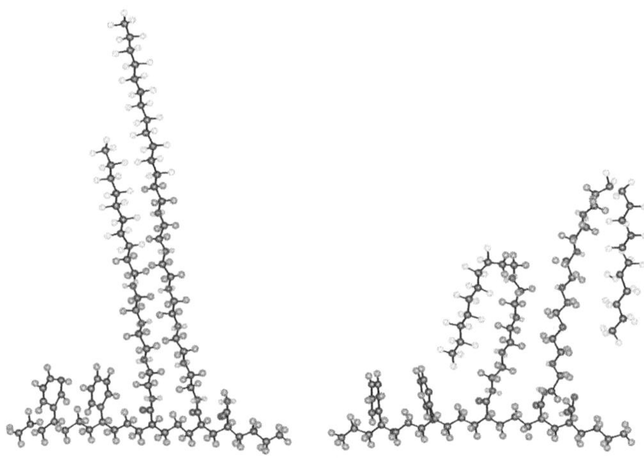

FIG. 11.22 Proposed mechanism of surface reconstruction of the ethoxylated fluoroalkyl side chains upon immersion of the surface in water. The image on the left indicates the orientation of side chains in air and the one on the right shows the effect of water immersion. Reprinted from a previous work with permission Ref. 93. © 2006 American Chemical Society.

biomacromolecules, which are amphiphilic in character themselves, and so reduces the adhesion strength of the whole organism [1, 93, 98].

Besides chemical confusion by the surface, the second key point in constructing effective AF amphiphilic surfaces is the capability of undergoing surface reconstruction after being immersed in an aquatic environment [93, 96, 99]. In most amphiphilic AF coatings, PEG or oligoethylene glycol (OEG) segments are the major hydrophilic constitutions. In order to produce certain mechanical properties applicable to practical applications, the amount of hydrophilic moieties is usually no more than 50 wt% while the hydrophobic segments dominate the whole system as the reinforcing network to complement the poor mechanical properties of the hydrophilic parts. In air,

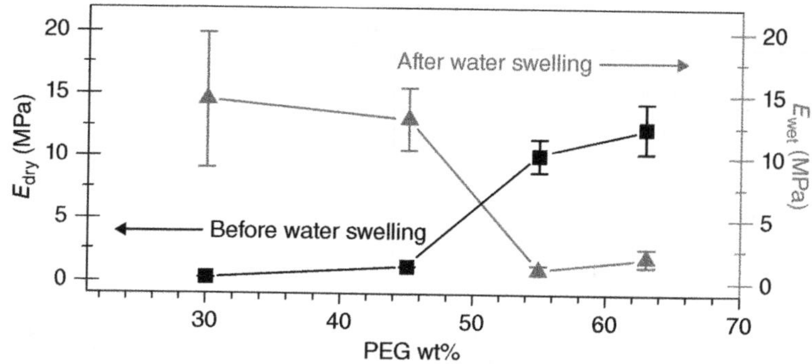

FIG. 11.23 Elastic moduli of HBFP(I)/PEG networks with varying amounts of PEG before and after swelling in water. Reproduced with kind permission from Ref. 96. © 2007 WILEY-VCH Verlag GmbH & Co. KGaA, Weinheim.

hydrophobic parts dominate the surface, which reduces the interfacial energy with water, and hydrophobic moieties retreat back into the body and meanwhile hydrophilic components are enriched at the water's surface (Fig. 11.22). Nevertheless, a certain amount of chain mobility is needed to enable the reconstruction of the surface. Unlike more rigid networks, which restrain the migration of hydrophilic polymers to the surface, flexible networks are available for surface reconstruction by adjusting the chain density, and chain length exhibits better AF properties. The mobility is actually the balance between induction of hydrophilic segments and the adsorption of water molecules into the hydrophilic domains. Bartels et al. pointed out that as long as absorbed water caused the network to swell but not burst (PEG content <50 wt%), there were remarkable enhancements in the mechanical properties (Fig. 11.23) [96]. These might also contribute to its enhanced AF and FR performance against several fouling species [96].

Despite the fact that amphiphilic coatings with smooth topographies parallel to the substrate could deter marine fouling, Wooley et al. pointed out that the ordered uneven physical topographies with discrete domains were unfavorable to marine foulants [35, 94, 96]. For this reason, submicro- and nanoscale physical topographies are a third factor contributing to the improvement of AF properties. Their principles can be considered analogous to those of the bioinspired AF coatings discussed in Section 6.1.

To conclude, it could be said that amphiphilic coatings combat marine fouling by integrating the essences of hydrophilic NF ones, hydrophobic FR ones, and bioinspired microtopographical dynamic coatings.

11.7.2 PEG-Fluoropolymers Amphiphilic Coatings

Wooley's group first prepared amphiphilic coatings, based on hyperbranched fluoropolymers (HBFP) and linear PEGs, involving the nucleophilic aromatic substitution of the pentafluorophenylpara-fluorines in HBFP by the amino groups on the telechelic PEG linear segments [94, 100] (Fig. 11.24). The fabricated surface coatings were

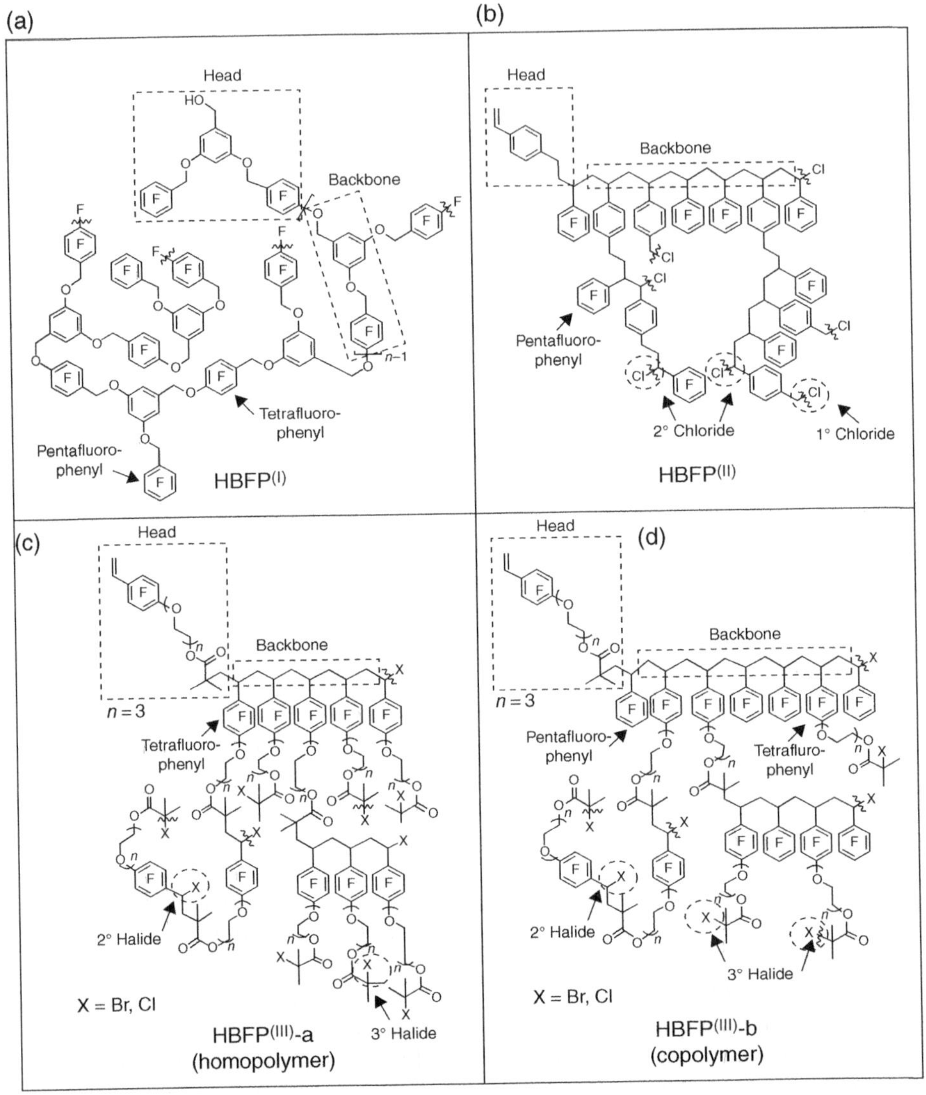

FIG. 11.24 Representative structures of (a) first-type HBFP (HBFP(I), by polycondensation); (b) second-type HBFP (HBFP(II), by ATR-SCVCP); (c) third-type HBFP homopolymer (HBFP(III)-a, by ATR-SCVP); and (d) third-type HBFP copolymer (HBFP(III)-b, by ATR-SCVCP). Reprinted with permission from Ref. 96. © 2007 WILEY-VCH Verlag GmbH & Co. KGaA, Weinheim.

reported to exhibit nano- or microscopic domains, the dimensions of which were strongly influenced by the composition of two components and the effect of water immersion. The deterrent effect against biomacromolecules and *Ulva*, which was caused by surface complexity and reconstruction, became more obvious with the increasing PEG content and reached an optimum at 45 wt% PEG [95]. The HBFP was first built up from the A_2B structured monomer, 3,5-bis[(pentafluorobenzyl)oxy] benzyl alcohol, by polycondensation (Fig. 11.24 HBFP$^{(I)}$). Later, this group developed two other precursors, HBFP$^{(II)}$ and HBFP$^{(III)}$, via atom transfer radical self-condensing vinyl (co)polymerization (ATR-SCVCP) [96, 101]. For HBFP$^{(II)}$, the one-pot reaction with all commercially available reactants renders it the advantages of being a more economical material [96], and HBFP$^{(III)}$ was prepared from a fluorinated amphiphilic inimer, 4-{oxy[tri(ethylene glycol)]-bromoisobutyryl}-2,3,5,6-tetrafluorostyrene, with increased hydrophilicity due to the incorporation of tri(ethylene glycol) in advance [96, 101].

In addition to hyperbranched amphiphilic polymers, a series of block copolymers and terpolymers with perfluoroalkyl tagged moieties as side chains and OEG segments either in side chains or in backbones were designed. Krishnan et al. synthesized comb-structured block copolymer by grafting hydrophilic oligoethylene glycol and hydrophobic perfluoroalkyl side chains to a poly(styrene-block-acrylic acid) backbone (Fig. 11.25, Structure **1**) [93]. The copolymer demonstrated universal AF properties for the algal species that adhered strongly to both hydrophobic surface and hydrophilic surface [93]. Similar copolymers with the same amphiphilic side chains grafted onto different backbones were prepared by Weinman et al. [102] (Fig. 11.25, Structure **2**) and Martinelli et al. [103–105] (Fig. 11.25. Structure **5**, **6**) using controlled ATRP. In addition, Zoelen et al. attached two types of peptoid monomers with opposite wettabilities to the PS-*b*-P(EO-*co*-AGE)) scaffold (Fig. 11.25, Structure **4**) via thiol-ene click chemistry [106, 107]. The polymer (Fig. 11.25, Structure **3**) was used as a platform to study the effect of sequences of hydrophilic and hydrophobic moieties on both AF and FR performance of *U. linza* [106]. Specifically, the position of the fluorinated units in the peptoid chain changes the surface chemistry and the AF behavior, while the number of fluorinated residues affects the fouling-release properties [106]. Polymer coatings with identical compositions but fluorinated groups at different positions showed different surface composition because the fluorinated group can spontaneously drag whatever the portion of the segment it links to the surface [106].

In most aforementioned cases, a bilayer strategy was used, where the prepared surface active block copolymers are sedimented as a film upon a thick layer of poly(styrene-block-ethylene-random-butylene)-block-polystyrene (SEBS) thermoplastic elastomer [93, 104, 107]. Weinman et al. evaluated the performances of two different commercially available SEBSs (Kraton MD6945 and G1652) whose modulus values differ by an order of magnitude [102]. Results showed that MD6945, which had a lower modulus, also had further reduced attachment strength of the sporelings [102]. This in another perspective verified the statement that surface with low modulus can enhance FR properties.

The hydrophobic segments within the above amphiphilic block copolymers are long perfluoroalkyl chains (C_nF_{2n+1}) [5]. However, their degradation becomes an environmental problem because it produces perfluorooctanic acid, which is resistant to degradation with long bio half-life and bioaccumulative in living organisms [5]. Because of this, PFPE was used by Wang et al. as an alternative to perfluoroalkyl groups [99, 108].

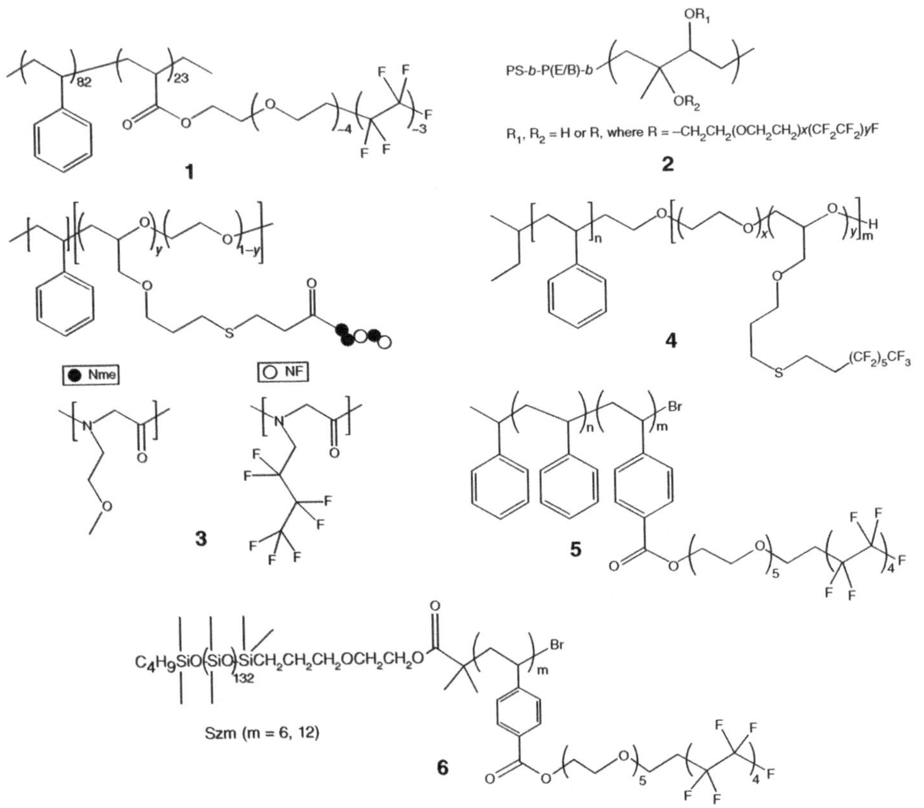

FIG. 11.25 Representative AF amphiphilic polymer structures with perfluoroalkyl chains.

PFPE polymers are reported to be biodegradable and meanwhile exhibit low surface energy (<13 mJ/m^2) [5]. Because suitable surface mechanical properties could enhance AF and FR properties, Hu et al. first tuned the cross-linking density using a series of (di)(meth)acryloxy-terminated photocurable PFPE precursors [109]. Wang et al. [99, 108] embedded two PEG, dimethacryloxy-functionalized PEG (550 g/mol) (PEG550-DMA) and monomethacryloxy-functionalized PEG macromonomer (454 g/mol) (PEG454-MA), into the polymer matrix to weaken fouling adhesion [99]. Results showed that because the PFPE/PEG454-MA coatings had one untethered chain end in PEG segment it could enable the drive of more polar groups to the surface. These displayed better spore and barnacle release performance than PFPE/PEG550-DMA coatings [99]. To overcome the energy barrier and drive more flexible hanging PEG brushes onto the surface, Wang et al. cured the coating under different moisture conditions (Fig. 11.26). Contact angle analysis showed that the density of PEG could be enhanced when cured in high humidity. The PEG-enriched surface differed from that of *Ulva* sporelings on the coatings cured in 57% humidity even after 6 days of growth [108]. The biomass of sporelings accumulated on coatings cured in 24% humidity was comparable to that of PDMS elastomer surfaces [108].

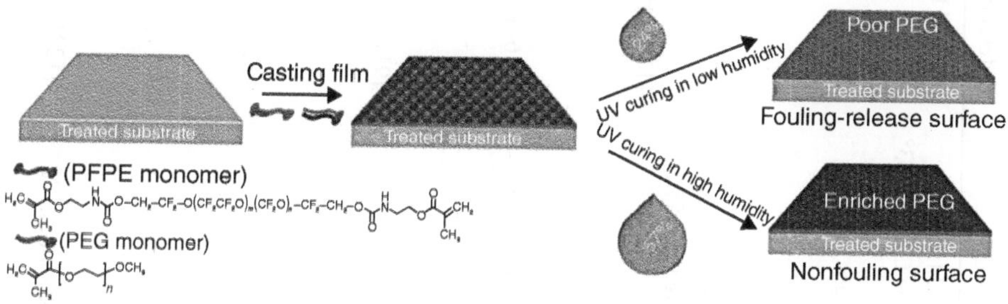

FIG. 11.26 Schematic illustration of fabrication process of a kind of amphiphilic co-networks with moisture-induced surface segregation for high-performance NF coatings. Reprinted from a previous work with kind permission Ref. 99. © 2011 American Chemical Society.

Besides its more environmentally benign precursors, this method also had other advantages such as that the coating can be UV-cured in one step, it is highly efficient, costs little energy, and is more and environmentally friendly than traditional thermo-curing procedures.

11.7.3 Other Amphiphilic AF Polymers

In addition to the PEG-fluoropolymers, many miscellaneous amphiphilic polymer coatings with different hydrophilic and hydrophobic parts have been evaluated. Table 11.1 lists some amphiphilic polymer coatings that may be able to weaken fouling adhesion. These polymers are divided into two categories: ionic and nonionic. Despite the emergence of this variety of amphiphilic surfaces in the field of marine AF research, they are still confined to the laboratory level. Amphiphilic AF coatings up to date are a conceptual model rather than a product ready for practical use.

11.7.4 Characterization Techniques

To study the amphiphilic coatings' AF principle, the two methods NEXAFS and angle-resolved X-ray photoelectron spectroscopy are commonly used in *in vacuo* techniques [1, 91, 93, 102, 106, 107]. However, the information they provide on surface structure in the dry state is not as meaningful as when surfaces are immersed in water. For better understanding, *in situ* characterization techniques for the surface properties underwater are needed. Atomic force microscopy in fluid mode, neutron reflectivity measurement, and sum frequency generation (SFG) spectroscopy make it possible to monitor the changes in immersed surface topographies, thickness, and surface chemistry over time [42, 113, 121]. These advancing physical techniques for interfacial characterization are crucial to further exploration. Amphiphilic coatings are still believed to have the potential to be developed into a universal AF coating capable of resisting a wide range of foulants and releasing them under hydrodynamic pressure.

TABLE 11.1 Other Amphiphilic AF Polymer Coating Systems

Coating Composition	Fabrication Method	AF Performances	Contact Angle	References
PEG-4,4′-diphenylmethanediisocyanate–PDMS multiblock copolymers	Two-step growth polymerization, thermal anneal	↓Human fibrinogen adsorption	Water CA ≈ 106–110°, Diiodomethane CA ≈ 59–63°	[110, 111]
Amphiphilic PES/F127-b-PDMS membrane surface	Free radical polymerization of F127-b-PDMS, nonsolvent-induced phase-inversion (NIPS) method	↓Bovine serum albumin (BSA), sodium alginate (SA) and yeast	Water CA ≈ 50–90°	[112]
Self-repairable high density PEG brushes in PDMS matrix	Hydrosilation of PDMS-b-PEG copolymer	—	Water advancing CA = 93.6°, water receding CA = 55.5°	[113]
Amphiphilic PEG/PDMS conetworks	Co-hydrosilation and condensation/cross-linking, ($SiPh(SiH)_2OEt$)) as extender/cross-linker	—	—	[114, 115]
Siloxane–polyurethane	Aminopropyl terminated PDMS macromers via hydrosilylation, ambient curing	↓Pseudobarnacle removal forces, ↑removal of marine bacteria, diatoms, algal sporelings	Water CA ≈ 90–110°	[116–118]
PEG/hydrocarbon grafted triblock copolymers	—	↓Settlement of zoospores of *Ulva linza*, cells of *Navicula incerta*	—	[91]
PMMA/PEG binary brushes	"Grafting-to" technique	↓Bovine serum albumin (BSA) adsorption	Water CA ≈ 30–70°	[119]
Polyanion with perfluoroalkyl polyethylene glycol (fPEG) side chains	Electrostatic layer-by-layer (LbL) deposition	↓Marine bacterium *Pseudomonas* adsorption	Water CA = 47 ± 2°, Oil CA = 41 ± 2°	[120]
Three amphiphilic polybetaines: ZI-OEG, ZI-C8, ZI-F13	—	Good AF properties for ZI-OEG and ZI-F13, poor AF properties for ZI-C8	—	[121]
PDMS network containing QAS-functional trimethoxysilanes	Moisture-curable	Effective in bacterium *Cellulophaga lytica*, diatom *Navicula incerta*, macrofouling	Water CA hysterisis ≈ 5–25°	[122]

11.8 SUMMARY

Development of non-biocidal AF coatings is becoming a priority. Currently, two general strategies, biodegrading coatings and biopreventing coatings, are used. However, the biocide-containing SPCs still dominate the AF coatings market. Biocide-free AF coatings hold only a small proportion of the total marine AF coatings market, being mainly contributed from the nontoxic FR coatings. Nevertheless, nontoxic AF coatings have emerged as the trend of novel AF technologies. The major difficulties that hinder the commercialization of biopreventing coatings are the high cost, impracticality for large-scale production, and low durability. Further efforts must be made toward enhancing the efficiency of action against the formation of slime and FR at low vehicle speeds and during idle periods.

Currently, different types of novel AF coatings keep springing up from laboratories worldwide. The examples chosen in this chapter are abundant but not exhaustive. There are still a few other ways to eradicate fouling microbes and organisms. These include coatings consisting of nanocomposites. Researches have shown that incorporation of organomodified montmorillonite sheets [75], carbon nanotubes [123, 124], natural sepiolite needles [123], and vanadium pentoxide (V_2O_5) nanowires [125] into polymer matrixes could all largely enhance the FR properties in laboratory assays with major fouling organisms (*Ulva* and/or barnacles) compared with pure PDMS controls.

However, due to the leaching of nanofillers in the polymer matrix, the challenge for the nanocomposite coatings must be associated with the current debate on the issues of nanoparticle toxicity [1]. In the design of future prospective coatings, nature is an excellent teacher. Artificial coatings inspired by the skins of NF or self-cleaning flora and fauna may emerge as real candidates for practical application, so long as the methods of fabrication make it possible for the large-scale production. The complexity and diversity of marine fouling opportunists makes it difficult to control biofouling by introducing a single factor into the coatings. Multifunctional coatings, therefore, are perceived to have a special niche in the field of future AF coatings technologies. A coating of many functions means the incorporation of a mixture of attributes, such as an appropriate topography combined with a suitable amphiphilic or zwitterionic surface chemistry and environmentally benign compounds that deter settlement of biofoulers. Smart coatings are also a promising concept in AF applications. They could spontaneously heal themselves upon external stimulations (pH changes, cracks/puncture, magnetic fields, etc.). The aforementioned self-repairable SLIPS are one possible case [31, 126]. The ideas of smart coatings for AF might as well have been borrowed from that of self-healing anticorrosion coatings. Inspired by biological systems where self-healing is common, capsule-based polymer composites and nanocontainer-based self-healing coatings have been reported [127–130]. More recently, Wooley et al. designed an amphiphilic cross-linked polymer as AF coatings are capable of self-repair with the thermally reversible Diels–Alder (DA) chemistry [131].

Successful design of AF coatings requires a thorough understanding of detailed mechanisms. Even though we have demonstrated a few examples of remarkable correlations between biological performance and specific types of topographical or chemical surface structuring, tackling problems such as how the chain arrangements of amphiphilic diblocks affect the AF performance against zoospores, the manner in which AF properties of silicon coatings vary with surface patterns of different engineered micro-topography gradients, and what are the critical dimensions of the hydrophilic/hydrophobic surface domains that are able to effectively confuse foulants might be far from sufficiently explained [1, 79, 106, 107]. Most AF property tests barely remain at the macro level. They involve assessments of fouled appearance after a certain period. Barriers still exist in understanding how these AF coatings function microscopically.

Advances are needed to understand the biological and molecular mechanisms underlying the prevention of biofilm formation. Callow pointed out that adhesion mechanisms and detailed molecular characteristics of the glues are largely unknown [2]. Deep understanding of fouling mechanisms is required. This is because biopreventing AF coatings is built upon the fact that fouling organisms are influenced in their initial adhesion behavior and the following adhesion strength by surface physicochemical properties and nano- or microscale topographies. In the field of fabrication techniques, living polymerization enables the controlled propagation of polymers and hence the defined polymer structures with narrow distribution. This largely facilitates the regulation of the surface AF properties in lab and so enables more thorough understanding of the relationship between polymer composition, surface density, and AF performance. In addition, exploration of mechanisms can also be put forward using sophisticated analysis techniques, such as SFG, QCM, NEXAFS, OWLS, and laser scanning confocal microscope (LSCM). These not only clearly show the performance of the AF coatings but also identify the underlying structural elements responsible for certain properties.

Interdisciplinary cooperation in AF technologies mirrors the increasing awareness of materials scientists and researchers that progresses could never go without a good command of biological principles of the biofouling process. The subject of AF must be evaluated in a systematic and comprehensive way. In other words, information regarding materials engineering, chemistry, and interdisciplinary subjects such as molecular biology, ecotoxicology, nanofabrication techniques, and more crucially the understanding of regulatory science, are necessary, as depicted in Fig. 11.27.

Trial and error in experiments of molecular design is the conventional route for identification of effective AF properties in polymer coatings. However, few designs are based upon the establishment of predictive models. This provides guidance for design rules and can so greatly shorten the time required for research. The past few decades have shown a swift increase in the use of computer simulations alongside experiments in dealing with problems in materials science. Mesoscopic simulation is suitable for the study of large molecules, allowing computer simulations to be run on length and time scales that are two to three orders of magnitude larger than atomistic simulations. This could, for instance, help predict the micro phase structures in amphiphilic coatings and explain their relationships to their AF performance.

In conclusion, quite a few strategies for combating fouling organisms are reviewed here. Even though most of them have considerable shortcomings concerning stability,

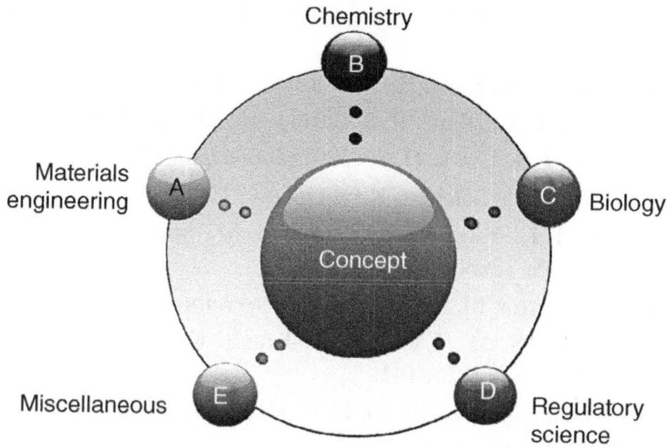

FIG. 11.27 Interdisciplinary cooperation in AF technologies.

toxicity, and methods of fabrication, they can be modified and tailored to the industrial (technical, economic, and ecological) requirements of particular coatings. There is no shortage of creative ideas. The constant pursuit of enhancement suggests advent of better AF materials in the future.

REFERENCES

1. Callow, J. A., Callow, M. E. (2011) *Nat. Commun.*, 2: 244.
2. Callow, M. E., Callow, J. A. (2002) *Biologist*, 49: 1.
3. Smith, A. M. ed. (2006) *Biological adhesives*, Springer: Berlin/Heidelberg.
4. Costlow, J. D., Tipper, R. C. (1984) *Marine biodeterioration: An interdisciplinary study: Proceedings of the symposium on marine biodeterioration, Uniformed Services University of Health Sciences, 20–23 April 1981*. Naval Institute Press, Annapolis.
5. Lejars, M. N, Margaillan, A, Bressy, C (2012) *Chem. Rev.*, 112: 4347.
6. Hellio, C., Yebra, D. M. eds. (2009) *Advances in marine antifouling coatings and technologies*, Woodhead Publishing Ltd., Cambridge.
7. Dürr, S., Thomason, J. C. eds. (2010) *Biofouling*, Blackwell Publishing Ltd., Massachusetts.
8. Rodrigues, R. C, Berenguer-Murcia, Á., Fernandez-Lafuente, R. (2011) *Adv. Synth. Catal.*, 353: 2216.
9. Noel, R. (1985) Antisoiling composition for addition to the coatings of immersed bodies and coating containing it. France Patent, FR2562554.
10. Olsen, S. M., Pedersen, L. T., Laursen, M., Kiil, S., Dam-Johansen, K. (2007) *Biofouling*, 23: 369.
11. Kristensen, J. B., Meyer, R. L., Laursen, B. S., Shipovskov, S., Besenbacher, F., Poulsen, C. H. (2008) *Biotechnol. Adv.*, 26: 471.
12. Pettitt, M., Henry, S., Callow, M., Callow, J., Clare, A. (2004) *Biofouling*, 20: 299.
13. Aldred, N., Phang, I. Y., Conlan, S. L., Clare, A. S., Vancso, G. J. (2008) *Biofouling*, 24: 97.

14. Tasso, M., Pettitt, M. E., Cordeiro, A. L., Callow, M. E., Callow, J. A., Werner, C. (2009) *Biofouling*, 25: 505.
15. Asuri, P., Karajanagi, S. S., Kane, R. S., Dordick, J. S. (2007) *Small*, 3: 50.
16. Banerjee, I., Pangule, R. C., Kane, R. S. (2011) *Adv. Mater.*, 23: 690.
17. Kim, Y. D., Dordick, J. S., Clark, D. S. (2001) *Biotechnol. Bioeng.*, 72: 475.
18. Tasso, M., Cordeiro, A. L., Salchert, K., Werner, C. (2009) *Macromol. Biosci.*, 9: 922.
19. Liu, Y., Ogorzalek, T. L., Yang, P., Schroeder, M. M., Marsh, E. N. G., Chen, Z. (2013) *J. Am. Chem. Soc.*, 135: 12660.
20. Messersmith, P. B., Textor, M. (2007) *Nat. Nanotechnol.*, 2: 138.
21. Kristensen, J., Olsen, S. M., Laursen, B., Kragh, K., Poulsen, C., Besenbacher, F., Meyer, R. (2010) *Biofouling*, 26: 141.
22. Schultz, M. P., Kavanagh, C. J., Swain, G. W. (1999) *Biofouling*, 13: 323.
23. Brady, R. F. (2005) *Def. Sci. J.*, 55: 75.
24. Baier, R. E. (2006) *J. Mater. Sci. Mater. Med.*, 17: 1057.
25. Brady, R. F.,Singer, I. L. (2000) *Biofouling*, 15: 73.
26. Singer, I. L., Kohl, J. G., Patterson, M. (2000) *Biofouling*, 16: 301.
27. Holland, R., Dugdale, T. M., Wetherbee, R., Brennan, A. B., Finlay, J. A., Callow, J. A., Callow, M. E. (2004) *Biofouling*, 20: 323.
28. Bullock, S., Johnston, E. E., Willson, T., Gatenholm, P., Wynne, K. J. (1999) *J. Colloid Interface Sci.*, 210: 18.
29. Griffith, J. R., Bultman, J. D. (1978) *Ind. Eng. Chem. Prod. Res. Dev.*, 17: 8.
30. Grinthal, A., Aizenberg, J. (2014) *Chem. Mater.*, 26: 698.
31. Wong, T.-S., Kang, S. H., Tang, S. K., Smythe, E. J., Hatton, B. D., Grinthal, A., Aizenberg, J. (2011) *Nature*, 477: 443.
32. Yao, X., Dunn, S. S., Kim, P., Duffy, M., Alvarenga, J., Aizenberg, J. (2014) *Angew. Chem. Int. Ed.*, 53: 4418.
33. Lin, P., Lin, C.-W., Mansour, R., Gu, F. (2013) *Biosens. Bioelectron.*, 47: 451.
34. Parrott, M. C., DeSimone, J. M. (2012) *Nat. Chem..*, 4: 13.
35. Krishnan, S., Weinman, C. J., Ober, C. K. (2008) *J. Mater. Chem.*, 18: 3405.
36. Morra, M. (2000) *J. Biomat. Sci., Polym. Ed.*, 11: 547.
37. Llanos, G. R., Sefton, M. V. (1993) *J. Biomater. Sci., Polym. Ed.*, 4: 381.
38. Yandi, W., Mieszkin, S., Martin-Tanchereau, P., Callow, M. E., Callow, J. A., Tyson, L., Liedberg, B., Ederth, T. (2014) *ACS Appl. Mater. Interfaces*, 6: 11448.
39. Zhao, C. W., Lin, Z. F., Yin, H. B., Ma, Y. H., Xu, F. J., Yang, W. T. (2014) *Sci. Rep.*, 4: 4982.
40. Kim, D.-G., Kang, H., Han, S., Lee, J.-C. (2012) *J. Mater. Chem.*, 22: 8654.
41. Kim, D.-G., Kang, H., Han, S., Lee, J.-C. (2012) *ACS Appl. Mater. Interfaces*, 4: 5898.
42. Kim, D. G., Kang, H., Choi, Y. S., Han, S., Lee, J. C. (2013) *Polym. Chem.*, 4: 5065.
43. Balachandran, V. S., Jadhav, S. R., Vemula, P. K., John, G. (2013) *Chem. Soc. Rev.*, 42: 427.
44. Gillich, T., Benetti, E. M., Rakhmatullina, E., Konradi, R., Li, W., Zhang, A., Schluter, A. D., Textor, M. (2011) *J. Am. Chem. Soc.*, 133: 10940.
45. Huang, K., Lee, B. P., Ingram, D. R., Messersmith, P. B. (2002) *Biomacromolecules*, 3: 397.
46. Lee, B. P., Dalsin, J. L., Messersmith, P. B. (2002) *Biomacromolecules*, 3: 1038.

47. Mizrahi, B., Khoo, X., Chiang, H. H., Sher, K. J., Feldman, R. G., Lee, J.-J., Irusta, S., Kohane, D. S. (2013) *Langmuir*, 29: 10087.
48. Dalsin, J. L., Messersmith, P. B. (2005) *Mater. Today*, 8: 38.
49. Zhang, Z., Finlay, J. A., Wang, L., Gao, Y., Callow, J. A., Callow, M. E., Jiang, S. (2009) *Langmuir*, 25: 13516.
50. Chen, S., Jiang, S. (2008) *Adv. Mater.*, 20: 335.
51. Keefe, A. J.,Jiang, S. (2012) *Nat. Chem.*, 4: 59.
52. Chen, S., Li, L., Zhao, C., Zheng, J. (2010) *Polymer*, 51: 5283.
53. Kitano, H., Mori, T., Takeuchi, Y., Tada, S., Gemmei-Ide, M., Yokoyama, Y., Tanaka, M. (2005) *Macromol. Biosci.*, 5: 314.
54. Aldred, N.,Li, G. Z., Gao, Y., Clare, A. S., Jiang, S. Y. (2010) *Biofouling*, 26: 673.
55. Kuang, J., Messersmith, P. B. (2012) *Langmuir*, 28: 7258.
56. Xue, L., Lu, X., Wei, H., Long, P., Xu, J., Zheng, Y. (2014) *J. Colloid Interface Sci.*, 421: 178.
57. Xie, L., Hong, F., He, C., Ma, C., Liu, J., Zhang, G., Wu, C. (2011) *Polymer*, 52: 3738.
58. Hong, F., Xie, L., He, C., Liu, J., Zhang, G., Wu, C. (2013) *Polymer*, 54: 2966.
59. Hong, F., Xie, L., He, C., Liu, J., Zhang, G., Wu, C. (2013) *J. Mater. Chem. B*, 1: 2048.
60. Simon, R. J., Kania, R. S., Zuckermann, R. N., Huebner, V. D., Jewell, D. A., Banville, S., Ng, S., Wang, L., Rosenberg, S., Marlowe, C. K. (1992) *PNAS*, 89: 9367.
61. Statz, A. R., Meagher, R. J., Barron, A. E., Messersmith, P. B. (2005) *J. Am. Chem. Soc.*, 127: 7972.
62. Lin, S., Zhang, B., Skoumal, M. J., Ramunno, B., Li, X., Wesdemiotis, C., Liu, L., Jia, L. (2011) *Biomacromolecules*, 12: 2573.
63. Lau, K. H. A., Ren, C., Sileika, T. S., Park, S. H., Szleifer, I., Messersmith, P. B. (2012) *Langmuir*, 28: 16099.
64. Bixler, G. D., Bhushan, B. (2012) *Soft Matter*, 8: 11271.
65. Liu, K., Yao, X., Jiang, L. (2010) *Chem. Soc. Rev.*, 39: 3240.
66. Liu, K., Jiang, L. (2012) *Annu. Rev. Mater. Res.*, 42: 231.
67. Autumn, K., Sitti, M., Liang, Y. A., Peattie, A. M., Hansen, W. R., Sponberg, S., Kenny, T. W., Fearing, R., Israelachvili, J. N., Full, R. J. (2002) *PNAS*, 99: 12252.
68. Parker, A. R., Lawrence, C. R. (2001) *Nature*, 414: 33.
69. Liu, K., Tian, Y., Jiang, L. (2013) *Prog. Mater. Sci.*, 58: 503.
70. Scardino, A., Harvey, E., De Nys, R. (2006) *Biofouling*, 22: 55.
71. Railkin, A. I. (2003) *Marine biofouling: Colonization processes and defenses*, CRC press, Boca Raton.
72. Bixler, G. D., Bhushan, B. (2012) *Philos. Trans. R. Soc. A*, 370: 2381.
73. Xiao, L. (2014) PhD dissertation, the Ruprecht-Karls University of Heidelberg, Heidelberg.
74. Scardino, A., Guenther, J., De Nys, R. (2008) *Biofouling*, 24: 45.
75. Beigbeder, A., Labruyère, C., Viville, P., Pettitt, M. E., Callow, M. E., Callow, J. A., Bonnaud, L., Lazzaroni, R., Dubois, P. (2011) *J. Adhes. Sci. Technol.*, 25: 1689.
76. Friedlander, R. S., Vlamakis, H., Kim, P., Khan, M., Kolter, R., Aizenberg, J. (2013) *PNAS*, 110: 5624.
77. Schumacher, J. F., Carman, M. L., Estes, T. G., Feinberg, A. W., Wilson, L. H., Callow, M. E., Callow, J. A., Finlay, J. A., Brennan, A. B. (2007) *Biofouling*, 23: 55.

78. Long, C. J., Schumacher, J. F., Robinson, P. A., Finlay, J. A., Callow, M. E., Callow, J. A., Brennan, A. B. (2010) *Biofouling*, 26: 411.
79. Schumacher, J. F., Long, C. J., Callow, M. E., Finlay, J. A., Callow, J. A., Brennan, A. B. (2008) *Langmuir*, 24: 4931.
80. Magin, C. M., Long, C. J., Cooper, S. P., Ista, L. K., López, G. P., Brennan, A. B. (2010) *Biofouling*, 26: 719.
81. Decker, J. T., Kirschner, C. M., Long, C. J., Finlay, J. A., Callow, M. E., Callow, J. A., Brennan, A. B. (2013) *Langmuir*, 29: 13023.
82. Schumacher, J. F., Aldred, N., Callow, M. E., Finlay, J. A., Callow, J. A., Clare, A. S., Brennan, A. B. (2007) *Biofouling*, 23: 307.
83. Liu, D., Bastiaansen, C. W., den Toonder, J. M., Broer, D. J. (2012) *Macromolecules*, 45: 8005.
84. Efimenko, K., Finlay, J., Callow, M. E., Callow, J. A., Genzer, J. (2009) *ACS Appl. Mater. Interfaces*, 1: 1031.
85. Evensen, H., Jiang, H., Gotrik, K., Denes, F., Carpick, R. (2009) *Nano Lett.*, 9: 2884.
86. Woltman, S. J., Jay, G. D., Crawford, G. P. (2007) *Nat. Mater.*, 6: 929.
87. Liu, D., Bastiaansen, C. W., den Toonder, J. M., Broer, D. J. (2012) *Angew. Chem. Int. Ed.*, 51: 892.
88. Elias, A., Harris, K., Bastiaansen, C., Broer, D., Brett, M. (2006) *J. Mater. Chem.*, 16: 2903.
89. Sousa, M. E., Broer, D. J., Bastiaansen, C. W., Freund, L., Crawford, G. P. (2006) *Adv. Mater.*, 18: 1842.
90. Epstein, A. K., Hong, D., Kim, P., Aizenberg, J. (2013) *New J. Phys.*, 15: 095018.
91. Zhou, Z., Calabrese, D. R., Taylor, W., Finlay, J. A., Callow, M. E., Callow, J. A., Fischer, D., Kramer, E. J., Ober, C. K. (2014) *Biofouling*, 1.
92. Kim, J., Kim, S., Kim, H., Akaike, T., Kim, S. C. (2002) *J. Biomed. Mater. Res.*, 62: 613.
93. Krishnan, S., Ayothi, R., Hexemer, A., Finlay, J. A., Sohn, K. E., Perry, R., Ober, C. K., Kramer, E. J., Callow, M. E., Callow, J. A. (2006) *Langmuir*, 22: 5075.
94. Gudipati, C. S., Greenlief, C. M., Johnson, J. A., Prayongpan, P., Wooley, K. L. (2004) *J. Polym. Sci., Part A: Polym. Chem.*, 42: 6193.
95. Gudipati, C. S., Finlay, J. A., Callow, J. A., Callow, M. E., Wooley, K. L. (2005) *Langmuir*, 21: 3044.
96. Bartels, J. W., Cheng, C., Powell, K. T., Xu, J., Wooley, K. L. (2007) *Macromol. Chem. Phys.*, 208: 1676.
97. Finlay, J. A., Krishnan, S., Callow, M. E., Callow, J. A., Dong, R., Asgill, N., Wong, K., Kramer, E. J., Ober, C. K. (2008) *Langmuir*, 24: 503.
98. Lin, F. Y., Chen, W. Y., Hearn, M. T. (2002) *J. Mol. Recognit.*, 15: 55.
99. Wang, Y., Finlay, J. A., Betts, D. E., Merkel, T. J., Luft, J. C., Callow, M. E., Callow, J. A., DeSimone, J. M. (2011) *Langmuir*, 27: 10365.
100. Gan, D., Mueller, A., Wooley, K. L. (2003) *J. Polym. Sci., Part A: Polym. Chem.*, 41: 3531.
101. Powell, K. T., Cheng, C., Wooley, K. L. (2007) *Macromolecules*, 40: 4509.
102. Weinman, C. J., Finlay, J. A., Park, D., Paik, M. Y., Krishnan, S., Sundaram, H. S., Dimitriou, M., Sohn, K. E., Callow, M. E., Callow, J. A. (2009) *Langmuir*, 25: 12266.
103. Martinelli, E., Agostini, S., Galli, G., Chiellini, E., Glisenti, A., Pettitt, M. E., Callow, M. E., Callow, J. A., Graf, K., Bartels, F. W. (2008) *Langmuir*, 24: 13138.

104. Martinelli, E., Menghetti, S., Galli, G., Glisenti, A., Krishnan, S., Paik, M. Y., Ober, C. K., Smilgies, D. M., Fischer, D. A. (2009) *J. Polym. Sci., Part A: Polym. Chem.*, 47: 267.
105. Martinelli, E., Suffredini, M., Galli, G., Glisenti, A., Pettitt, M. E., Callow, M. E., Callow, J. A., Williams, D., Lyall, G. (2011) *Biofouling*, 27: 529.
106. van Zoelen, W., Buss, H. G., Ellebracht, N. C., Lynd, N. A., Fischer, D. A., Finlay, J., Hill, S., Callow, M. E., Callow, J. A., Kramer, E. J. (2014) *ACS Macro Lett.*, 3: 364.
107. Dimitriou, M. D., Zhou, Z., Yoo, H.-S., Killops, K. L., Finlay, J. A., Cone, G., Sundaram, H. S., Lynd, N. A., Barteau, K. P., Campos, L. M. (2011) *Langmuir*, 27: 13762.
108. Wang, Y., Betts, D. E., Finlay, J. A., Brewer, L., Callow, M. E., Callow, J. A., Wendt, D. E., DeSimone, J. M. (2011) *Macromolecules*, 44: 878.
109. Hu, Z., Finlay, J. A., Chen, L., Betts, D. E., Hillmyer, M. A., Callow, M. E., Callow, J. A., DeSimone, J. M. (2009) *Macromolecules*, 42: 6999.
110. Fang, H., Zhou, S., Wu, L. (2006) *Appl. Surf. Sci.*, 253: 2978.
111. Fang, H., Feng, L., You, B., Wu, L. (2007) *J. Polym. Sci., Part B: Polym. Phys.*, 45: 208.
112. Zhao, X., Su, Y., Li, Y., Zhang, R., Zhao, J., Jiang, Z. (2014) *J. Membr. Sci.*, 450: 111.
113. Inutsuka, M., Yamada, N. L., Ito, K., Yokoyama, H. (2013) *ACS Macro Lett.*, 2: 265.
114. Erdodi, G., Kennedy, J. P. (2005) *J. Polym. Sci., Part A: Polym. Chem.*, 43: 4953.
115. Erdodi, G.,Kennedy, J. (2005) J. Polym. Sci. Part A: *Polym. Chem.*, 43: 4965.
116. Majumdar, P., Webster, D. C. (2005) *Macromolecules*, 38: 5857.
117. Majumdar, P., Webster, D. C. (2006) *Polymer*, 47: 4172.
118. Sommer, S., Ekin, A., Webster, D. C., Stafslien, S. J., Daniels, J., VanderWal, L. J., Thompson, S. E., Callow, M. E., Callow, J. A. (2010) *Biofouling*, 26: 961.
119. Feng, L., Gu, G., Chen, M., Wu, L. (2007) *Macromol. Mater. Eng.*, 292: 754.
120. Zhu, X., Guo, S., Jańczewski, D., Parra Velandia, F. J., Teo, S. L.-M., Vancso, G. J. (2013) *Langmuir*, 30: 288.
121. Leng, C., Gibney, K. A., Liu, Y., Tew, G. N., Chen, Z. (2013) *ACS Macro Lett.*, 2: 1011.
122. Majumdar, P., Lee, E., Patel, N., Ward, K., Stafslien, S. J., Daniels, J., Chisholm, B. J., Boudjouk, P., Callow, M. E., Callow, J. A. (2008) *Biofouling*, 24: 185.
123. Beigbeder, A., Degee, P., Conlan, S. L., Mutton, R. J., Clare, A. S., Pettitt, M. E., Callow, M. E., Callow, J. A., Dubois, P. (2008) *Biofouling*, 24: 291.
124. Kim, H. J., Baek, Y., Choi, K., Kim, D.-G., Kang, H., Choi, Y.-S., Yoon, J., Lee, J.-C. (2014) *RSC Adv.*, 4: 32802.
125. Natalio, F., André, R., Hartog, A. F., Stoll, B., Jochum, K. P., Wever, R., Tremel, W. (2012) *Nat. Nanotechnol.*, 7: 530.
126. Grinthal, A., Aizenberg, J. (2013) *Chem. Mater.*, 26: 698.
127. White, S. R., Sottos, N., Geubelle, P., Moore, J., Kessler, M. R., Sriram, S., Brown, E., Viswanathan, S. (2001) *Nature*, 409: 794.
128. Blaiszik, B., Kramer, S., Olugebefola, S., Moore, J. S., Sottos, N. R., White, S. R. (2010) *Annu. Rev. Mater. Res.*, 40: 179.
129. Shchukin, D., Möhwald, H. (2013) *Science*, 341: 1458.
130. Zheng, Z., Huang, X., Schenderlein, M., Borisova, D., Cao, R., Möhwald, H., Shchukin, D. (2013) *Adv. Funct. Mater.*, 23: 3307.
131. Imbesi, P. M., Fidge, C., Raymond, J. E., Cauët, S. N. I., Wooley, K. L. (2012) *ACS Macro Lett.*, 1: 473.

Index

Note: Page numbers in *italics* refer to Figures; those in **bold** to Tables.

Functional Polymer Coatings: Principles, Methods, and Applications, First Edition.
Edited by Limin Wu and Jamil Baghdachi.

Printed and bound by CPI Group (UK) Ltd, Croydon, CR0 4YY
16/09/2026
14963420-0008